LONDON MATHEMATICAL SOCIETY LECTURE NOTE SERIES

Managing Editor: Professor M. Reid, Mathematics Institute,
University of Warwick, Coventry CV4 7AL, United Kingdom

The titles below are available from booksellers, or from Cambridge University Press at
www.cambridge.org/mathematics

287 Topics on Riemann surfaces and Fuchsian groups, E. BUJALANCE, A.F. COSTA & E. MARTÍNEZ (eds)
288 Surveys in combinatorics, 2001, J.W.P. HIRSCHFELD (ed)
289 Aspects of Sobolev-type inequalities, L. SALOFF-COSTE
290 Quantum groups and Lie theory, A. PRESSLEY (ed)
291 Tits buildings and the model theory of groups, K. TENT (ed)
292 A quantum groups primer, S. MAJID
293 Second order partial differential equations in Hilbert spaces, G. DA PRATO & J. ZABCZYK
294 Introduction to operator space theory, G. PISIER
295 Geometry and integrability, L. MASON & Y. NUTKU (eds)
296 Lectures on invariant theory, I. DOLGACHEV
297 The homotopy category of simply connected 4-manifolds, H.-J. BAUES
298 Higher operads, higher categories, T. LEINSTER (ed)
299 Kleinian groups and hyperbolic 3-manifolds, Y. KOMORI, V. MARKOVIC & C. SERIES (eds)
300 Introduction to Möbius differential geometry, U. HERTRICH-JEROMIN
301 Stable modules and the D(2)-problem, F.E.A. JOHNSON
302 Discrete and continuous nonlinear Schrödinger systems, M.J. ABLOWITZ, B. PRINARI & A.D. TRUBATCH
303 Number theory and algebraic geometry, M. REID & A. SKOROBOGATOV (eds)
304 Groups St Andrews 2001 in Oxford I, C.M. CAMPBELL, E.F. ROBERTSON & G.C. SMITH (eds)
305 Groups St Andrews 2001 in Oxford II, C.M. CAMPBELL, E.F. ROBERTSON & G.C. SMITH (eds)
306 Geometric mechanics and symmetry, J. MONTALDI & T. RATIU (eds)
307 Surveys in combinatorics 2003, C.D. WENSLEY (ed.)
308 Topology, geometry and quantum field theory, U.L. TILLMANN (ed)
309 Corings and comodules, T. BRZEZINSKI & R. WISBAUER
310 Topics in dynamics and ergodic theory, S. BEZUGLYI & S. KOLYADA (eds)
311 Groups: topological, combinatorial and arithmetic aspects, T.W. MÜLLER (ed)
312 Foundations of computational mathematics, Minneapolis 2002, F. CUCKER *et al* (eds)
313 Transcendental aspects of algebraic cycles, S. MÜLLER-STACH & C. PETERS (eds)
314 Spectral generalizations of line graphs, D. CVETKOVIĆ, P. ROWLINSON & S. SIMIĆ
315 Structured ring spectra, A. BAKER & B. RICHTER (eds)
316 Linear logic in computer science, T. EHRHARD, P. RUET, J.-Y. GIRARD & P. SCOTT (eds)
317 Advances in elliptic curve cryptography, I.F. BLAKE, G. SEROUSSI & N.P. SMART (eds)
318 Perturbation of the boundary in boundary-value problems of partial differential equations, D. HENRY
319 Double affine Hecke algebras, I. CHEREDNIK
320 L-functions and Galois representations, D. BURNS, K. BUZZARD & J. NEKOVÁŘ (eds)
321 Surveys in modern mathematics, V. PRASOLOV & Y. ILYASHENKO (eds)
322 Recent perspectives in random matrix theory and number theory, F. MEZZADRI & N.C. SNAITH (eds)
323 Poisson geometry, deformation quantisation and group representations, S. GUTT *et al* (eds)
324 Singularities and computer algebra, C. LOSSEN & G. PFISTER (eds)
325 Lectures on the Ricci flow, P. TOPPING
326 Modular representations of finite groups of Lie type, J.E. HUMPHREYS
327 Surveys in combinatorics 2005, B.S. WEBB (ed)
328 Fundamentals of hyperbolic manifolds, R. CANARY, D. EPSTEIN & A. MARDEN (eds)
329 Spaces of Kleinian groups, Y. MINSKY, M. SAKUMA & C. SERIES (eds)
330 Noncommutative localization in algebra and topology, A. RANICKI (ed)
331 Foundations of computational mathematics, Santander 2005, L.M PARDO, A. PINKUS, E. SÜLI & M.J. TODD (eds)
332 Handbook of tilting theory, L. ANGELERI HÜGEL, D. HAPPEL & H. KRAUSE (eds)
333 Synthetic differential geometry (2nd Edition), A. KOCK
334 The Navier–Stokes equations, N. RILEY & P. DRAZIN
335 Lectures on the combinatorics of free probability, A. NICA & R. SPEICHER
336 Integral closure of ideals, rings, and modules, I. SWANSON & C. HUNEKE
337 Methods in Banach space theory, J.M.F. CASTILLO & W.B. JOHNSON (eds)
338 Surveys in geometry and number theory, N. YOUNG (ed)
339 Groups St Andrews 2005 I, C.M. CAMPBELL, M.R. QUICK, E.F. ROBERTSON & G.C. SMITH (eds)
340 Groups St Andrews 2005 II, C.M. CAMPBELL, M.R. QUICK, E.F. ROBERTSON & G.C. SMITH (eds)
341 Ranks of elliptic curves and random matrix theory, J.B. CONREY, D.W. FARMER, F. MEZZADRI & N.C. SNAITH (eds)
342 Elliptic cohomology, H.R. MILLER & D.C. RAVENEL (eds)
343 Algebraic cycles and motives I, J. NAGEL & C. PETERS (eds)
344 Algebraic cycles and motives II, J. NAGEL & C. PETERS (eds)

345 Algebraic and analytic geometry, A. NEEMAN
346 Surveys in combinatorics 2007, A. HILTON & J. TALBOT (eds)
347 Surveys in contemporary mathematics, N. YOUNG & Y. CHOI (eds)
348 Transcendental dynamics and complex analysis, P.J. RIPPON & G.M. STALLARD (eds)
349 Model theory with applications to algebra and analysis I, Z. CHATZIDAKIS, D. MACPHERSON, A. PILLAY & A. WILKIE (eds)
350 Model theory with applications to algebra and analysis II, Z. CHATZIDAKIS, D. MACPHERSON, A. PILLAY & A. WILKIE (eds)
351 Finite von Neumann algebras and masas, A.M. SINCLAIR & R.R. SMITH
352 Number theory and polynomials, J. MCKEE & C. SMYTH (eds)
353 Trends in stochastic analysis, J. BLATH, P. MÖRTERS & M. SCHEUTZOW (eds)
354 Groups and analysis, K. TENT (ed)
355 Non-equilibrium statistical mechanics and turbulence, J. CARDY, G. FALKOVICH & K. GAWEDZKI
356 Elliptic curves and big Galois representations, D. DELBOURGO
357 Algebraic theory of differential equations, M.A.H. MACCALLUM & A.V. MIKHAILOV (eds)
358 Geometric and cohomological methods in group theory, M.R. BRIDSON, P.H. KROPHOLLER & I.J. LEARY (eds)
359 Moduli spaces and vector bundles, L. BRAMBILA-PAZ, S.B. BRADLOW, O. GARCÍA-PRADA & S. RAMANAN (eds)
360 Zariski geometries, B. ZILBER
361 Words: Notes on verbal width in groups, D. SEGAL
362 Differential tensor algebras and their module categories, R. BAUTISTA, L. SALMERÓN & R. ZUAZUA
363 Foundations of computational mathematics, Hong Kong 2008, F. CUCKER, A. PINKUS & M.J. TODD (eds)
364 Partial differential equations and fluid mechanics, J.C. ROBINSON & J.L. RODRIGO (eds)
365 Surveys in combinatorics 2009, S. HUCZYNSKA, J.D. MITCHELL & C.M. RONEY-DOUGAL (eds)
366 Highly oscillatory problems, B. ENGQUIST, A. FOKAS, E. HAIRER & A. ISERLES (eds)
367 Random matrices: High dimensional phenomena, G. BLOWER
368 Geometry of Riemann surfaces, F.P. GARDINER, G. GONZÁLEZ-DIEZ & C. KOUROUNIOTIS (eds)
369 Epidemics and rumours in complex networks, M. DRAIEF & L. MASSOULIÉ
370 Theory of *p*-adic distributions, S. ALBEVERIO, A.YU. KHRENNIKOV & V.M. SHELKOVICH
371 Conformal fractals, F. PRZYTYCKI & M. URBAŃSKI
372 Moonshine: The first quarter century and beyond, J. LEPOWSKY, J. MCKAY & M.P. TUITE (eds)
373 Smoothness, regularity, and complete intersection, J. MAJADAS & A. G. RODICIO
374 Geometric analysis of hyperbolic differential equations: An introduction, S. ALINHAC
375 Triangulated categories, T. HOLM, P. JØRGENSEN & R. ROUQUIER (eds)
376 Permutation patterns, S. LINTON, N. RUŠKUC & V. VATTER (eds)
377 An introduction to Galois cohomology and its applications, G. BERHUY
378 Probability and mathematical genetics, N. H. BINGHAM & C. M. GOLDIE (eds)
379 Finite and algorithmic model theory, J. ESPARZA, C. MICHAUX & C. STEINHORN (eds)
380 Real and complex singularities, M. MANOEL, M.C. ROMERO FUSTER & C.T.C WALL (eds)
381 Symmetries and integrability of difference equations, D. LEVI, P. OLVER, Z. THOMOVA & P. WINTERNITZ (eds)
382 Forcing with random variables and proof complexity, J. KRAJÍČEK
383 Motivic integration and its interactions with model theory and non-Archimedean geometry I, R. CLUCKERS, J. NICAISE & J. SEBAG (eds)
384 Motivic integration and its interactions with model theory and non-Archimedean geometry II, R. CLUCKERS, J. NICAISE & J. SEBAG (eds)
385 Entropy of hidden Markov processes and connections to dynamical systems, B. MARCUS, K. PETERSEN & T. WEISSMAN (eds)
386 Independence-friendly logic, A.L. MANN, G. SANDU & M. SEVENSTER
387 Groups St Andrews 2009 in Bath I, C.M. CAMPBELL *et al* (eds)
388 Groups St Andrews 2009 in Bath II, C.M. CAMPBELL *et al* (eds)
389 Random fields on the sphere, D. MARINUCCI & G. PECCATI
390 Localization in periodic potentials, D.E. PELINOVSKY
391 Fusion systems in algebra and topology M. ASCHBACHER, R. KESSAR & B. OLIVER
392 Surveys in combinatorics 2011, R. CHAPMAN (ed)
393 Non-abelian fundamental groups and Iwasawa theory, J. COATES *et al* (eds)
394 Variational Problems in Differential Geometry, R. BIELAWSKI, K. HOUSTON & M. SPEIGHT (eds)
395 How groups grow, A. MANN
396 Arithmetic Differential Operators over the p-adic Integers, C.C. RALPH & S.R. SIMANCA
397 Hyperbolic geometry and applications in quantum Chaos and cosmology, J. BOLTE & F. STEINER (eds)
398 Mathematical models in contact mechanics, M. SOFONEA & A. MATEI
399 Circuit double cover of graphs, C.-Q. ZHANG
400 Dense sphere packings: a blueprint for formal proofs, T. HALES
401 A double Hall algebra approach to affine quantum Schur–Weyl theory, B. DENG, J. DU & Q. FU
402 Mathematical aspects of fluid mechanics, J. ROBINSON, J. L. RODRIGO & W. SADOWSKI
403 Foundations of computational mathematics, Budapest 2011, F. CUCKER, T. KRICK, A. PINKUS & A. SZANTO
404 Operator methods for boundary value problems, S. HASSI, H. S. V. DE SNOO & F. H. SZAFRANIEC

London Mathematical Society Lecture Note Series: 406

# Appalachian Set Theory 2006–2012

JAMES CUMMINGS
*Carnegie Mellon University, Pennsylvania*

ERNEST SCHIMMERLING
*Carnegie Mellon University, Pennsylvania*

CAMBRIDGE
UNIVERSITY PRESS

CAMBRIDGE
UNIVERSITY PRESS

University Printing House, Cambridge CB2 8BS, United Kingdom

One Liberty Plaza, 20th Floor, New York, NY 10006, USA

477 Williamstown Road, Port Melbourne, VIC 3207, Australia

314-321, 3rd Floor, Plot 3, Splendor Forum, Jasola District Centre, New Delhi - 110025, India

103 Penang Road, #05-06/07, Visioncrest Commercial, Singapore 238467

Cambridge University Press is part of the University of Cambridge.

It furthers the University's mission by disseminating knowledge in the pursuit of education, learning and research at the highest international levels of excellence.

www.cambridge.org
Information on this title: www.cambridge.org/9781107608504



First published 2013

*A catalogue record for this publication is available from the British Library*

ISBN 978-1-107-60850-1 Paperback

# Contents

# Contributors

## Workshop lecturers

Todd Eisworth
*Department of Mathematics*
*Ohio University*
*Athens, Ohio, USA*

Ilijas Farah
*Department of Mathematics and Statistics*
*York University*
*Toronto, Ontario, Canada*
&
*Matematički Institut*
*Beograd, Serbia*

Moti Gitik
*School of Mathematical Sciences*
*Tel Aviv University*
*Tel Aviv, Israel*

Alexander S. Kechris
*Department of Mathematics*
*California Institute of Technology*
*Pasadena, California, USA*

Paul B. Larson
*Department of Mathematics*
*Miami University*
*Oxford, Ohio, USA*

Menachem Magidor
*Einstein Institute of Mathematics*
*The Hebrew University of Jerusalem*
*Jerusalem, Israel*

Justin Tatch Moore
*Department of Mathematics*
*Cornell University*
*Ithaca, New York, USA*

Itay Neeman
*Department of Mathematics*
*University of California Los Angeles*
*Los Angeles, California, USA*

Vladimir G. Pestov
*Department of Mathematics and Statistics*
*University of Ottawa*
*Ottawa, Ontario, Canada*

Simon Thomas
*Department of Mathematics*
*Rutgers University*
*Piscataway, New Jersey, USA*

Asger Törnquist
*Department of Mathematics*
*University of Copenhagen*
*Copenhagen, Denmark*

Boban Veličković
*Équipe de Logique Mathématique*
*Université Denis-Diderot Paris*
*Paris, France*

W. Hugh Woodin
*Department of Mathematics*
*University of California, Berkeley*
*Berkeley, California, USA*

## Student assistants

In cases where two addresses are given for a student contributor, the first is the address at the time of writing and the second is the current address (as of June 2012).

Jacob Davis
*Department of Mathematical Sciences*
*Carnegie Mellon University*
*Pittsburgh, Pennsylvania, USA*

Aleksandra Kwiatkowska
*Department of Mathematics*
*University of Illinois at Urbana-Champaign*
*Urbana, Illinois, USA*

*Department of Mathematics*
*University of California Los Angeles*
*Los Angeles, California, USA*

Chris Lambie-Hanson
*Department of Mathematical Sciences*
*Carnegie Mellon University*
*Pittsburgh, Pennsylvania, USA*

Peter Lumsdaine
*Department of Mathematical Sciences*
*Carnegie Mellon University*
*Pittsburgh, Pennsylvania, USA*

*Department of Mathematics and Statistics*
*Dalhousie University*
*Halifax, Nova Scotia, Canada*

Martino Lupini
*Department of Mathematics and Statistics*
*York University*
*Toronto, Ontario, Canada*

David Milovich
*Department of Mathematics*
*University of Wisconsin*
*Madison, Wisconsin, USA*

*Department of Engineering, Mathematics and Physics*
*Texas A&M International University*
*Laredo, Texas, USA*

Daniel Rodríguez
*Department of Mathematical Sciences*
*Carnegie Mellon University*
*Pittsburgh, Pennsylvania, USA*

Scott Schneider
*Department of Mathematics*
*Rutgers University*
*Piscataway, New Jersey, USA*

*Department of Mathematics*
*University of Michigan*
*Ann Arbor, Michigan, USA*

Robin D. Tucker-Drob
*Department of Mathematics*
*California Institute of Technology*
*Pasadena, California, USA*

Spencer Unger
*Department of Mathematical Sciences*
*Carnegie Mellon University*
*Pittsburgh, Pennsylvania, USA*

Giorgio Venturi
*Équipe de Logique Mathématique*
*Université Denis-Diderot Paris*
*Paris, France*

Eric Wofsey
*Department of Mathematics*
*Washington University in St. Louis*
*St. Louis, Missouri, USA*

*Department of Mathematics*
*Harvard University*
*Cambridge, Massachusetts, USA*

Yimu Yin
*Department of Philosophy*
*Carnegie Mellon University*
*Pittsburgh, Pennsylvania, USA*

*Équipe Analyse Algébrique*
*Institut de Mathématiques de Jussieu*
*Université Pierre et Marie Curie*
*Paris, France*

# Introduction

This volume collects lecture notes from talks given in the Appalachian Set Theory workshop series (supported by the National Science Foundation) during the period 2006–2012.

This workshop series grew out of an informal series of expository lectures held at Carnegie Mellon University and attended by set theorists from universities in Appalachian states before 2006. The success of these earlier gatherings inspired the editors to formalize the series and seek funding to help more people attend. Participants from other universities were invited to host workshops as well. Typically there are three meetings a year with one taking place at Carnegie Mellon University and the remaining two elsewhere. Several of the workshops have been held in neighbouring regions but the series retains its Appalachian flavour.

At each workshop a leading researcher lectures for six hours on an important topic or technique in modern set theory. Students are engaged to assist in writing notes based on the lectures, and these notes are disseminated on the web. This provides a learning opportunity for the students and makes the notes universally available.

The papers collected here represent more polished versions of the lecture notes from most of the workshops to date. They were prepared collaboratively by the lecturers and the student assistants. The lecturers are the principal authors and their names appear first, followed by the names of the assistants. One workshop (represented in Chapter 7) had two lecturers, and two workshops (represented in Chapters 1 and 13) had two assistants. All the papers in this volume were refereed, several by two referees; many of the referee reports were outstandingly detailed and helpful.

The main goals of the series are:

- To disseminate important ideas (some of which have been known to experts for a long time) to a wider audience, and in particular to make them accessible to graduate students.
- To bring researchers in set theory into contact with each other, and strengthen the set-theoretic research community.

The workshops have covered a wide range of topics including forcing and large cardinals, descriptive set theory, and applications of set-theoretic ideas in group theory and analysis. In line with our goals, about half of the workshop audiences have been students and postdoctoral researchers. We have also attracted regular faculty mathematicians at all levels of seniority. Most of the participants work in various parts of set theory but a significant number are experts in other areas of mathematics.

Information about past and future workshops in the series can be found at the URL

http://www.math.cmu.edu/~eschimme/Appalachian/Index.html

James Cummings
Ernest Schimmerling
Pittsburgh, June 2012

**Acknowledgements:**

We are truly and lastingly grateful to the lecturers and student assistants for the very large commitment of time and energy which they have made, and to the referees of the papers in this volume for their diligent and painstaking work.

Many thanks are also due to the local organisers of the meetings: Dietmar Bisch, Elizabeth Brown, John Clemens, Alan Dow, Todd Eisworth, George Elliott, Ilijas Farah, Paul Larson, Justin Moore, Christian Rosendal, Steve Simpson, and Juris Steprāns.

The Appalachian Set Theory workshops have been generously supported by the National Science Foundation under grants DMS-0631446 and DMS-0903845. Thanks are due to the staff of the Foundations program in the NSF Division of Mathematical Sciences, and in particular to the Foundations program director Tomek Bartoszyński.

Thanks also to the office staff at CMU (particularly Ferna Hartman, Jeff Moreci, and Nancy Watson) for their work on administering the NSF grants and their help with the CMU-based workshops.

We have received excellent support from the staff at Cambridge University Press during the preparation of this volume. Special thanks to Sam Harrison and the TEX support group for solving some knotty LaTeX problems at the last minute.

# 1

# An introduction to $\mathbb{P}_{max}$ forcing

Paul B. Larson[a], Peter Lumsdaine and Yimu Yin

---

The first Appalachian Set Theory workshop was held at Carnegie Mellon University in Pittsburgh on September 9, 2006. The lecturer was Paul Larson. As graduate students Peter Lumsdaine and Yimu Yin assisted in writing this chapter, which is based on the workshop lectures.

---

## 1 Introduction

The forcing construction $\mathbb{P}_{max}$ was invented by W. Hugh Woodin in the early 1990's in the wake of his result that the saturation of the nonstationary ideal on $\omega_1$ ($NS_{\omega_1}$) plus the existence of a measurable cardinal implies that there is a definable counterexample to the Continuum Hypothesis (in particular, it implies that $\delta^1_2 = \omega_2$, which implies $\neg$CH). These notes outline a proof of the $\Pi_2$ maximality of the $\mathbb{P}_{max}$ extension, which we can state as follows.

**Theorem 1.1** ([9]) *Suppose that there exist proper class many Woodin cardinals, $A \subseteq \mathbb{R}$, $A \in L(\mathbb{R})$, $\varphi$ is $\Pi_2$ in the extended language containing two additional unary predicates, and in some set forcing extension*

$$\langle H(\omega_2), \in, NS_{\omega_1}, A^* \rangle \models \varphi$$

*(where $A^*$ is the reinterpretation of $A$ in this extension). Then*

$$L(\mathbb{R})^{\mathbb{P}_{max}} \models [\langle H(\omega_2), \in, NS_{\omega_1}, A \rangle \models \varphi] .$$

[a] Research supported by NSF grants DMS-0401603 and DMS-0801009.

Forcing with $\mathbb{P}_{\max}$ does not add reals, so there is no need to reinterpret $A$ in the last line of the theorem. The theorem says that any such $\Pi_2$ statement that we can force in any extension must hold in the $\mathbb{P}_{\max}$ extension of $L(\mathbb{R})$, so $H(\omega_2)$ of $L(\mathbb{R})^{\mathbb{P}_{\max}}$ is maximal, or complete, in a certain sense; among other things, it models ZFC, Martin's Axiom, certain fragments of Martin's Maximum [1], and the negation of the Continuum Hypothesis. The reinterpretation $A^*$ will be defined later, in terms of tree representations for sets of reals. We will not give the definition of Woodin cardinals (but see [4]).

We have reworked the standard proof of Theorem 1.1 in order to minimize the prerequisites. In particular, the need for (mentioning) sharps has been eliminated. However, they and much more will need to be reintroduced to go any further than the material presented here.

Woodin's book on $\mathbb{P}_{\max}$, *The axiom of determinacy, forcing axioms and the nonstationary ideal* [9] runs to around 1000 pages. The first author's article for the Handbook of Set Theory [6], introducing $\mathbb{P}_{\max}$, has about 65. The advance notes for these lectures were about 30 pages, and previous lecture courses have taken about 12-15 hours to cover $\mathbb{P}_{\max}$. Today we have about six hours, so we will, of course, have to be brief...

## 2 Setup: iterations and the definition of $\mathbb{P}_{\max}$

Suppose that $M \models$ ZFC and that $I \in M$ is a normal ideal on $\omega_1^M$ in $M$. Force over $M$ with $((\mathcal{P}(\omega_1) \setminus I)/I)^M$. The resulting generic $G$ is now an $M$-normal ultrafilter on $\omega_1^M$; so we may form the corresponding ultrapower and elementary embedding

$$j : M \to \mathrm{Ult}(M, G) := \{f : \omega_1^M \to M \mid f \in M\}/ =_G .$$

(We will use this construction many times today.) Via Fodor's Lemma, we get that $\mathrm{crit}(j) = \omega_1^M$, and for $A \in \mathcal{P}(\omega_1)^M$,

$$A \in G \iff \omega_1^M \in j(A).$$

Via consideration of surjections from $\omega_1^M$ onto each ordinal below $\omega_2^M$, we get that $j(\omega_1^M) \geq \omega_2^M$. There is no need to assume that $M$ is well-founded, though it will be in the cases we are interested in. When the model $\mathrm{Ult}(M, G)$ is well-founded, we identify it with its Mostowski collapse. It is not hard to see that in this case $\mathrm{Ord}^{\mathrm{Ult}(M,G)} = \mathrm{Ord}^M$.

**Definition 2.1** $I$ is *precipitous* if $\mathrm{Ult}(M, G)$ thus constructed is well-founded from the point of view of $M[G]$, for all $M$-generic $G$. (N.B. this is definable in $M$ via forcing.)

We need a pair of theories satisfying the following conditions.

- $T_0$, a theory consistent with ZFC and strong enough to make sense of the generic ultrapower construction above and prove that $j : M \to \mathrm{Ult}(M, G)$ is elementary.
- $T_1$, a theory consistent with ZFC and at least as strong as $T_0$ + "every set lies in some $H(\kappa) \models T_0$."

In [6], we take $T_0$ (which we call ZFC°) to be ZFC - Replacement - Powerset plus "$\mathcal{P}(\mathcal{P}(\omega_1))$ exists" plus the scheme saying that definable trees of height $\omega_1$ have maximal branches. Then we let

$$T_1 = T_0 + \text{Powerset} + \text{Choice} + \Sigma_1\text{–Replacement}$$

(though we don't express it in these terms). In [9], Woodin has an even weaker fragment of ZFC (which he calls ZFC*) playing the role of $T_0$. Today we may as well let $T_0$ be ZFC and $T_1$ be ZFC plus the existence of a proper class of strongly inaccessible cardinals. From now on we will just use the terms $T_0$ and $T_1$.

We now extend the generic ultrapower construction given above to iterated ultrapowers. Suppose we have $(M_0, I_0)$, $G_0 \subsetneq (\mathcal{P}(\omega_1)/I_0)^{M_0}$,

$$j_0 : (M_0, I_0) \to \mathrm{Ult}(M_0, G_0),$$

all as before; let $M_1 = \mathrm{Ult}(M_0, G_0)$, $I_1 = j_0(I_0)$. Now we can take the generic ultrapower of $M_1$ by $I_1$, and iterate. At limit stages, we have a directed system of elementary embeddings, so we can just take the direct limit, so we can keep going up to length $\omega_1$. (No further, as if we force again there, we collapse $\omega_1^V$, so are back to countable length!) Note that the final model of the iteration, $M_{\omega_1}$, is an element of $H(\omega_2)$.

**Definition 2.2** An *iteration* of $(M, I)$ (as above, with $\mathcal{P}(\mathcal{P}(\omega_1))^M$ countable) of length $\gamma$ consists of $M_\alpha$, $I_\alpha$ ($\alpha \leq \gamma$), $G_\eta$ ($\eta < \gamma$), and $j_{\alpha,\beta}$ ($\alpha \leq \beta \leq \gamma$), satisfying

- $M_0 = M$, $I_0 = I$.
- $G_\eta$ is $M_\eta$-generic for $(P(\omega_1)/I_\eta)^{M_\eta}$.
- $j_{\eta,\eta+1}$ is the canonical embedding of $M_\eta$ into $\mathrm{Ult}(M_\eta, G_\eta) = M_{\eta+1}$.
- $j_{\alpha,\beta} : M_\alpha \to M_\beta$ are a commuting family of elementary embeddings.

- $I_\beta = j_{0,\beta}(I)$.
- For limit $\beta$, $M_\beta$ is the direct limit of $\{M_\alpha \mid \alpha < \beta\}$ under the embeddings $j_{\alpha,\eta}$ ( $\alpha \leq \eta < \beta$).

In practice, we almost always have $\gamma = \omega_1^N$ for some larger $N \supseteq M$. We will generally write $\langle M_\alpha, I_\alpha, G_\eta, j_{\alpha,\beta} \mid \alpha \leq \beta \leq \gamma,\ \eta < \gamma\rangle$ for the iteration, or just "$j$ is an iteration" to mean that $j$ is the $j_{0,\gamma}$ of an iteration, with $j(M)$ for $M_\gamma$. (As we will see, in some circumstances, if we know $M_0$, $M_\gamma$, $j_{0,\gamma}$, we can (with slight assumptions) recover the full iteration.) We say that the $M_\alpha$'s are *iterates* of $(M, I)$; $(M, I)$ is *iterable* if all iterates are well-founded; and $(M, I)$ is an *iterable pair* if $M$ is a transitive model of $T_0$ with $\mathcal{P}(\mathcal{P}(\omega_1))^M$ countable, $I$ a normal ideal on $\mathcal{P}(\omega_1)$ in $M$, and $(M, I)$ is iterable.

If $M$ is well-founded and $M \models$ "$I$ is precipitous", then certainly $(M, I)$ is finitely iterable (i.e., its finite-length iterations produce wellfounded models); and in fact, we will show that in this case $(M, I)$ is iterable to any $\alpha \in \mathrm{Ord}^M$.

The proof of the following lemma is left as an exercise (the proof is by induction on the length of the iteration). In a typical application, $M$ is $H(\kappa)^N$ for some suitable $\kappa$.

**Lemma 2.3** *Suppose that $M \in N$ are models of $T_0$, $M$ is closed under $\omega_1$-sequences from $N$, and $\mathcal{P}(\mathcal{P}(\omega_1))^M = \mathcal{P}(\mathcal{P}(\omega_1))^N$. Let $I \in M$ be an $M$-normal ideal on $\omega_1^M$. Then the following hold.*

- *For each iteration $\langle M_\alpha, I_\alpha, G_\eta, j_{\alpha,\beta} \mid \alpha \leq \beta \leq \gamma,\ \eta < \gamma\rangle$ of $(M, I)$ there is a unique iteration $\langle N_\alpha, I_\alpha, G_\eta, j^*_{\alpha,\beta} \mid \alpha \leq \beta \leq \gamma,\ \eta < \gamma\rangle$ of $(N, I)$ such that for all $\beta \leq \gamma$: $j^*_{0,\beta}(M) = M_\beta$, $M_\beta$ is closed under $\omega_1$-sequences from $N_\beta$, $\mathcal{P}(\mathcal{P}(\omega_1))^{M_\beta} = \mathcal{P}(\mathcal{P}(\omega_1))^{N_\beta}$, and $j^*_{\alpha,\beta} \restriction M_\alpha = j_{\alpha,\beta}$.*
- *For each iteration $\langle N_\alpha, I_\alpha, G_\eta, j^*_{\alpha,\beta} \mid \alpha \leq \beta \leq \gamma,\ \eta < \gamma\rangle$ of $(N, I)$ there is a unique iteration $\langle M_\alpha, I_\alpha, G_\eta, j_{\alpha,\beta} \mid \alpha \leq \beta \leq \gamma,\ \eta < \gamma\rangle$ of $(M, I)$ such that for all $\beta \leq \gamma$: $j^*_{0,\beta}(M) = M_\beta$, $M_\beta$ is closed under $\omega_1$-sequences from $N_\beta$, $\mathcal{P}(\mathcal{P}(\omega_1))^{M_\beta} = \mathcal{P}(\mathcal{P}(\omega_1))^{N_\beta}$, and $j^*_{\alpha,\beta} \restriction M_\alpha = j_{\alpha,\beta}$.*

**Corollary 2.4** *In the context of Lemma 2.3, if $(M, I)$ has an ill-founded iterate by an iteration of length $\alpha$, then so does $(N, I)$.*

Lemma 2.5 below then shows that $(M, I)$ is iterable if $N$ contains $\omega_1$ (recall that iterations can have length at most $\omega_1$, and note that an illfounded iteration of length $\omega_1$ must be illfounded at some countable stage).

First we fix a coding of elements of $H(\omega_1)$ by reals. Fix a recursive bijection $\pi : \omega \times \omega \to \omega$, and say $X \subseteq \omega$ *codes* $a \in H(\omega_1)$ if

$$(\mathrm{tc}(\{a\}), \in) \cong (\omega, \{(i, j) \mid \pi(i, j) \in X\}),$$

where $tc(b)$ denotes the transitive closure of $b$. Then $\in$ and $=$ are $\Sigma^1_1$ (as permutations of $\omega$ induce different codes for the same object).

**Lemma 2.5** *Suppose that $N$ is a transitive model of $T_1$, $\gamma \in \mathrm{Ord}^N$, and $I$ is a normal precipitous ideal on $\omega_1^N$ in $N$. Then any iterate of $(N, I)$ by an iteration of length $\gamma$ is well-founded.*

*Proof* It suffices to prove that iterations of the form $(H(\kappa)^N, I)$ produce wellfounded models for all $\kappa \in N$ such that $H(\kappa)^N \models T_0$; for if any iterate of $N$ is ill-founded, then some ordinal in $N$ is large enough to witness this (i.e., $\sup(\mathrm{rge}(f))$, where $f$ witnesses ill-foundedness) and by assumption (as $N \models T_1$) this is contained in some $H(\kappa)^N$ that models $T_0$.

Let $(\bar{\gamma}, \bar{\kappa}, \bar{\eta})$ be the lexicographically minimal triple $(\gamma, \kappa, \eta)$ satisfying (with $N$) the formula $\varphi(N, \gamma, \kappa, \eta)$ defined by "$H(\kappa)^N \models T_0$ and there exists an iteration of $(H(\kappa)^N, I)$ of length $\gamma$ which is ill-founded below the image of $\eta$".

Using our fixed coding of elements of $H(\omega_1)$ by reals there is a $\Sigma^1_1$ formula $\varphi'(x, y, z)$ saying "$x$ codes a model of $T_0$ and a normal ideal in the model on the $\omega_1$ of the model and there exists an iteration of this pair whose length is coded by $y$ and which is illfounded below the image of the element of this model coded by $z$."

For all cardinals $\kappa, \rho \in N$ and all ordinals $\gamma, \eta \in N$, if $\rho \in N$ is larger than $|H(\kappa)|^N$, $|\eta|^N$ and $|\gamma|^N$, then there exist reals coding $H(\kappa)^N$, $\eta$, and $\gamma$ in any forcing extension of $N$ by $\mathrm{Coll}(\omega, \rho)$. Such an extension is correct about whether these reals satisfy $\varphi'$. However, this is a homogeneous forcing extension of $N$; so there is a formula $\psi(\gamma, \kappa, \eta)$ saying that in every forcing extension in which $H(\kappa)$ (of the ground model), $\eta$ and $\gamma$ are all countable there exist reals coding $H(\kappa)$, $\eta$ and $\gamma$ which satisfy $\varphi'$. It follows that that $N \models \psi(\gamma, \kappa, \eta)$ if and only if $\varphi(N, \gamma, \kappa, \eta)$ holds, and furthermore, since $\varphi'$ is $\Sigma^1_1$, for all well-founded iterates $N^*$ of $N$, and all $\gamma, \kappa, \eta \in N^*$, $N^* \models \psi(\gamma, \kappa, \eta)$ if and only if $\varphi(N^*, \gamma, \kappa, \eta)$ holds.

Since $I$ is precipitous in $N$, $\bar{\gamma}$ is a limit ordinal, and clearly $\bar{\eta}$ is a limit ordinal as well. Fix an iteration $\langle M_\alpha, G_\beta, j_{\alpha,\delta} : \beta < \alpha \leq \delta \leq \bar{\gamma}\rangle$ of $(H(\bar{\kappa})^N, I)$ such that $j_{0,\bar{\gamma}}(\bar{\eta})$ is not wellfounded, and let $\langle N_\alpha, G_\beta, j'_{\alpha,\delta} : \beta < \alpha \leq \delta \leq \bar{\gamma}\rangle$ be the corresponding iteration of $N$ as in Lemma 2.3. By the minimality of $\bar{\gamma}$ we have that $N_\alpha$ is wellfounded for all $\alpha < \bar{\gamma}$. Since

$M_{\bar{\gamma}}$ is the direct limit of the iteration leading up to it, we may fix $\gamma^* < \bar{\gamma}$ and $\eta^* < j_{0,\gamma^*}(\bar{\eta})$ such that $j_{\gamma^*,\bar{\gamma}}(\eta^*)$ is not wellfounded. By Lemma 2.3, $j'_{\gamma^*,\bar{\gamma}}(\eta^*) = j_{\gamma^*,\bar{\gamma}}(\eta^*)$ and $j'_{\gamma^*,\bar{\gamma}}(\bar{\eta}) = j_{\gamma^*,\bar{\gamma}}(\bar{\eta})$.

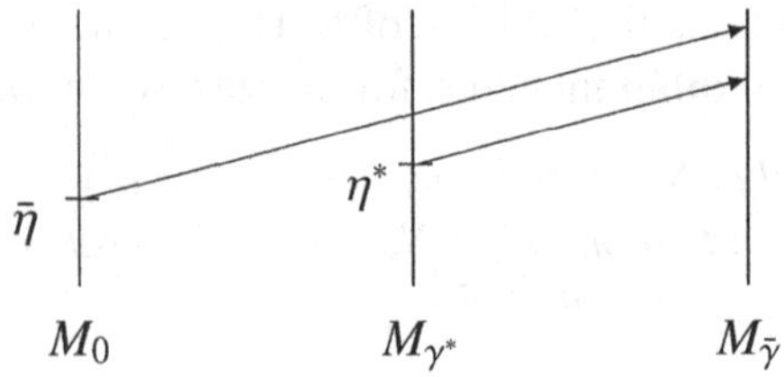


But now, $N_{\gamma^*} \models \psi(\bar{\gamma} - \gamma^*, j_{0,\gamma^*}(\bar{\kappa}), \eta^*)$, $\bar{\gamma} - \gamma^* \leq \bar{\gamma}$, and $\eta^* < j_{0,\gamma^*}(\bar{\eta})$, contradicting the minimality of $(j_{0,\gamma^*}(\bar{\gamma}), j_{0,\gamma^*}(\bar{\kappa}), j_{0,\gamma^*}(\bar{\eta}))$ in $N_{\gamma^*}$. □

We note that ZFC does not imply the existence of iterable pairs. However, by Lemma 2.5, if there exist a normal, precipitous ideal $J$ on $\omega_1$, and a measurable cardinal $\kappa$ with a $\kappa$-complete ultrafilter $\mu$, then there exist iterable pairs. The main point here is that if $\theta > \kappa$ is a regular cardinal and $X$ is a countable elementary submodel of $H(\theta)$ with $\kappa, J \in X$, then $X$ can be end-extended below $\kappa$ by taking $\gamma$ to be any member of $\bigcap(X \cap \mu)$, and letting $X[\gamma]$ be the set of values $f(\gamma)$ for all functions $f$ in $X$ with domain $\kappa$. Applying this fact $\omega_1$ many times, we get that the transitive collapse $M$ of $X \cap V_\kappa$ is a countable model which is a rank initial segment of a model containing $\omega_1$. Letting $I$ be the image of $J$ under the transitive collapse, then, $(M, I)$ is an iterable pair. This is a special case of the proof of Lemma 4.6, and a key point in Woodin's proof (which appears in Chapter 3 of [9]) that if there exists a measurable cardinal and the nonstationary ideal on $\omega_1$ is saturated, then CH fails.

If there is a precipitous ideal on $\omega_1$, then sharps exist for subsets of $\omega_1$, and a countable iterable model will be correct about these sharps. We will work around this today to avoid having to talk about sharps.

**Lemma 2.6** *If $(M, I)$ is an iterable pair and $A$ is an element of $\mathcal{P}(\omega_1)^M$, then $(\omega_1^{L[A]})^M = \omega_1^{L[A]}$.*

*Proof* Let $\langle M_\alpha, I_\alpha, G_\eta, j_{\alpha,\beta} \mid \alpha \leq \beta \leq \omega_1,\ \eta < \omega_1 \rangle$ be an iteration of $(M, I)$. The ordinals of $M$ and $M_1$ are the same, so $L[A]^M = L[A]^{M_1}$. The critical point of $j_{1,\omega_1}$ is greater than $\omega_1^M$, and thus greater than the $\omega_1$ of $L[A]^M$. The restriction of $j_{1,\omega_1}$ to $L[A]^M$ embeds $L[A]^M$ elementarily into $L[A]^{M_{\omega_1}}$, which means that $L[A]^M$ and $L[A]^{M_{\omega_1}}$ have the same $\omega_1$. Since $\omega_1 \subset M_{\omega_1}$, $(\omega_1^{L[A]})^{M_{\omega_1}} = \omega_1^{L[A]}$. □

Now we can define $\mathbb{P}_{\max}$.

**Definition 2.7** The partial order $\mathbb{P}_{\max}$ is the set of pairs $\langle (M, I), a\rangle$ such that

1. $M$ is a countable transitive model of $T_0 + \mathrm{MA}_{\aleph_1}$,
2. $(M, I)$ is an iterable pair,
3. $a \in \mathcal{P}(\omega_1)^M$ and there exists $x \in \mathcal{P}(\omega)^M$ such that $\omega_1^{L[x,a]} = \omega_1^M$,

ordered by: $p < q$ (where $p = \langle (M, I), a\rangle$, $q = \langle (N, J), b\rangle$) if there is some iteration $j : (N, J) \to (N^*, J^*)$ such that

1. $j \in M$,
2. $j(b) = a$,
3. $J^* = N^* \cap I$ ( and hence $j(\omega_1^N) = \omega_1^M$),
4. $q \in H(\omega_1)^M$.

Note that since $j \in M$ in definition of the $\mathbb{P}_{\max}$ order above, $N$ and $N^*$ are both in $M$ as well.

**Definition 2.8** We say that $(M, I)$ is a $\mathbb{P}_{\max}$ *precondition* if there exists an $a$ such that $\langle (M, I), a\rangle \in \mathbb{P}_{\max}$, or equivalently just if $(M, I)$ satisfies conditions 1 and 2 in the definition of $\mathbb{P}_{\max}$ conditions above.

Suppose that $(M, I)$ is an iterable pair, and $j$: $(M, I) \to (M', I')$ is an iteration. Then for any $A \in \mathcal{P}(\omega_1)^M$ which is bounded in $\omega_1^M$, $j(A) = A$. By Lemma 2.6, it follows then that $\omega_1^{L[A]} < \omega_1^M$, since $j(\omega_1^M) > \omega_1^M$ if $j$ is nontrivial. Therefore, the set $a$ from a $\mathbb{P}_{\max}$ condition $\langle (M, I), a\rangle$ must always be unbounded in $\omega_1^M$ to make $\omega_1^{L[x,a]} = \omega_1^M$ possible.

If $p_0 < p_1 < p_2$ ($p_i = \langle (M_i, I_i), a_i\rangle$), and these are witnessed by $j_{1,0}$, $j_{2,1}$, then $p_0 < p_2$ is witnessed by $j_{1,0}(j_{2,1})$: $j_{1,0} \in H(\omega_2)^{M_0}$, $j_{2,1} \in H(\omega_2)^{M_1}$; $j_{2,1}$ is an iteration of $(M_2, I_2)$, and $j_{1,0}((M_2, I_2)) = (M_2, I_2)$.

Under our fixed coding, "$(M, I)$ is iterable" is $\Pi^1_2$ in a code for $(M, I)$: roughly, "for anything satisfying the first-order properties of being an iteration, either there is no infinite descending sequence in the ordinals of the final model, or there is an infinite descending sequence in the indices of the iteration." Since iterable models embed elementarily into models containing $\omega_1$, they are $\mathbf{\Pi}^1_2$-correct. It follows that "$(M, I)$ is iterable" is absolute to iterable models containing a code for $(M, I)$.

So now we see that $\mathbb{P}_{\max} \in L(\mathbb{R})$ — all constructions involved are nicely codable.

## 3 First properties of $\mathbb{P}_{\max}$

The requirement that the models in $\mathbb{P}_{\max}$ conditions satisfy $\mathrm{MA}_{\aleph_1}$ is used for a particular consequence of $\mathrm{MA}_{\aleph_1}$ known as *almost disjoint coding* [2]. That is, it follows from $\mathrm{MA}_{\aleph_1}$ that if $Z = \{z_\alpha : \alpha < \omega_1\}$ is a collection of infinite subsets of $\omega$ whose pairwise intersections are finite (i.e., $Z$ is an *almost disjoint family*), then for each $B \subseteq \omega_1$ there exists a $y \subseteq \omega$ such that for all $\alpha < \omega_1$, $\alpha \in B$ if and only if $y \cap z_\alpha$ is infinite. This is used to show that if $\langle (M, I), a\rangle$ is a $\mathbb{P}_{\max}$ condition, then any iteration of $(M, I)$ is uniquely determined by the image of $a$ (Lemma 3.1 below), so the order on each comparable pair of conditions is witnessed by a unique iteration.

**Lemma 3.1** *Let $\langle (M, I), a\rangle$ be a $\mathbb{P}_{\max}$ condition and let $A$ be a subset of $\omega_1$. Then there is at most one iteration of $(M, I)$ for which $A$ is the image of $a$.*

*Proof* Fix a real $x$ in $M$ such that $\omega_1^M = \omega_1^{L[a,x]}$, and let

$$Z = \langle z_\alpha : \alpha < \omega_1^M\rangle$$

be the almost disjoint family defined recursively from the constructibility order in $L[a, x]$ on $\mathcal{P}(\omega)^{L[a,x]}$ (using $a$ and $x$ as parameters) by letting $\langle z_i : i < \omega\rangle$ be the constructibly least partition of $\omega$ into infinite pieces, and, for each $\alpha \in [\omega, \omega_1^M)$, letting $z_\alpha$ be the constructibly least infinite $z \subset \omega$ almost disjoint from each $z_\beta$ $(\beta < \alpha)$. Suppose that

$$\mathcal{I} = \langle M_\alpha, G_\beta, j_{\alpha,\delta} : \beta < \alpha \leq \delta \leq \gamma\rangle$$

and

$$\mathcal{I}' = \langle M'_\alpha, G'_\beta, j'_{\alpha,\delta} : \beta < \alpha \leq \delta \leq \gamma'\rangle$$

are two iterations of $(M, I)$ such that $j_{0,\gamma}(a) = A = j'_{0\gamma'}(a)$. Then

$$j_{0,\gamma}(\omega_1^M) = \omega_1^{L[x,A]} = j_{0,\gamma}(\omega_1^{M'})$$

and $j_{0,\gamma}(Z) = j'_{0,\gamma'}(Z)$ (this uses Lemma 2.6 to see that the constructibility order on reals in $L[A, x]$ is computed correctly in $M_\gamma$ and $M'_{\gamma'}$). Let $\langle z_\alpha : \alpha < j_{0,\gamma}(\omega_1^M)\rangle$ enumerate $j_{0,\gamma}(Z)$.

Without loss of generality, $\gamma \leq \gamma'$. We show by induction on $\alpha < \gamma$ that, for each such $\alpha$, $G_\alpha = G'_\alpha$. This will suffice. Fix $\alpha$ and suppose that

$$\{G_\beta : \beta < \alpha\} = \{G'_\beta : \beta < \alpha\}.$$

Then $M_\alpha = M'_\alpha$. For each $B \in \mathcal{P}(\omega_1)^{M_\alpha}$, $B \in G_\alpha$ if and only if $\omega_1^{M_\alpha} \in$

$j_{\alpha,\alpha+1}(B)$, and $B \in G'_\alpha$ if and only if $\omega_1^{M_\alpha} \in j'_{\alpha,\alpha+1}(B)$. Fixing B and applying almost disjoint coding, let $y \in \mathcal{P}(\omega)^{M_\alpha}$ be such that for all $\eta < \omega_1^{M_\alpha}$, $\eta \in B$ if and only if $y \cap z_\eta$ is infinite. Then $B \in G_\alpha$ if and only if $y \cap z_{\omega_1^{M_\alpha}}$ is infinite, which holds if and only if $B \in G'_\alpha$. □

**Lemma 3.2** *($T_0$) Suppose that $(M, I)$ is an iterable pair, and $J$ is a normal ideal on $\omega_1$. Then there exists an iteration $j\colon (M, I) \to (M^*, I^*)$ of length $\omega_1$ such that $I^* = M^* \cap J$.*

*Proof* Note that $I^* \subseteq M^* \cap J$ holds for any such $\omega_1$-length iteration. To see this, first note that the critical sequence of an iteration of length $\omega_1$ is a club. Every element $B$ of $I^*$ is $j_{\alpha,\omega_1}(b)$ for some $\alpha < \omega_1$ and $b \in I_\alpha$. Then for all $\beta \in [\alpha, \omega_1)$, $j_{\alpha,\beta}(b) \notin G_\beta$, so $\omega_1^{M_\beta} \notin j_{\alpha,\omega_1}(b) = B$; thus $B \in \mathrm{NS}_{\omega_1}$, but $J$ is normal, so $NS_{\omega_1} \subseteq J$.

Conversely, for $\supseteq$: as $J$ is normal, we may let $\langle E_i^\alpha \mid \alpha < \omega_1, i < \omega\rangle$ be a partition of $\omega_1$ into $J$-positive pieces. Now, as we construct an iteration, let $\{e_i^\alpha \mid i < \omega\}$ enumerate $\mathcal{P}(\omega_1)^{M_\alpha} \setminus I_\alpha$, and build each $G_\beta$ in such a way that if $\omega_1^{N_\beta} \in E_i^\alpha$ for some $\alpha \leq \beta$ and $i < \omega$, then $j_{\alpha,\beta}(e_i^\alpha)$ is in $G_\beta$.

Now, for all $B \in \mathcal{P}(\omega_1)^{M_\alpha} \setminus I_{\omega_1}$, there exist $\alpha < \omega_1$, $i < \omega$ such that $B = j_{\alpha,\omega_1}(e_i^\alpha)$ and for all $\beta \in [\alpha, \omega_1)$,

$$\omega_1^{M_\beta} \in E_i^\alpha \Rightarrow \omega_1^{M_\beta} \in j_{0,\beta+1}(e_i^\alpha) \Rightarrow \omega_1^{M_\beta} \in B.$$

So in particular, we have a club $C \subseteq \omega_1$ such that $C \cap E_i^\alpha \subseteq B$, so $B \notin J$. □

We may consider this as an *iteration game* $G((M, I), J, B)$: two players collaborate on building an iteration of $(M, I)$, and play is as follows at each round $\alpha$:

- if $\alpha \notin B$, player I does nothing, and player II chooses $G_\alpha$;
- if $\alpha \in B$, player I specifies some element for $G_\alpha$, and player II must choose some $G_\alpha$ containing it.

Player I wins if $I_{\omega_1} = M_{\omega_1} \cap J$. The above proof shows that player I has a winning strategy iff $B \notin J$. (More precisely, it shows $\Leftarrow$; $\Rightarrow$ is because if $B \in J$ then II may choose some $I$-positive set and keep its images out of every $G_\alpha$.)

The following lemma shows that $\mathbb{P}_{\max}$ satisfies a homogeneity property strong enough to imply that the theory of the generic extension can be computed in the ground model. Since the existence of a proper class of Woodin cardinals implies that the theory of $L(\mathbb{R})$ is generically absolute, it implies that the theory of the $\mathbb{P}_{\max}$ extension of $L(\mathbb{R})$ is generically absolute as well.

**Lemma 3.3** *Suppose that for each $x \in H(\omega_1)$ there exists a $\mathbb{P}_{\max}$ precondition $(M, I)$ such that $x \in M$. Then for all $p_0, p_1 \in \mathbb{P}_{\max}$, there exist $q_0, q_1 \in \mathbb{P}_{\max}$ such that each $q_i \leq p_i$, and $\mathbb{P}_{\max} \restriction q_0 \cong \mathbb{P}_{\max} \restriction q_1$.*

*Proof* Take $p_i = \langle (M_i, I_i), a_i \rangle$ for $i = 0, 1$. Then let $(N, J)$ be a $\mathbb{P}_{\max}$ precondition such that $p_0, p_1 \in H(\omega_1)^N$. Now take $j_i \colon (M_i, I_i) \to (M_i^*, I_i^*)$ $(i = 0, 1)$ to be iterations in $N$ such that $I_i^* = M_i^* \cap J$ (we may do so, by the previous theorem applied in $N$).

Now set $q_i = \langle (N, J), j_i(a_i) \rangle \in \mathbb{P}_{\max}$. Certainly these satisfy $q_i < p_i$ as desired. (To see that these $q_i$ are indeed conditions, note that the witnessing $x_i$ for $p_i$ (i.e., the $x \in \mathcal{P}(\omega)^{M_i}$ such that $\omega_1^{L[x,a_i]} = \omega_1^{M_i}$) still works for $q_i$.) But now $\mathbb{P}_{\max} \restriction q_0 \cong \mathbb{P}_{\max} \restriction q_1$, for given any $r_0 = \langle (N', I'), b \rangle < q_0$, there is unique $j : (N, J) \to (N^*, J^*)$ witnessing this (and we have $b = j(j_0(a_0))$); now take $r_0$ to $r_1 := \langle (N', J'), j(j_1(a_1)) \rangle < q_1$, also witnessed by $j$. □

Given $\gamma \in [\omega_1, \omega_2)$, a *canonical function* for $\gamma$ is a function $f \colon \omega_1 \to \omega_1$ such that for some (equivalently, every) bijection $\pi : \omega_1 \to \gamma$,

$$\{\alpha < \omega_1 \mid \mathrm{ot}(\pi[\alpha]) = f(\alpha)\}$$

contains a club. In a normal ultrapower context, we then have: $f \colon \omega_1 \to \mathrm{Ord}$, $[f] = j(f)(\omega_1) = \mathrm{ot}(\pi[\omega_1]) = \gamma$.

Suppose that $\langle M_\alpha, I_\alpha, G_\eta, j_{\alpha,\beta} \mid \alpha \leq \beta \leq \omega_1,\ \eta < \omega_1 \rangle$ is an iteration of length $\omega_1$ of some $(M, I)$, and let $\pi : \omega_1 \to \mathrm{Ord}^{M_{\omega_1}}$ be a bijection; then $\pi$ induces a canonical function $g$. For club-many $\alpha < \omega_1$, $\omega_1^{M_\alpha} = \alpha$, and also

$$\pi[\alpha] = j_{\alpha,\omega_1}[\mathrm{Ord}^{M_\alpha}]$$

(recall that we take direct limits at limit stages of iterations); so there is a club $C \subset \omega_1$ such that for all $\alpha \in C$, $\mathrm{ot}(\pi[\alpha]) = \mathrm{Ord}^{M_\alpha}$. If $f \in ({\omega_1}^{\omega_1})^{M_\beta}$, for some $\beta < \alpha$, then for any $\alpha \in [\beta, \omega_1)$,

$$j_{\beta,\alpha+1}(f)(\omega_1^{M_\alpha}) < \omega_1^{M_{\alpha+1}} < \mathrm{Ord}^{M_{\alpha+1}} = \mathrm{Ord}^{M_\alpha},$$

which equals $g(\alpha)$ if $\alpha \in C$. It follows that for any $f \in ({\omega_1}^{\omega_1})^{M_{\omega_1}}$, $g(\alpha) > f(\alpha)$ for club-many $\alpha$. We will use this fact to show that $\mathbb{P}_{\max}$ is $\sigma$-closed (this is an alternate proof avoiding sharps; some of what follows can be done more easily and in more generality with sharps).

**Lemma 3.4** *Suppose that for each $x \in H(\omega_1)$ there exists a $\mathbb{P}_{\max}$ pre-condition $(M, I)$ such that $x \in M$. If $p_i \in \mathbb{P}_{\max}$ $(i < \omega)$ are such that for all $i$ $p_{i+1} < p_i$, then there exists $q \in \mathbb{P}_{\max}$ such that $q < p_i$ for all $i$.*

The proof of Lemma 3.4 involves some new notions. Say that

$$p_i = \langle (M_i, I_i), a_i \rangle,$$

and for each $j < i < \omega$, let $k_{i,j} : M_j \to M_j^* \in M_i$ be the unique witness for $p_i < p_j$. By the uniqueness of witnesses, the $k'_{i,j}s$ commute, so let $N_i$, $J_i$ be the images of $M_i$, $I_i$ in the limit of the directed system given by the embeddings $k_{i,j}$. Each $(N_i, J_i)$ is an iterate of the corresponding $(M_i, I_i)$ by an iteration of length $\sup\{\omega_1^{M_i} : i < \omega\}$, so each $N_i$ is wellfounded. Let $b = \bigcup_{i<\omega} a_i$. Then:

1. Each $(N_i, J_i)$ is iterable;
2. for all $i$, $\omega_1^{N_i} = \sup_{j<\omega} \omega_1^{M_j}$;
3. $i < j \Rightarrow N_i \in H(\omega_2)^{N_j}$;
4. for all $i$, $J_i = J_{i+1} \cap N_i$;
5. for each $i$ there exists some iteration $j_i : (M_i, I_i) \to (N_i, J_i)$ in $N_{i+1}$ such that $j_1(a_i) = b$ (and so in $N_{i+1}$, there is a canonical function for $\mathrm{Ord}^{N_i}$ that dominates on a club every member of $(\omega_1^{\omega_1})^{N_i}$).

We call a sequence $\langle (N_i, J_i) : i < \omega \rangle$ satisfying (1.) to (5.) above a $\mathbb{P}_{\max}$ *limit sequence*. An $\langle (N_i, J_i) \mid i < \omega \rangle$*-normal ultrafilter* is a filter $G \subseteq \bigcup_i (\mathcal{P}(\omega_1)^{N_i} \setminus J_i)$ such that for all $i < \omega$, and for all regressive $f \in ({\omega_1}^{\omega_1})^{N_i}$ there exists $e \in G$ such that $f$ is constant on $e$. The fact that $J_i = J_{i+1} \cap N_i$ for all $i$, plus the countability of the models, implies that such ultrafilters exist. Then we have

$$\mathrm{Ult}(\langle (N_i, J_i) : i < \omega \rangle, G),$$

i.e., a sequence of models whose $j$th model $[\mathrm{Ult}(\langle (N_i, J_i) : i < \omega \rangle, G)]_j$ is

$$\{f : \omega_1^{N_0} \to N_j \mid f \in \bigcup \{N_i \mid i < \omega\}\} / =_G .$$

This sequence is also a $\mathbb{P}_{\max}$ limit sequence, and there is an induced embedding $j$ whose restriction to $N_i$ maps to the $i$th model of the sequence.

Now we will iterate this operation.

**Definition 3.5** An *iteration* of $\langle (N_i, J_i) : i < \omega \rangle$ of length $\gamma$ is some

$$\langle \langle (N_i^\alpha, J_i^\alpha) : i < \omega \rangle, G_\eta, j_{\alpha,\beta} \mid \alpha \le \beta \le \gamma, \eta < \gamma \rangle$$

in which:

- each $G_\eta$ is a $\langle (N_i^\eta, J_i^\eta) \mid i < \omega\rangle$-normal ultrafilter contained in
$$\bigcup_i (\mathcal{P}(\omega_1)^{N_i^\eta} \setminus J_i^\eta)$$
such that for each $i < \omega$, $j_{\eta,\eta+1} \restriction N_i^\eta \to N_i^{\eta+1}$ is the induced ultrapower
- the embeddings $j_{\alpha,\beta}$ commute, and, for limit $\beta$, $N_i^\beta$ is the direct limit of $N_i^\alpha$ $(\alpha < \beta)$ under $j_{\alpha,\rho} \restriction N_i^\alpha$ $(\alpha \leq \rho < \beta)$.

In an iteration of this form, for each pair $i, \alpha$, there is in $N_{i+1}^\alpha$ an iteration of $(M_i, I_i)$ of length $\omega_1^{N_0^\alpha}$, with final model $N_i^\alpha$. Since each $(M_i, I_i)$ is iterable, the wellfoundedness of each $N_i^\alpha$ will follow from the wellfoundedness of $\omega_1^{N_0^\alpha}$.

For each $\alpha < \gamma$,
$$\omega_1^{N_0^{\alpha+1}} = \sup\{\mathrm{Ord}^{N_i^\alpha} \mid i < \omega\}.$$

To see this, fix for each $i \in \omega$ a canonical function $g_i \in N_{i+1}^\alpha$ for $\mathrm{Ord}^{N_i}$. Then each $g_i$ dominates on a club every member of $(\omega_1^{\omega_1})^{N_i^\alpha}$. The $g_i$ are cofinal under mod-$NS_{\omega_1}$ domination in $\bigcup_i (\omega_1{}^{\omega_1})^{N_i^\alpha}$, and each $g_i$ represents an ordinal in $N_{i+1}^\alpha$ in this ultrapower, which shows that $\omega_1^{N_0^{\alpha+1}} = \sup\{\mathrm{Ord}^{N_i^\alpha} \mid i < \omega\}$. For limit $\beta$,
$$\omega_1^{N_0^\beta} = \sup\{\omega_1^{N_0^\alpha} \mid \alpha < \beta\}.$$
It follows that each $N_0^\alpha$ is wellfounded.

We have shown the following.

*Fact* 3.6 All iterations of $\mathbb{P}_{\max}$ pre-limit sequences give well-founded models.

Again this can be rephrased in terms of games. Let
$$G_\omega(\langle (N_i, J_i) \mid i < \omega\rangle, I, B)$$
be the game of length $\omega_1$ in which I and II collaborate to build an iteration of $\langle (N_i, J_i) : i < \omega\rangle$ of length $\omega_1$, in which at stage $\alpha$:

- if $\alpha \in B$, player I chooses $e \in \bigcup_i (\mathcal{P}(\omega_1)^{N_i^\alpha} \setminus J_i^\alpha)$, and player II chooses $G_\alpha$, a $\langle (N_i^\alpha, J_\alpha^i) \mid i < \omega\rangle$-normal ultrafilter contained in $\bigcup_i (\mathcal{P}(\omega_1)^{N_i^\alpha} \setminus J_i^\alpha)$, with $e \in G_\alpha$;
- if $\alpha \notin B$, player I does nothing, and player II chooses any suitable $G_\alpha$.

Player I wins if for all $i$ $j_{0,\omega_1}(J_i) = I \cap M_i^{\omega_1}$. The argument just given (along with the argument for Lemma 3.2) shows the following.

**Lemma 3.7** *Suppose $\langle (N_i, J_i) \mid i < \omega \rangle$ is a $\mathbb{P}_{\max}$ pre-limit-sequence, $I$ is a normal ideal on $\omega_1$ and $B \subseteq \omega_1$. Then Player I has a winning strategy in $G_\omega(\langle (N_i, J_i) \mid i < \omega \rangle, I, B)$ if and only if $B$ is not in $I$.*

We now return to the proof of Lemma 3.4. We have that the limit sequence $\langle (N_i, J_i) : i < \omega \rangle$ induced by the descending sequence $p_i$ ($i \in \omega$) is iterable. Fix a $\mathbb{P}_{\max}$ precondition $(M', I')$ such that this sequence is in $H(\omega_1)^{M'}$. Apply a winning strategy for player I in $M'$ for $G_{\omega_1}(\langle (N_i, J_i) \mid i < \omega \rangle, I', \omega_1)$ to get an iteration $j$ of $\langle (N_i, J_i) : i < \omega \rangle$ of length $\omega_1^{M'}$. Then for all $i < \omega$, $j(j_i)$ witnesses that $p_i > \langle (M', I'), j(b) \rangle$.

Thus $\mathbb{P}_{\max}$ forcing is $\sigma$-closed, so it does not add any reals; so

$$L(\mathbb{R})^{V^{\mathbb{P}_{\max}}} = L(\mathbb{R})^V.$$

## 4 Existence of $\mathbb{P}_{\max}$ conditions

**Definition 4.1** Given $A \subseteq \mathbb{R}$, and an iterable pair $(M, I)$, we say that $(N, I)$ is *$A$-iterable* if $A \cap M \in M$ and for any iteration $j : (M, I) \to (M^*, I^*)$,

$$j(A \cap M) = A \cap M^*.$$

In this section we will work through a proof of the following existence theorem for $\mathbb{P}_{\max}$ conditions. The same conclusion can be reached from Woodin's axiom $AD^+$.

**Lemma 4.2** (Main existence lemma) *Suppose there are infinitely many Woodin cardinals below some measurable cardinal, and let $A \in \mathcal{P}(\mathbb{R}) \cap L(\mathbb{R})$. Then there exists an $A$-iterable $\mathbb{P}_{\max}$ precondition $(M, I)$ such that for every set forcing extension $M^+$ of $M$ and every precipitous ideal $I^+ \in M^+$ on $\omega_1^{M^+}$, $(M^+, I^+)$ is $A$-iterable.*

We need to introduce towers of measures and homogeneous trees. See [8] for a detailed discussion of this material

**Definition 4.3** Given $Z \neq \phi$, a *tower of measures* on $Z$ is a sequence

$$\langle \mu_i \mid i < \omega \rangle$$

such that each $\mu_i \subseteq \mathcal{P}(Z^i)$ is an ultrafilter, and for all $k < i < j$ and all $A \in \mu_i$, we have $\{b \in Z^j \mid b \restriction i \in A\} \in \mu_j$ and $\{b \restriction k \mid b \in A\} \in \mu_k$.

Such a tower is *countably complete* if whenever $\langle A_i \mid i < \omega \rangle$ is such that each $A_i \in \mu_i$, there is $a \in Z^\omega$ such that for all $i$ $a \upharpoonright i \in A_i$.

We note briefly that countable completeness is equivalent to: the direct limit of $\mathrm{Ult}(V, \mu_i)$ is well-founded.

**Definition 4.4** A *tree on* $\omega \times Z$ is a set $T \subseteq (\omega \times Z)^{<\omega}$ such that for all $i < \omega, t \in T$ we have $t \upharpoonright i \in T$. The projection of $T$ is

$$p[T] := \{y \in \omega^\omega \mid \exists c \in Z^\omega \, \forall i < \omega \, (y \upharpoonright i, c \upharpoonright i) \in T\}.$$

Such a tree is *weakly $\kappa$-homogeneous* (for $\kappa$ a cardinal) if there exist $\kappa$-complete ultrafilters $\mu_{a,b} \subseteq \mathcal{P}(Z^{|a|})$ (for all $a, b \in \omega^{<\omega}$ with $|a| = |b|$) such that

$$\{c \in Z^{|a|} \mid (a, c) \in T\} \in \mu_{a,b},$$

and such that for each $x \in p[T]$ there exists a $b \in \omega^\omega$ such that

$$\langle \mu_{x \upharpoonright i, b \upharpoonright i} \mid i < \omega \rangle$$

is a countably complete tower.

Weakly homogeneous trees originated from work of Kechris, Martin and Solovay. The following fact is due to Woodin. A proof appears in [5].

*Fact* 4.5 If $\delta$ is a limit of Woodin cardinals and there is a measurable cardinal above $\delta$, then for each $A \in \mathcal{P}(\mathbb{R}) \cap L(\mathbb{R})$ and $\gamma < \delta$, there exists a $\gamma$-weakly-homogeneous tree $T$ such that $p[T] = A$.

**Lemma 4.6** *Suppose that $T \subset (\omega \times Z)^{<\omega}$ is a $\gamma^+$-weakly-homogeneous tree, $\theta > (2^{|T|})^+$ is regular, $X \prec H(\theta)$, $T, \gamma \in X$, $|X| < \gamma$, and $\bar{a} \in p[T]$. Then there exists $Y \prec H(\theta)$ with $X \subseteq Y$, $X \cap \gamma = Y \cap \gamma$, $|X| = |Y|$, and $\bar{a} \in p[T \cap Y]$.*

*Proof* Fix $\langle \mu_{a,b} \mid a, b \in \omega^{<\omega} \rangle \in X$ witnessing the $\gamma^+$-weak-homogeneity of T. Since $\bar{a} \in p[T]$, there exists $\bar{b}$ such that $\langle \mu_{\bar{a} \upharpoonright i, \bar{b} \upharpoonright i} \mid i < \omega \rangle$ is a countably complete tower. Now let $A_i = \bigcap (X \cap \mu_{\bar{a} \upharpoonright i, \bar{b} \upharpoonright i})$. Each $A_i$ is in $\mu_{\bar{a} \upharpoonright i, \bar{b} \upharpoonright i}$; so there is $\bar{c} \in Z^\omega$ such that for all $i$ $\bar{c} \upharpoonright i \in A_i$. Take

$$Y = X[\{\bar{c} \upharpoonright i \mid i < \omega\}] := \{f(\bar{c} \upharpoonright i) \mid f \in X, \mathrm{dom}(f) = Z^{<\omega}, i < \omega\}.$$

Elementarity of $Y$ follows from an argument similar to the proof of Łós's Theorem (see Theorem 1.1.13 of [5]). To see that $Y \cap \gamma = X \cap \gamma$, note that if $\alpha \in Y \cap \gamma$, then $\alpha = f(c \upharpoonright i)$ for some $f \in X$, $\mathrm{dom}\, f = Z^i$; but $\mu_{a \upharpoonright i, b \upharpoonright i}$ is $\gamma^+$-complete, so $f$ is constant on a set in $\mu_{a \upharpoonright i, b \upharpoonright i}$. But then this constant value is in $X$, and $f(c \restriction i)$ is this value, since $c \restriction i = \bigcap(\mu_{a\restriction i, b\restriction i} \cap X)$. □

Note that $Y$ in the proof above is in some sense a limit ultrapower of the transitive collapse of $X$.

The following was first proved by Foreman, Magidor and Shelah from a supercompact cardinal, and later improved by Woodin.

*Fact* 4.7 If $\delta$ is Woodin, then $\mathrm{Coll}(\omega_1, <\delta)$ forces that $\mathrm{NS}_{\omega_1}$ is presaturated, and hence precipitous.

Recall that an ideal $I$ on $\omega_1$ is *presaturated* if for every sequence of maximal antichains $\{Q_i \mid i < \omega\} \subset \mathcal{P}(\omega_1) \setminus I$, for every $A \in I^+$, there is $B \subseteq A$, $B \in I^+$ such that for all $i < \omega$,

$$|\{E \in Q_i \mid E \cap B \in I^+\}| \leq \aleph_1.$$

The following was proved by Kakuda and Magidor independently [3, 7].

*Fact* 4.8 If $\mathrm{NS}_{\omega_1}$ is precipitous, then it remains precipitous after any c.c.c. forcing.

Recall the hypotheses of the main existence lemma: $\delta$ is a limit of Woodin cardinals, there exists a measurable cardinal greater than $\delta$, and $A$ is in $\mathcal{P}(\mathbb{R}) \cap L(\mathbb{R})$. To prove the lemma, let $\kappa$ be the least Woodin cardinal, and $\gamma$ the least strong inaccessible above $\kappa$. Fix $\gamma^+$-weakly-homogeneous trees $S, T$, with $p[S] = A$, $p[T] = \mathbb{R} \setminus A$. Fix a regular $\theta > (2^{|S|})^+, (2^{|T|})^+$. Let $X$ be a countable elementary submodel of $H(\theta)$, with $S, T, \gamma, \kappa \in X$. Repeatedly apply Lemma 4.6 above to obtain $Y \prec H(\theta)$ such that $X \subseteq Y$, $X \cap \gamma = Y \cap \gamma$, $A = p[S \cap Y]$, $\mathbb{R} \cap A = p[T \cap Y]$. (Then $|Y \cap \mathrm{Ord}| = 2^\omega$.) Now let $N$ be the transitive collapse of $Y$, and let $\bar{S}, \bar{T}, \bar{\gamma}, \bar{\kappa}$ be the images of $S, T, \gamma, \kappa$ therein. Let $h$ be $N$-generic for $\mathrm{Coll}(\omega_1, <\bar{\kappa})$ followed by a c.c.c. poset of size $2^{\omega_1}$ to make $\mathrm{MA}_{\aleph_1}$ hold. Then

$$N[h] \models \mathrm{MA}_{\aleph_1} + \text{“}\mathrm{NS}_{\omega_1} \text{is precipitous''}.$$

Then $N[h]$ is iterable, by Lemma 2.5. Let $M$ be $(V_{\bar{\gamma}})^{N[h]}$, and let

$$j : (M, \mathrm{NS}_{\omega_1}^M) \to (M^*, \mathrm{NS}_{\omega_1}^*)$$

be an iteration. By Lemma 2.3, this induces an iteration of $(N[h], \mathrm{NS}_{\omega_1}^{N[h]})$ with final model $(N^*, I^*)$ (which we'll also call $j$). Now, $N^*$ is well-founded, and $p[\bar{S}] \subseteq p[j(\bar{S})]$ and $p[\bar{T}] \subseteq p[j(\bar{T})]$. But by elementarity,

$$N^* \models p[j(\bar{S})] \cap p[j(\bar{T})] = \emptyset,$$

and since $N^*$ is well-founded it is correct about this. Then $p[\bar{S}] = p[j(\bar{S})]$ and $p[\bar{T}] = p[j(\bar{T})]$, so $j(A \cap M) = p[j(\bar{S})] \cap M^* = A \cap M^*$.

*Remark* 4.9   Instead of $\mathrm{Coll}(\omega_1, < \bar{\kappa})$, we could have taken $h$ to be $N$-generic for any poset in $V^N_{\bar{\gamma}}$ such that

$$N[h] \models \text{"there exists precipitous } I \text{ on } \omega_1\text{"},$$

and the rest of the proof would have still gone through.

The main existence lemma gives not only $A$-iterable preconditions for any $A \in \mathcal{P}(\mathbb{R}) \cap L(\mathbb{R})$, but also $A$-iterable preconditions containing any given real $x$, for any $A \in \mathcal{P}(\mathbb{R}) \cap L(\mathbb{R})$, applying the lemma to the set $\{y \oplus x \mid y \in A\}$. Thus we have shown: if there exist infinitely many Woodin cardinals below a measurable, then for all $x \in \mathbb{R}$, for all $A \in \mathcal{P}(\mathbb{R}) \cap L(\mathbb{R})$, there exists some $\mathbb{P}_{\max}$ condition $\langle(M, I), a\rangle$, with $x \in M$, and $(M, I)$ $A$-iterable.

We didn't quite show that $\langle H(\omega_1)^M, \in, A \cap M\rangle \prec \langle H(\omega_1), \in, A\rangle$. We can do this using $A^\sharp$, or by using not just $S, T$ as above but similar trees for all sets projective in $A$.

Given a filter $G \subset \mathbb{P}_{\max}$, $A_G$ denotes the set $\bigcup\{e \mid \exists\langle(N, J), e\rangle \in G\}$. We also omit a proof of the following:

*Fact* 4.10 ("The combinatorial heart of the $\mathbb{P}_{\max}$ analysis")   Suppose that for each $A \in \mathcal{P}(\mathbb{R}) \cap L(\mathbb{R})$ there exists an $A$-iterable $\mathbb{P}_{\max}$ precondition $(N, I)$ such that

$$\langle H(\omega_1)^M, \in, A \cap M\rangle \prec \langle H(\omega_1), \in, A\rangle,$$

and suppose that $G \subseteq \mathbb{P}_{\max}$ is an $L(\mathbb{R})$-generic filter. Then for all $B \in \mathcal{P}(\omega_1)^{L(\mathbb{R})[G]}$, there exists $\langle(M, I), a\rangle \in G$ such that $B = j(b)$ for some $b \in \mathcal{P}(\omega_1)^M$, where $j$ is the unique iteration of $(M, I)$ satisfying $j(a) = A_G$.

In other words, all subsets of $\omega_1$ in extensions come from models in the conditions, and $L(\mathbb{R})[G] = L(\mathbb{R})[A_G]$.

**Corollary 4.11**   *Suppose that for each $A \in \mathcal{P}(\mathbb{R}) \cap L(\mathbb{R})$ there exists an $A$-iterable $\mathbb{P}_{\max}$ precondition $(N, I)$ such that*

$$\langle H(\omega_1)^M, \in, A \cap M\rangle \prec \langle H(\omega_1), \in, A\rangle,$$

*and suppose that $G \subseteq \mathbb{P}_{\max}$ is an $L(\mathbb{R})$-generic filter. Then $\mathrm{NS}_{\omega_1}^{L(\mathbb{R})[G]}$ is the collection of all sets of the form $j(e)$, where for some $\langle(M, I), a\rangle \in G$, $e \in I$, and $j$ is the iteration of $(M, I)$ sending $a$ to $A_G$.*

Woodin has shown that the hypotheses of Fact 4.10 are equivalent to the assertion that AD holds in $L(\mathbb{R})$.

## 5 $\Pi_2$ maximality

*Proof of Goal 1* Fix a $\Pi_2$ sentence $\varphi = \forall x \exists y \psi(x, y)$ (in the extended language with two new unary predicates), and an $A \in \mathcal{P}(\mathbb{R}) \cap L(\mathbb{R})$. To show that

$$\langle H(\omega_2), \in, A, \mathrm{NS}_{\omega_1} \rangle^{L(\mathbb{R})^{\mathbb{P}_{\max}}} \models \varphi,$$

it is sufficient to show that for each $\langle (M, I), a \rangle \in \mathbb{P}_{\max}$ and each $b \in H(\omega_2)^M$, there exist $\langle (N, \mathrm{NS}^N_{\omega_1}), e \rangle \in \mathbb{P}_{\max}$ and $j : (M, I) \to (M^*, I^*)$ in $N$ such that $j(a) = e$, $I^* = M^* \cap \mathrm{NS}^N_{\omega_1}$, and

$$\langle H(\omega_2)^N, \in, A \cap N, \mathrm{NS}^N_{\omega_1} \rangle \models \exists d\, \psi(j(b), d).$$

The argument is like the one for existence of conditions.

So suppose $\langle (M, I), a \rangle$ is given. Fix $P$ forcing $\varphi$. Let $\delta$ be the least Woodin cardinal with $p \in V_\delta$; let $\kappa$ be the least strong inaccessible above $\delta$. Let $S, T$ be $\kappa^+$-weakly-homogeneous trees projecting to $A$, $\mathbb{R} \setminus A$. Let $\theta > (2^{|S|})^+, (2^{|T|})^+$ be regular. Fix $Y \prec H(\theta)$ with $Y \cap \kappa$ countable, $p[S \cap Y] = A$, $p[T \cap Y] = \mathbb{R} \setminus A$ and $\langle (M, I), a \rangle \in Y$.

Let $N$ be the transitive collapse of $Y$, and let $\bar{P}$, $\bar{S}$, $\bar{\delta}$, $\bar{\kappa}$ be the respective images of $P$, $S$, $\delta$, $\kappa$ under this collapse. Let $h_0$ be $\bar{P}$-generic for $N$. Note that since $P \in V_\delta$, $\bar{\delta}$ remains Woodin in $N[h_0]$. The reinterpretation of $A$ is the projection of $\bar{S}$ in the extension. Thus

$$\langle H(\omega_2)^{N[h_0]}, \in, (p[\bar{S}])^{N[h_0]}, \mathrm{NS}^{N[h_0]}_{\omega_1} \rangle \models \varphi.$$

Pick an iteration $j$ of $(M, I)$ in $N$ such that $j(I) = j(M) \cap \mathrm{NS}^{N[h_0]}_{\omega_1}$. Then there exists a $d \in H(\omega_2)^{N[h_0]}$ such that

$$\langle H(\omega_2)^{N[h_0]}, \in, (p[\bar{S}])^{N[h_0]}, \mathrm{NS}^{N[h_0]}_{\omega_1} \rangle \models \psi(j(b), d).$$

Let $h_1$ be $N[h_0]$-generic for $\mathrm{Coll}(\omega_1, < \bar{\delta})^{N[h_0]}$ followed by some c.c.c. forcing making $\mathrm{MA}_{\aleph_1}$ hold. Now $\langle ((V_{\bar{\kappa}})^{N[h_0][h_1]}, \mathrm{NS}^{N[h_0][h_1]}_{\omega_1}), j(a) \rangle$ is the desired condition. □

## 6 Discussion

*Question* 6.1 You've shown that under these conditions, any forceable $\Pi_2$ statement must hold in the $\mathbb{P}_{max}$ extension. Can you give us some cool examples?

*Answer.* One example is $\varphi_{AC}$: "For every stationary, costationary $A, B \subseteq \omega_1$, there is some $\gamma \in [\omega_1, \omega_2)$, some bijection $\pi : \omega_1 \to \gamma$ such that

$$\{\alpha < \omega_1 \mid \alpha \in A \iff (\pi[\alpha]) \in B\}$$

contains a club." This can be used to get an injection $\mathcal{P}(\omega_1) \hookrightarrow \omega_2$, which shows that the Axiom of Choice holds in the $\mathbb{P}_{max}$ extension of $L(\mathbb{R})$.

Also, in some cases one can use $\mathbb{P}_{max}$ to get $\Pi_2$ maximality relative to a given $\Sigma_2$ statement; that is, for a given $\Sigma_2$ statement for $H(\omega_2)$, you can simultaneously get all $\Pi_2$ statements forceably consistent with it.

Another useful aspect: often, the combinatorics of forcing to kill off one thing while preserving another are not clear; the combinatorics of doing the same by an iteration may be much clearer. For instance, an analysis of iterations may help answer the question of whether there exists a Dowker space on $\omega_1$.

## References

[1] Matthew Foreman, Menachem Magidor, and Saharon Shelah. Martin's Maximum, saturated ideals, and nonregular ultrafilters. I. *Ann. of Math. (2)*, 127(1):1–47, 1988.

[2] Ronald B. Jensen and Robert M. Solovay. Some applications of almost disjoint sets. In *Mathematical Logic and Foundations of Set Theory (Proc. Internat. Colloq., Jerusalem, 1968)*, pages 84–104. North-Holland, Amsterdam, 1970.

[3] Yuzuru Kakuda. On a condition for Cohen extensions which preserve precipitous ideals. *J. Symbolic Logic*, 46(2):296–300, 1981.

[4] Akihiro Kanamori. *The higher infinite*. Springer Monographs in Mathematics. Springer-Verlag, Berlin, second edition, 2003. Large cardinals in set theory from their beginnings.

[5] Paul B. Larson. *The stationary tower*, volume 32 of *University Lecture Series*. American Mathematical Society, Providence, RI, 2004. Notes on a course by W. Hugh Woodin.

[6] Paul B. Larson. Forcing over models of determinacy. In *Handbook of Set Theory*, pages 2121–2177. Springer, New York, 2010.

[7] Menachem Magidor. Precipitous ideals and $\Sigma^1_4$ sets. *Israel J. Math.*, 35(1-2):109–134, 1980.

[8] Donald A. Martin and John R. Steel. A proof of projective determinacy. *J. Amer. Math. Soc.*, 2(1):71–125, 1989.

[9] W. Hugh Woodin. *The axiom of determinacy, forcing axioms, and the nonstationary ideal*, volume 1 of *de Gruyter Series in Logic and its Applications*. Walter de Gruyter & Co., Berlin, 1999.

# 2

# Countable Borel Equivalence Relations

Simon Thomas[a] and Scott Schneider

The fourth Appalachian Set Theory workshop was held at Ohio University in Athens on November 17, 2007. The lecturer was Simon Thomas. As a graduate student Scott Schneider assisted in writing this chapter, which is based on the workshop lectures.

*Dedicated to the memory of Greg Hjorth*

## First lecture

In these lectures, we will present an introduction to the theory of countable Borel equivalence relations, focusing on some recent applications of the Popa and Ioana Superrigidity Theorems.

### 1.1 Standard Borel spaces and Borel equivalence relations

A topological space is said to be *Polish* if it admits a complete, separable metric. If $\mathcal{B}$ is a $\sigma$-algebra of subsets of a given set $X$, then the pair $(X, \mathcal{B})$ is called a *standard Borel space* if there exists a Polish topology $\mathcal{T}$ on $X$ that generates $\mathcal{B}$ as its Borel $\sigma$-algebra; in which case, we write $\mathcal{B} = \mathcal{B}(\mathcal{T})$. For example, each of the sets $\mathbb{R}$, $[0, 1]$, $\mathbb{N}^{\mathbb{N}}$, and $2^{\mathbb{N}} = \mathcal{P}(\mathbb{N})$ is Polish in its natural topology, and so may be viewed, equipped with its corresponding Borel structure, as a standard Borel space.

The abstraction involved in passing from a topology to its associated

[a] Research partially supported by NSF Grant 0600940.

Borel structure is analogous to that of passing from a metric to its induced topology. Just as distinct metrics on a space may induce the same topology, distinct topologies may very well generate the same Borel $\sigma$-algebra. In a standard Borel space, then, one "remembers" only the Borel sets, and forgets which of them were open; it is natural therefore to imagine that *any* of them might have been, and indeed this is the case:

**Theorem 1.1** *Let $(X, \mathcal{T})$ be a Polish space and $Y \subseteq X$ any Borel subset. Then there exists a Polish topology $\mathcal{T}_Y \supseteq \mathcal{T}$ such that $\mathcal{B}(\mathcal{T}_Y) = \mathcal{B}(\mathcal{T})$ and $Y$ is clopen in $(X, \mathcal{T}_Y)$.*

It follows that if $(X, \mathcal{B})$ is a standard Borel space and $Y \in \mathcal{B}$, then $(Y, \mathcal{B} \upharpoonright Y)$ is also a standard Borel space. In fact, so much structural information is "forgotten" in passing from a Polish space to its Borel structure that we obtain the following theorem of Kuratowski [22].

**Theorem 1.2** *There exists a unique uncountable standard Borel space up to isomorphism.*

It turns out that many classification problems from diverse areas of mathematics may be viewed as definable equivalence relations on standard Borel spaces. For example, consider the problem of classifying all countable graphs up to graph isomorphism. Let $C$ be the set of graphs of the form $\Gamma = \langle \mathbb{N}, E \rangle$. Then identifying each graph $\Gamma \in C$ with its edge relation $E \in 2^{\mathbb{N}^2}$, one easily checks that $C$ is a Borel subset of $2^{\mathbb{N}^2}$ and hence is a standard Borel space. Moreover, the isomorphism relation on $C$ is simply the orbit equivalence relation arising from the natural action of $\mathrm{Sym}(\mathbb{N})$ on $C$. More generally, if $\sigma$ is a sentence of $\mathcal{L}_{\omega_1,\omega}$, then

$$\mathrm{Mod}(\sigma) = \{ \mathcal{M} = \langle \mathbb{N}, \cdots \rangle \mid \mathcal{M} \models \sigma \}$$

is a standard Borel space, and the isomorphism relation on $\mathrm{Mod}(\sigma)$ is the orbit equivalence relation arising from the natural $\mathrm{Sym}(\mathbb{N})$-action. However, while the isomorphism relation on $\mathrm{Mod}(\sigma)$ is always an analytic subset of $\mathrm{Mod}(\sigma) \times \mathrm{Mod}(\sigma)$, it is not in general a Borel subset; for example, the graph isomorphism relation on $C$ is not Borel. On the other hand, the restriction of graph isomorphism to the standard Borel space of connected locally finite graphs is Borel; and more generally, the isomorphism relation on a standard Borel space of countable structures will be Borel if each structure is "finitely generated" in some broad sense. With these examples in mind, we make the following definitions.

**Definition 1.3** If $X$ is a standard Borel space, then a *Borel equivalence* *relation* on $X$ is an equivalence relation $E \subseteq X^2$ which is a Borel subset of $X^2$.

**Definition 1.4** If $G$ is a Polish group, then a *standard Borel $G$-space* is a standard Borel space $X$ equipped with a Borel $G$-action $(g, x) \mapsto g \cdot x$. The corresponding $G$-orbit equivalence relation is denoted by $E_G^X$.

We observe that if $G$ is a countable group and $X$ is a standard Borel $G$-space, then $E_G^X$ is a Borel equivalence relation. As further examples, we will next consider the standard Borel space $R(\mathbb{Q}^n)$ of torsion-free abelian groups of rank $n$ and the Polish space $\mathcal{G}$ of finitely generated groups.

For each $n \geq 1$, let $\mathbb{Q}^n = \bigoplus_{1 \leq i \leq n} \mathbb{Q}$. Then the standard Borel space of torsion-free abelian groups of rank $n$ is defined to be

$$R(\mathbb{Q}^n) = \{A \leq \mathbb{Q}^n \mid A \text{ contains a basis of } \mathbb{Q}^n\}.$$

Notice that if $A, B \in R(\mathbb{Q}^n)$, then we have that

$$A \cong B \quad \text{iff} \quad \text{there exists } \varphi \in \mathrm{GL}_n(\mathbb{Q}) \text{ such that } \varphi(A) = B,$$

and hence the isomorphism relation on $R(\mathbb{Q}^n)$ is the Borel equivalence relation arising from the natural action of $\mathrm{GL}_n(\mathbb{Q})$ on $R(\mathbb{Q}^n)$.

As a step towards defining the Polish space $\mathcal{G}$ of finitely generated groups, for each $m \in \mathbb{N}$, let $\mathbb{F}_m$ be the free group on the $m$ generators $\{x_1, \ldots, x_m\}$ and let $2^{\mathbb{F}_m}$ be the compact space of all functions $\varphi : \mathbb{F}_m \to 2$. Then, identifying each subset $S \subseteq \mathbb{F}_m$ with its characteristic function $\chi_S \in 2^{\mathbb{F}_m}$, it is easily checked that the collection $\mathcal{G}_m$ of normal subgroups of $\mathbb{F}_m$ is a closed subset of $2^{\mathbb{F}_m}$. In particular, $\mathcal{G}_m$ is a compact Polish space. Next, as each $m$-generator group can be realized as a quotient $\mathbb{F}_m/N$ for some $N \in \mathcal{G}_m$, we can identify $\mathcal{G}_m$ with the space of $m$-generator groups. Finally, there exists a natural embedding $\mathcal{G}_m \hookrightarrow \mathcal{G}_{m+1}$ defined by

$$N \mapsto \text{the normal closure of } N \cup \{ x_{m+1} \} \text{ in } \mathbb{F}_{m+1};$$

and so we can define the space of finitely generated groups by $\mathcal{G} = \bigcup_{m \geq 1} \mathcal{G}_m$.

By a theorem of Tietze, if $N, M \in \mathcal{G}_m$, then $\mathbb{F}_m/N \cong \mathbb{F}_m/M$ if and only if there exists $\pi \in \mathrm{Aut}(\mathbb{F}_{2m})$ such that $\pi(N) = M$.[1] In particular, it

[1] It is probably worth pointing out that this is *not* a misprint. For example, by Dunwoody-Pietrowski [9], there exist normal subgroups $N, M \leqslant \mathbb{F}_2$ with $\mathbb{F}_2/N \cong \mathbb{F}_2/M$ such that $\theta(N) \neq M$ for all $\theta \in \mathrm{Aut}(\mathbb{F}_2)$. However, if we identify $N, M$ with the corresponding normal subgroups of $\mathbb{F}_4$ via the natural embedding $\mathcal{G}_2 \hookrightarrow \mathcal{G}_3 \hookrightarrow \mathcal{G}_4$, then there exists $\pi \in \mathrm{Aut}(\mathbb{F}_4)$ such that $\pi(N) = M$.

follows that the isomorphism relation $\cong$ on the space $\mathcal{G}$ of finitely generated groups is the orbit equivalence relation arising from the action of the countable group $\mathrm{Aut}_f(\mathbb{F}_\infty)$ of finitary automorphisms of the free group $\mathbb{F}_\infty$ on $\{x_1, x_2, \cdots, x_m, \cdots\}$. (For more details, see either Champetier [6] or Thomas [36].)

## 1.2 Borel reducibility

We have seen that many naturally occurring classification problems may be viewed as Borel equivalence relations on standard Borel spaces. In particular, the complexity of the problem of finding complete invariants for such classification problems can be measured to some extent by the "structural complexity" of the associated Borel equivalence relations. Here the crucial notion of comparison is that of a *Borel reduction.*

**Definition 1.5** If $E$ and $F$ are Borel equivalence relations on the standard Borel spaces $X$, $Y$ respectively, then we say that $E$ is *Borel reducible* to $F$, and write $E \leq_B F$, if there exists a Borel map $f : X \to Y$ such that $xEy \leftrightarrow f(x)Ff(y)$. Such a map is called a *Borel reduction* from $E$ to $F$. We say that $E$ and $F$ are *Borel bireducible*, and write $E \sim_B F$, if both $E \leq_B F$ and $F \leq_B E$; and we write $E <_B F$ if both $E \leq_B F$ and $F \nleq_B E$.

If $E$ and $F$ are Borel equivalence relations, then we interpret $E \leq_B F$ as meaning that the classification problem associated with $E$ is at most as complicated as that associated with $F$, in the sense that an assignment of complete invariants for $F$ would, via composition with the Borel reduction from $E$ to $F$, yield one for $E$ as well. Additionally we observe that if $f$ is a Borel reduction from $E$ to $F$, then the induced map $\tilde{f} : X/E \to Y/F$ is an embedding of quotient spaces, the existence of which is sometimes interpreted as meaning that $X/E$ has "Borel cardinality" less than or equal to that of $Y/F$.

This notion of Borel reducibility imposes a partial (pre)-order on the collection of Borel equivalence relations, and much of the work currently taking place in the theory of Borel equivalence relations concerns determining the structure of this partial ordering. As a first step towards describing the $\leq_B$-hierarchy, we introduce the so-called *smooth* and *hyperfinite* Borel equivalence relations. Writing $\mathrm{id}_{\mathbb{R}}$ for the identity relation on $\mathbb{R}$, the following result is a special case of a more general result of Silver [31] concerning co-analytic equivalence relations.

**Theorem 1.6** (Silver) *If $E$ is a Borel equivalence relation with uncountably many classes, then $id_{\mathbb{R}} \leq_B E$.*

Hence $id_{\mathbb{R}}$ – and any Borel equivalence relation bireducible with it – is a $\leq_B$-minimal element in the partial ordering of Borel equivalence relations with uncountably many classes.

**Definition 1.7** The Borel equivalence relation $E$ is *smooth* iff $E \leq_B id_Z$ for some (equivalently every) uncountable standard Borel space $Z$.

For example, it is easily checked that if the Borel equivalence relation $E$ on the standard Borel space $X$ admits a Borel transversal, then $E$ is smooth. (Here a Borel transversal is a Borel subset $T \subseteq X$ which intersects every $E$-class in a single point.) While the converse does not hold for arbitrary Borel equivalence relations, we will later see that a countable Borel equivalence relation $E$ is smooth iff $E$ admits a Borel transversal.

The isomorphism relation on the space of countable divisible abelian groups is an example of a smooth equivalence relation. Similarly, if $\equiv$ is the equivalence relation defined on the space $\mathcal{G}$ of finitely generated groups by $G \equiv H$ iff $Th(G) = Th(H)$, then $\equiv$ is also smooth. For an example of a *non*-smooth Borel equivalence relation, we turn to the following:

**Definition 1.8** $E_0$ is the Borel equivalence relation defined on $2^{\mathbb{N}}$ by $xE_0y$ iff $x(n) = y(n)$ for all but finitely many $n$.

To see that $E_0$ is not smooth, suppose $f : 2^{\mathbb{N}} \to [0, 1]$ is a Borel reduction from $E_0$ to $id_{[0,1]}$ and let $\mu$ be the usual product probability measure on $2^{\mathbb{N}}$. Then $f^{-1}([0, \frac{1}{2}])$ and $f^{-1}([\frac{1}{2}, 1])$ are Borel tail events, so by Kolmogorov's zero-one law, either $\mu(f^{-1}([0, \frac{1}{2}])) = 1$ or $\mu(f^{-1}([\frac{1}{2}, 1])) = 1$. Continuing to cut intervals in half in this manner, we obtain that $f$ is $\mu$-a.e. constant, which is a contradiction.

## 1.3 Countable Borel equivalence relations

An important subclass of Borel equivalence relations consists of those with countable equivalence classes.

**Definition 1.9** A Borel equivalence relation on a standard Borel space is called *countable* if each of its equivalence classes is countable.

The importance of this subclass stems in large part from the fact that each such equivalence relation can be realized as the orbit equivalence relation of a Borel action of a countable group. Of course, if $G$ is a countable group and $X$ a standard Borel $G$-space, then the corresponding orbit equivalence relation $E_G^X$ is a countable Borel equivalence relation. But by a remarkable result of Feldman-Moore [10], the converse is also true:

**Theorem 1.10** (Feldman-Moore) *If $E$ is a countable Borel equivalence relation on the standard Borel space $X$, then there exists a countable group $G$ and a Borel action of $G$ on $X$ such that $E = E_G^X$.*

*Sketch Proof* (For more details, see Srivastava [33, 5.8.13]). Let $E$ be a countable Borel equivalence relation on the standard Borel space $X$. Since $E \subseteq X^2$ has countable sections, the Lusin-Novikov Uniformization Theorem [21, 18.10] implies that we can write $E$ as a countable union of graphs of injective partial Borel functions, $f_n : \operatorname{dom} f_n \to X$. Each $f_n$ is easily modified into a Borel bijection $g_n : X \to X$ with the same "orbits." But then $E$ is simply the orbit equivalence relation arising from the resulting Borel action of the group $G$ generated by $\{g_n \mid n \in \mathbb{N}\}$. □

*Remark* 1.11 The Lusin-Novikov Uniformization Theorem also implies that if $E$ is a smooth countable Borel equivalence relation on the standard Borel space $X$, then $E$ admits a Borel transversal. To see this, notice that if $f : X \to \mathbb{R}$ is a Borel reduction from $E$ to $\mathrm{id}_{\mathbb{R}}$, then $f$ is countable-to-one. Applying the Lusin-Novikov Uniformization Theorem to the Borel relation $R = \{(f(x), x) \mid x \in X\}$, it follows that $f(X)$ is Borel and that there exists an injective Borel function $g : f(X) \to X$ such that $f(g(y)) = y$ for all $y \in f(X)$. Hence $T = g(f(X))$ is a Borel transversal for $E$.

Unfortunately, the countable group action given by the Feldman-Moore Theorem is by no means canonical. For example, let us define the *Turing equivalence relation* $\equiv_T$ on $\mathcal{P}(\mathbb{N})$ by

$$A \equiv_T B \quad \text{iff} \quad A \leq_T B \text{ and } B \leq_T A,$$

where $\leq_T$ denotes Turing reducibility. Then $\equiv_T$ is clearly a countable Borel equivalence relation; and hence by the Feldman-Moore Theorem, it must arise as the orbit equivalence relation induced by a Borel action of some countable group $G$ on $\mathcal{P}(\mathbb{N})$. However, the proof of the Feldman-Moore Theorem gives us no information about either the group $G$ or its action, and so it is reasonable to ask:

*Vague question* Can $\equiv_T$ be realized as the orbit equivalence relation of a "nice" Borel action of some countable group?

We have earlier seen that there is a $\leq_B$-minimal Borel equivalence relation on an uncountable standard Borel space. On the other hand, by Friedman-Stanley [11], there does not exist a maximal equivalence relation in the setting of arbitrary Borel equivalence relations. However, by Dougherty-Jackson-Kechris [7], the subclass of countable Borel equivalence relations does admit a universal element.

**Definition 1.12** A countable Borel equivalence relation $E$ is *universal* iff $F \leq_B E$ for every countable Borel equivalence relation $F$.

This universal countable Borel equivalence relation can be realized as follows. Let $\mathbb{F}_\omega$ be the free group on infinitely many generators and define a Borel action of $\mathbb{F}_\omega$ on

$$(2^{\mathbb{N}})^{\mathbb{F}_\omega} = \{p \mid p : \mathbb{F}_\omega \to 2^{\mathbb{N}}\}$$

by setting

$$(g \cdot p)(h) = p(g^{-1}h), \quad p \in (2^{\mathbb{N}})^{\mathbb{F}_\omega}.$$

Let $E_\omega$ be the resulting orbit equivalence relation.

*Claim* 1.13 $E_\omega$ is a universal countable Borel equivalence relation.

*Proof* Let $X$ be a standard Borel space and let $E$ be any countable Borel equivalence relation on $X$. Since every countable group is a homomorphic image of $\mathbb{F}_\omega$, the Feldman-Moore Theorem implies that $E$ is the orbit equivalence relation of a Borel action of $\mathbb{F}_\omega$. Let $\{U_i\}_{i\in\mathbb{N}}$ be a sequence of Borel subsets of $X$ which separates points and define $f : X \to (2^{\mathbb{N}})^{\mathbb{F}_\omega}$ by $x \mapsto f_x$, where

$$f_x(h)(i) = 1 \ \text{ iff } \ x \in h(U_i).$$

Then $f$ is injective and

$$\begin{aligned}(g \cdot f_x)(h)(i) = 1 \ &\text{ iff } \ f_x(g^{-1}h)(i) = 1\\ &\text{ iff } \ x \in g^{-1}h(U_i)\\ &\text{ iff } \ g \cdot x \in h(U_i)\\ &\text{ iff } \ f_{g\cdot x}(h)(i) = 1\end{aligned}$$

□

Dougherty-Jackson-Kechris [7] have also shown that the orbit equivalence relation $E_\infty$ arising from the translation action of the free group $\mathbb{F}_2$ on its powerset is a universal countable Borel equivalence relation.

(Of course, any two universal countable Borel equivalence relations are Borel bireducible.)

We have now seen that within the class of countable Borel equivalence relations, there exist $\leq_B$-least and $\leq_B$-greatest such equivalence relations, up to $\sim_B$, with realizations given by $\mathrm{id}_{\mathbb{R}}$ and $E_\infty$, respectively. It turns out that the minimal $\mathrm{id}_{\mathbb{R}}$ has an immediate $\leq_B$-successor:

**Theorem 1.14** (Harrington-Kechris-Louveau [14]) *If $E$ is a nonsmooth Borel equivalence relation, then $E_0 \leq_B E$.*

A Borel equivalence relation $E$ is said to be *hyperfinite* if it can be written as an increasing union $E = \cup_n F_n$ of a sequence of finite Borel equivalence relations. (Here a Borel equivalence relation $F$ is said to be *finite* if every $F$-class is finite.) It is easily shown that $E_0$ is hyperfinite; and it turns out that every nonsmooth hyperfinite countable Borel equivalence relation is Borel bireducible with $E_0$. Furthermore, by a result of Dougherty-Jackson-Kechris [7], if $E$ is a countable Borel equivalence relation, then $E$ can be realized as the orbit equivalence relation of a Borel $\mathbb{Z}$-action if and only if $E \leq_B E_0$. (There will be a further discussion of the class of hyperfinite equivalence relations in Section 5.1.) For many years, it was unknown whether there were infinitely many countable Borel equivalence relations up to Borel bireducibility. This question was finally settled in 2000 when Adams-Kechris [2] proved that the partial ordering of Borel subsets of $2^{\mathbb{N}}$ under inclusion embeds into the $\leq_B$ ordering on the class of countable Borel equivalence relations.Combining these fundamental results, we obtain the picture of the universe of countable Borel equivalence relations given in Figure 2.1.

Given this picture, one can ask where a particular countable Borel equivalence relation lies relative to the known benchmarks. In the following section, we shall consider this question for the Turing equivalence relation $\equiv_T$. Here it is interesting to note that Martin has conjectured that $\equiv_T$ is *not* universal, while Kechris has conjectured that it is. However, despite some progress, which we will discuss below, this important problem remains open.

## 1.4 Turing equivalence and the Martin conjectures

We first define the set of *Turing degrees* to be the collection

$$\mathcal{D} = \{\, \mathbf{a} = [A]_{\equiv_T} \mid A \in \mathcal{P}(\mathbb{N})\}$$

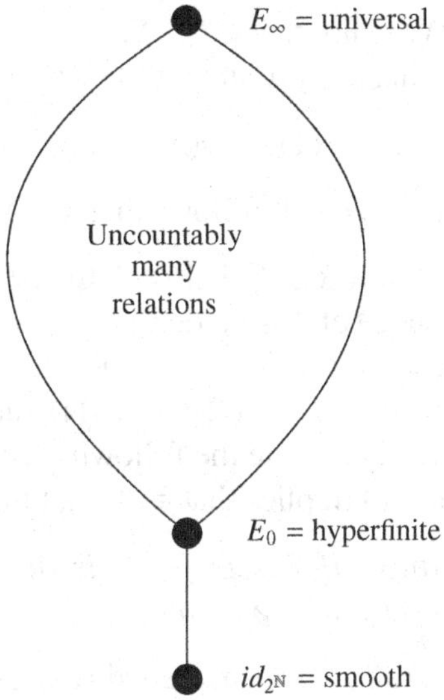


Figure 2.1 Countable Borel equivalence relations

of $\equiv_T$-classes. A subset $X \subseteq \mathcal{D}$ is said to be *Borel* iff $X^* = \bigcup\{\mathbf{a} \mid \mathbf{a} \in X\}$ is a Borel subset of $\mathcal{P}(\mathbb{N})$. It is well-known that if $E$ is a countable Borel equivalence relation on a standard Borel space $X$, then the quotient Borel space $X/E = \{[x]_E \mid x \in X\}$ is a standard Borel space if and only if $E$ is smooth. (If $X/E$ is a standard Borel space, then the map $x \mapsto [x]_E$ witnesses that $E$ is smooth. Conversely, if $E$ is smooth, then $E$ admits a Borel transversal $T$ and $X/E$ is isomorphic to the standard Borel space $T$.) In particular, since $\equiv_T$ is not smooth, it follows that $\mathcal{D}$ is *not* a standard Borel space.

For $\mathbf{a}, \mathbf{b} \in \mathcal{D}$, we define $\mathbf{a} \leq \mathbf{b}$ iff $A \leq_T B$ for each $A \in \mathbf{a}$ and $B \in \mathbf{b}$; and for each $\mathbf{a} \in \mathcal{D}$, we define the corresponding *cone* to be $C_{\mathbf{a}} = \{\mathbf{b} \in \mathcal{D} \mid \mathbf{a} \leq \mathbf{b}\}$. Of course, each cone $C_{\mathbf{a}}$ is a Borel subset of $\mathcal{D}$.

**Theorem 1.15** (Martin) *If $X \subseteq \mathcal{D}$ is Borel, then for some $\mathbf{a} \in \mathcal{D}$, either $C_{\mathbf{a}} \subseteq X$ or $C_{\mathbf{a}} \subseteq \mathcal{D} \setminus X$.*

*Proof* Let $X \subseteq \mathcal{D}$ be Borel and consider the 2-player game $G(X^*)$

$$a = a(0)a(1)a(2)\cdots, \quad \text{where each } a(n) \in 2,$$

such that Player 1 wins iff $a \in X^*$. Then $G(X^*)$ is Borel and hence is determined. Suppose, for example, that $\varphi : 2^{<\mathbb{N}} \to 2$ is a winning strategy for Player 1. We claim that $C_\varphi \subseteq X$.

To see this, suppose that $\varphi \leq_T x$ and let Player 2 play

$$x = a(1)a(3)a(5)\cdots.$$

Then $y = \varphi(x) \in X^*$ and $x \equiv_T y$. It follows that $x \in X^*$. □

For later use, notice that if $X \subseteq \mathcal{D}$ is Borel, then $X$ contains a cone iff $X$ is $\leq_T$-cofinal in the set $\mathcal{D}$ of Turing degrees.

In a similar fashion, we define a function $f : \mathcal{D} \to \mathcal{D}$ to be *Borel* iff there exists a Borel function $\varphi : \mathcal{P}(\mathbb{N}) \to \mathcal{P}(\mathbb{N})$ such that $f([A]_{\equiv_T}) = [\varphi(A)]_{\equiv_T}$. We are now ready to state the following conjecture of Martin, which (as we will soon see) implies that $\equiv_T$ is not universal.

**Conjecture 1.16** (Martin) *If $f : \mathcal{D} \to \mathcal{D}$ is Borel, then either $f$ is constant on a cone or else $f(\mathbf{a}) \geq \mathbf{a}$ on a cone.*

While this conjecture remains open, there do exist some partial results of Slaman-Steel [32] that point in its direction:

**Theorem 1.17** (Slaman-Steel) *If $f : \mathcal{D} \to \mathcal{D}$ is Borel and $f(\mathbf{a}) < \mathbf{a}$ on a cone, then $f$ is constant on a cone.*

**Theorem 1.18** (Slaman-Steel) *If the Borel map $f : \mathcal{D} \to \mathcal{D}$ is uniformly invariant, then either $f$ is constant on a cone or else $f(\mathbf{a}) \geq \mathbf{a}$ on a cone.*

(The definition of a uniformly invariant map can be found in Slaman-Steel [32].) Next, following Dougherty-Kechris [8], we will show that the Martin conjecture implies that $\equiv_T$ is not universal. First recall that, by Dougherty-Jackson-Kechris [7], if $E$, $F$ are countable Borel equivalence relations on the standard Borel spaces $X$, $Y$ respectively, then $E \sim_B F$ iff there exist Borel complete sections $A \subseteq X$, $B \subseteq Y$ such that $E \restriction A \cong F \restriction B$ via a Borel isomorphism. (Here a Borel subset $A \subseteq X$ is said to be a *complete section* if $A$ intersects every $E$-class.) In particular, if $\equiv_T$ is universal, then $(\equiv_T \times \equiv_T) \sim_B \equiv_T$; and hence there exist Borel complete sections $Y \subseteq \mathcal{P}(\mathbb{N}) \times \mathcal{P}(\mathbb{N})$ and $Z \subseteq \mathcal{P}(\mathbb{N})$ such that $(\equiv_T \times \equiv_T) \restriction Y \cong \equiv_T \restriction Z$ via a Borel isomorphism $\varphi$. Let $f : \mathcal{D} \times \mathcal{D} \to \mathcal{D}$ be the Borel pairing function induced by $\varphi$. Then fixing $\mathbf{d}_0 \neq \mathbf{d}_1 \in \mathcal{D}$, we can define Borel maps $f_i : \mathcal{D} \to \mathcal{D}$ by $f_i(\mathbf{a}) = f(\mathbf{d}_i, \mathbf{a})$. By the Martin Conjecture, $f_i(\mathbf{a}) \geq \mathbf{a}$ on a cone and so each ran $f_i$ is a cofinal Borel subset of $\mathcal{D}$. But this means that each ran $f_i$ contains a cone, which is impossible since ran $f_0 \cap$ ran $f_1 = \emptyset$.

In contrast, let $\equiv_A$ be the *arithmetic equivalence relation* defined on $\mathcal{P}(\mathbb{N})$ by

$$B \equiv_A C \text{ iff } B \leq_A C \text{ and } C \leq_A B,$$

where $\leq A$ denotes arithmetic reducibility. Then Slaman-Steel [27] have shown that $\equiv_A$ is a universal countable Borel equivalence relation. One might take this as evidence that $\equiv_T$ is also universal. However, as Slaman has pointed out, an important difference between the two cases is that the arithmetic degrees have less closure with respect to arithmetic equivalences than the Turing degrees do with respect to recursive equivalences.

# Second lecture

## 2.1 The fundamental question in the theory of countable Borel equivalence relations

We have already seen that, by the Feldman-Moore Theorem, every countable Borel equivalence relation on a standard Borel space arises as the orbit equivalence relation of a Borel action of a suitable countable group. However, we have also seen that this action is not canonically determined, and that it is sometimes difficult to express a given countable Borel equivalence relation as the orbit equivalence relation arising from a "natural" group action. Since many of the techniques currently available for analyzing countable Borel equivalence relations involve properties of the groups and actions from which they arise, one of the fundamental questions in the theory concerns the extent to which an orbit equivalence relation $E_G^X$ determines the group $G$ and its action on $X$. Ideally one would hope for the complexity of $E_G^X$ to reflect the complexity of $G$, so that relations $E_G^X$ and $E_H^X$ can be distinguished (in the sense of $\leq_B$) by distinguishing $G$ from $H$.

Of course, strong hypotheses on a countably infinite group $G$ and its action on a standard Borel space $X$ must be made if there is to be any hope of recovering $G$ and its action from $E_G^X$. For example, let $G$ be any countable group and consider the Borel action of $G$ on $G \times [0, 1]$ defined by $g \cdot (h, r) = (gh, r)$. Then the Borel map $(h, r) \mapsto (1_G, r)$ selects a point in each $G$-orbit, and so the corresponding orbit equivalence relation is smooth. Notice, however, that this action does not admit an invariant probability measure. In fact, we have the following simple but important observation.

**Definition 2.1** A Borel action of a countable group $G$ on a standard Borel space $X$ is said to be *free* iff $g \cdot x \neq x$ for all $1 \neq g \in G$ and $x \in X$. In this case, we say that $X$ is a *free standard Borel G-space.*

**Proposition 2.2** *If a countably infinite group $G$ acts freely on $X$ and preserves a probability measure $\mu$, then $E_G^X$ is not smooth.*

*Proof* If $E$ is smooth, then $E$ admits a Borel transversal $T \subseteq X$. But since $G$ acts freely, it follows that $X$ can be expressed as the disjoint union $X = \bigsqcup_{g \in G} g(T)$, which means that $T$ is not $\mu$-measurable, which is a contradiction. □

The following two theorems show that if we are serious about recovering the group $G$ and its action from $E_G^X$, then it is necessary to assume that $G$ satisfies *both* of the hypotheses of Proposition 2.2

**Theorem 2.3** (Dougherty-Jackson-Kechris [7]) *Let $G$ be a countable group and let $X$ be a standard Borel G-space. If $X$ does not admit a G-invariant probability measure, then for every countable group $H \supset G$, there exists a Borel action of $H$ on $X$ such that $E_H^X = E_G^X$. Furthermore, if $G$ acts freely on $X$, then there exists a free Borel action of $H$ on $X$ such that $E_H^X = E_G^X$.*

In order to see that it is also necessary to assume that the action of $G$ on $X$ is free, consider the associated homomorphism $\pi : G \to \mathrm{Sym}(X)$. Of course, if $\ker \pi \neq 1$, then we cannot recover $G$ from its action on $X$. Thus it is certainly necessary to assume that $G$ acts faithfully on $X$. Following Miller [28], the action of $G$ on $X$ is said to be *everywhere faithful* if $G$ acts faithfully on every $G$-orbit. The following is an easy consequence of a much more general result of Miller [28].

**Theorem 2.4** (Miller) *Suppose that $E$ is a countable Borel equivalence relation such that every E-class is infinite. Then there exists an uncountable family $\mathcal{F}$ of pairwise non-embeddable countable groups such that $E$ can be realized as the orbit equivalence relation of an everywhere faithful Borel action of each group $G \in \mathcal{F}$.*

**Definition 2.5** A countable Borel equivalence relation in which every $E$-class is infinite is called *aperiodic*.

Consequently, we shall be especially concerned with *free*, *measure-preserving* Borel actions of countable groups on standard Borel probability spaces. A natural question, then, is whether we can *always* hope for this setting:

*Question* 2.6 Let $E$ be a nonsmooth countable Borel equivalence relation. Does there necessarily exist a countable group $G$ with a free measure-preserving Borel action on a standard probability space $(X, \mu)$ such that $E \sim_B E_G^X$?

We first observe that half of this question is easily answered: namely, if $E$ is a countable Borel equivalence relation on an uncountable standard Borel space $Y$, then there exists a countable group $G$ and a standard Borel $G$-space $X$ such that $G$ preserves a nonatomic probability measure $\mu$ on $X$, and $E \sim_B E_G^X$. To see this, let $G$ be a countable group with a Borel action on $Y$ such that $E_G^Y = E$. Then we can regard $X = Y \sqcup [0, 1]$ as a standard Borel $G$-space by letting $G$ act trivially on $[0, 1]$. If we regard the usual probability measure $\mu$ on $[0, 1]$ as a probability measure on $X$ which concentrates on $[0, 1]$, then $E_G^X$ satisfies our requirements. At this point, it is convenient to introduce two more definitions.

**Definition 2.7** The countable Borel equivalence relation $E$ on $X$ is *free* iff there exists a countable group $G$ with a free Borel action on $X$ such that $E_G^X = E$.

**Definition 2.8** The countable Borel equivalence relation $E$ is *essentially free* iff there exists a free countable Borel equivalence relation $F$ such that $E \sim_B F$.

In view of the above discussion, it is clear that we should replace Question 2.6 by the following question (which no longer mentions an invariant measure).

*Question* 2.9 (Jackson-Kechris-Louveau [19]) Is every countable Borel equivalence relation essentially free?

## 2.2 Essentially free countable Borel equivalence relations

Before answering Question 2.9, we first list some closure properties of essential freeness, which we will state without proof.

**Theorem 2.10** (Jackson-Kechris-Louveau [19]) *Let $E, F$ be countable Borel equivalence relations on the standard Borel spaces $X, Y$ respectively.*

- *If $E \leq_B F$ and F is essentially free, then so is E.*
- *If $E \subseteq F$ and F is essentially free, then so is E.*

It follows that *every* countable Borel equivalence relation is essentially free iff the universal countable Borel equivalence relation $E_\infty$ is essentially free. The following result will be proved in Section 2.3.

**Theorem 2.11** (Thomas 2006, [37]) *The class of essentially free countable Borel equivalence relations does not admit a universal element. In particular, $E_\infty$ is* not *essentially free.*

Thus, unfortunately, the answer to Question 2.6 is negative. As a corollary to 2.11 and 2.10, we observe that $\equiv_T$ is not essentially free; for identifying the free group $\mathbb{F}_2$ with a suitably chosen group of recursive permutations of $\mathbb{N}$, we have that $E_\infty \subseteq \equiv_T$.

Thus we obtain the map of the universe of nonsmooth countable Borel equivalence relations given in Figure 2.2.

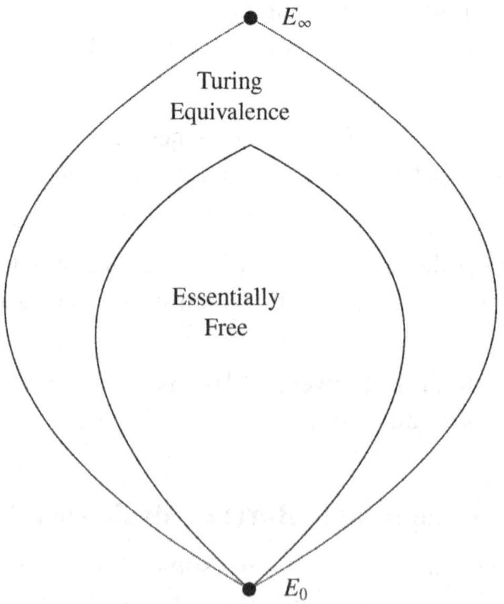


Figure 2.2 Nonsmooth countable Borel equivalence relations

## 2.3 Bernoulli actions, Popa superrigidity, and the proof of Theorem 2.11

In this section, we will state a striking consequence of Popa's Superrigidity Theorem, which easily implies Theorem 2.11. We will begin with a short discussion of *Bernoulli actions.*

By a Bernoulli action, we mean the shift action of a countably infinite discrete group $G$ on its powerset $\mathcal{P}(G) = 2^G$, defined by $g \cdot x(h) = x(g^{-1}h)$. (This is a special case of the notion as it appears in Popa [30]). Under this action, the usual product probability measure $\mu$ on $2^G$ is $G$-invariant and the *free part*

$$\mathcal{P}^*(G) = (2)^G = \{x \in 2^G \mid g \cdot x \neq x \text{ for all } 1 \neq g \in G\}$$

has $\mu$-measure 1. We let $E_G$ denote the corresponding orbit equivalence relation on $(2)^G$ and make the following observation:

**Proposition 2.12** *If $G \leq H$, then $E_G \leq_B E_H$.*

*Proof* The inclusion map $\mathcal{P}^*(G) \hookrightarrow \mathcal{P}^*(H)$ is a Borel reduction from $E_G$ to $E_H$. □

Now we just need a few more preliminary definitions before we can state the consequence of Popa's theorem which we will need to prove Theorem 2.11.

**Definition 2.13** If $E$, $F$ are Borel equivalence relations on the standard Borel spaces $X$, $Y$ respectively, then a Borel map $f : X \to Y$ is said to be a *homomorphism* from $E$ to $F$ if

$$x \, E \, y \quad \Longrightarrow \quad f(x) \, F \, f(y)$$

for all $x, y \in X$.

**Definition 2.14** If $\mu$ is an $E$-invariant probability measure on $X$, then the Borel homomorphism $f : X \to Y$ from $E$ to $F$ is said to be *$\mu$-trivial* if there exists a Borel subset $Z \subseteq X$ with $\mu(Z) = 1$ such that $f$ maps $Z$ into a single $F$-class.

**Definition 2.15** If $G$ and $H$ are countable groups, then the homomorphism $\pi : G \to H$ is a *virtual embedding* if $|\ker \pi| < \infty$.

Now we are finally ready to state the consequence of Popa's Cocycle Superrigidity Theorem [30] that we shall use to prove Theorem 2.11. We shall discuss Popa's theorem and sketch the proof of the following result in Section 3.2.

**Theorem 2.16** *Let $G = SL_3(\mathbb{Z}) \times S$, where $S$ is any countable group. Let $H$ be any countable group and let $Y$ be a free standard Borel $H$-space. If there exists a $\mu$-nontrivial Borel homomorphism from $E_G$ to $E_H^Y$, then there exists a virtual embedding $\pi : G \to H$.*

We observe that, in particular, this conclusion holds if there exists a Borel subset $Z \subseteq (2)^G$ with $\mu(Z) = 1$ such that $E_G \restriction Z \leq_B E_H^Y$. Theorem 2.11 is then an immediate corollary of the following:

**Theorem 2.17** *If $E$ is an essentially free countable Borel equivalence relation, then there exists a countable group $G$ such that $E_G \nleq_B E$.*

*Proof* We can suppose that $E = E_H^X$ is realized by a free Borel action of the countable group $H$ on the standard Borel space $X$. Let $L$ be a finitely generated group which does not embed into $H$. Let $S = L * \mathbb{Z}$ and let $G = SL_3(\mathbb{Z}) \times S$. Then $G$ has no finite normal subgroups and so there does not exist a virtual embedding $\pi : G \to H$. It follows that $E_G \nleq_B E_H^X$. □

## 2.4 Free and non-essentially free countable Borel equivalence relations

We will now use 2.16 to show that there are continuum many free countable Borel equivalence relations. For each prime $p \in \mathbb{P}$, let $A_p = \bigoplus_{i=0}^{\infty} C_p$, where $C_p$ is the cyclic group of order $p$ ; and for each subset $C \subseteq \mathbb{P}$, let $G_C = SL_3(\mathbb{Z}) \times \bigoplus_{p \in C} A_p$. Then the desired result is an immediate consequence of the following:

**Theorem 2.18** *If $C, D \subseteq \mathbb{P}$, then $E_{G_C} \leq_B E_{G_D}$ iff $C \subseteq D$.*

*Proof* If $C \subseteq D$, then $G_C \leq G_D$, and hence $E_{G_C} \leq_B E_{G_D}$. Conversely, applying 2.16, if $E_{G_C} \leq_B E_{G_D}$, then there exists a virtual embedding $\pi : G_C \to G_D$. Since $SL_3(\mathbb{Z})$ contains a torsion-free subgroup of finite index, it follows that for each $p \in C$, the cyclic group $C_p$ embeds into $\bigoplus_{q \in D} A_q$ and this implies that $p \in D$. □

We will now show that there also exist continuum many *non*-essentially free countable Borel equivalence relations. We begin by introducing the notion of *ergodicity*.

**Definition 2.19** Let $G$ be a countable group and let $X$ be a standard Borel $G$-space with $G$-invariant probability measure $\mu$. Then the action of $G$ on $(X, \mu)$ is said to be *ergodic* if $\mu(A) = 0$ or $\mu(A) = 1$ for every $G$-invariant Borel subset $A \subseteq X$.

For example, every countable group $G$ acts ergodically on $((2)^G, \mu)$. (This is a consequence of Theorem 3.2.) The following characterization of ergodicity is well-known.

**Theorem 2.20** *If $\mu$ is a $G$-invariant probability measure on the standard Borel $G$-space $X$, then the following statements are equivalent.*

- *The action of $G$ on $(X, \mu)$ is ergodic.*
- *If $Y$ is a standard Borel space and $f : X \to Y$ is a $G$-invariant Borel function, then there exists a $G$-invariant Borel subset $M \subseteq X$ with $\mu(M) = 1$ such that $f \restriction M$ is a constant function.*

Finally we need just one more definition.

**Definition 2.21** The countable groups $G, H$ are said to be *virtually isomorphic* if there exist finite normal subgroups $N \lhd G$, $M \lhd H$ such that $G/N \cong H/M$.

The groups given by the following lemma will be used below to construct the desired examples of non-essentially free countable Borel equivalence relations. (The proof of Lemma 2.22 can be found in Thomas [37].)

**Lemma 2.22** *There exists a Borel family $\{S_x \mid x \in 2^{\mathbb{N}}\}$ of finitely generated groups such that if $G_x = SL_3(\mathbb{Z}) \times S_x$, then the following conditions hold:*

- *If $x \neq y$, then $G_x$ and $G_y$ are not virtually isomorphic.*
- *If $x \neq y$, then $G_x$ does not virtually embed in $G_y$.*

Now, for each Borel subset $A \subseteq 2^{\mathbb{N}}$, let $E_A = \bigsqcup_{x \in A} E_{G_x}$ be the corresponding *smooth disjoint union*; i.e. $E_A$ is the countable Borel equivalence relation defined on the standard Borel space

$$\bigsqcup_{x \in A} (2)^{G_x} = \{ (x, r) \mid x \in A, r \in (2)^{G_x} \}$$

defined by

$$(x, r)\ E_A\ (y, s) \quad \Longleftrightarrow \quad x = y \text{ and } r\ E_{G_x}\ s.$$

**Lemma 2.23** *If the Borel subset $A \subseteq 2^{\mathbb{N}}$ is uncountable, then $E_A$ is not essentially free.*

*Proof* Suppose that $E_A \leq_B E_H^Y$, where $H$ is a countable group and $Y$ is a free standard Borel $H$-space. Then for each $x \in A$, we have that $E_{G_x} \leq_B E_H^Y$ and so there exists a virtual embedding $\pi_x : G_x \to H$. Since $A$ is uncountable and each $G_x$ is finitely generated, there exist $x \neq y \in A$ such that $\pi_x[G_x] = \pi_y[G_y]$. But then $G_x, G_y$ are virtually isomorphic, which is a contradiction. □

**Lemma 2.24** *$E_A \leq_B E_B$ iff $A \subseteq B$.*

*Proof* It is clear that if $A \subseteq B$, then $E_A \leq_B E_B$. Conversely, suppose that $E_A \leq_B E_B$ and that $A \nsubseteq B$. Let $x \in A \setminus B$. Then there exists a Borel reduction

$$f : (2)^{G_x} \to \bigsqcup_{y \in B} (2)^{G_y}$$

from $E_{G_x}$ to $E_B$. By ergodicity, there exists a $\mu_x$-measure 1 subset of $(2)^{G_x}$ which maps to a fixed $(2)^{G_y}$. This yields a $\mu_x$-nontrivial Borel homomorphism from $E_{G_x}$ to $E_{G_y}$ and so $G_x$ virtually embeds into $G_y$, which is a contradiction. □

Of course, the existence of uncountably many non-essentially free countable Borel equivalence relations is an immediate consequence of Lemmas 2.23 and 2.24.

# Third lecture

## 3.1 Ergodicity, strong mixing and Borel cocycles

In this section, we will discuss some of the background material which is necessary in order to understand the statement of Popa's Cocycle Superrigidity Theorem and the proof of Theorem 2.16. As usual, if a countable group $G$ acts on a standard probability space $(X, \mu)$, then we assume that the action is both free and measure-preserving, so that we may stand some chance of recovering the group $G$ and its action on $X$ from the orbit equivalence relation $E_G^X$.

Recall now that a measure-preserving action of a countable group $G$ on a standard Borel probability $G$-space $(X, \mu)$ is *ergodic* iff every $G$-invariant Borel subset of $X$ is null or conull; equivalently, the action of $G$ on $(X, \mu)$ is ergodic iff whenever $Y$ is a standard Borel space and $f : X \to Y$ is a $G$-invariant Borel function, then there exists a $G$-invariant Borel subset $M \subseteq X$ with $\mu(M) = 1$ such that $f \restriction M$ is a constant function. In particular, ergodicity is a natural obstruction to smoothness: if $G$ acts ergodically on the standard Borel probability $G$-space $(X,\mu)$, then the corresponding orbit equivalence relation $E_G^X$ is not smooth.

**Definition 3.1** The action of $G$ on the standard probability space $(X,\mu)$ is *strongly mixing* if for any Borel subsets $A, B \subseteq X$, we have that

$$\mu(g(A) \cap B) \to \mu(A) \cdot \mu(B) \quad \text{as } g \to \infty.$$

In other words, if $\langle g_n \mid n \in \mathbb{N} \rangle$ is *any* sequence of distinct elements of $G$, then

$$\lim_{n\to\infty} \mu(g_n(A) \cap B) = \mu(A) \cdot \mu(B).$$

Mixing is a strong form of ergodicity. Indeed, suppose that the action of $G$ on $(X,\mu)$ is strongly mixing and let $A \subseteq X$ be a $G$-invariant Borel subset. Then

$$\mu(A)^2 = \lim_{g\to\infty} \mu(g(A) \cap A) = \lim_{g\to\infty} \mu(A) = \mu(A),$$

which implies that $\mu(A) = 0$ or 1. Hence strongly mixing actions are ergodic. However, unlike ergodicity, strong mixing is a property that passes to infinite subgroups.

*Observation* If the action of $G$ on $(X,\mu)$ is strongly mixing and $H \leq G$ is an infinite subgroup of $G$, then the action of $H$ on $(X,\mu)$ is also strongly mixing.

That the above observations actually apply to our setting is given by the following:

**Theorem 3.2** *The action of $G$ on $((2)^G, \mu)$ is strongly mixing.*

*Sketch Proof* Consider the special case when there exist finite subsets $S, T \subseteq G$ and subsets $\mathcal{F} \subseteq 2^S$, $\mathcal{G} \subseteq 2^T$ such that $A = \{f \in (2)^G \mid f \upharpoonright S \in \mathcal{F}\}$ and $B = \{f \in (2)^G \mid f \upharpoonright T \in \mathcal{G}\}$. (Of course, the "cylinder" sets of this form generate the measure $\mu$.) If $\langle g_n \mid n \in \mathbb{N} \rangle$ is a sequence of distinct elements of $G$, then $g_n(S) \cap T = \emptyset$ for all but finitely many $n$. This means that $g_n(A)$ and $B$ are independent events and so

$$\mu(g_n(A) \cap B) = \mu(g_n(A)) \cdot \mu(B) = \mu(A) \cdot \mu(B).$$

It follows that $\lim_{n\to\infty} \mu(g_n(A) \cap B) = \mu(A) \cdot \mu(B)$. □

The final important concept which we must introduce before stating Popa's Theorem is that of a Borel cocycle. Let $G$, $H$ be countable discrete groups and let $X$ be a standard Borel $G$-space with invariant Borel probability measure $\mu$. Then a Borel map $\alpha : G \times X \to H$ is a *cocycle* iff $\alpha$ satisfies the cocycle identity

$$\forall g, h \in G \quad \alpha(hg, x) = \alpha(h, gx)\, \alpha(g, x) \quad \mu\text{-a.e.}(x).$$

If $\beta : G \times X \to H$ is another cocycle into $H$, then we say that $\alpha$ and $\beta$ are *equivalent*, and write $\alpha \sim \beta$, iff there is a Borel map $b : X \to H$ such that

$$\forall g \in G \quad \beta(g, x) = b(gx)\, \alpha(g, x)\, b(x)^{-1} \quad \mu\text{-a.e.}(x).$$

It is easily checked that $\sim$ is an equivalence relation on the set of cocycles $G \times X \to H$.

In these lectures, cocycles $\alpha : G \times X \to H$ will always arise from Borel homomorphisms into free standard Borel $H$-spaces in the following way. Suppose that $Y$ is a free standard Borel $H$-space and that $f$ is a Borel homomorphism from $E_G^X$ to $E_H^Y$. Then we can define a corresponding cocycle $\alpha : G \times X \to H$ by

$$\alpha(g, x) = \text{ the unique } h \in H \text{ such that } h \cdot f(x) = f(g \cdot x).$$

Furthermore, if $\alpha$ is the cocycle corresponding in this manner to the Borel homomorphism $f : X \to Y$ and if $b : X \to H$ is any Borel function, then the map $f' : X \to Y$ defined by $f'(x) = b(x)f(x)$ is also a Borel homomorphism, and the corresponding cocycle $\beta$ is equivalent to $\alpha$ via the the equation

$$\beta(g, x) = b(gx)\, \alpha(g, x)\, b(x)^{-1}.$$

Equivalence of cocycles can be easily visualized with the aid of the following diagram:

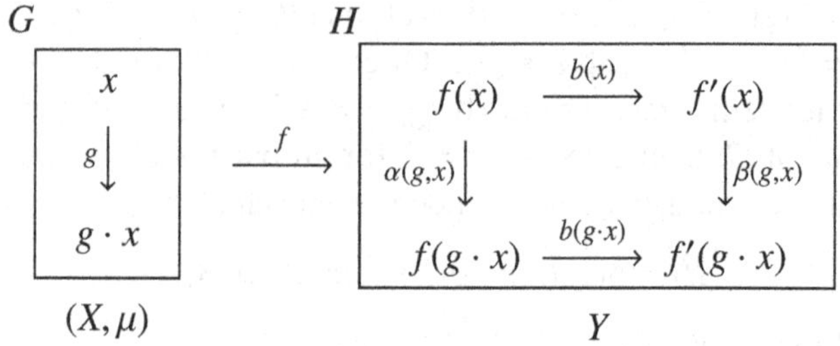


Notice that if the cocycle $\alpha : G \times X \to H$ is actually a function of only one variable, i.e. the value of $\alpha(g, x) = \alpha(g)$ is independent of $x$, then $\alpha$ is a group homomorphism from $G$ to $H$; and if $f : X \to Y$ is the corresponding Borel homomorphism, then $(G, X) \xrightarrow{(\alpha,f)} (H, Y)$ is a permutation group homomorphism.

## 3.2 Popa's Cocycle Superrigidity Theorem and the proof of Theorem 2.16

We are almost ready to state Popa's Cocycle Superrigidity Theorem [30]. But first we need to present a short discussion concerning the notions of amenable, nonamenable and Kazhdan groups.

A countable (discrete) group $G$ is *amenable* if there exists a finitely additive $G$-invariant probability measure $\nu : \mathcal{P}(G) \to [0, 1]$ defined on every subset of $G$. For example, finite groups are amenable and abelian groups are amenable. Furthermore, the class of amenable is closed under taking subgroups, forming extensions and taking direct limits. In particular, solvable groups are also amenable. On the other hand, nonabelian free groups are nonamenable; and for many years, it was a open problem whether every countable nonamenable group contained a nonabelian free subgroup, until Ol'shanskii [29] constructed a periodic nonamenable group in 1980. (An excellent introduction to the theory of amenable groups can be found in Wagon [38].)

In many senses, the opposite of the notion of an amenable group is that of a Kazhdan group. For our purposes in these notes, it is enough to know that if $m \geq 3$, then $\mathrm{SL}_m(\mathbb{Z})$ is a Kazhdan group. However, for the sake of completeness, we will provide the formal definition. So let $G$ be a countably infinite group and let $\pi : G \to U(\mathcal{H})$ be a unitary representation of $G$ on the separable Hilbert space $\mathcal{H}$. Then $\pi$ *almost admits invariant vectors* if for every $\varepsilon > 0$ and every finite subset $K \subseteq G$, there exists a unit vector $v \in \mathcal{H}$ such that $\|\pi(g) \cdot v - v\| < \varepsilon$ for all $g \in K$. We say that $G$ is a *Kazhdan group* if for every unitary representation $\pi$ of $G$, if $\pi$ almost admits invariant vectors, then $\pi$ has a non-zero invariant vector. (An excellent introduction to the theory of Kazhdan groups can be found in Lubotzky [24].)

We are finally ready to state (a special case of) Popa's Cocycle Superrigidity Theorem [30].

**Theorem 3.3** (Popa) *Let $\Gamma$ be a countably infinite Kazhdan group and let $G$ be a countable group such that $\Gamma \lhd G$. If $H$ is any countable group, then every Borel cocycle*

$$\alpha : G \times (2)^G \to H$$

*is equivalent to a group homomorphism of G into H.*

For example, we may let $\Gamma = SL_n(\mathbb{Z})$ for any $n \geq 3$ and $G = \Gamma \times S$, where $S$ is any countable group. We are now ready to prove Theorem 2.16.

*Proof of Theorem 2.16* Let $G = SL_3(\mathbb{Z}) \times S$ and let $Y$ be a free standard Borel $H$-space, where $S$ and $H$ are any countable groups. Suppose the $f : (2)^G \to Y$ is a $\mu$-nontrivial Borel homomorphism from $E_G$ to $E_H^Y$, where $E_G$ denotes the orbit equivalence relation of the Bernoulli action of $G$ on $((2)^G, \mu)$. Then we can define a Borel cocycle $\alpha : G \times (2)^G \to H$ by

$$\alpha(g, x) = \text{ the unique } h \in H \text{ such that } h \cdot f(x) = f(g \cdot x).$$

By Theorem 3.3, after deleting a null set of $(2)^G$ and adjusting $f$ if necessary, we can suppose that $\alpha : G \to H$ is a group homomorphism.

Suppose that $K = \ker \alpha$ is infinite. Note that if $k \in K$, then $f(k \cdot x) = \alpha(k) \cdot x = f(x)$ and so $f : (2)^G \to X$ is $K$-invariant. Also since the action of $G$ is strongly mixing, it follows that $K$ acts ergodically on $((2)^G, \mu)$. But then the $K$-invariant function $f : (2)^G \to X$ is $\mu$-a.e. constant, which is a contradiction. □

### 3.3 Torsion-free abelian groups of finite rank

Recall that an additive subgroup $G \leq \mathbb{Q}^n$ has rank $n$ iff $G$ contains $n$ linearly independent elements; and that we have previously defined the standard Borel space $R(\mathbb{Q}^n)$ of torsion-free abelian groups of rank $n$ to be

$$R(\mathbb{Q}^n) = \{A \leq \mathbb{Q}^n \mid A \text{ contains a basis of } \mathbb{Q}^n\}.$$

Recall also that for $A, B \in R(\mathbb{Q}^n)$, we have that

$$A \cong B \quad \text{iff} \quad \text{there exists } g \in GL_n(\mathbb{Q}) \text{ such that } g(A) = B.$$

Thus the isomorphism relation $\cong_n$ on $R(\mathbb{Q}^n)$ is the orbit equivalence relation arising from the natural action of $GL_n(\mathbb{Q})$ on $R(\mathbb{Q}^n)$.

In 1937, Baer [4] gave a satisfactory classification of the rank 1 groups, which showed that $\cong_1$ is hyperfinite. In 1938, Kurosh [23] and Malcev [25] independently gave *unsatisfactory* classifications of the higher rank groups. In light of this failure to classify even the rank 2 groups in a satisfactory way, Hjorth-Kechris [17] conjectured in 1996 that the isomorphism relation for the torsion-free abelian groups of rank 2 was countable universal. As an initial step towards establishing this result, Hjorth [15] then proved in 1998 that the classification problem for the rank 2 groups is strictly harder than that for the rank 1 groups; that is, Hjorth proved that $\cong_1 <_B \cong_2$. Soon afterwards, making essential use of the techniques of Hjorth [15] and Adams-Kechris [2], Thomas obtained the following [35]:

**Theorem 3.4** (Thomas 2000) *The complexity of the classification problem for the torsion-free abelian groups of rank n increases strictly with the rank n.*

Of course, this implies that none of the relations $\cong_n$ is countable universal. It remained open, however, whether the isomorphism relation on the space of torsion-free abelian groups of *finite* rank was countable universal. In 2006 [37], making use of Popa's Cocycle Superrigidity Theorem, Thomas was finally able to show that it is not.

**Theorem 3.5** (Thomas 2006) *The isomorphism relation on the space of torsion-free abelian groups of finite rank is not countable universal.*

In the next two sections, we shall present an outline of the proof of Theorem 3.5. We will begin by introducing the notion of $E_0$-ergodicity, which will play an important role at the end of the proof.

## 3.4 $E_0$-ergodicity

The following is another useful strengthening of ergodicity.

**Definition 3.6** Let $E, F$ be countable Borel equivalence relations on the standard Borel spaces $X, Y$ and let $\mu$ be an $E$-invariant probability measure on $X$. Then $E$ is said to be *$F$-ergodic* iff every Borel homomorphism $f : X \to Y$ from $E$ to $F$ is $\mu$-trivial.

Thus $\mathrm{id}_{\mathbb{R}}$-ergodicity coincides with the usual notion of ergodicity. Furthermore, observe that if $E$ is $F$-ergodic and $F' \leq_B F$, then $E$ is also $F'$-ergodic. The following characterization of $E_0$-ergodicity is due to Jones-Schmidt [20].

**Definition 3.7** Let $E = E_G^X$ be a countable Borel equivalence relation and let $\mu$ be an $E$-invariant probability measure on $X$. Then $E$ has *nontrivial almost invariant subsets* iff there exists a sequence of Borel subsets $\langle A_n \subseteq X \mid n \in \mathbb{N} \rangle$ satisfying the following conditions:

- $\mu(g \cdot A_n \triangle A_n) \to 0$ for all $g \in G$.

- There exists $\delta > 0$ such that $\delta < \mu(A_n) < 1 - \delta$ for all $n \in \mathbb{N}$.

**Theorem 3.8** (Jones-Schmidt) *Suppose that $E$ is a countable Borel equivalence relation on the standard Borel space $X$ and that $\mu$ is an ergodic $E$-invariant probability measure. Then $E$ is $E_0$-ergodic iff $E$ has no nontrivial almost invariant subsets.*

This can in turn be used to prove the following:

**Theorem 3.9** (Jones-Schmidt) *If $G$ is a countable group and $H \leq G$ is a nonamenable subgroup, then the shift action of $H$ on $((2)^G, \mu)$ is $E_0$-ergodic.*

Finally, we remark for later use that if $E$ is $E_0$-ergodic and $F$ is hyperfinite, then $E$ is also $F$-ergodic. We are now ready to commence our sketch of the proof of the non-universality of the isomorphism relation on the space of torsion-free abelian groups of finite rank.

## 3.5 The non-universality of the isomorphism relation for torsion-free abelian groups of finite rank

Roughly speaking, the strategy of our proof will be as follows. The results of Jackson-Kechris-Louveau [19, Section 5.2] easily imply that a smooth disjoint union of *countably* many essentially free countable Borel equivalence relations is itself essentially free; and we already know that the class of essentially free countable Borel equivalence relations does not admit a universal element. Since the isomorphism relation on the space of torsion-free abelian groups of finite rank is the smooth disjoint union of the $\cong_n$ relations, $n \geq 1$, it would thus suffice to show that each $\cong_n$ is essentially free. Unfortunately, it appears to be difficult to determine whether this is true even for the case when $n = 2$. However, we shall show that the coarser quasi-isomorphism relation is "(hyperfinite)-by-(essentially free)", and this will turn out be enough. We will now proceed with the details.

Let $G = SL_3(\mathbb{Z}) \times S$, where $S$ is a suitably chosen countable group that we shall specify at a later stage in the proof. Let $E = E_G$ be the orbit equivalence relation arising from the action of $G$ on $((2)^G, \mu)$. Suppose that

$$f : (2)^G \to \bigsqcup_{n \geq 1} R(\mathbb{Q}^n)$$

is a Borel reduction from $E$ to the isomorphism relation for the torsion-free abelian groups of finite rank. After deleting a null set of $(2)^G$ if necessary, we may assume that $f$ takes values in $R(\mathbb{Q}^n)$ for some *fixed* $n \geq 1$.

At this point, we would like to define a Borel cocycle corresponding to $f$, but unfortunately $GL_n(\mathbb{Q})$ does not act freely on $R(\mathbb{Q}^n)$. In fact, the stabilizer of each $B \in R(\mathbb{Q}^n)$ under the action of $GL_n(\mathbb{Q})$ is precisely its automorphism group $\mathrm{Aut}(B)$. We shall overcome this difficulty by shifting our focus from the isomorphism relation on $R(\mathbb{Q}^n)$ to the coarser quasi-isomorphism relation.

**Definition 3.10** If $A, B \in R(\mathbb{Q}^n)$, then $A$ and $B$ are said to be *quasi-equal*, written $A \approx_n B$, if $A \cap B$ has finite index in both $A$ and $B$.

**Definition 3.11** If $A, B \in R(\mathbb{Q}^n)$, then $A$ and $B$ are said to be *quasi-isomorphic* if there exists $\varphi \in GL_n(\mathbb{Q})$ such that $\varphi(A) \approx_n B$.

The following result will play a key role in the proof of Theorem 3.5.

**Theorem 3.12** (Thomas [35]) *The quasi-equality relation* $\approx_n$ *is a hyperfinite countable Borel equivalence relation.*

For each $A \in R(\mathbb{Q}^n)$, let $[A]$ be the $\approx_n$-class containing $A$. We shall consider the induced action of $GL_n(\mathbb{Q})$ on the set $X = \{[A] \mid A \in R(\mathbb{Q}^n)\}$ of $\approx_n$-classes. Of course, since $\approx_n$ is not smooth, $X$ is not a standard Borel space; but fortunately this will not pose a problem in what follows. In order to describe the setwise stabilizer in $GL_n(\mathbb{Q})$ of each $\approx_n$-class $[A]$, we now make some further definitions.

**Definition 3.13** For each $A \in R(\mathbb{Q}^n)$, the ring of *quasi-endomorphisms* is

$$QE(A) = \{\varphi \in Mat_n(\mathbb{Q}) \mid (\exists m \geq 1)\, m\varphi \in End(A)\}.$$

Clearly $\mathrm{QE}(A)$ is a $\mathbb{Q}$-subalgebra of $\mathrm{Mat}_n(\mathbb{Q})$, and so there are only countably many possibilities for $\mathrm{QE}(A)$, a fact which will be of crucial importance below.

**Definition 3.14** *QAut(A)* is the group of units of the $\mathbb{Q}$-algebra *QE(A)*.

**Lemma 3.15** (Thomas [35]) *If* $A \in R(\mathbb{Q}^n)$, *then* $\mathrm{QAut}(A)$ *is the setwise stabilizer of* $[A]$ *in* $GL_n(\mathbb{Q})$.

For each $x \in (2)^G$, let $A_x = f(x) \in R(\mathbb{Q}^n)$. Since there are only countably many possibilities for the group $\mathrm{QAut}(A_x)$, there exists a *fixed* subgroup $L \leq GL_n(\mathbb{Q})$ and a Borel subset $X \subseteq (2)^G$ with $\mu(X) > 0$ such that $\mathrm{QAut}(A_x) = L$ for all $x \in X$. Since $G$ acts ergodically on $((2)^G, \mu)$, it follows that $\mu(G \cdot X) = 1$. In order to simplify notation, we shall assume that $G \cdot X = (2)^G$. Then, after slightly adjusting $f$ if necessary, we can suppose that $\mathrm{QAut}(A_x) = L$ for all $x \in (2)^G$.

Notice that the quotient group $H = N_{\mathrm{GL}_n(\mathbb{Q})}(L)/L$ acts freely on the corresponding set $Y = \{ [A] \mid \mathrm{QAut}(A) = L\}$ of $\approx_n$-classes. Furthermore, if $x \in (2)^G$ and $g \in G$, then there exists $\varphi \in \mathrm{GL}_n(\mathbb{Q})$ such that $\varphi(A_x) = A_{g \cdot x}$ and it follows that $\varphi([A_x]) = [A_{g \cdot x}]$. Hence we can define a corresponding cocycle

$$\alpha : G \times (2)^G \to H$$

by setting

$$\alpha(g, x) = \text{ the unique } h \in H \text{ such that } h \cdot [A_x] = [A_{g \cdot x}].$$

Now let $S$ be a countable simple nonamenable group which does not embed into any of the countably many possibilities for $H$. Applying Theorem 3.3, after deleting a null set and slightly adjusting $f$ if necessary, we can suppose that

$$\alpha : G = \mathrm{SL}_3(\mathbb{Z}) \times S \to H$$

is a group homomorphism. Since $S \leq \ker \alpha$, it follows that $f : (2)^G \to R(\mathbb{Q}^n)$ is a Borel homomorphism from the $S$-action on $(2)^G$ to the hyperfinite quasi-equality $\approx_n$-relation. Since $S$ is nonamenable, the $S$-action on $(2)^G$ is $E_0$-ergodic and hence $\mu$-almost all $x \in (2)^G$ are mapped to a single $\approx_n$-class, which is a contradiction. This completes the proof of Theorem 3.5.

# Fourth lecture

## 4.1 Containment vs. Borel reducibility

Our next goal will be to present some applications of Ioana's Cocycle Superrigidity Theorem. We shall focus on a problem that was initially raised in the context of the Kechris Conjecture that the Turing equivalence relation $\equiv_T$ is universal. Recall that the translation action of the free group $\mathbb{F}_2$ on its power set gives rise to a universal countable Borel equivalence relation, which is denoted by $E_\infty$. If we identify $\mathbb{F}_2$ with a suitably chosen group of recursive permutations of $\mathbb{N}$, then we see that $E_\infty$ can be realized as a subset of $\equiv_T$. Thus the following conjecture of Hjorth [3] implies that $\equiv_T$ is universal.

**Conjecture 4.1** (Hjorth) *If $F$ is a universal countable Borel equivalence relation on the standard Borel space $X$ and $E$ is a countable Borel equivalence relation such that $F \subseteq E$, then $E$ is also universal.*

In [35], Thomas pointed out that it was not even known whether there existed a pair $F \subseteq E$ of countable Borel equivalence relations for which $F \not\leq_B E$. Soon afterwards, Adams [1] constructed a pair of countable Borel equivalence relations $F \subseteq E$ which were incomparable with respect to Borel reducibility. Most of this lecture will be devoted to a sketch of the proof of the following application of Ioana's Cocycle Superrigidity Theorem:

**Theorem 4.2** (Thomas [34] 2002) *There exists a pair of countable Borel equivalence relations $F \subseteq E$ on a standard Borel space $X$ such that $E <_B F$.*

Here $E$ and $F$ will arise from the actions of $\mathrm{SL}_n(\mathbb{Z})$ and a suitably chosen congruence subgroup on $\mathrm{SL}_n(\mathbb{Z}_p)$. We shall first need to recall some basic facts about the ring $\mathbb{Z}_p$ of $p$-adic integers.

**Definition 4.3** The ring $\mathbb{Z}_p$ of $p$-adic integers is the inverse limit of the system

$$\cdots \xrightarrow{\varphi_{n+1}} \mathbb{Z}/p^{n+1}\mathbb{Z} \xrightarrow{\varphi_n} \mathbb{Z}/p^n\mathbb{Z} \xrightarrow{\varphi_{n-1}} \cdots \xrightarrow{\varphi_1} \mathbb{Z}/p\mathbb{Z},$$

where $x + p^{n+1}\mathbb{Z} \overset{\varphi_n}{\longmapsto} x + p^n\mathbb{Z}$.

It is useful to think of the $p$-adic integers as formal sums

$$z = a_0 + a_1 p + a_2 p^2 + \cdots + a_n p^n + \cdots$$

where each $0 \leq a_n < p$. We define the *$p$-adic norm* $|\ |_p$ by

$$|z|_p = p^{-\operatorname{ord}_p(z)}, \qquad \operatorname{ord}_p(z) = \min\{n \mid a_n \neq 0\},$$

and the *$p$-adic metric* by

$$d_p(x, y) = |x - y|_p.$$

With this metric, $\mathbb{Z}_p$ is a compact Polish space having the integers $\mathbb{Z}$ as a dense subring. It follows that $\mathrm{SL}_n(\mathbb{Z}_p)$ is a compact Polish group with dense subgroup $\mathrm{SL}_n(\mathbb{Z}) \leq \mathrm{SL}_n(\mathbb{Z}_p)$. Note that $\mathrm{SL}_n(\mathbb{Z}_p)$ is the inverse limit of the system

$$\cdots \xrightarrow{\theta_{n+1}} \mathrm{SL}_n(\mathbb{Z}/p^{n+1}\mathbb{Z}) \xrightarrow{\theta_n} \mathrm{SL}_n(\mathbb{Z}/p^n\mathbb{Z}) \xrightarrow{\theta_{n-1}} \cdots \xrightarrow{\theta_1} \mathrm{SL}_n(\mathbb{Z}/p\mathbb{Z}),$$

where $\theta_n$ is the map induced by $\varphi_n$.

Since $\mathrm{SL}_n(\mathbb{Z}_p)$ is compact, there exists a unique Haar probability measure on $\mathrm{SL}_n(\mathbb{Z}_p)$; i.e. a unique probability measure $\mu_p$ which is invariant under the left translation action of $\mathrm{SL}_n(\mathbb{Z}_p)$ on itself.[2] In fact, $\mu_p$ is simply the inverse limit of the counting measures on

$$\cdots \xrightarrow{\theta_{n+1}} \mathrm{SL}_n(\mathbb{Z}/p^{n+1}\mathbb{Z}) \xrightarrow{\theta_n} \mathrm{SL}_n(\mathbb{Z}/p^n\mathbb{Z}) \xrightarrow{\theta_{n-1}} \cdots \xrightarrow{\theta_1} \mathrm{SL}_n(\mathbb{Z}/p\mathbb{Z}).$$

Observe that if $H \le \mathrm{SL}_n(\mathbb{Z}_p)$ is an open subgroup, then $H$ has finite index in $\mathrm{SL}_n(\mathbb{Z}_p)$ and

$$\mu_p(H) = \frac{1}{[\mathrm{SL}_n(\mathbb{Z}_p) : H]}.$$

**Theorem 4.4** *$\mu_p$ is the unique $\mathrm{SL}_n(\mathbb{Z})$-invariant probability measure on $\mathrm{SL}_n(\mathbb{Z}_p)$.*

*Proof* First note that $\mathrm{SL}_n(\mathbb{Z}_p)$ acts continuously on the space $\mathcal{M}$ of probability measures on $\mathrm{SL}_n(\mathbb{Z}_p)$. It follows that if $\nu$ is any probability measure on $\mathrm{SL}_n(\mathbb{Z}_p)$, then

$$S_\nu = \{g \in \mathrm{SL}_n(\mathbb{Z}_p) \mid \nu \text{ is } g\text{-invariant}\}$$

is a closed subgroup of $\mathrm{SL}_n(\mathbb{Z}_p)$. Hence, since $\mathrm{SL}_n(\mathbb{Z})$ is a dense subgroup of $\mathrm{SL}_n(\mathbb{Z}_p)$, any $\mathrm{SL}_n(\mathbb{Z})$-invariant probability measure is actually $\mathrm{SL}_n(\mathbb{Z}_p)$-invariant and thus must be $\mu_p$. □

## 4.2 Unique ergodicity and ergodic components

An action of a group $G$ on a standard Borel $G$-space $X$ is said to be *uniquely ergodic* iff there exists a unique $G$-invariant probability measure $\mu$ on $X$. In this case, it is well-known that $\mu$ must be ergodic. To see this, suppose that $A \subseteq X$ is a $G$-invariant Borel set with $0 < \mu(A) < 1$. Then we can define distinct $G$-invariant probability measures by

$$\begin{aligned} \nu_1(Z) &= \mu(Z \cap A)/\mu(A) \\ \nu_2(Z) &= \mu(Z \setminus A)/\mu(X \setminus A), \end{aligned}$$

which is a contradiction. Note that Theorem 4.4 simply states that the action of $\mathrm{SL}_n(\mathbb{Z})$ on $\mathrm{SL}_n(\mathbb{Z}_p)$ is uniquely ergodic.

Next suppose that $\Gamma$ is a countable group and that $\Lambda \le \Gamma$ is a subgroup of finite index. Let $X$ be a standard Borel $\Gamma$-space with an invariant ergodic probability measure $\mu$. Then a $\Lambda$-invariant Borel set $Z \subseteq X$ with

[2] The Haar measure on a compact group is also invariant under the right translation action.

$\mu(Z) > 0$ is said to be an *ergodic component* for the action of $\Lambda$ on $X$ iff $\Lambda$ acts ergodically on $(Z, \mu_Z)$, where $\mu_Z$ is the normalized probability measure on $Z$ defined by $\mu_Z(A) = \mu(A)/\mu(Z)$. It is easily checked that there exists a partition $Z_1 \sqcup \cdots \sqcup Z_d$ of $X$ into finitely many ergodic components and that the collection of ergodic components is uniquely determined up to $\mu$-null sets. Furthermore, if the action of $\Gamma$ on $X$ is uniquely ergodic, then the action of $\Lambda$ on each ergodic component is also uniquely ergodic.

Now let $n \geq 3$ and fix some prime $p$. Consider the left translation action of the subgroup $\mathrm{SL}_n(\mathbb{Z})$ on $\mathrm{SL}_n(\mathbb{Z}_p)$. Then we have already seen that this action is uniquely ergodic. Let $\Lambda = \ker \varphi$ and $H = \ker \psi$ be the kernels of the homomorphisms

$$\varphi : \mathrm{SL}_n(\mathbb{Z}) \to \mathrm{SL}_n(\mathbb{Z}/p\mathbb{Z})$$

and

$$\psi : \mathrm{SL}_n(\mathbb{Z}_p) \to \mathrm{SL}_n(\mathbb{Z}_p/p\mathbb{Z}_p) \cong \mathrm{SL}_n(\mathbb{Z}/p\mathbb{Z}).$$

Then $H$ is the closure of $\Lambda$ in $\mathrm{SL}_n(\mathbb{Z}_p)$ and the ergodic decomposition of the $\Lambda$-action coincides with the coset decomposition

$$\mathrm{SL}_n(\mathbb{Z}_p) = Hg_1 \sqcup \cdots \sqcup Hg_d, \qquad d = |\,\mathrm{SL}_n(\mathbb{Z}/p\mathbb{Z})|.$$

**Theorem 4.5** *Let $n \geq 3$ and let $F \subseteq E$ be the orbit equivalence relations of the actions of $\Lambda$ and $\mathrm{SL}_n(\mathbb{Z})$ on $\mathrm{SL}_n(\mathbb{Z}_p)$. Then $E <_B F$.*

We shall devote the next section to a proof of this result.

## 4.3 The proof of Theorem 4.5

By considering the ergodic decomposition of the $\Lambda$-action,

$$\mathrm{SL}_n(\mathbb{Z}_p) = Hg_1 \sqcup \cdots \sqcup Hg_d, \qquad d = |\,\mathrm{SL}_n(\mathbb{Z}/p\mathbb{Z})|,$$

we see that

$$F = E_1 \oplus \cdots \oplus E_d, \qquad \text{where } E_i = F \restriction Hg_i.$$

We claim that $E \sim_B E_i$ for each $1 \leq i \leq d$.

To see that $E_i \leq_B E$, we check that the inclusion map $Hg_i \to \mathrm{SL}_n(\mathbb{Z}_p)$ is a Borel reduction. Suppose that $x, y \in Hg_i$. Clearly if $xE_iy$ then $xEy$, since $F \subseteq E$. Conversely, if $xEy$, then there exists $\gamma \in \mathrm{SL}_n(\mathbb{Z})$ such that $\gamma x = y$, whence $\emptyset \neq \gamma Hg_i \cap Hg_i = H\gamma g_i \cap Hg_i$ and so $\gamma \in \mathrm{SL}_n(\mathbb{Z}) \cap H = \Lambda$.

To show that $E \leq_B E_i$, we choose the coset representatives $g_k$ so that each $g_k \in \mathrm{SL}_n(\mathbb{Z})$. Then for each $1 \leq k \leq d$, define $h_k : Hg_k \to Hg_i$ by $h_k(x) = g_i g_k^{-1} x$. We claim that $h = h_1 \cup \cdots \cup h_d$ is a Borel reduction from $E$ to $E_i$. To see this, note that if $x, y \in \mathrm{SL}_n(\mathbb{Z}_p)$, then

$$\begin{aligned} xEy \quad &\text{iff} \quad h(x)Eh(y) \\ &\text{iff} \quad h(x)E_i h(y), \end{aligned}$$

where this last equivalence holds because $h(x), h(y) \in Hg_i$. This completes the proof that $E \sim_B E_i$ for each $1 \leq i \leq d$; and hence we have that

$$F \sim_B \underbrace{E \oplus \cdots \oplus E}_{d \text{ times}}.$$

Therefore it will be enough to prove the following:

**Theorem 4.6** (Thomas [34] 2002) *If $n \geq 3$, then*

$$E <_B E \oplus E <_B \cdots <_B \underbrace{E \oplus \cdots \oplus E}_{m \text{ times}} <_B \cdots$$

*Proof* Let $\Gamma = \mathrm{SL}_n(\mathbb{Z})$ and let $(K, \mu) = (\mathrm{SL}_n(\mathbb{Z}_p), \mu_p)$, so that $E$ is the orbit equivalence relation arising from the action of $\Gamma$ on $K$. It clearly suffices to show that if $f : K \to K$ is a Borel reduction from $E$ to $E$, then $\mu(\Gamma \cdot f(K)) = 1$.

So suppose that $f : K \to K$ is a Borel reduction from $E$ to $E$. Since $\Gamma$ acts freely on $K$, we can define a corresponding Borel cocycle $\alpha : \Gamma \times K \to \Gamma$ by

$$\alpha(g, x) = \text{ the unique } h \in \Gamma \text{ such that } h \cdot f(x) = f(g \cdot x).$$

By Ioana's Superrigidity Theorem [18] (which we will state and discuss in the next section), there exists a subgroup $\Delta \leq \Gamma$ of finite index and an ergodic component $X \subseteq K$ for the $\Delta$-action such that $\alpha \restriction (\Delta \times X)$ is equivalent to a group homomorphism

$$\psi : \Delta \to \mathrm{SL}_n(\mathbb{Z}).$$

After slightly adjusting $f$ if necessary, we can suppose that $\alpha \restriction (\Delta \times X) = \psi$ and hence that

$$\psi(g) \cdot f(x) = f(g \cdot x) \quad \text{for all } g \in \Delta \text{ and } x \in X.$$

Furthermore, since $\Delta$ is residually finite, after passing to a subgroup of finite index if necessary, we can also suppose that $\Delta \cap Z(\mathrm{SL}_n(\mathbb{Z})) = 1$.

*Claim* 4.7 Either $\psi(\Delta)$ is finite, or else $\psi$ is an embedding and $\psi(\Delta)$ is a subgroup of finite index in $\mathrm{SL}_n(\mathbb{Z})$.

*Proof of Claim 4.7* Suppose that $\psi$ is not an embedding and let $N = \ker \psi$. Then the Margulis Normal Subgroup Theorem [26, Chapter VIII] implies that $[\Delta : N] < \infty$ and hence $\psi(\Delta)$ is finite. Thus we can suppose that $\psi$ is an embedding. Let $\pi : SL_n(\mathbb{Z}) \to PSL_n(\mathbb{Z})$ be the canonical surjective homomorphism and let $\theta = \pi \circ \psi$. Then, arguing as above, we see that $\theta : \Delta \to PSL_n(\mathbb{Z})$ is also an embedding. Applying Margulis [26, Chapter VII], it follows that $\theta$ extends to an $\mathbb{R}$-rational homomorphism $\Theta : SL_n(\mathbb{R}) \to PSL_n(\mathbb{R})$ and it is easily seen that $\Theta$ is surjective. Since $\Delta$ is a lattice in $SL_n(\mathbb{R})$, it follows that $\theta(\Delta) = \Theta(\Delta)$ is a lattice in $PSL_n(\mathbb{R})$ and this implies that $\theta(\Delta)$ is a subgroup of finite index in $PSL_n(\mathbb{Z})$. Hence $\psi(\Delta)$ is a subgroup of finite index in $SL_n(\mathbb{Z})$. □

First suppose that $\psi(\Delta)$ is finite. Then we can define a $\Delta$-invariant map $\phi : X \to [K]^{<\omega}$ by

$$\phi(x) = \{f(g \cdot x) \mid g \in \Delta\};$$

and since $\Delta$ acts ergodically on $X$, it follows that $\phi$ is constant on a $\mu$-conull subset of $X$, which is a contradiction.

Thus $\psi$ is an embedding and $\psi(\Delta)$ is a subgroup of finite index in $SL_n(\mathbb{Z})$. Let $Y_1, \dots, Y_d$ be the ergodic components for the action of $\psi(\Delta)$ on $K$. Since $\Delta$ acts ergodically on $X$, we can suppose that there exists a fixed $Y = Y_i$ such that $f : X \to Y$. Recalling that $\psi(g) \cdot f(x) = f(g \cdot x)$, we can now define a $\psi(\Delta)$-invariant probability measure $\nu$ on $Y$ by

$$\nu(Z) = \mu(f^{-1}(Z))/\mu(X).$$

Since the action of $\psi(\Delta)$ on $Y$ is uniquely ergodic, it follows that $\nu(Z) = \mu(Z)/\mu(Y)$. Hence $\mu(f(X)) = \mu(Y) > 0$ and so $\mu(\Gamma \cdot f(K)) = 1$, as desired. This completes the proof of Theorem 4.6, and hence also that of Theorem 4.5. □

## 4.4 Profinite actions and Ioana superrigidity

**Definition 4.8** Suppose that $\Gamma$ is a countable group and that $X$ is a standard Borel $\Gamma$-space with invariant probability measure $\mu$. Then the action of $\Gamma$ on $(X, \mu)$ is said to be *profinite* if there exists a directed system of finite $\Gamma$-spaces $X_n$ with invariant probability measures $\mu_n$ such that

$$(X, \mu) = \varprojlim (X_n, \mu_n).$$

For example, suppose that $K$ is a profinite group and that $\Gamma \leq K$ is a countable dense subgroup. If $L \leq K$ is a closed subgroup, then the action of $\Gamma$ on $K/L$ is profinite. In particular, if $\Gamma$ is a residually finite group and

$$\Gamma = \Gamma_0 > \Gamma_1 > \cdots > \Gamma_n > \cdots$$

is a decreasing sequence of finite index normal subgroups such that $\bigcap \Gamma_n = 1$, then $\Gamma$ is a dense subgroup of the profinite group $\varprojlim \Gamma/\Gamma_n$ and its action as a subgroup will be profinite. Of course, this example covers the situation discussed above; i.e. the action of $\mathrm{SL}_n(\mathbb{Z})$ on $\mathrm{SL}_n(\mathbb{Z}_p)$ is profinite.

We are now ready to state Ioana's Cocycle Superrigidity Theorem [18], which was used in our proof of Theorem 4.5.

**Theorem 4.9** (Ioana) *Let $\Gamma$ be a countably infinite Kazhdan group and let $(X, \mu)$ be a free ergodic profinite $\Gamma$-space. Suppose that $H$ is any countable group and that $\alpha : \Gamma \times X \to H$ is a Borel cocycle. Then there exists a subgroup $\Delta \leq \Gamma$ of finite index and an ergodic component $Y \subseteq X$ for the $\Delta$-action such that $\alpha \restriction (\Delta \times Y)$ is equivalent to a homomorphism $\psi : \Delta \to H$.*

To conclude this section, we shall present a final application of Ioana's theorem.

**Theorem 4.10** (Thomas [34] 2002) *Fix $n \geq 3$. For each nonempty set $S$ of primes, regard $\mathrm{SL}_n(\mathbb{Z})$ as a subgroup of*

$$G(S) = \prod_{p \in S} \mathrm{SL}_n(\mathbb{Z}_p)$$

*via the diagonal embedding and let $E_S$ be the corresponding orbit equivalence relation. If $S \neq T$, then $E_S$ and $E_T$ are incomparable with respect to Borel reducibility.*

*Sketch Proof* For simplicity, suppose that $S = \{p\}$ and $T = \{q\}$, where $p \neq q$ are distinct primes. Suppose that $f : \mathrm{SL}_n(\mathbb{Z}_p) \to \mathrm{SL}_n(\mathbb{Z}_q)$ is a Borel reduction from $E_{\{p\}}$ to $E_{\{q\}}$. Then applying Ioana Superrigidity and arguing as in the proof of Theorem 4.6, after passing to subgroups of finite index and ergodic components if necessary, we can suppose that

$$(\mathrm{SL}_n(\mathbb{Z}), \mathrm{SL}_n(\mathbb{Z}_p), \mu_p) \cong (\mathrm{SL}_n(\mathbb{Z}), \mathrm{SL}_n(\mathbb{Z}_q), \mu_q)$$

as measure-preserving permutation groups. Hence it only remains to detect the prime $p$ in $(\mathrm{SL}_n(\mathbb{Z}), \mathrm{SL}_n(\mathbb{Z}_p), \mu_p)$.

Towards this end, recall that $\mathrm{Aut}(\mathrm{SL}_n(\mathbb{Z}), \mathrm{SL}_n(\mathbb{Z}_p), \mu_p)$ consists of the measure-preserving bijections $\varphi : \mathrm{SL}_n(\mathbb{Z}_p) \to \mathrm{SL}_n(\mathbb{Z}_p)$ such that for all $\gamma \in \mathrm{SL}_n(\mathbb{Z})$,

$$\varphi(\gamma \cdot x) = \gamma \cdot \varphi(x) \quad \text{for } \mu_p\text{-a.e. } x,$$

where we identify two such maps if they agree $\mu_p$-a.e. Notice that for each $g \in \mathrm{SL}_n(\mathbb{Z}_p)$, we can define a corresponding automorphism $\varphi \in \mathrm{Aut}(\mathrm{SL}_n(\mathbb{Z}), \mathrm{SL}_n(\mathbb{Z}_p), \mu_p)$ by $\varphi(x) = xg$. (Here we have made use of the fact that the Haar measure $\mu_p$ on the compact group $\mathrm{SL}_n(\mathbb{Z}_p)$ is also invariant under the right translation action.) The following proposition shows that there are no other automorphisms.

**Proposition 4.11** (Gefter-Golodets [13]) *$Aut(\mathrm{SL}_n(\mathbb{Z}), \mathrm{SL}_n(\mathbb{Z}_p), \mu_p) = \mathrm{SL}_n(\mathbb{Z}_p)$.*

*Proof* Let $\varphi \in \mathrm{Aut}(\mathrm{SL}_n(\mathbb{Z}), \mathrm{SL}_n(\mathbb{Z}_p), \mu_p)$. For each $x \in \mathrm{SL}_n(\mathbb{Z}_p)$, let $h(x) \in \mathrm{SL}_n(\mathbb{Z}_p)$ be such that $\varphi(x) = xh(x)$. If $\gamma \in \mathrm{SL}_n(\mathbb{Z})$, then for $\mu_p$-a.e. $x$,

$$\varphi(\gamma \cdot x) = \gamma \cdot \varphi(x) = \gamma \cdot xh(x)$$

and so $h(\gamma \cdot x) = h(x)$. Since $\mathrm{SL}_n(\mathbb{Z})$ acts ergodically on $(\mathrm{SL}_n(\mathbb{Z}_p), \mu_p)$, there exists a fixed $g \in \mathrm{SL}_n(\mathbb{Z}_p)$ such that $h(x) = g$ for $\mu_p$-a.e. $x$. □

Thus we have reduced our problem to that of detecting the prime $p$ in the topological group $\mathrm{SL}_n(\mathbb{Z}_p)$. But this is easy, since $\mathrm{SL}_n(\mathbb{Z}_p)$ is virtually a pro-$p$ group. More precisely, if $H$ is any open subgroup of $\mathrm{SL}_n(\mathbb{Z}_p)$, then

$$[\mathrm{SL}_n(\mathbb{Z}_p) : H] = bp^l$$

for some $l \geq 0$ and some divisor $b$ of $|\,\mathrm{SL}_n(\mathbb{Z}/p\mathbb{Z})|$. This completes our sketch of a proof of Theorem 4.10. □

## Open problems

In this closing section, we shall point out some of the many open problems in the field of countable Borel equivalence relations.

## 5.1 Hyperfinite relations.

Recall that a countable Borel equivalence relation $E$ on a standard Borel space $X$ is said to be *hyperfinite* if $E$ can be written as the union of an increasing sequence of *finite* Borel equivalence relations. A theorem of Dougherty-Jackson-Kechris [7] provides two additional characterizations:

**Theorem 5.1** (Dougherty-Jackson-Kechris) *If $E$ is a countable Borel equivalence relation on a standard Borel space $X$, then the following are equivalent:*

- *$E$ is hyperfinite.*
- *$E \leq_B E_0$.*
- *There exists a Borel action of $\mathbb{Z}$ on $X$ such that $E = E_{\mathbb{Z}}^X$.*

In particular, *every* $\mathbb{Z}$-action on a standard Borel $\mathbb{Z}$-space $X$ yields a hyperfinite orbit equivalence relation; and by a recent theorem of Gao-Jackson [12], even more is true.

**Theorem 5.2** (Gao-Jackson) *If $G$ is a countable abelian group and $X$ is a standard Borel $G$-space, then $E_G^X$ is hyperfinite.*

An important question concerns how much further this result can be extended. By a theorem of Jackson-Kechris-Louveau [19], if $G$ is a countable, nonamenable group, then the orbit equivalence relation $E_G$ arising from the free action of $G$ on $((2)^G, \mu)$ is *not* hyperfinite. However, the following problem remains open:

*Question* 5.3 (Weiss [39]) Suppose that $G$ is a countable amenable group and that $X$ is a standard Borel $G$-space. Does it follow that $E_G^X$ is hyperfinite?

As a partial answer, we have the following theorem of Connes, Feldman and Weiss[5].

**Theorem 5.4** (Connes-Feldman-Weiss) *Suppose that $G$ is a countable amenable group and that $X$ is a standard Borel $G$-space. If $\mu$ is* any *Borel probability measure on $X$, then there exists a Borel subset $Y \subseteq X$ with $\mu(Y) = 1$ such that $E \restriction Y$ is hyperfinite.*

## 5.2 Treeable relations.

**Definition 5.5** The countable Borel equivalence relation $E$ on $X$ is said to be *treeable* iff there is an acyclic Borel graph $(X, R)$ whose connected components are the $E$-classes.

For example, if a countable free group $\mathbb{F}$ acts freely on a standard Borel $\mathbb{F}$-space $X$, then the corresponding orbit equivalence relation $E_{\mathbb{F}}^{X}$ is treeable. Conversely, by a theorem of Jackson-Kechris-Louveau [19], if $E$ is treeable, then there exists a free Borel action of a countable free group $\mathbb{F}$ on a standard Borel space $Y$ such that $E \sim_B E_{\mathbb{F}}^{Y}$. It is easily seen that every hyperfinite countable Borel equivalence relation is treeable; and it is known that the universal countable Borel equivalence relation $E_\infty$ is *not* treeable. On the other hand, there exist countable Borel equivalence relations which are treeable but not hyperfinite. For example, the orbit equivalence relation $E_{\infty T}$ arising from the free action of $\mathbb{F}_2$ on $(2)^{\mathbb{F}_2}$ is not hyperfinite.

**Theorem 5.6** (Jackson-Kechris-Louveau [19]) *$E_{\infty T}$ is universal for treeable countable Borel equivalence relations.*

For many years, it was an important open problem whether there existed infinitely many treeable countable Borel equivalence relations up to Borel bireducibility. This question has recently been solved by Hjorth:

**Theorem 5.7** (Hjorth [16]) *There exist uncountably many treeable countable Borel equivalence relations which are pairwise incomparable with respect to Borel reducibility.*

An intriguing aspect of Hjorth's proof is that it does not provide an *explicit example* of a single pair $E$, $F$ of incomparable treeable countable Borel equivalence relations. However, there is a natural candidate for an explicit family of uncountably many treeable countable Borel equivalence relations which are pairwise incomparable with respect to Borel reducibility. For each nonempty set $S$ of primes, regard $SL_2(\mathbb{Z})$ as a subgroup of

$$G(S) = \prod_{p \in S} SL_2(\mathbb{Z}_p)$$

via the diagonal embedding and let $E_S$ be the corresponding orbit equivalence relation. Then $E_S$ is a non-hyperfinite profinite treeable Borel equivalence relation.

**Conjecture 5.8** (Thomas) *If $S \neq T$, then $E_S$ and $E_T$ are incomparable with respect to Borel reducibility.*

We will finish this subsection with an attractive (but almost certainly false) conjecture of Kechris. Recall that in 1980, Ol'shanskii [29] refuted the so-called "von Neumann conjecture" (which is actually due to Day) by constructing a periodic nonamenable group, which clearly had no free nonabelian subgroups. However, the following analogous problem remains open:

**Conjecture 5.9** (Kechris) *If $E$ is a non-hyperfinite countable Borel equivalence relation, then there exists a non-hyperfinite treeable countable Borel equivalence relation $F$ such that $F \leq_B E$.*

## 5.3 Universal relations.

There are many basic open problems concerning universal countable Borel equivalence relations, including the following:

**Conjecture 5.10** (Hjorth) *If $E$ is a universal countable Borel equivalence relation on the standard Borel space $X$ and $F$ is a countable Borel equivalence relation such that $E \subseteq F$, then $F$ is also universal.*

**Conjecture 5.11** (Kechris) *The Turing equivalence relation $\equiv_T$ is countable universal.*

*Question* 5.12 (Jackson-Kechris-Louveau [19]) Suppose that $E$ is a universal countable Borel equivalence relation on the standard Borel space $X$ and that $Y \subseteq X$ is an $E$-invariant Borel subset. Does it follow that either $E \restriction Y$ or $E \restriction (X \setminus Y)$ is universal? [3]

Finally we conclude with two questions concerning the notion of a *minimal cover* of an equivalence relation.

**Definition 5.13** If $E$, $E'$ are countable Borel equivalence relations, then $E'$ is a minimal cover of $E$ if:

- $E <_B E'$; and
- if $F$ is a countable Borel equivalence relation such that $E \leq_B F \leq_B E'$, then either $E \sim_B F$ or $F \sim_B E'$.

*Open problem* (Thomas) Find an example of a nonsmooth countable Borel equivalence relation which has a minimal cover.

[3] Andrew Marks [27] has recently shown that this is indeed the case.

*Open problem* (Thomas) Find an example of a nonuniversal countable Borel equivalence relation which does *not* have a minimal cover.

## References


[1] S. Adams, *Containment does not imply Borel reducibility*, in: Set Theory: The Hajnal Conference (Ed: S. Thomas), DIMACS Series, vol. 58, American Mathematical Society, 2002, pp. 1-23.

[2] S. Adams and A. S. Keckris, *Linear algebraic groups and countable Borel equivalence relations*, J. Amer. Math. Soc. **13** (2000), 909-943.

[3] A. Andretta, R. Camerlo, and G. Hjorth, *Conjugacy equivalence relations on subgroups*, Fund. Math. **167** (2001), 189-212.

[4] R. Baer, *Abelian groups without elements of finite order*, Duke Math. Journal **3** (1937), 68-122.

[5] A. Connes, J. Feldman, and B. Weiss, *An amenable equivalence relation is generated by a single transformation*, Ergodic Theory and Dynamical Systems **1** (1981), 430-450.

[6] C. Champetier, *L'espace des groupes de type fini*, Topology **39** (2000), 657–680.

[7] R. Dougherty, S. Jackson, and A. S. Kechris, *The structure of hyperfinite Borel equivalence relations*, Trans. Amer. Math. Soc. **341 (1)** (1994), 193-225.

[8] R. Dougherty and A. S. Kechris, *How many Turing degrees are there?*, in: Computability Theory and its Applications (Boulder, CO, 1999), Contemp. Math. **257**, Amer. Math. Soc., 2000, 83–94.

[9] M. J. Dunwoody and A. Pietrowski, *Presentations of the trefoil group*, Canad. Math. Bull. **16** (1973), 517–520.

[10] J. Feldman and C. C. Moore, *Ergodic equivalence relations and Von Neumann algebras I*, Trans. Amer. Math. Soc. **234** (1977), 289-324.

[11] H. Friedman and L. Stanley, *A Borel reducibility theory for classes of countable structures* J. Symbolic Logic **54** (1989), 894–914.

[12] S. Gao and S. Jackson, *Countable abelian group actions and hyperfinite equivalence relations*, to appear in Inventiones Mathematicae.

[13] S. L. Gefter and V. Ya. Golodets, *Fundamental groups for ergodic actions and actions with unit fundamental groups*, Publ. RIMS, Kyoto Univ. **24** (1988), 821-847.

[14] L. Harrington, A. S. Kechris, and A. Louveau, *A Glimm-Effros dechotomy for Borel equivalence relations*, J. Amer. Math. Soc. **3 (4)** (1990), 903-928.

[15] G. Hjorth, *Around nonclassifiability for countable torsion-free abelian groups*, in Abelian groups and modules (Dublin, 1998), Trends Math., Birkhäuser, Basel (1999), 269-292.

[16] G. Hjorth, *Treeable equivalence relations*, preprint, 2008.

[17] G. Hjorth and A. S. Kechris, *Borel equivalence relations and classification of countable models*, Annals of Pure and Applied Logic **82** (1996), 221-272.

[18] A. Ioana, *Cocycle superrigidity for profinite actions of property (T) groups*, Duke Math. J. **157** (2011), 337–367.

[19] S. Jackson, A. S. Kechris, and A. Louveau, *Countable Borel equivalence relations*, J. Math. Logic, **2** (2002), 1-80.

[20] V. F. R. Jones and K. Schmidt, *Asymptotically invariant sequences and approximate finiteness*, Amer. J. Math. **109** (1987), 91-114.

[21] A. S. Kechris, *Classical Descriptive Set Theory*, Graduate Texts in Mathematics **156** Springer-Verlag, New York, 1994.

[22] K. Kuratowski, *Sur une généralisation de la notion d'homéomorphie*, Fund. Math. **22** (1934), 206-220.

[23] A. G. Kurosh, *Primitive torsionsfreie abelsche Gruppen vom endlichen Range*, Ann. Math. 38 (1937), 175-203.

[24] A. Lubotzky, *Discrete Groups, Expanding Graphs and Invariant Measures*, Progress in Mathematics **125**, Birkhäuser, 1994.

[25] A. I. Malcev, *Torsion-free abelian groups of finite rank* (Russian), Mat. Sbor. **4** (1938), 45-68.

[26] G. A. Margulis, *Discrete Subgroups of Semisimple Lie Groups*, Erg. der Math. und ihrer Grenz. **17**, Springer-Verlag, 1991.

[27] A. Marks, T. Slaman and S. Steel, *Martin's Conjecture, Arithmetic Equivalence and Countable Borel Equivalence Relations*, preprint, 2011.

[28] B. D. Miller, *Borel equivalence relations and everywhere faithful actions of free products*, preprint, 2006.

[29] A. Yu. Ol'shanskii, *On the question of the existence of an invariant mean on a group* (Russian), Uspekhi Mat. Nauk, **35** (1980), no. 4, 199-200.

[30] S. Popa, *Cocycle and orbit equivalence superrigidity for malleable actions of $\omega$-rigid groups*, Inventiones Mathematicae **170 (2)** (2007), 243-295.

[31] J. Silver, *Counting the number of equivalence classes of Borel and co-analytic equivalence relations*, Ann. Math. Logic **18** (1980), 1-28.

[32] T. Slaman and J. Steel, *Definable functions on degrees*, in: Cabal Seminar 81-85, Lecture Notes in Mathematics, **1333**, Springer-Verlag (1998), 37-55

[33] S. M. Srivastava, *A Course on Borel Sets*, Graduate Texts in Mathematics **180**, Springer-Verlag, 1998.

[34] S. Thomas, *Superrigidity and countable Borel equivalence relations*, Annals Pure Appl. Logic **120** (2003), 237-262.

[35] S. Thomas, *The classification problem for torsion-free abelian groups of finite rank*, J. Amer. Math. Soc. **16** (2003), 233-258.

[36] S. Thomas, *On the complexity of the quasi-isometry and virtual isomorphism problems for finitely generated groups*, Groups Geom. Dyn. **2** (2008), 281–307.

[37] S. Thomas, *Popa superrigidity and countable Borel equivalence relations*, Ann. Pure Appl. Logic **158** (2009), 175-189.

[38] S. Wagon, *The Banach-Tarski Paradox*, Encyclopedia of Mathematics and its Applications **24**, Cambridge University Press, 1985.

[39] B. Weiss, *Measureable dynamics*, Conference in modern analysis and probability (R. Beals et al., editors), Contemporary Mathematics, vol. 26, American Mathematical Society, Providence, Rhode Island, 1984, 395-421.

# 3

# Set theory and operator algebras

Ilijas Farah and Eric Wofsey

---

The fifth Appalachian Set Theory workshop was held at Carnegie Mellon University on February 9, 2008. The lecturer was Ilijas Farah. As an undergraduate student Eric Wofsey assisted in writing this chapter, which is based on the workshop lectures.

---

These notes are based on the six-hour Appalachian Set Theory workshop given by Ilijas Farah on February 9th, 2008 at Carnegie Mellon University. The first half of the workshop (Sections 1–4) consisted of a review of Hilbert space theory and an introduction to C*-algebras, and the second half (Sections 5–6) outlined a few set-theoretic problems relating to C*-algebras. The number and variety of topics covered in the workshop was unfortunately limited by the available time.

Good general references on Hilbert spaces and C*-algebras include [8], [13], [17], [51], [61], and [67]. An introduction to spectral theory is given in [9]. Most of the omitted proofs can be found in most of these references. For a survey of applications of set theory to operator algebras, see [64].

## Acknowledgments

We owe greatest thanks to Ernest Schimmerling, without whose enthusiasm and support this paper would not have existed. We would like to thank Nik Weaver for giving us a kind permission to include some of his unpublished results. I.F. would like to thank George Elliott, N. Christopher Phillips, Efren Ruiz, Juris Steprāns, Nik Weaver and the second author for many conversations related to the topics presented here. He would also like to thank Sam Coskey, Paul Ellis, Saeed Ghasemi, and Ernest Schimmerling for a number of remarks about the early version of the present paper, and to the anonymous referee for one of the longest and most painstaking reports in the history of mathematics. Special thanks to Gala for her indispensable help with the bibliography.

The early draft of these notes was produced for a series of lectures that the first author gave at the Kobe University in November 2006. He would like to thank Jörg Brendle for organizing this visit and for his hospitality. Some of the results included here were obtained during the first author's stay at the IHES in July 2008. I.F. is partially supported by NSERC.

## 1 Introduction

A more accurate title for the present paper would be 'Set theory and C*-algebras' but this title was already taken by Weaver's excellent survey ([64]). Apart from C*-algebras, set theory (both combinatorial and descriptive) has strong connections to the theory of von Neumann algebras, and in particular to $\mathrm{II}_1$ factors (but see §1.2 and §1.4 below). This subject will not be touched upon in the present paper. A very intuitive and approachable introduction to von Neumann algebras can be found in [43].

The fact that there were very few interactions between set theory and theory of operator algebras may be somewhat surprising, not only because John von Neumann played a role in the development of both subjects. Apart from the work of Joel Anderson in the 1970's and some attempts at developing 'noncommutative' set theory, there was virtually no interaction between the two areas until recently.

This situation has dramatically changed in the last decade to the extent that we will not even be able to outline the entire subject in the present paper.[1] Let us instead outline what is covered here. A set-theorist-friendly introduction to operators on Hilbert spaces, continuous function calculus, C*-algebras, and their representation theory is given in sections §1–4. In §5 we consider the Calkin algebra as a quantized version of $\mathcal{P}(\mathbb{N})/\mathrm{Fin}$ and consider some problems about the former which are direct translations of theorems (and problems) about the latter. In §6 we return to the representation theory of C*-algebras and consider two

[1] It has changed even more drastically during the four years since February 2008, when this paper was written.

of the most interesting recent applications of set theory to C*-algebras, both due to Akemann and Weaver. These are the construction of a counterexample to Naimark's problem using Jensen's $\diamondsuit$ and a construction of a pure state on $\mathcal{B}(H)$ that is not diagonalizable, using the Continuum Hypothesis. In the latter case we present an unpublished result of the first author and Weaver, showing that a substantial weakening of the Continuum Hypothesis suffices. It is not known whether either of these two results can be proved from ZFC alone.

Applications of set theory to the theory of operator algebras fall into several categories, and we shall now describe (a part of) what is being omitted.

## 1.1 Nonseparable C*-algebras

Some long-standing open problems in theory of C*-algebras were recently solved in ZFC, by using rather elementary set theory to construct nonseparable C*-algebras with properties not present in separable C*-algebras. We should mention Weaver's construction of a prime C*-algebra that is not primitive ([62], see also [18] and [45] for simpler constructions). In [31] and [24] it was demonstrated that even direct limits of full matrix C*-algebras (the nonseparable analogues of UHF algebras, see §3.4) can have rather exotic properties. Curiously, each of these results answers a (different) long-standing open problem posed by Jacques Dixmier.

## 1.2 Ultrapowers

Ultrapowers are an indispensable tool both in model theory and in operator algebras, yet until [35] the two theories were developed essentially independently. This can be contrasted to the fact that ultraproducts of Banach spaces were well-studied by logicians. Largely motivated by some questions of Eberhard Kirchberg, a few papers appeared recently showing that the structure of ultrapowers and relative commutants of C*-algebras and $\mathrm{II}_1$ factors can depend on the choice of the ultrafilter ([32], [23], [30]). A model-theoretic logic suitable for study of C*-algebras and $\mathrm{II}_1$ factors, adapted from [10], was developed in [29].

### 1.3 Structure of corona algebras

The question whether the Calkin algebra has outer automorphisms was asked in the seminal Brown–Douglas–Fillmore paper [15]. The answer to this question is independent from ZFC by [52] and [26]. Further results on rigidity of corona algebras can be found in [28], [36] and [25]

### 1.4 Classification and descriptive set theory

While the present survey is exclusively concerned with applications of combinatorial set theory, some of the most exciting interactions between operator algebras and set theory were purely descriptive. The abstract classification theory was recently applied to determine lower bounds of classification problems for von Neumann algebras ([56], [57], [58]), of spectra of C*-algebras ([46], [27]), and for classification of C*-algebras ([34], [33]). Also, Popa superrigidity developed in the context of $\mathrm{II}_1$ factors was indispensable in proving some of the most interesting recent results on countable Borel equivalence relations. But this is another story (see [1]).

## 2 Hilbert spaces and operators

We begin with a review of the basic properties of operators on a Hilbert space. Throughout we let $H$ denote a complex infinite-dimensional separable Hilbert space, and we let $(e_n)$ be an orthonormal basis for $H$ (see Example 2.1). For $\xi, \eta \in H$, we denote their inner product by $(\xi \mid \eta)$. We recall that

$$(\eta \mid \xi) = \overline{(\xi \mid \eta)}$$

and the norm defined by

$$\|\xi\| = \sqrt{(\xi \mid \xi)}.$$

The Cauchy–Schwartz inequality says that

$$|(\xi \mid \eta)| \leq \|\xi\| \cdot \|\eta\|.$$

**Example 2.1** The space

$$\ell^2(\mathbb{N}) = \left\{(\alpha_k)_{k\in\mathbb{N}} : \alpha_k \in \mathbb{C}, \|\alpha\|^2 = \sum |\alpha_k|^2 < \infty\right\}$$

(sometimes denoted simply by $\ell^2$) is a Hilbert space under the inner product $(\alpha \mid \beta) = \sum \alpha_k \overline{\beta_k}$. If we define $e^n \in \ell^2(\mathbb{N})$ by $e^n_k = \delta_{nk}$ (the Kronecker $\delta$), then $(e^n)$ is an orthonormal basis for $\ell^2$. For any $\alpha \in \ell^2$, $\alpha = \sum \alpha_n e^n$.

Any Hilbert space has an orthonormal basis, and this can be used to prove that all separable infinite-dimensional Hilbert spaces are isomorphic. Moreover, any two infinite-dimensional Hilbert spaces with the same character density (the minimal cardinality of a dense subset) are isomorphic.

**Example 2.2** If $(X, \mu)$ is a measure space,

$$L^2(X,\mu) = \left\{ f : X \to \mathbb{C} \text{ measurable} : \int |f|^2 d\mu < \infty \right\} / \{f : f = 0 \text{ a.e.}\}$$

is a Hilbert space under the inner product $(f \mid g) = \int f\overline{g}d\mu$ and with the norm defined by $\|f\|^2 = \int |f|^2 d\mu$.

We will let $a, b, \ldots$ denote linear operators $H \to H$. We recall that

$$\|a\| = \sup\{\|a\xi\| : \xi \in H, \|\xi\| = 1\}.$$

If $\|a\| < \infty$, we say $a$ is *bounded*. An operator is bounded if and only if it is continuous. We denote the algebra of all bounded operators on $H$ by $\mathcal{B}(H)$ (some authors use $L(H)$), and throughout the paper all of our operators will be bounded. We define the *adjoint* $a^*$ of $a$ to be the unique operator satisfying

$$(a\xi \mid \eta) = (\xi \mid a^*\eta)$$

for all $\xi, \eta \in H$. Note that since an element of $H$ is determined by its inner products with all other elements of $H$ (e.g., take an orthonormal basis), an operator $a$ is determined by the values of $(a\xi \mid \eta)$ for all $\xi, \eta$ or even by the values $(ae_m \mid e_n)$ for $m$ and $n$ in $\mathbb{N}$.

**Lemma 2.3** *For all $a, b$ we have*

1. $(a^*)^* = a$
2. $(ab)^* = b^*a^*$
3. $\|a\| = \|a^*\|$
4. $\|ab\| \leq \|a\| \cdot \|b\|$
5. $\|a^*a\| = \|a\|^2$

*Proof* These are all easy calculations. For example, for (5.), for $\|\xi\| = 1$,

$$\|a\xi\|^2 = (a\xi \mid a\xi) = (\xi \mid a^*a\xi) \le \|\xi\| \cdot \|a^*a\xi\| \le \|a^*a\|,$$

the first inequality holding by Cauchy–Schwartz. Taking the sup over all $\xi$, we obtain $\|a\|^2 \le \|a^*a\|$. Conversely,

$$\|a^*a\| \le \|a^*\|\|a\| = \|a\|^2$$

by (3.) and (4.). □

The first four parts of this say that $\mathcal{B}(H)$ is a *Banach *-algebra* (or a *Banach algebra with involution* *) and (5.) (sometimes called the C*-*equality*) says that $\mathcal{B}(H)$ is a C*-algebra.

**Definition 2.4** An (*abstract*) *C*-algebra* is a Banach *-algebra satisfying the C*-equality, $\|a^*a\| = \|a\|^2$ for all $a$.

## 2.1 Normal operators and the spectral theorem

In this section we introduce some distinguished classes of operators in $\mathcal{B}(H)$, such as normal and self-adjoint operators (cf. §3).

**Example 2.5** Assume $(X, \mu)$ is a probability measure space. If $H_0 = L^2(X, \mu)$ and $f: X \to \mathbb{C}$ is bounded and measurable, then

$$H_0 \ni g \stackrel{m_f}{\longmapsto} fg \in H_0$$

is a bounded linear operator. We have $\|m_f\| = \|f\|_\infty$ and

$$m_f^* = m_{\bar{f}}.$$

Hence $m_f^* m_f = m_f m_f^* = m_{|f|^2}$. We call operators of this form *multiplication operators.*

An operator $a$ is *normal* if $aa^* = a^*a$. Clearly, all multiplication operators are normal. Normal operators have a nice structure theory, which is summarized in the following theorem.

**Theorem 2.6** (Spectral Theorem) *If $a$ is a normal operator then there is a probability measure space $(X, \mu)$, a measurable function $f$ on $X$, and a Hilbert space isomorphism $\Phi: L^2(X, \mu) \to H$ such that $\Phi a \Phi^{-1} = m_f$.*

*Proof* For an elegant proof using Corollary 3.13 see [9, Theorem 2.4.5]. □

That is, every normal operator is a multiplication operator for some identification of $H$ with an $L^2$ space. Conversely, every multiplication operator is clearly normal. If $X$ is discrete and $\mu$ is counting measure, the characteristic functions of the points of $X$ form an orthonormal basis for $L^2(X,\mu)$ and the spectral theorem says that $a$ is diagonalized by this basis. In general, the spectral theorem says that normal operators are "measurably diagonalizable".

If $\Phi\colon H_1 \to H_2$ is an isomorphism between Hilbert spaces, then

$$a \mapsto \operatorname{Ad}\Phi(a) = \Phi a \Phi^{-1}$$

is an isomorphism between $\mathcal{B}(H_1)$ and $\mathcal{B}(H_2)$. The operator $\operatorname{Ad}\Phi(a)$ is just $a$ with its domain and range identified with $H_2$ via $\Phi$.

Our stating of the Spectral Theorem is rather premature in the formal sense since we are going to introduce some of the key notions used in its proof later on, in §2.2 and §3.2. This was motivated by the insight that the Spectral Theorem provides to theory of C*-algebras.

An operator $a$ is *self-adjoint* if $a = a^*$. Self-adjoint operators are obviously normal. For any $b \in \mathcal{B}(H)$, the "real" and "imaginary" parts of $b$, defined by $b_0 = (b + b^*)/2$ and $b_1 = (b - b^*)/2i$, are self-adjoint and satisfy $b = b_0 + ib_1$. Thus any operator is a linear combination of self-adjoint operators. It is easy to check that an operator is normal if and only if its real and imaginary parts commute, so the normal operators are exactly the linear combinations of commuting self-adjoint operators.

**Example 2.7** The real and imaginary parts of a multiplication operator $m_f$ are $m_{\operatorname{Re} f}$ and $m_{\operatorname{Im} f}$. A multiplication operator $m_f$ is self-adjoint if and only if $f$ is real (a.e.). By the spectral theorem, all self-adjoint operators are of this form.

It is easy to verify that for any $a \in \mathcal{B}(H)$ and $\xi, \eta \in H$ the following so-called *polarization identity* holds

$$(a\xi \mid \eta) = \frac{1}{4}\sum_{k=0}^{3} i^k (a(\xi + i^k\eta) \mid \xi + i^k\eta).$$

**Proposition 2.8** *An operator $a$ is self-adjoint if and only if $(a\xi \mid \xi)$ is real for all $\xi$.*

*Proof* First, note that

$$((a-a^*)\xi \mid \xi) = (a\xi \mid \xi) - (a^*\xi \mid \xi) = (a\xi \mid \xi) - (\xi \mid a\xi) = (a\xi \mid \xi) - \overline{(a\xi \mid \xi)}.$$

Thus $(a\xi \mid \xi)$ is real for all $\xi$ if and only if $((a - a^*)\xi \mid \xi) = 0$ for all

$\xi$. But by polarization, the operator $a - a^*$ is entirely determined by the values $((a - a^*)\xi \mid \xi)$, so this is equivalent to $a - a^* = 0$. □

A self-adjoint operator $b$ is called *positive* if $(b\xi \mid \xi) \geq 0$ for all $\xi \in H$. In this situation we write $b \geq 0$. By Proposition 2.8, positive operators are self-adjoint. For instance, a multiplication operator $m_f$ is positive if and only if $f \geq 0$ (a.e.). By the spectral theorem, all positive operators are of the form $m_f$ for $f \geq 0$ a.e.

*Exercise* 2.9 For any self-adjoint $a \in \mathcal{B}(H)$ we can write $a = a_0 - a_1$ for some positive operators $a_0$ and $a_1$. (Hint: Use the spectral theorem.)

**Proposition 2.10** *An operator $b$ is positive if and only if $b = a^*a$ for some (non-unique) $a$. This $a$ may be chosen to be positive.*

*Proof* For the converse implication note that $(a^*a\xi \mid \xi) = (a\xi \mid a\xi) = \|a\xi\|^2 \geq 0$. For the direct implication, assume $b$ is positive. By the spectral theorem we may assume $b = m_f$ for $f \geq 0$. Let $a = m_{\sqrt{f}}$. □

We say that $p \in \mathcal{B}(H)$ is a *projection* if $p^2 = p^* = p$.

**Lemma 2.11** *$p$ is a projection if and only if it is the orthogonal projection onto a closed subspace of $H$.*

*Proof* Any linear projection $p$ onto a closed subspace of $H$ satisfies $p = p^2$, and orthogonal projections are exactly those that also satisfy $p = p^*$. Conversely, suppose $p$ is a projection. Then $p$ is self-adjoint, so we can write $p = m_f$ for $f : X \to \mathbb{C}$, and we have $f = f^2 = \bar{f}$. Hence $f(x) \in \{0, 1\}$ for (almost) all $x$. We then set $A = f^{-1}(\{1\})$, and it is easy to see that $p$ is the orthogonal projection onto the closed subspace $L^2(A) \subseteq L^2(X)$. □

If $E \subseteq H$ is a closed subspace, we denote the projection onto $E$ by $\mathrm{proj}_E$.

We denote the identity operator on $H$ by $I$ (some authors use 1). An operator $u$ is *unitary* if $uu^* = u^*u = I$. This is equivalent to $u$ being invertible and satisfying

$$(\xi \mid \eta) = (u^*u\xi \mid \eta) = (u\xi \mid u\eta)$$

for all $\xi, \eta \in H$. That is, an operator is unitary if and only if it is a Hilbert space automorphism of $H$. Unitary operators are obviously normal. For instance, a multiplication operator $m_f$ is unitary if $m_f(m_f)^* = I$, or equivalently, if $|f|^2 = 1$ (a.e.). By the spectral theorem, all unitaries are of this form.

An operator $v$ is a *partial isometry* if

$$p = vv^* \text{ and } q = v^*v$$

are both projections. Partial isometries are essentially isomorphisms (isometries) between closed subspaces of $H$: For every partial isometry $v$ there is a closed subspace $H_0$ of $H$ such that $v \restriction H_0$ is an isometry into a closed subspace of $H$ and $v \restriction H_0^\perp \equiv 0$. (As usually, $H_0^\perp$ is the *orthogonal complement* of $H_0$, $\{\xi : (\xi \mid \eta) = 0 \text{ for all } \eta \in H_0\}$.) However, as the following example shows, partial isometries need not be normal.

**Example 2.12** Let $(e_n)$ be an orthonormal basis of $H$. We define the *unilateral shift* $S$ by $S(e_n) = e_{n+1}$ for all $n$. Then $S^*(e_{n+1}) = e_n$ and $S^*(e_0) = 0$. We have $S^*S = I$ but $SS^* = \text{proj}_{\overline{\text{span}\{e_n\}_{n\geq 1}}}$.

Any complex number $z$ can be written as $z = re^{i\theta}$ for $r \geq 0$ and $|e^{i\theta}| = 1$. Considering $\mathbb{C}$ as the set of operators on a one-dimensional Hilbert space, there is an analogue of this on an arbitrary Hilbert space.

**Theorem 2.13** (Polar Decomposition) *Any $a \in \mathcal{B}(H)$ can be written as $a = bv$ where $b$ is positive and $v$ is a partial isometry.*

*Proof* See e.g., [51, Theorem 3.2.17 and Remark 3.2.18]. □

However, this has less value as a structure theorem than than one might think, since $b$ and $v$ may not commute. While positive operators and partial isometries are both fairly easy to understand, polar decomposition does not always make arbitrary operators easy to understand. For example, it is easy to show that positive operators and partial isometries always have nontrivial closed invariant subspaces, but it is a famous open problem whether this is true for all operators.

## 2.2 The spectrum of an operator

The *spectrum* of an operator $a$ is

$$\sigma(a) = \{\lambda \in \mathbb{C} : a - \lambda I \text{ is not invertible}\}.$$

For a finite-dimensional matrix, the spectrum is the set of eigenvalues.

**Example 2.14** A multiplication operator $m_f$ is invertible if and only if there is some $\epsilon > 0$ such that $|f| > \epsilon$ (a.e.). Thus since $m_f - \lambda I = m_{f-\lambda}$, $\sigma(m_f)$ is the essential range of $f$ (the set of $\lambda \in \mathbb{C}$ such that for every neighborhood $U$ of $\lambda$, $f^{-1}(U)$ has positive measure).

**Lemma 2.15** *If* $\|a\| < 1$ *then* $I - a$ *is invertible in* $\mathcal{B}(H)$.

*Proof* The series $b = \sum_{n=0}^{\infty} a^n$ is convergent and hence in $\mathcal{B}(H)$. By considering partial sums one sees that $(I - a)b = b(I - a) = I$. □

The following Lemma is an immediate consequence of the Spectral Theorem. However, since its part (1.) is used in the proof of the latter, we provide its proof.

**Lemma 2.16** *Let* $a \in \mathcal{B}(H)$.

1. $\sigma(a)$ *is a compact subset of* $\mathbb{C}$.
2. $\sigma(a^*) = \{\overline{\lambda} : \lambda \in \sigma(a)\}$.
3. *If* $a$ *is normal, then* $a$ *is self-adjoint if and only if* $\sigma(a) \subseteq \mathbb{R}$.
4. *If* $a$ *is normal, then* $a$ *is positive if and only if* $\sigma(a) \subseteq [0, \infty)$.

*Proof of* (1.) If $|\lambda| > \|a\|$ then $a - \lambda \cdot I = \lambda(\frac{1}{\lambda}a - I)$ is invertible by Lemma 2.15, and therefore $\sigma(a)$ is bounded.

We shall now show that the set of invertible elements is open. Fix an invertible $a$. Since the multiplication is continuous, we can find $\epsilon > 0$ such that for every $b$ in the $\epsilon$-ball centered at $a$ there is $c$ such that both $\|I - bc\| < 1$ and $\|I - cb\| < 1$. By Lemma 2.15 there are $d_1$ and $d_2$ such that $bcd_1 = d_2cb = I$. Then we have

$$cd_1 = I \cdot cd_1 = d_2cbcd_1 = d_2c \cdot I = d_2c$$

and therefore $cd_1 = d_2c$ is the inverse of $b$.

Let $a$ be an arbitrary operator. If $\lambda \notin \sigma(a)$ then by the above there is an $\epsilon > 0$ such that every $b$ in the $\epsilon$-ball centered at $a - \lambda I$ is invertible. In particular, if $|\lambda' - \lambda| < \epsilon$ then $\lambda' \notin \sigma(a)$, concluding the proof that $\sigma(a)$ is compact. □

## 3 C*-algebras

We say that a *concrete* C*-*algebra* is a norm-closed *-subalgebra of $\mathcal{B}(H)$. For $X \subseteq \mathcal{B}(H)$ by $C^*(X)$ we denote the *C*-algebra generated by* $X$, i.e., the norm-closure of the algebra of all *-polynomials in elements of $X$. Equivalently, $C^*(X)$ is the intersection of all C*-subalgebras of $\mathcal{B}(H)$ including $X$.

When talking about C*-algebras, everything is 'starred': subalgebras are *-subalgebras (i.e. closed under involution), homomorphisms are *-homomorphisms (i.e. preserve the involution), etc.

**Definition 3.1** An *(abstract)* C*-*algebra* is a Banach algebra with involution that satisfies the C*-equality $\|aa^*\| = \|a\|^2$ for all $a$. That is, it is a Banach space with a product and involution satisfying Lemma 2.3.

A C*-algebra is *unital* if it has a unit (multiplicative identity). For unital C*-algebras, we can talk about the spectrum of an element.

**Lemma 3.2** *Every* C*-*algebra $A$ is contained in a unital* C*-*algebra* $\tilde{A} \cong A \oplus \mathbb{C}$.

*Proof* On $A \times \mathbb{C}$ define the operations as follows: $(a, \lambda)(b, \xi) = (ab + \lambda b + \xi a, \lambda\xi)$, $(a, \lambda)^* = (a^*, \bar{\lambda})$ and $\|(a, \lambda)\| = \sup_{\|b\| \leq 1} \|ab + \lambda b\|$ and check that this is still a C*-algebra.

A straightforward calculation shows that $(0, 1)$ is the unit of $\tilde{A}$ and that $A \ni a \mapsto (a, 0) \in \tilde{A}$ is an isomorphic embedding of $A$ into $\tilde{A}$. □

*Exercise* 3.3 Work out the details of the proof of Lemma 3.2.

We call $\tilde{A}$ the *unitization* of $A$. By passing to the unitization, we can talk about the spectrum of an element of a nonunital C*-algebra. The unitization retains many of the properties of the algebra $A$, and many results are proved by first considering the unitization. However, some caution is advised; for example, the unitization is never a simple algebra. (Recall that an algebra is *simple* if it has no nontrivial (two-sided) ideals. In case of C*-algebras, an algebra is simple if it has no nontrivial closed ideals. Such ideals are automatically self-adjoint.)

If $A$ and $B$ are unital and $A \subseteq B$ we say $A$ is a *unital subalgebra* of $B$ if the unit of $B$ belongs to $A$ (that is, $B$ has the same unit as $A$).

## Types of operators in C*-algebras

We import all of our terminology for distinguished classes of operators in $\mathcal{B}(H)$ (normal, self-adjoint, projections, etc.) to describe elements of any C*-algebra (cf. §2.1). More precisely, for an operator $a$ in a C*-algebra $A$ we say that

1. $a$ is *normal* if $aa^* = a^*a$,
2. $a$ is *self-adjoint* (or *Hermitian*) if $a = a^*$,
3. $a$ is a *projection* if $a^2 = a^* = a$,
4. $a$ is *positive* (or $a \geq 0$) if $a = b^*b$ for some $b$,
5. If $A$ is unital then $a$ is *unitary* if $aa^* = a^*a = I$.

Note that a positive element is automatically self-adjoint. For self-adjoint elements $a$ and $b$ write $a \leq b$ if $b - a$ is positive.

## 3.1 Some examples of C*-algebras

Let us consider several important C*-algebras and classes of C*-algebras (see also [19]).

$C_0(X)$

Let $X$ be a locally compact Hausdorff space. Then

$$C_0(X) = \{f : X \to \mathbb{C} : f \text{ is continuous and vanishes at } \infty\}$$

is a C*-algebra with the involution $f^* = \overline{f}$. Here "vanishes at $\infty$" means that $f$ extends continuously to the one-point compactification $X \cup \{\infty\}$ of $X$ such that the extension vanishes at $\infty$. Equivalently, for any $\epsilon > 0$, there is a compact set $K \subseteq X$ such that $|f(x)| < \epsilon$ for $x \notin K$. In particular, if $X$ itself is compact, all continuous functions vanish at $\infty$, and we write $C_0(X) = C(X)$.

$C_0(X)$ is abelian, so in particular every element is normal. $C_0(X)$ is unital if and only if $X$ is compact (iff the constant function 1 vanishes at $\infty$). The unitization of $C_0(X)$ is $C(X^*)$, where $X^*$ is the one-point compactification of $X$. For $f \in C_0(X)$, we have:

| | | |
|---|---|---|
| $f$ is self-adjoint | if and only if | $\text{range}(f) \subseteq \mathbb{R}$. |
| $f$ is positive | if and only if | $\text{range}(f) \subseteq [0, \infty)$. |
| $f$ is a projection | if and only if | $f^2(x) = f(x) = \overline{f(x)}$ |
| | if and only if | $\text{range}(f) \subseteq \{0, 1\}$ |
| | if and only if | $f = \chi_U$ for a clopen $U \subseteq X$. |

For any $f \in C_0(X)$, $\sigma(f) = \text{range}(f)$.

**Full matrix algebras**

$M_n(\mathbb{C})$, the set of $n \times n$ complex matrices is a unital C*-algebra. In fact, $M_n(\mathbb{C}) \cong \mathcal{B}(\mathbb{C}^n)$, where $\mathbb{C}^n$ is the $n$-dimensional complex Hilbert space.

| | |
|---|---|
| adjoint, unitary: | the usual meaning. |
| self-adjoint: | Hermitian. |
| positive: | positive semidefinite. |
| $\sigma(a)$: | the set of eigenvalues. |

The spectral theorem on $M_n(\mathbb{C})$ is the spectral theorem of elementary linear algebra: normal matrices are diagonalizable.

$L^\infty(X, \mu)$

If $(X, \mu)$ is a measure space, then the space $L^\infty(X, \mu)$ of all essentially bounded $\mu$-measurable functions on $X$ can be identified with the space of all multiplication operators (see Example 2.5). Then $L^\infty(X, \mu)$ is a concrete C*-algebra acting on $L^2(X, \mu)$. It is easy to see that $\|m_f\|$ is equal to the essential supremum of $f$,

$$\|f\|_\infty = \sup\{t \geq 0 : \mu\{x : |f(x)| > t\} > 0\}.$$

### The algebra of compact operators

It is equal to[2]

$$\begin{aligned}\mathcal{K}(H) &= C^*(\{a \in \mathcal{B}(H) : a[H] \text{ is finite-dimensional}\})\\ &= \{a \in \mathcal{B}(H) : a[\text{unit ball}] \text{ is precompact}\}\\ &= \{a \in \mathcal{B}(H) : a[\text{unit ball}] \text{ is compact}\}.\end{aligned}$$

(Note that $\mathcal{K}(H)$ is denoted by $C(H)$ in [50] and by $\mathbf{B}_0(H)$ in [51], by analogy with $C_0(X)$.) We write $r_n = \mathrm{proj}_{\overline{\mathrm{span}}\{e_j | j \leq n\}}$ for a fixed basis $\{e_n\}$ of $H$. Then for $a \in \mathcal{B}(H)$, the following are equivalent:

1. $a \in \mathcal{K}(H)$,
2. $\lim_n \|a(I - r_n)\| = 0$,
3. $\lim_n \|(I - r_n)a\| = 0$.

Note that if $a$ is self-adjoint then

$$\|a(I - r_n)\| = \|(a(I - r_n))^*\| = \|(I - r_n)a\|.$$

In the following exercises and elsewhere, 'ideal' always stands for a two-sided, closed, self-adjoint ideal. Actually, the las property follows from the previous ones since every two-sided closed ideal in a C*-algebra is automatically self-adjoint.

*Exercise* 3.4 Prove $\mathcal{K}(H)$ is an ideal of $\mathcal{B}(H)$. That is, prove that for $a \in \mathcal{K}(H)$ and $b \in \mathcal{B}(H)$ both $ab$ and $ba$ belong to $\mathcal{K}(H)$, that $\mathcal{K}(H)$ is norm-closed and that $b \in \mathcal{K}(H)$ if and only if $b^* \in \mathcal{K}(H)$.

*Exercise* 3.5

1. Prove that $\mathcal{K}(H)$ is the unique ideal of $\mathcal{B}(H)$ when $H$ is a separable Hilbert space.
2. Assume $\kappa$ is an infinite cardinal. Show that the number of proper ideals of $\mathcal{B}(\ell^2(\kappa))$ is equal to the number of infinite cardinals $\leq \kappa$.

[2] The third equality is a nontrivial fact specific to the Hilbert space; see [51, Theorem 3.3.3 (iii)]

### The Calkin algebra

This is an example of an abstract C*-algebra. The quotient $C(H) = \mathcal{B}(H)/\mathcal{K}(H)$ is called the *Calkin algebra*. It is sometimes denoted by $Q$ or $Q(H)$. We write $\pi : \mathcal{B}(H) \to C(H)$ for the quotient map. The norm on $C(H)$ is the usual quotient norm for Banach spaces:

$$\|\pi(a)\| = \inf\{\|b\| : \pi(b) = \pi(a)\}$$

The Calkin algebra turns out to be a very "set-theoretic" C*-algebra, analogous to the Boolean algebra $\mathcal{P}(\mathbb{N})/\mathrm{Fin}$.

We shall give more examples of C*-algebras in §3.4, after proving a fundamental result.

## 3.2 Automatic continuity and the Gelfand transform

In this section we prove two important results. First, any *-homomorphism between C*-algebras is norm-decreasing (Lemma 3.9) and second, every unital abelian C*-algebra is of the form $C(X)$ for some compact Hausdorff space $X$ (Theorem 3.10).

**Lemma 3.6** *If $a$ is normal then $\|a^{2^n}\| = \|a\|^{2^n}$ for all $n \in \mathbb{N}$.*

*Proof* Repeatedly using the C*-equality and normality of $a$ we have

$$\|a^2\| = (\|(a^*)^2a^2\|)^{1/2} = (\|(a^*a)^*(a^*a)\|)^{1/2} = \|a^*a\| = \|a\|^2.$$

The Lemma now follows by a straightforward induction. □

*Exercise* 3.7 Find $a \in \mathcal{B}(H)$ such that $\|a\| = 1$ and $a^2 = 0$. (Hint: Choose $a$ to be a partial isometry.)

It can be proved that a C*-algebra is abelian if and only if it contains no nonzero element $a$ such that $a^2 = 0$ (see [13, II.6.4.14]).

The *spectral radius* of an element $a$ of a C*-algebra is defined as

$$r(a) = \max\{|\lambda| : \lambda \in \sigma(a)\}.$$

**Lemma 3.8** *Let $A$ be a C*-algebra and $a \in A$ be normal. Then $\|a\| = r(a)$.*

*Sketch of a proof* It can be proved (see [9, Theorem 1.7.3], also the first line of the proof of Lemma 2.16) that for an arbitrary $a$ we have

$$\lim_n \|a^n\|^{1/n} = r(a),$$

in particular, the limit on the left hand side exists. By Lemma 3.6, for a normal $a$ this limit is equal to $\|a\|$. □

**Lemma 3.9** *Any *-homomorphism $\Phi : A \to B$ between* C**-algebras is a contraction (in particular, it is continuous). Therefore, any (algebraic) isomorphism between C*-algebras is an isometry.*

*Proof* By passing to the unitizations, we may assume $A$ and $B$ are unital and $\Phi$ is unital as well (i.e., $\Phi(I_A) = I_B$).

Note that for any $a \in A$, $\sigma(\Phi(a)) \subseteq \sigma(a)$ (by the definition of the spectrum). Thus for $a$ normal, using Lemma 3.8,

$$\begin{aligned} \|a\| &= \sup\{|\lambda| : \lambda \in \sigma(a)\} \\ &\geq \sup\{|\lambda| : \lambda \in \sigma(\Phi(a))\} \\ &= \|\Phi(a)\|. \end{aligned}$$

For general $a$, $aa^*$ is normal so by the C*-equality we have

$$\|a\| = \sqrt{\|aa^*\|} \geq \sqrt{\|\Phi(aa^*)\|} = \|\Phi(a)\|.$$

□

The reader may want to compare the last sentence of Lemma 3.9 with the situation in Banach space theory, where isomorphism and isometry drastically differ—even in the case of the Hilbert space (see [49]).

For a unital abelian C*-algebra $A$ consider its *spectrum*

$$\hat{A} = \{\phi\colon A \to \mathbb{C} : \phi \text{ is a nonzero algebra homomorphism}\}.$$

It is not difficult to see that every homomorphism from $A$ into $\mathbb{C}$ is a *-homomorphism. By Lemma 3.9 each $\phi \in \hat{A}$ is a contraction. Also $\phi(I) = 1$, and therefore $\hat{A}$ is a subset of the unit ball of the Banach space dual $A^*$ of $A$. Since it is obviously closed, it is weak*-compact by the Banach–Alaoglu theorem.

**Theorem 3.10** (Gelfand–Naimark) *If $A$ is unital and abelian* C**-algebra and $\hat{A}$ is its spectrum, then $A \cong C(\hat{A})$.*

*Proof* For $a \in A$ the map $f_a\colon \hat{A} \to \mathbb{C}$ defined by

$$f_a(\phi) = \phi(a)$$

is continuous in the weak*-topology. The transformation

$$A \ni a \longmapsto f_a \in C(X)$$

is the *Gelfand transform* of $a$. An easy computation shows that the Gelfand transform is a *-homomorphism, and therefore by Lemma 3.9 continuous. We need to show it is an isometry.

For $b \in A$ we claim that $b$ is not invertible if and only if $\phi(b) = 0$ for some $\phi \in \hat{A}$. Only the forward direction requires a proof. Fix a non-invertible $b$. The $J_b = \{xb : x \in A\}$ is a proper (two-sided since $A$ is abelian) ideal containing $b$. Let $J \supseteq J_b$ be a maximal proper two sided (not necessarily closed and not necessarily self-adjoint) ideal. Lemma 2.15 implies that $\|I - b\| \geq 1$ for all $b \in J$. Hence the closure of $J$ is still proper, and by maximality $J$ is a closed ideal. Every closed two-sided ideal in a C*-algebra is automatically self-adjoint (see [8, p.11]). Therefore the quotient map $\phi_J$ from $A$ to $A/J$ is a *-homomorphism. Since $A$ is abelian, by the maximality of $J$ the algebra $A/J$ is a field. For any $a \in A/J$, Lemma 3.8 implies that $\sigma(a)$ is nonempty, and for any $\lambda \in \sigma(a)$, $a - \lambda I = 0$ since $A/J$ is a field. Thus $A/J$ is generated by $I$ and therefore isomorphic to $\mathbb{C}$, so $\phi_J \in X$. Clearly $\phi_J(b) = 0$.

Therefore range$(f_a) = \sigma(a)$ for all $a$. Lemma 3.8 implies

$$\|a\| = \max\{|\lambda| : \lambda \in \sigma(a)\} = \|f_a\|.$$

Thus the algebra $B = \{f_a : a \in A\}$ is isometric to $A$. Since it separates the points in $X$, by the Stone–Weierstrass theorem (e.g., [51, Theorem 4.3.4]) $B$ is norm-dense in $C(X)$. Being isometric to $A$, it is closed and therefore equal to $C(X)$. □

The following exercise shows that the category of abelian unital C*-algebras is contravariantly equivalent to the category of compact Hausdorff spaces.

*Exercise* 3.11 Assume $X$ and $Y$ are compact Hausdorff spaces and $\Phi\colon C(X) \to C(Y)$ is a *-homomorphism.

1. Prove that there exists a unique continuous $f\colon Y \to X$ such that $\Phi(a) = a \circ f$ for all $a \in C(X)$.
2. Prove that $\Phi$ is a surjection if and only if $f$ is an injection.
3. Prove that $\Phi$ is an injection if and only if $f$ is a surjection.
4. Prove that for every $f\colon Y \to X$ there exists a unique $\Phi\colon C(X) \to C(Y)$ such that (1)–(3) above hold.

The following exercise provides an alternative construction of a Čech–Stone compactification $\beta X$ of a completely regular space $X$. (Recall that a topological space is *completely regular* if it is homeomorphic to a subspace of some Hilbert cube, $[0, 1]^J$.)

*Exercise* 3.12 Let $X$ be a completely regular space and let $A$ be the *-algebra $C_b(X, \mathcal{B})$ of all continuous bounded functions from $X$ into $\mathbb{C}$, equipped with the sup norm.

1. Prove that $A$ is a C*-algebra.
2. Let $\gamma X$ denote the compact Hausdorff space such that $C(\gamma X)$ is isomorphic to $C_b(X, \mathcal{B})$. Show that there is a homeomorphic embedding $f: X \to \gamma X$ such that $\Phi(a) = a \circ f$ represents the isomorphism of $C(\gamma X)$ and $C_b(X, \mathcal{B})$.
3. Prove that $f[X]$ is dense in $\gamma X$.
4. Prove that every continuous real-valued function on $f[X]$ has unique continuous extension with domain $\gamma X$.

## 3.3 Continuous functional calculus

Recall that $\sigma(a)$ is always a compact subset of $\mathbb{C}$ (Lemma 2.16). Theorem 2.6 (Spectral Theorem) is a consequence of the following Corollary and some standard manipulations; see [9, Theorem 2.4.5].

**Corollary 3.13** *If $a \in \mathcal{B}(H)$ is normal then $C^*(a, I) \cong C(\sigma(a))$.*

*Proof* We first prove that $C^*(a, I)$ is isomorphic to $C(\sigma_0(a))$, where $\sigma_0(a)$ denotes the spectrum of $a$ as defined in $C^*(a, I)$. Let $C^*(a, I) \cong C(X)$ as in Theorem 3.10. For any $\lambda \in \sigma(a)$, $a - \lambda I$ is not invertible so there exists $\phi_\lambda \in X$ such that $\phi_\lambda(a - \lambda I) = 0$, or $\phi_\lambda(a) = \lambda$. Conversely, if there is $\phi \in X$ such that $\phi(a) = \lambda$, then $\phi(a - \lambda I) = 0$ so $\lambda \in \sigma(a)$. Since any nonzero homomorphism to $\mathbb{C}$ is unital, an element $\phi \in X$ is determined entirely by $\phi(a)$. Since $X$ has the weak* topology, $\phi \mapsto \phi(a)$ is thus a continuous bijection from $X$ to $\sigma(a)$, which is a homeomorphism since $X$ is compact.

It remains to show that $\sigma_0(a) = \sigma(a)$. Since an element invertible in the smaller algebra is clearly invertible in the larger algebra, we only need to check that $\sigma(a) \supseteq \sigma_0(a)$. Pick $\lambda \in \sigma_0(a)$. We need to prove that $a - \lambda I$ is not invertible in $\mathcal{B}(H)$. Assume the contrary and let $b$ be the inverse of $a - \lambda I$. Fix $\epsilon > 0$ and let $U \subseteq \sigma_0(a)$ be the open ball around $\lambda$ od radius $\epsilon$. Let $g \in C(\sigma_0(a))$ be a function supported by $U$ such that $\|g\| = 1$. Then $g = b(a - \lambda I)g$, hence $\|b(a - \lambda I)g\| = 1$. On the other hand, $(a - \lambda I)g = f \in C(\sigma_0(a))$ so that $f$ vanishes outside of $U$ and $\|f(x)\| < \epsilon$ for $x \in U$, hence $\|(a - \lambda I)g\| < \epsilon$. Thus $\|b\| > 1/\epsilon$ for every $\epsilon > 0$, a contradiction. □

A spectrum $\sigma_A(a)$ of an element $a$ of an arbitrary C*-algebra $A$ can be defined as

$$\sigma_A(a) = \{\lambda \in \mathbb{C} | a - \lambda I \text{ is not invertible in } A\}.$$

**Lemma 3.14** *Suppose $A$ is a unital subalgebra of $B$ and $a \in A$ is normal. Then $\sigma_A(a) = \sigma_B(a)$, where $\sigma_A(a)$ and $\sigma_B(a)$ denote the spectra of $a$ as an element of $A$ and $B$, respectively.*

*Proof* See e.g., [51, Corollary 4.3.16] or [9, Corollary 2 on p. 49]. □

Note that the isomorphism above is canonical and maps $a$ to the identity function on $\sigma(a)$. It follows that for any polynomial $p$, the isomorphism maps $p(a)$ to the function $z \mapsto p(z)$. More generally, for any continuous function $f : \sigma(a) \to \mathbb{C}$, we can then define $f(a) \in C^*(a, I)$ as the preimage of $f$ under the isomorphism. For example, we can define $|a|$ and if $a$ is self-adjoint then it can be written as a difference of two positive operators as

$$a = \frac{|a| + a}{2} - \frac{|a| - a}{2}.$$

If $a \geq 0$, then we can also define $\sqrt{a}$. Here is another application of the "continuous functional calculus" of Corollary 3.13.

**Lemma 3.15** *Every $a \in \mathcal{B}(H)$ is a linear combination of unitaries.*

*Proof* By decomposing an arbitrary operator into the positive and negative parts of its real and imaginary parts, it suffices to prove that each positive operator $a$ of norm $\leq 1$ is a linear combination of two unitaries, $u = a + i\sqrt{I - a^2}$ and $v = a - i\sqrt{I - a^2}$. Clearly $a = \frac{1}{2}(u + v)$. Since $u = v^*$ and $uv = vu = I$, the conclusion follows. □

*Exercise* 3.16 For $a \in \mathcal{B}(H)$ let $a = bu$ be its polar decomposition (see Theorem 2.13).

1. Show that $b \in C^*(a, I)$.
2. Give an example of $a$ such that $u \notin C^*(a, I)$.

(Hint: For (1) use $b = \sqrt{aa^*}$. For (2) take $a$ which is compact, but not of finite rank.)

## 3.4 More examples of C*-algebras

We are now equipped to describe another construction of C*-algebras and more examples.

### Direct limits

We now return to giving examples of C*-algebras.

**Definition 3.17** If $\Omega$ is a directed set, $A_i$, $i \in \Omega$ are C*-algebras and

$$\varphi_{i,j} : A_i \to A_j \qquad \text{for } i < j$$

is a commuting family of homomorphisms, we define the *direct limit* (also called the *inductive limit*) $A = \varinjlim_i A_i$ by taking the algebraic direct limit and completing it. We define a norm on $A$ by saying that if $a \in A_i$,

$$\|a\|_A = \lim_j \|\varphi_{i,j}(a)\|_{A_j}.$$

This limit makes sense because the $\varphi_{i,j}$ are all contractions by Lemma 3.9.

### UHF (uniformly hyperfinite) algebras

For each natural number $n$, define a *-homomorphism $\Phi_n : M_{2^n}(\mathbb{C}) \to M_{2^{n+1}}(\mathbb{C})$ by

$$\Phi_n(a) = \begin{pmatrix} a & 0 \\ 0 & a \end{pmatrix}.$$

We then define the CAR (Canonical Anticommutation Relations) algebra (aka the Fermion algebra, aka $M_{2^\infty}$ UHF algebra) as the direct limit $M_{2^\infty} = \varinjlim(M_{2^n}(\mathbb{C}), \Phi_n)$. Alternatively, $M_{2^\infty} = \bigotimes_{n \in \mathbb{N}} M_2(\mathbb{C})$, since $M_{2^{n+1}}(\mathbb{C}) = M_{2^n}(\mathbb{C}) \otimes M_2(\mathbb{C})$ for each $n$ and $\Phi_n(a) = a \otimes 1_{M_2(\mathbb{C})}$.

Note $\Phi_n$ maps diagonal matrices to diagonal matrices, so we can talk about the diagonal elements of $M_{2^\infty}$. These turn out to be isomorphic to the algebra $C(K)$, where $K$ is the Cantor set. Thus we can think of $M_{2^\infty}$ as a "noncommutative Cantor set."

It is not difficult to see that for $m$ and $n$ in $\mathbb{N}$ there is a unital homomorphism from $M_m$ into $M_n(\mathbb{C})$ if and only if $m$ divides $n$. If it exists, then this map is unique up to conjugacy. Direct limits of full matrix algebras are called *UHF algebras* and they were classified by Glimm (the unital case) and Dixmier (the general case) in the 1960s. This was the start of the Elliott classification program of separable unital C*-algebras (see [54], [20]).

The following somewhat laborious exercise is intended to introduce representation theory of the CAR algebra.

*Exercise* 3.18 Fix $x \in 2^{\mathbb{N}}$ and let $D_x = \{y \in 2^{\mathbb{N}} : (\forall^\infty n) y(n) = x(n)\}$. Enumerate a basis of a complex, infinite-dimensional separable Hilbert space $H$ as $\xi_y$, $y \in D_x$. Let $s, t$ range over functions from a finite subset of $\mathbb{N}$ into $\{0, 1\}$. For such $s$ define a partial isometry of $H$ as follows. If $y(m) \neq s(m)$ for some $m \in \mathrm{dom}(s)$ then let $u_s(\xi_y) = 0$. Otherwise, if $y \restriction \mathrm{dom}(s) = s$, then let $z \in 2^{\mathbb{N}}$ be such that $z(n) = 1 - y(n)$ for $n \in \mathrm{dom}(s)$ and $z(n) = y(n)$ for $n \notin \mathrm{dom}(s)$ and set $u_s(\xi_y) = \xi_z$.

1. Prove that $u_s^* = u_{\bar{s}}$, where $\mathrm{dom}(\bar{s}) = \mathrm{dom}(s)$ and $\bar{s}(n) = 1 - s(n)$ for all $n \in \mathrm{dom}(s)$.
2. Prove that $u_s u_s^*$ is the projection to $\overline{\mathrm{span}}\{\xi_y : y \restriction \mathrm{dom}(s) = \bar{s}\}$ and $u_s^* u_s$ is the projection to $\overline{\mathrm{span}}\{\xi_y : y \restriction \mathrm{dom}(s) = s\}$.
3. Let $A_x$ be the C*-algebra generated by $u_s$ as defined above. Prove that $A_x$ is isomorphic to $M_{2^\infty}$.
4. Show that the intersection of $A_x$ with the atomic masa (see §5.1) diagonalized by $\xi_y$, $y \in D_x$, consists of all operators of the form $\sum_y \alpha_y \xi_y$ where $y \mapsto \alpha_y$ is a continuous function.
5. Show that for $x$ and $y$ in $2^{\mathbb{N}}$ there is a unitary $v$ of $H$ such that $\mathrm{Ad}\, v$ sends $A_x$ to $A_y$ if and only if $(\forall^\infty n) x(n) = y(n)$. (Hint: cf. Example 4.24.)

## AF (approximately finite) algebras

Let us start with an exercise. A *direct sum* of C*-algebras $A$ and $B$ is the algebra $A \oplus B$ whose elements are sums $a + b$ for $a \in A$ and $b \in B$ (assuming $A \cap B = \{0\}$), with the convention that $ab = ba = 0$ whenever $a \in A$ and $b \in B$. One similarly defines a direct sum of a family of C*-algebras.

*Exercise* 3.19 Show that a C*-algebra $A$ is a finite-dimensional vector space if and only if it is the direct sum of finitely many full matrix algebras.

A C*-algebra is AF, or *approximately finite*, if it is a direct limit of finite-dimensional C*-algebras. This class of C*-algebras is much more extensive than the class of UHF algebras. Elliott's classification of unital separable AF algebras by K-theoretic invariant $K_0$ (see [54]) marked the beginning of Elliott program for classification of C*-algebras.

*Exercise* 3.20 Show that an abelian C*-algebra is AF if and only if it is of the form $C_0(X)$ for a zero-dimensional, locally compact, Hausdorff, space $X$.

**Even more examples**

Giving an exhaustive treatment of techniques for building C*-algebras is beyond the scope of this article. Tensor products, group algebras, and crossed products are indispensable tools in theory of C*-algebras. Some of these constructions were blended with set-theoretic methods in [31] and [24] to construct novel examples of nonseparable C*-algebras.

## 4 Positivity, states and the GNS construction

The following is a generalization of the spectral theorem to abstract C*-algebras.

**Theorem 4.1** (Gelfand–Naimark) *Every abelian* C*-*algebra is isomorphic to* $C_0(X)$ *for a unique locally compact Hausdorff space* $X$. *The algebra is unital if and only if* $X$ *is compact.*

*Proof* By Theorem 3.10, the unitization $B$ of $A$ is isomorphic to $C(\hat{B})$ for a compact Hausdorff space $\hat{B}$, the spectrum of $A$. If $\phi \in \hat{B}$ is the unique map whose kernel is equal to $A$, then $A \cong C_0(\hat{B} \setminus \{\phi\})$. The uniqueness of $X = \hat{B} \setminus \{\phi\}$ follows from Theorem 4.17 below. □

In fact, the Gelfand–Naimark theorem is functorial: the category of abelian C*-algebras is contravariantly isomorphic to the category of locally compact Hausdorff spaces (cf. Exercise 3.11). The space $X$ is a natural generalization of the spectrum of a single element of a C*-algebra.

*Exercise* 4.2

1. If $C^*(a)$ is unital and isomorphic to $C(X)$, then $0 \notin \sigma(a)$ and $\sigma(a)$ is homeomorphic to $X$.
2. If $C^*(a)$ is not unital and it is isomorphic to $C_0(X)$, then $0 \in \sigma(a)$ and $\sigma(a) \setminus \{0\}$ is homeomorphic to $X$.

Recall that an element $a$ of a C*-algebra $A$ is *positive* if $a = b^*b$ for some $b \in A$. It is not difficult to see that for projections $p$ and $q$ we have $p \leq q$ if and only if $pq = p$ if and only if $qp = p$ (see Lemma 5.5).

*Exercise* 4.3 Which of the following are true for projections $p$ and $q$ and positive $a$ and $b$?

1. $pqp \leq p$?
2. $a \leq b$ implies $ab = ba$?
3. $p \leq q$ implies $pap \leq qaq$?

4. $p \leq q$ implies $prp \leq qrq$ for a projection $r$?
5. $prp \leq p$ for a projection $r$?

(Hint 1: Only one of the above is true. Hint 2: Formula (1) is easy to prove. Hint 3: For (2) note that $a \geq 0$ implies $a \leq a + c$ for every $c \geq 0$. Hint 4: There is a counterexample for (5) on the two-dimensional Hilbert space.)

**Definition 4.4** Let $A$ be a C*-algebra. A continuous linear functional $\varphi : A \to \mathbb{C}$ is *positive* if $\varphi(a) \geq 0$ for all positive $a \in A$. It is a *state* if it is positive and of norm 1. We denote the space of all states on $A$ by $\mathbb{S}(A)$.

*Exercise* 4.5 Assume $\phi$ is a positive functional on a C*-algebra. Show that $\phi(a)$ is a real whenever $a$ is self-adjoint and show that $\phi(b^*) = \overline{\phi(b)}$ for all $b$ . (Hint: Use the continuous function calculus.)

**Example 4.6** If $\xi \in H$ is a unit vector, define a functional $\omega_\xi$ on $\mathcal{B}(H)$ by

$$\omega_\xi(a) = (a\xi \mid \xi).$$

Then $\omega_\xi(a) \geq 0$ for a positive $a$ and $\omega_\xi(I) = 1$; hence it is a state. We call a state of this form a *vector state*.

**Lemma 4.7** *Each positive functional $\phi$ satisfies a Cauchy–Schwartz inequality:*

$$|\varphi(b^*a)|^2 \leq \varphi(a^*a)\varphi(b^*b).$$

*Proof* The proof is similar to the proof of the standard Cauchy–Schwartz inequality. Let $\lambda$ be a complex number. Since $(\lambda a+b)^*(\lambda a+b)$ is positive, we have

$$0 \leq \phi((\lambda a + b)^*(\lambda a + b)) = |\lambda|^2\phi(a^*a) + \bar{\lambda}\phi(a^*b) + \lambda\phi(b^*a) + \phi(b^*b).$$

We may assume $|\phi(b^*a)| \neq 0$ since the inequality is trivial otherwise. Let $\lambda = t\phi(a^*b)/|\phi(b^*a)|$ for a real $t$. Noting that $\phi(a^*b) = \overline{\phi(b^*a)}$ (Exercise 4.5), we obtain

$$0 \leq t^2\frac{|\phi(a^*b)|^2}{|\phi(b^*a)|^2}\phi(a^*a) + t\frac{\overline{\phi(a^*b)}}{|\phi(b^*a)|}\phi(a^*b) + t\frac{\phi(a^*b)}{|\phi(b^*a)|}\phi(b^*a) + \phi(b^*b)$$

or equivalently

$$0 \leq t^2\phi(a^*a) + 2t|\phi(b^*a)| + \phi(b^*b).$$

The discriminant of this equation is nonpositive, and the inequality follows. □

**Lemma 4.8** *If $\varphi$ is a state on $A$ and $0 \leq a \leq I$ is such that $\varphi(a) = 1$, then $\varphi(b) = \varphi(aba)$ for all $b$.*

*Proof* By the Cauchy–Schwartz inequality for states (Lemma 4.7)

$$|\varphi((I-a)b)| \leq \sqrt{\varphi(I-a)\varphi(b^*b)} = 0.$$

Since $b = ab + (I-a)b$, we have $\varphi(b) = \varphi(ab) + \varphi((I-a)b) = \varphi(ab)$. By applying the same argument to $ab$ and multiplying by $I - a$ on the right one proves that $\varphi(ab) = \varphi(aba)$. □

The basic reason we care about states is that they give us representations of abstract C*-algebras as concrete C*-algebras.

**Theorem 4.9** (The GNS construction) *Let $\varphi$ be a state on $A$. Then there is a Hilbert space $H_\varphi$, a representation $\pi_\varphi : A \to \mathcal{B}(H_\varphi)$, and a unit vector $\xi = \xi_\varphi$ in $H_\varphi$ such that $\varphi = \omega_\xi \circ \pi_\varphi$.*

*Sketch of the proof* We define an "inner product" on $A$ by

$$(a \mid b) = \varphi(b^* a).$$

We let $J = \{a : (a \mid a) = 0\}$, so that Lemma 4.7 implies $(\cdot \mid \cdot)$ is actually an inner product on the quotient space $A/J$. We then define $H_\varphi$ to be the completion of $A/J$ under the induced norm. For any $a \in A$, $\pi_\varphi(a)$ is then the operator that sends $b + J$ to $ab + J$, and $\xi_\varphi$ is $I + J$. □

## 4.1 Irreducible representations and pure states

In this section we introduce a particularly important class of states called pure states. The following exercise focuses on a state that is not pure.

*Exercise* 4.10 Assume $\psi_1$ and $\psi_2$ are states on $A$ and $0 < t < 1$ and let

$$\phi = t\psi_1 + (1-t)\psi_2.$$

1. Show that $\phi$ is a state.
2. Show that $H_\phi \cong H_{\psi_1} \oplus H_{\psi_2}$, with $\pi_\phi(a) = \pi_{\psi_1}(a) + \pi_{\psi_2}(a)$, and $\xi_\phi = \sqrt{t}\xi_{\psi_1} + \sqrt{1-t}\xi_{\psi_2}$. In particular, projections to $H_{\psi_1}$ and $H_{\psi_2}$ commute with $\pi_\varphi(a)$ for all $a \in A$.

Hence states form a convex subset of $A^*$. We say that a state is *pure* if it is an extreme point of $\mathbb{S}(A)$. That is, $\varphi$ is pure iff

$$\varphi = t\psi_0 + (1-t)\psi_1, \qquad 0 \leq t \leq 1$$

for $\psi_0, \psi_1 \in \mathbb{S}(A)$ implies $\varphi = \psi_0$ or $\varphi = \psi_1$. We denote the set of all pure states on $A$ by $\mathbb{P}(A)$.

*Exercise* 4.11 Prove the following.

1. If $\phi$ is a pure state on $M_n(\mathbb{C})$ then there is a rank one projection $p$ such that $\phi(a) = \phi(pap)$ for all $a$.
2. Identify $M_n(\mathbb{C})$ with $\mathcal{B}(\ell_2^n)$. Show that all pure states of $M_n(\mathbb{C})$ are vector states.

Recall that the Krein–Milman theorem states that every compact convex subset of a locally convex topological vector space is the closed convex hull of its extreme points. Since the dual space of a C*-algebra is locally convex and since the convex hull of $\mathbb{S}(A) \cup \{0\}$ is compact, we conclude that $\mathbb{S}(A)$ is the weak* closure of the convex hull of $\mathbb{P}(A)$. Since one can show that a C*-algebra has an ample supply of states (see Lemma 4.25) the same is true for pure states.

The space $\mathbb{P}(A)$ is weak*-compact only for a very restrictive class of C*-algebras, including $\mathcal{K}(H)$ and abelian algebras (see Definition 6.8). For example, for UHF algebras the pure states form a dense subset in the compactum of all states ([39, Theorem 2.8]).

*Exercise* 4.12 Let $A$ be a separable C*-algebra. Prove that $\mathbb{P}(A)$ is Polish in the weak*-topology.

(Hint: Show that $\mathbb{P}(A)$ is a $G_\delta$ subset of $\mathbb{S}(A)$.)

**Definition 4.13** A representation $\pi : A \to \mathcal{B}(H)$ of a C*-algebra is *irreducible* (sometimes called an *irrep*) if there is no nontrivial subspace $H_0 \subset H$ such that $\pi(a)H_0 \subseteq H_0$ for all $a \in A$. Such a subspace is said to be *invariant* for $\pi[A]$ or *reducing* for $\pi$.

**Theorem 4.14** *A state $\varphi$ is pure if and only if $\pi_\varphi$ is irreducible. Every irreducible representation is of the form $\pi_\varphi$ for some pure state $\varphi$.*

*Proof* The easy direction is Exercise 4.10. For the other direction see e.g., [8, Theorem 1.6.6] or [50, (i) $\Leftrightarrow$ (vi) of Theorem 3.13.2]. □

**Example 4.15** If $A = C(X)$, then by the Riesz representation theorem states are in a bijective correspondence with the Borel probability measures on $X$ (writing $\mu(f) = \int f d\mu$).

*Exercise* 4.16 For a state $\varphi$ of $C(X)$ the following are equivalent:

1. $\varphi$ is pure,
2. for a unique $x_\varphi \in X$ we have $\varphi(f) = f(x_\varphi)$
3. $\varphi : C(X) \to \mathbb{C}$ is a homomorphism ($\varphi$ is "multiplicative").

(Hint: Use Example 4.15 and see the proof of Theorem 3.10.)

**Theorem 4.17** *If $X$ is a compact Hausdorff space then $\mathbb{P}(C(X))$ with respect to the weak*-topology is homeomorphic to $X$.*

*Proof* By (2.) in Exercise 4.16, there is a natural map $F : \mathbb{P}(C(X)) \to X$. By (3.), it is not hard to show that $F$ is surjective, and it follows from Urysohn's lemma that $F$ is a homeomorphism. □

**Proposition 4.18** *For any unit vector $\xi \in H$ the vector state $\omega_\xi \in \mathbb{S}(\mathcal{B}(H))$ is pure.*

*Proof* Immediate from Theorem 4.14. □

**Definition 4.19** We say $\varphi \in \mathbb{S}(\mathcal{B}(H))$ is *singular* if $\varphi[\mathcal{K}(H)] = \{0\}$.

*Exercise* 4.20 Prove that a nonsingular state on $\mathcal{B}(H)$ is pure if and only if it is a vector state. (Hint: First show that a nontrivial linear combination of vector states is never pure.)

By factoring through the quotient map $\pi : \mathcal{B}(H) \to \mathcal{C}(H)$, the space of singular states is isomorphic to the space of states on the Calkin algebra $\mathcal{C}(H)$.

**Theorem 4.21** *Each state of $\mathcal{B}(H)$ is a weak*-limit of vector states. A pure state is singular if and only if it is not a vector state.*

*Proof* The first sentence is a special case of [37, Lemma 9] when $\mathfrak{A} = \mathcal{B}(H)$. The second sentence is trivial (modulo Exercise 4.20). □

We now take a closer look at the relationship between states and representations of a C*-algebra.

**Definition 4.22** Let $A$ be a C*-algebra and $\pi_i : A \to \mathcal{B}(H_i)$ $(i = 1, 2)$ be representations of $A$. We say $\pi_1$ and $\pi_2$ are *(unitarily) equivalent* and write $\pi_1 \sim \pi_2$ if there is a unitary (Hilbert space isomorphism)

$u\colon H_1 \to H_2$ such that the following commutes:

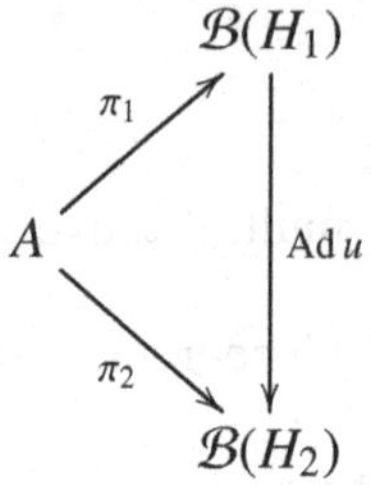


$\mathrm{Ad}\, u(a) = uau^*$

Similarly, if $\varphi_i \in \mathbb{P}(A)$, we say $\varphi_1 \sim \varphi_2$ if there is a unitary $u \in \tilde{A}$ such that the following commutes:

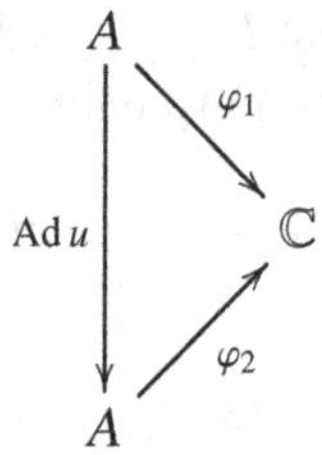


**Proposition 4.23** *For $\varphi_i \in \mathbb{P}(A)$, $\varphi_1 \sim \varphi_2$ if and only if $\pi_{\varphi_1} \sim \pi_{\varphi_2}$.*

*Proof* The direct implication is easy and the converse is a consequence of the remarkable Kadison's Transitivity Theorem. For the proof see e.g., [50, the second sentence of Proposition 3.13.4]. □

## 4.2 On the existence of states

States on an abelian C*-algebra $C(X)$ correspond to probability Borel measures on $X$ (see Example 4.15).

**Example 4.24** On $M_2(\mathbb{C})$, the following are pure states:

$$\varphi_0 : \quad \begin{pmatrix} a_{11} & a_{12} \\ a_{21} & a_{22} \end{pmatrix} \mapsto a_{11}$$

$$\varphi_1 : \quad \begin{pmatrix} a_{11} & a_{12} \\ a_{21} & a_{22} \end{pmatrix} \mapsto a_{22}$$

For any $f \in 2^{\mathbb{N}}$, $\varphi_f = \bigotimes_n \varphi_{f(n)}$ is a pure state on $\bigotimes M_2(\mathbb{C}) = M_{2^\infty}$. Furthermore, one can show that $\varphi_f$ and $\varphi_g$ are equivalent if and only if $f$ and $g$ differ at only finitely many points, and that $\|\varphi_f - \varphi_g\| = 2$ for $f \neq g$. See [50, §6.5] for a more general setting and proofs.

**Lemma 4.25** *If $\phi$ is a linear functional of norm 1 on a unital* C*-*algebra then $\phi$ is a state if and only if $\phi(I) = 1$.*

*Proof* First assume $\phi$ is a state. Since $\|\phi\| \leq 1$ and since $\phi$ is positive we have $0 \leq \phi(I) \leq 1$. For positive operators $a \leq b$ we have $\phi(a) \leq \phi(b)$. Since for a positive operator $a$ we have $a \leq \|a\| \cdot I$, we must have $\phi(I) = 1$.

Now assume $\phi(I) = \|\phi\| = 1$ and fix $a \geq 0$. The algebra $C^*(a, I)$ is abelian, and by the Riesz representation theorem the restriction of $\phi$ to this algebra is given by a Borel measure $\mu$ on $\sigma(a)$. The assumption that $\phi(I) = \|\phi\|$ translates as $|\mu| = \mu$, hence $\mu$ is a positive probability measure. Since $a$ corresponds to the identity function on $\sigma(a) \subseteq [0, \infty)$ we have $\phi(a) \geq 0$. □

**Lemma 4.26** *If A is a subalgebra of B then every state of B restricts to a state of A. Also, every (pure) state of A can be extended to a (pure) state of B.*

*Proof* The first statement is trivial. Now assume $\phi$ is a state on $A \subseteq B$. We shall extend $\phi$ to a state of $B$ under an additional assumption that $A$ is a unital subalgebra of $B$; the general case is then a straightforward exercise (see Lemma 3.2).

By the Hahn–Banach theorem we can extend $\phi$ to a functional $\psi$ on $B$ of norm 1. By Lemma 4.25, $\psi$ is a state of $B$.

Note that the (nonempty) set of extensions of $\phi$ to a state of $B$ is weak*-compact and convex. If we start with a pure state $\varphi$, then by Krein–Milman the set of extensions of $\varphi$ to $B$ has an extreme point, which can then be shown to be a pure state on $B$. □

**Lemma 4.27** *For every normal $a \in A$ there is a pure state $\phi$ such that $|\phi(a)| = \|a\|$.*

*Proof* The algebra $C^*(a)$ is by Corollary 3.13 isomorphic to $C(\sigma(a))$. Consider its state $\phi_0$ defined by $\phi_0(f) = f(\lambda)$, where $\lambda \in \sigma(a)$ is such that $\|a\| = |\lambda|$. This is a pure state and satisfies $|\phi(a)| = \|a\|$.

By Lemma 4.25 extend $\phi_0$ to a pure state $\phi$ on $A$. □

*Exercise* 4.28 Show that there is a C*-algebra $A$ and $a \in A$ such that $|\phi(a)| < \|a\|$ for every state $\phi$ of $A$.

(Hint: First do Exercise 4.11. Then consider $\begin{pmatrix} 0 & 1 \\ 0 & 0 \end{pmatrix}$ in $M_2(\mathbb{C})$.)

**Theorem 4.29** (Gelfand–Naimark–Segal) *Every* C*-*algebra A is isomorphic to a concrete* C*-*algebra.*

*Proof* By taking the unitization, we may assume $A$ is unital. Each state $\varphi$ on $A$ gives a representation $\pi_\varphi$ on a Hilbert space $H_\varphi$, and we take the product of all these representations to get a single representation $\pi = \bigoplus_{\varphi \in \mathbb{S}(A)} \pi_\varphi$ on $H = \bigoplus H_\varphi$.

We need to check that this representation is faithful. By Lemma 3.9 this is equivalent to $\pi$ being an isometry. By the same Lemma 3.9 we have $\|\pi(a)\| \leq \|a\|$. By Lemma 4.27 for every self-adjoint $a$ we have $|\phi(a)| = \|a\|$.

We claim that $a \neq 0$ implies $\pi(a) \neq 0$. We have that $a = b + ic$ for self-adjoint $b$ and $c$, at least one of which is nonzero. Therefore $\pi(a) = \pi(b) + i\pi(c)$ is nonzero. Thus $A$ is isomorphic to its image $\pi(A) \subseteq \mathcal{B}(H)$, a concrete C*-algebra. □

*Exercise* 4.30 Prove that a separable abstract C*-algebra can be faithfully represented on a separable Hilbert space. (Hint for logicians: the Löwenheim–Skolem theorem.)

Note that the converse of last exercise is false, since $\mathcal{B}(H)$ itself is nonseparable in norm topology.

*Exercise* 4.31

1. Prove that for every C*-algebra $A$ and every $a \in A$ we have
$$\|a\|^2 = \sup_\varphi \sup_b \varphi(b^* a^* ab)$$
where the supremum is taken over all (pure) states $\phi$ and over all $b \in A$ of norm 1.
2. If a C*-algebra $A$ is simple, then for every state $\phi$ and every $a \in A$ we have
$$\|a\|^2 = \sup_b \varphi(b^* a^* ba)$$
where the supremum is taken over all $b \in A$ of norm 1.

## 5 Projections in the Calkin algebra

Recall that $\mathcal{K}(H)$ (see Example 3.1) is a (norm-closed two-sided) ideal of $\mathcal{B}(H)$, and the quotient $C(H) = \mathcal{B}(H)/\mathcal{K}(H)$ is the *Calkin algebra* (see §3.1). We write $\pi : \mathcal{B}(H) \to C(H)$ for the quotient map.

Many instances of the question whether an element in a quotient C*-algebra can be lifted to an element with similar properties are well-studied. We shall now consider some of them.

**Lemma 5.1** *If* $\mathbf{a} \in C(H)$ *is self-adjoint, then there is a self-adjoint* $a \in \mathcal{B}(H)$ *such that* $\mathbf{a} = \pi(a)$.

*Proof* Fix any $a_0$ such that $\pi(a_0) = \mathbf{a}$. Let $a = (a_0 + a_0^*)/2$. Then $a$ is self-adjoint and $a - a_0$ is compact. Therefore $a$ is as required. □

*Exercise* 5.2 Assume $f: A \to B$ is a *-homomorphism between C*-algebras and $p$ is a projection in the range of $f$. Is there necessarily a projection $q \in A$ such that $f(q) = p$? (Hint: Consider the natural *-homomorphism from $C([0, 1])$ to $C([0, 1/3] \cup [2/3, 1])$.)

The following lemma, showing that the answer to question in Exercise 5.2 is sometimes positive, is taken from [64].

**Lemma 5.3** *If* $\mathbf{p} \in C(H)$ *is a projection, then there is a projection* $p \in \mathcal{B}(H)$ *such that* $\mathbf{p} = \pi(p)$.

*Proof* Fix a self-adjoint $a$ such that $\mathbf{p} = \pi(a)$. Represent $a$ as a multiplication operator $m_f$. Since $\pi(m_f)$ is a projection, $m_{f^2-f} \in \mathcal{K}(H)$ Let

$$h(x) = \begin{cases} 1, & f(x) \geq 1/2 \\ 0, & f(x) < 1/2. \end{cases}$$

Then $m_h$ is a projection. Also, if $(x_\alpha)$ is such that $f(x_\alpha)^2 - f(x_\alpha) \to 0$, then $h(x_\alpha) - f(x_\alpha) \to 0$. One can show that this implies that since $m_{f^2-f}$ is compact, so is $m_{h-f}$. Hence $\pi(m_h) = \pi(m_f) = \mathbf{p}$. □

Thus self-adjoints and projections in $C(H)$ are just self-adjoints and projections in $\mathcal{B}(H)$ modded out by compacts. However, the same is not true for unitaries.

**Example 5.4** Let $S \in \mathcal{B}(H)$ be the unilateral shift (Example 2.12). Then $S^*S = I$ and $SS^* = I - \mathrm{proj}_{\overline{\mathrm{span}(\{e_0\})}} = I - p$. Since $p$ has finite-dimensional range, it is compact, so $\pi(S)^*\pi(S) = I = \pi(S)\pi(S^*)$. That is, $\pi(S)$ is unitary.

If $\pi(a)$ is invertible, one can define the *Fredholm index* of $a$ by

$$\mathrm{index}(a) = \dim\ker a - \dim\ker a^*.$$

The Fredholm index is (whenever defined) invariant under compact perturbations of $a$ ([51, Theorem 3.3.17]). Since $\mathrm{index}(u) = 0$ for any unitary $u$ and $\mathrm{index}(S) = -1$, there is no unitary $u \in \mathcal{B}(H)$ such that $\pi(u) = \pi(S)$.

For a C*-algebra $A$ we write $\mathcal{P}(A)$ for the set of projections in $A$. We partially order $\mathcal{P}(A)$ by saying $p \le q$ if $pq = p$. This agrees with the restriction of the ordering on positive operators. If they exist, we denote joins and meets under this ordering by $p \vee q$ and $p \wedge q$. Note that every $p \in \mathcal{P}(A)$ has a canonical (orthogonal) complement $q = I - p$ such that $p \vee q = I$ and $p \wedge q = 0$.

**Lemma 5.5** *Let $p, q \in A$ be projections. Then $pq = p$ if and only if $qp = p$.*

*Proof* Since $p = p^*$ and $q = q^*$, if $pq = p$ then $pq = (pq)^* = q^* p^* = qp$. The converse is similar. □

**Lemma 5.6** *Let $p, q \in A$ be projections. Then $pq = qp$ if and only if $pq$ is a projection, in which case $pq = p \wedge q$ and $p + q - pq = p \vee q$.*

*Proof* If $pq = qp$, $(pq)^* = q^* p^* = qp = pq$ and $(pq)^2 = p(qp)q = p^2 q^2 = pq$. Conversely, if $pq$ is a projection then $qp = (pq)^* = pq$. Clearly then $pq \le p$ and $pq \le q$, and if $r \le p$ and $r \le q$ then $rpq = (rp)q = rq = r$ so $r \le pq$. Hence $pq = p \wedge q$. We similarly have $(1 - p)(1 - q) = (1 - p) \wedge (1 - q)$; since $r \mapsto 1 - r$ is an order-reversing involution it follows that $p + q - pq = 1 - (1 - p)(1 - q) = p \vee q$. □

For $A = \mathcal{B}(H)$, note that $p \le q$ if and only if $\text{range}(p) \subseteq \text{range}(q)$. Also, joins and meets always exist in $\mathcal{B}(H)$ and are given by

$$p \wedge q = \text{ the projection onto } \text{range}(p) \cap \text{range}(q),$$

$$p \vee q = \text{ the projection onto } \overline{\text{span}}(\text{range}(p) \cup \text{range}(q)).$$

That is, $\mathcal{P}(\mathcal{B}(H))$ is a lattice (in fact, it is a complete lattice, as the definitions of joins and meets above generalize naturally to infinite joins and meets).

Note that if $X$ is a connected compact Hausdorff space then $C(X)$ has no projections other than 0 and $I$.

**Proposition 5.7** $\mathcal{B}(H) = C^*(\mathcal{P}(\mathcal{B}(H)))$. *That is, $\mathcal{B}(H)$ is generated by its projections.*

*Proof* Since every $a \in \mathcal{B}(H)$ is a linear combination of self-adjoints $a + a^*$ and $i(a - a^*)$, it suffices to show that if $b$ is self-adjoint and $\epsilon > 0$ then there is a linear combination of projections $c = \sum_j \alpha_j p_j$ such that $\|b - c\| < \epsilon$. For this we may use the spectral theorem and approximate $m_f$ by a step function. □

**Corollary 5.8** *$C(H) = C^*(\mathcal{P}(C(H)))$. That is, $C(H)$ is generated by its projections.*

*Proof* Since a *-homomorphism sends projections to to projections, this is a consequence of Proposition 5.7 □

**Proposition 5.9** *Let $A$ be an abelian unital* C*-*algebra. Then $\mathcal{P}(A)$ is a Boolean algebra.*

*Proof* By Lemma 5.6, commuting projections always have joins and meets, and $p \mapsto I - p$ gives complements. It is then easy to check that this is actually a Boolean algebra using the formulas for joins and meets given by Lemma 5.6. □

## Stone duality

Let us recall the Stone duality for Boolean algebras. For a Boolean algebra $\mathbb{B}$ its Stone space Stone($\mathbb{B}$) is the compact Hausdorff space of all ultrafilters of $\mathbb{B}$ with the topology generated by its basic open sets $U_a = \{\mathcal{U} \in \text{Stone}(\mathbb{B}) : a \in \mathcal{U}\}$, for $a \in \mathbb{B} \setminus \{0_{\mathbb{B}}\}$. It is well-known that the algebra of clopen subsets of Stone($\mathbb{B}$) is isomorphic to $\mathbb{B}$. Also, to every Boolean algebra homomorphism $\Phi\colon \mathbb{B}_1 \to \mathbb{B}_2$ one associates a continuous map $f_\Phi\colon \text{Stone}(\mathbb{B}_2) \to \text{Stone}(\mathbb{B}_1)$, so that $f_\Phi(\mathcal{U}) = \Phi^{-1}(\mathcal{U})$ for all $\mathcal{U} \in \text{Stone}(\mathbb{B}_2)$. Conversely, if $f\colon \text{Stone}(\mathbb{B}_2) \to \text{Stone}(\mathbb{B}_1)$ is a continuous map, then $\Phi_f(a) = b$ for $b$ such that $f^{-1}(U_a) = U_b$ is a Boolean algebra homomorphism. It is straightforward to show that (i) the operations $f \mapsto \Phi_f$ and $\Phi \mapsto f_\Phi$ are inverses of one another, (ii) $f$ is a surjection if and only if $\Phi_f$ is an injection, and (iii) $f$ is an injection if and only if $\Phi_f$ is a surjection. Altogether this shows that the category of compact zero-dimensional Hausdorff spaces is contravariantly equivalent to the category of Boolean algebras.

By combining Stone duality with the Gelfand–Naimark theorem (see the remark after Theorem 4.1) one obtains isomorphism between the categories of Boolean algebras and abelian C*-algebras generated by their projections.

Note that if $A$ is nonabelian, then even if $\mathcal{P}(A)$ is a lattice it may be nondistributive and hence not a Boolean algebra. See also Proposition 5.26 below.

*Exercise* 5.10 Prove that the following are equivalent for a C*-algebra $A$.

1. The set of all invertible self-adjoint elements of $A$ is dense in the set of all self-adjoint elements of $A$.

2. The set of all linear combinations of projections is dense in $A$.

C*-algebras satisfying conditions of Exercise 5.10 are said to have *real rank zero*.

*Exercise* 5.11 Prove that $C(X)$ has real rank zero if and only if $X$ is zero-dimensional.

## 5.1 Maximal abelian subalgebras

In this section we will be interested in abelian (unital) subalgebras of $\mathcal{B}(H)$ and $C(H)$. In particular, we will look at maximal abelian subalgebras, or "masas." The acronym masa stands for 'Maximal Abelian SubAlgebra' or 'MAximal Self-Adjoint subalgebra.' Pedersen ([51]) uses MAÇA, for 'MAximal Commutative subAlgebra.'[3] Note that if $H = L^2(X, \mu)$, then $L^\infty(X, \mu)$ is an abelian subalgebra of $\mathcal{B}(H)$ (as multiplication operators).

**Theorem 5.12** *$L^\infty(X, \mu) \subset \mathcal{B}(L^2(X, \mu))$ is a masa.*

*Proof* See [9, Theorem 4.1.2] or [51, Theorem 4.7.7]. □

Conversely, every masa in $\mathcal{B}(H)$ is of this form. To prove this, we need a stronger form of the spectral theorem, which applies to abelian subalgebras rather than just single normal operators.

**Theorem 5.13** (General Spectral Theorem) *If $A$ is an abelian subalgebra of $\mathcal{B}(H)$ then there is a probability measure space $(X, \mu)$, a subalgebra $B$ of $L^\infty(X, \mu)$, and a Hilbert space isomorphism $\Phi : L^2(X, \mu) \to H$ such that* $\operatorname{Ad}\Phi[B] = A$.

*Proof* See [9, Theorem 4.7.13]. □

**Corollary 5.14** *For any masa $A \subset \mathcal{B}(H)$, there is a probability measure space $(X, \mu)$ and a Hilbert space isomorphism $\Phi : L^2(X, \mu) \to H$ such that*

$$\operatorname{Ad}\Phi[L^\infty(X, \mu)] = A.$$

*Proof* By maximality, $B$ must be all of $L^\infty(X)$ in the spectral theorem. □

Corollary 5.14 can be used to classify masas in $\mathcal{B}(H)$. The two most important examples of masas are given in the following two examples.

[3] This acronym acquires a whole new meaning in light of the related ASHCEFLC ([50, 6.2.13])

**Example 5.15** (Atomic masa in $\mathcal{B}(H)$) Fix an orthonormal basis $(e_n)$ for $H$, which gives an identification $H \cong \ell^2(\mathbb{N}) = \ell^2$. The corresponding masa is then $\ell^\infty$, or all operators that are diagonalized by the basis $(e_n)$. We call this an *atomic masa* because the corresponding measure space is atomic. The projections in $\ell^\infty$ are exactly the projections onto subspaces spanned by a subset of $\{e_n\}$. That is, $\mathcal{P}(\ell^\infty) \cong \mathcal{P}(\mathbb{N})$. In particular, if we fix a basis, then the Boolean algebra $\mathcal{P}(\mathbb{N})$ is naturally a sublattice of $\mathcal{P}(\mathcal{B}(H))$. Given $X \subseteq \mathbb{N}$, we write $P_X^{(\vec{e})}$ for the projection onto $\overline{\text{span}}\{e_n : n \in X\}$.

*Exercise* 5.16 Prove that the atomic masa is isomorphic to $C(\beta\mathbb{N})$, where $\mathbb{N}$ is the Čech–Stone compactification of $\mathbb{N}$ taken with the discrete topology. (Hint: Cf. Exercise 3.12.)

**Example 5.17** (Atomless masa in $\mathcal{B}(H)$) Let $(X, \mu)$ be any atomless probability measure space. Then if we identify $H$ with $L^2(X)$, $L^\infty(X) \subseteq \mathcal{B}(H)$ is the *atomless masa*. The projections in $L^\infty(X)$ are exactly the characteristic functions of measurable sets, so $\mathcal{P}(L^\infty(X))$ is the measure algebra of $(X, \mu)$ (modulo the null sets).

**Proposition 5.18** *Let $\mathcal{A} \subseteq \mathcal{B}(H)$ be an atomless masa. Then $\mathcal{P}(\mathcal{A})$ is isomorphic to the Lebesgue measure algebra of measurable subsets of* $[0, 1]$ *modulo null sets.*

*Proof* Omitted, but see the remark following Proposition 5.9. □

We now relate masas in $\mathcal{B}(H)$ to masas in $C(H)$.

**Theorem 5.19** (Johnson–Parrott, 1972 [42]) *If $\mathcal{A}$ is a masa in $\mathcal{B}(H)$ then $\pi[\mathcal{A}]$ is a masa in $C(H)$.*

*Proof* Assume $b \in \mathcal{B}(H)$ is such that $\pi(b)$ belongs to the commutant of $\pi[\mathcal{A}]$. We need to find $a \in \mathcal{A}$ such that $a - b$ is compact. Consider the map $\delta_b : \mathcal{A} \to \mathcal{B}(H)$ defined by

$$\delta_b(x) = bx - xb.$$

Then $\delta_b(x)$ is compact for every $a \in \mathcal{A}$. A straightforward computation shows that $\delta_b(xy) = \delta_b(x)y + x\delta_b(y)$. Such a map is called a *derivation*. By [42, Theorem 2.1], every derivation from the atomic masa into $\mathcal{B}(H)$ is *trivial*, i.e., of the form $\delta_a$ for some $a$ in the atomic masa. Then $a \in \mathcal{A}$ such that $\delta_a$ and $\delta_b$ agree on $\mathcal{A}$ is as required, by [42, Lemma 1.4]. □

**Theorem 5.20** (Akemann–Weaver [3]) *There exists a masa $A$ in $C(H)$ that is not of the form $\pi[\mathcal{A}]$ for any masa $\mathcal{A} \subset \mathcal{B}(H)$.*

*Proof* By Corollary 5.14, each masa in $\mathcal{B}(H)$ is induced by an isomorphism from $H$ to $L^2(X)$ for a probability measure space $X$. But the measure algebra of a probability measure space is countably generated, so there are only $2^{\aleph_0}$ isomorphism classes of probability measure spaces. Since $H$ is separable, it follows that there are at most $2^{\aleph_0}$ masas in $\mathcal{B}(H)$.

Now fix an almost disjoint (modulo finite) family $\mathbb{A}$ of infinite subsets of $\mathbb{N}$ of size $2^{\aleph_0}$. Recall that $P_X^{(\vec{e})}$ is the projection to the closed subspace spanned by $\{e_n : n \in X\}$. Then the projections $p_X = \pi(P_X^{(\vec{e})})$, for $X \in \mathbb{A}$, form a family of orthogonal projections in $C(H)$. Choose non-commuting projections $q_{X,0}$ and $q_{X,1}$ in $C(H)$ below $p_X$. To each $f \colon \mathbb{A} \to \{0, 1\}$ associate a family of orthogonal projections $\{q_{X,f(X)}\}$. Extending each of these families to a masa, we obtain $2^{2^{\aleph_0}}$ distinct masas in $C(H)$. Therefore some masa in $C(H)$ is not of the form $\pi[\mathcal{A}]$ for any masa in $\mathcal{B}(H)$. □

Anderson ([4]) asked whether there is a masa in the Calkin algebra that is generated by projections but not of the form $\pi[\mathcal{A}]$ for some masa $\mathcal{A}$ in $\mathcal{B}(H)$. Note that this question is not answered by Theorem 5.20 since masa constructed there are not necessarily generated by their projections. By [60] very mild set-theoretic assumptions imply the existence of such masa. It is not known whether this can be proved in ZFC.

**Lemma 5.21** *Let $\mathcal{A} \subset \mathcal{B}(H)$ be a masa. Then $J = \mathcal{P}(\mathcal{A}) \cap \mathcal{K}(H)$ is a Boolean ideal in $\mathcal{P}(\mathcal{A})$ and $\mathcal{P}(\pi[\mathcal{A}]) = \mathcal{P}(\mathcal{A})/J$.*

*Proof* It is easy to check that $J$ is an ideal since $\mathcal{K}(H) \subseteq \mathcal{B}(H)$ is an ideal. Let $a \in \mathcal{A}$ be such that $\pi(a)$ is a projection. Writing $\mathcal{A} = L^\infty(X)$, then in the proof of Lemma 5.3, we could have chosen to represent $a$ as a multiplication operator on $L^2(X)$, in which case the projection $p$ that we obtain such that $\pi(p) = \pi(a)$ is also a multiplication operator on $L^2(X)$. That is there is a projection $p \in \mathcal{A}$ such that $\pi(p) = \pi(a)$. Thus $\pi : \mathcal{P}(\mathcal{A}) \to \mathcal{P}(\pi[\mathcal{A}])$ is surjective. Furthermore, it is clearly a Boolean homomorphism and its kernel is $J$, so $\mathcal{P}(\pi[\mathcal{A}]) = \mathcal{P}(\mathcal{A})/J$. □

*Exercise* 5.22 Let $\mathcal{A}$ be the CAR algebra (§3.4) and let $\mathcal{D}$ be its subalgebra generated by the diagonal matrices. Show that $\mathcal{D}$ is a masa in $\mathcal{A}$.

## 5.2 Projections in the Calkin algebra

In the present section we study the poset of projections in the Calkin algebra. This structure is closely related to the Boolean algebra $\mathcal{P}(\mathbb{N})/\operatorname{Fin}$, although in many ways it is closer to quotients over analytic P-ideals such as the asymptotic density zero ideal, $\mathcal{Z}_0$.

**Lemma 5.23** *A projection $p \in \mathcal{B}(H)$ is compact if and only if its range is finite-dimensional.*

*Proof* If we let $B \subseteq H$ be the unit ball, $p$ is compact if and only if $p[B]$ is precompact. But $p[B]$ is just the unit ball in the range of $p$, which is (pre)compact if and only if the range is finite-dimensional. □

Let us now take a closer look at the images of the two distinguished masas in $\mathcal{B}(H)$.

If $\mathcal{A} = \ell^\infty$ is an atomic masa in $\mathcal{B}(H)$, then we obtain an "atomic" masa $\pi[\mathcal{A}]$ in $C(H)$. By Lemmas 5.21 and 5.23, $\mathcal{P}(\pi[\mathcal{A}]) \cong \mathcal{P}(\mathbb{N})/\mathrm{Fin}$, where Fin is the ideal of finite sets. In particular, if we fix a basis then $\mathcal{P}(\mathbb{N})/\mathrm{Fin}$ naturally embeds in $\mathcal{P}(C(H))$. For this reason, we can think of $\mathcal{P}(C(H))$ as a "noncommutative" version of $\mathcal{P}(\mathbb{N})/\mathrm{Fin}$. Moreover, one can show that $\mathcal{A} \cap \mathcal{K}(H) = c_0$, the set of sequences converging to 0, so that $\pi[\mathcal{A}] = \ell^\infty/c_0 = C(\beta\mathbb{N} \setminus \mathbb{N})$.

If $\mathcal{A}$ is an atomless masa in $\mathcal{B}(H)$, then all of its projections are infinite-dimensional. Thus $\mathcal{P}(\pi[\mathcal{A}]) = \mathcal{P}(\mathcal{A})$. Thus the Lebesgue measure algebra also embeds in $\mathcal{P}(C(H))$.

**Lemma 5.24** *For projections $p$ and $q$ in $\mathcal{B}(H)$, the following are equivalent:*

1. $\pi(p) \leq \pi(q)$,
2. $p(I - q)$ *is compact,*
3. *For any $\epsilon > 0$, there is a finite-dimensional projection $p_0 \leq I - p$ such that $\|q(I - p - p_0)\| < \epsilon$.*

*Proof* The equivalence of (1.) and (2.) is trivial. For the remaining part see [64, Proposition 3.3]. □

We write $p \leq_{\mathcal{K}} q$ if the conditions of Lemma 5.24 are satisfied. The poset $(\mathcal{P}(C(H)), \leq)$ is then isomorphic to the quotient $(\mathcal{P}(\mathcal{B}(H)), \leq_{\mathcal{K}})/\sim$, where $p \sim q$ if $p \leq_{\mathcal{K}} q$ and $q \leq_{\mathcal{K}} p$. In the strong operator topology, $\mathcal{P}(\mathcal{B}(H))$ is Polish, and (3) in Lemma 5.24 then implies that $\leq_{\mathcal{K}} \subset \mathcal{P}(\mathcal{B}(H)) \times \mathcal{P}(\mathcal{B}(H))$ is Borel.

**Lemma 5.25** *There are projections $p$ and $q$ in $\mathcal{B}(H)$ such that $\pi(p) = \pi(q) \neq 0$ but $p \wedge q = 0$.*

*Proof* Fix an orthonormal basis $(e_n)$ for $H$ and let $\alpha_n = 1 - \frac{1}{n}$ and $\beta_n = \sqrt{1-\alpha_n^2}$. Vectors $\xi_n = \alpha_n e_{2n} + \beta_n e_{2n+1}$ for $n \in \mathbb{N}$ are orthonormal and they satisfy $\lim_n(\xi_n \mid e_{2n}) = 1$. Projections $p = \mathrm{proj}_{\overline{\mathrm{span}}\{e_{2n}:n\in\mathbb{N}\}}$ and $q = \mathrm{proj}_{\overline{\mathrm{span}}\{\xi_n:n\in\mathbb{N}\}}$ are as required. □

Recall that $\mathcal{P}(\mathcal{B}(H))$ is a complete lattice, which is analogous to the fact that $\mathcal{P}(\mathbb{N})$ is a complete Boolean algebra. Since $\mathcal{P}(\mathbb{N})/\mathrm{Fin}$ is not a complete Boolean algebra, we would not expect $\mathcal{P}(C(H))$ to be a complete lattice. More surprisingly, however, the "noncommutativity" of $\mathcal{P}(C(H))$ makes it not even be a lattice at all.

**Proposition 5.26** (Weaver) *$\mathcal{P}(C(H))$ is not a lattice.*

*Proof* Enumerate an orthogonal basis of $H$ as $\{\xi_{mn}, \eta_{mn} : m \in \mathbb{N}, n \in \mathbb{N}\}$. Define

$$\zeta_{mn} = \frac{1}{n}\xi_{mn} + \frac{\sqrt{n-1}}{n}\eta_{mn}$$

and

$$K = \overline{\mathrm{span}}\{\eta_{mn} : m, n \in \mathbb{N}\}, \qquad p = \mathrm{proj}_K$$
$$L = \overline{\mathrm{span}}\{\zeta_{mn} : m, n \in \mathbb{N}\}, \qquad q = \mathrm{proj}_L\,.$$

For $f : \mathbb{N} \to \mathbb{N}$, define

$$M(f) = \overline{\mathrm{span}}\{\eta_{mn} : m \le f(n)\} \text{ and } r(f) = \mathrm{proj}_{M(f)}\,.$$

It is easy to show, using Lemma 5.24, that $r(f) \le p$ and $r(f) \le_{\mathcal{K}} q$ for all $f$, and that if $f < g$ then $r(f) <_{\mathcal{K}} r(g)$ strictly.

Now assume $r$ is a projection such that $r \le_{\mathcal{K}} p$ and $r \le_{\mathcal{K}} q$. Again using Lemma 5.24 one sees that $r \le_{\mathcal{K}} r(f)$ for some $f$. In particular, it follows that $p$ and $q$ cannot have a meet under $\le_{\mathcal{K}}$. □

## 5.3 Cardinal invariants

Since cardinal invariants can often be defined in terms of properties of subsets of $\mathcal{P}(\mathbb{N})/\mathrm{Fin}$ (see [14]), we can look for "noncommutative" (or "quantum") versions of cardinal invariants by looking at analogous properties of $\mathcal{P}(C(H))$.

Recall that $\mathfrak{a}$ denotes the minimal possible cardinality of a maximal infinite antichain in $\mathcal{P}(\mathbb{N})/\mathrm{Fin}$, or equivalently the minimal possible cardinality of an (infinite) maximal almost disjoint family in $\mathcal{P}(\mathbb{N})$.

**Definition 5.27** (Wofsey, [65]) A family $\mathbb{A} \subseteq \mathcal{P}(\mathcal{B}(H))$ is *almost orthogonal (ao)* if $pq$ is compact for $p \neq q$ in $\mathbb{A}$ but no $p \in \mathbb{A}$ is compact. We define $\mathfrak{a}^*$ to be the minimal possible cardinality of an infinite maximal ao family ("mao family").

Note that we require every $p \in \mathbb{A}$ to be noncompact since while Fin $\subset \mathcal{P}(\mathbb{N})$ is only countable, there are $2^{\aleph_0}$ compact projections in $\mathcal{P}(\mathcal{B}(H))$.

**Theorem 5.28** (Wofsey, [65]) *1. It is relatively consistent with ZFC that* $\aleph_1 = \mathfrak{a} = \mathfrak{a}^* < 2^{\aleph_0}$,
*2. MA implies* $\mathfrak{a}^* = 2^{\aleph_0}$.

*Proof* Omitted. □

*Question* 5.29 Is $\mathfrak{a} = \mathfrak{a}^*$? Is $\mathfrak{a} \geq \mathfrak{a}^*$? Is $\mathfrak{a}^* \geq \mathfrak{a}$?

It may seem easy to prove that $\mathfrak{a} \geq \mathfrak{a}^*$, since $\mathcal{P}(\mathbb{N})/\text{Fin}$ embeds in $\mathcal{P}(C(H))$ so any maximal almost disjoint family would give a mao family. However, it turns out that a maximal almost disjoint family can fail to be maximal as an almost orthogonal family. We now proceed to give an example of such a family.

An ideal $J$ on $\mathcal{P}(\mathbb{N})$ is a *p-ideal* if for every sequence $X_n$, $n \in \mathbb{N}$ of elements of $J$ there is $X \in J$ such that $X_n \setminus X$ is finite for all $n$.

**Lemma 5.30** (Steprāns, 2007) *Fix $a \in \mathcal{B}(H)$ and a basis $(e_n)$ for $H$. Then*

$$J_a = \{X \subseteq \mathbb{N} : P_X^{(\vec{e})} a \text{ is compact}\}$$

*is a Borel P-ideal.*

*Proof* Let $\varphi_a(X) = \|P_X a\|$. This is a lower semicontinuous submeasure on $\mathbb{N}$, and $P_X a$ is compact if and only if $\lim_n \varphi_a(X \setminus n) = 0$ (see equivalent conditions (1)–(3) in Example 3.1). Thus $J_a$ is $F_{\sigma\delta}$. Proving that it is a p-ideal is an easy exercise. □

**Proposition 5.31** (Wofsey, [65]) *There is a maximal almost disjoint family $\mathbb{A} \subset \mathcal{P}(\mathbb{N})$ whose image in $\mathcal{P}(\mathcal{B}(H))$ is not a mao family.*

*Proof* Let $\xi_n = 2^{-n/2} \sum_{j=2^n}^{2^{n+1}-1} e_j$. Then $\xi_n$, for $n \in \mathbb{N}$, are orthonormal and $q = \text{proj}_{\overline{\text{span}\{\xi_n\}}}$. Since $\lim_n \|qe_n\| = 0$ the ideal $J_q$ is *dense*: every infinite subset of $\mathbb{N}$ has an infinite subset in $J_q$ (choose a sparse enough subset $X$ such that $\sum_{n \in X} \|qe_n\| < \infty$). By density, we can find a maximal almost disjoint family $\mathbb{A}$ that is contained in $J_q$. Then $q$ is almost orthogonal to $P_X$ for all $X \in \mathbb{A}$, so $\{P_X : X \in \mathbb{A}\}$ is not a mao family. □

In some sense, this is the only way to construct such a counterexample. More precisely, we have the following:

**Theorem 5.32** *Let $\mathfrak{a}'$ denote the minimal possible cardinality of a maximal almost disjoint family that is not contained in any proper Borel P-ideal. Then $\mathfrak{a}' \geq \mathfrak{a}$ and $\mathfrak{a}' \geq \mathfrak{a}^*$.*

*Proof* The inequality $\mathfrak{a}' \geq \mathfrak{a}$ is trivial, and the inequality $\mathfrak{a}' \geq \mathfrak{a}^*$ follows by Lemma 5.30. □

One can also similarly define other quantum cardinal invariants: $\mathfrak{p}^*$, $\mathfrak{t}^*$, $\mathfrak{b}^*$, etc (see e.g., [14]). For example, recall that $\mathfrak{b}$ is the minimal cardinal $\kappa$ such that there exists a $(\kappa, \omega)$-gap in $\mathcal{P}(\mathbb{N})/\mathrm{Fin}$ and let $\mathfrak{b}^*$ be the minimal cardinal $\kappa$ such that there exists a $(\kappa, \omega)$-gap in $\mathcal{P}(C(H))$. Considerations similar to those needed in the proof of Proposition 5.26 lead to following.

**Theorem 5.33** (Zamora–Avilés, [66]) $\mathfrak{b} = \mathfrak{b}^*$.

*Proof* Omitted. □

Almost all other questions about the relationship between these and ordinary cardinal invariants are open. One should also note that equivalent definitions of standard cardinal invariants may lead to distinct quantum cardinal invariants.

## 5.4 A twist of projections

A question that may be related to cardinal invariants is when collections of commuting projections of $C(H)$ can be simultaneously lifted to $\mathcal{B}(H)$ such that the lifts still commute. Let $\mathfrak{l}$ (this symbol is `\mathfrak l`) be the minimal cardinality of such a collection that does not lift. From the proof of Theorem 5.20 it follows that such collections exist. Note that if instead of projections in the definition of $\mathfrak{l}$ we consider arbitrary commuting operators, then the value of a cardinal invariant defined in this way drops to 2. To see this, consider the unilateral shift and its adjoint (see Example 5.4).

**Lemma 5.34** *The cardinal $\mathfrak{l}$ is uncountable. Given any sequence $p_i$ of projections in $\mathcal{B}(H)$ such that $\pi(p_i)$ and $\pi(p_j)$ commute for all $i, j$, there is an atomic masa $\mathcal{A}$ in $\mathcal{B}(H)$ such that $\pi[\mathcal{A}]$ contains all $\pi(p_i)$.*

*Proof* Let $\zeta(i)$, $i \in \mathbb{N}$, be a norm-dense subset of the unit ball of $H$. We will recursively choose projections $q_i$ in $\mathcal{B}(H)$, orthonormal basis $e_i$, and

$k(j) \in \mathbb{N}$ so that for all $i \leq k(j)$ we have $\pi(q_i) = \pi(p_i)$, $q_i(e_j) \in \{e_j, 0\}$ and $\zeta(j)$ is in the span of $\{e_i : i < k(j)\}$. Assume $q_j$, $j < n$, and $e_i$, $i < k(n)$, have been chosen to satisfy these requirements. Let $r$ be the projection to the orthogonal complement of $\{e_i \mid i < k(n)\}$ and for each $\alpha \in \{1, \perp\}^n$ let $r_\alpha = r \prod_{i<n} q_i^{\alpha(i)}$. For each $\alpha \in \{1, \perp\}^n$ we have that $\pi(p_n)$ and $\pi(r_\alpha)$ commute, hence by Lemma 5.3 there is a projection $p_\alpha$ in $\mathcal{B}(r_\alpha[H])$ such that $\pi(p_\alpha) = \pi(p_n)\pi(r_\alpha)$, and $\pi(p_n) = \sum_\alpha \pi(p_n)\pi(r_\alpha)$. Note that we have

$$q_n = \sum_{\alpha \in \{1,\perp\}^n} p_\alpha.$$

Now pick $k(n+1)$ large enough and unit vectors $e_i$, $k(n) \leq i < k(n+1)$, each belonging in some $r_\alpha q_n[H]$, such that $e_i$, $i < k(n+1)$ span $\zeta(n)$.

This assures $(e_i)$ is a basis of $H$. Let $X(i) = \{n : n \geq k(i)$ and the unique $\alpha \in \{1, \perp\}^n$ such that $e_n \in r_\alpha(n)$ satisfies $\alpha(i) = 1\}$. Fix $i \in \mathbb{N}$. Clearly $q_i = P^{(\vec{e})}_{X(i)}$ satisfies $\pi(q_i) = \pi(p_i)$ and it is diagonalized by $(e_n)$. □

Note that it is not true that any countable collection of commuting projections in $\mathcal{B}(H)$ is simultaneously diagonalizable (e.g., take $H = L^2([0,1])$ and the projections onto $L^2([0,q])$ for each $q \in \mathbb{Q}$).

Theorem 5.35 below was inspired by [48]. In this paper Luzin proved the existence of an uncountable almost disjoint family $\{X_\xi : \xi < \omega_1\}$ of subsets of $\mathbb{N}$ with the property that for every $\mathcal{Z} \subseteq \omega_1$ such that both $\mathcal{Z}$ and $\omega_1 \setminus \mathcal{Z}$ are uncountable the families $\{X_\xi : \xi \in \mathcal{Z}\}$ and $\{X_\xi : \xi \in \omega_1 \setminus \mathcal{Z}\}$ cannot be separated, in the sense that there is no $Y \subseteq \mathbb{N}$ such that $X_\xi \setminus Y$ is finite for all $\xi \in \mathcal{Z}$ and $X_\xi \cap Y$ is finite for all $\xi \in \omega_1 \setminus \mathcal{Z}$ This family is one of the instances of incompactness of $\omega_1$ that are provable in ZFC, along with Hausdorff gaps, special Aronszajn trees, or nontrivial coherent families of partial functions.

**Theorem 5.35** (Farah, 2006 [22]) *There is a collection of $\aleph_1$ commuting projections in $C(H)$ that cannot be lifted to simultaneous diagonalizable projections in $\mathcal{B}(H)$.*

*Proof* Construct $p_\xi$ $(\xi < \omega_1)$ in $\mathcal{P}(\mathcal{B}(H))$ so that for $\xi \neq \eta$ (using the standard notation for the commutator of $a$ and $b$, $[a,b] = ab - ba$):

1. $p_\xi p_\eta$ is compact, and
2. $\|[p_\xi, p_\eta]\| > 1/4$.

Such a family can easily be constructed by repeatedly applying Lemma 5.34.

If there are lifts $P^{(\vec{e})}_{X_\xi}$ of $\pi(p_\xi)$ that are all diagonalized by a basis $(e_n)$, let $d_\xi = p_\xi - P^{(\vec{e})}_{X(\xi)}$. Write $r_n = P^{(\vec{e})}_{\{0,1,\dots,n-1\}}$, so $a$ is compact if and only if $\lim_n \|a(I - r_n)\| = 0$. By hypothesis, each $d_\xi$ is compact, so fix $\bar{n}$ such that $S = \{\xi : \|d_\xi(I - r_{\bar{n}})\| < 1/8\}$ is uncountable. Since the range of $I - r_{\bar{n}}$ is separable, there are distinct $\xi, \eta \in S$ such that $\|(d_\xi - d_\eta)r_{\bar{n}}\| < 1/8$. But then we can compute that

$$\|[p_\xi, p_\eta]\| \leq \|[P_{X(\xi)}, P_{X(\eta)}]\| + 1/8 = 1/4,$$

a contradiction. □

In the early draft of this paper it was conjectured that the projections constructed in Theorem 5.35 cannot be lifted to simultaneously commuting projections, and that in particular, $\mathfrak{l} = \aleph_1$. This conjecture was confirmed by Tristan Bice in [11, Theorem 2.4.18].

## 5.5 Maximal chains of projections in the Calkin algebra

A problem closely related to cardinal invariants is the description of isomorphism classes of maximal chains in $\mathcal{P}(\mathbb{N})/\mathrm{Fin}$ and $\mathcal{P}(C(H))$. The structure $(\mathcal{P}(\mathbb{N})/\mathrm{Fin}, \leq)$ is $\aleph_1$-saturated, in the model-theoretic sense: every consistent type over a countable set is realized in the structure (this was first noticed by Hadwin in [40]). Therefore under CH a back-and-forth argument shows that all maximal chains are order-isomorphic. Countable saturatedness of quotients $\mathcal{P}(\mathbb{N})/J$, for analytic ideals $J$, was well-studied. For example, by a result of Just and Krawczyk the quotient over every $F_\sigma$ ideal that includes Fin is countably saturated. Also, there are arbitrarily complex Borel ideals with countably saturated quotients. On the other hand, many well-studied $F_{\sigma\delta}$ ideals, for example the ideal $\mathcal{Z}_0$ of asymptotic density zero sets, don't have countably saturated quotients (see [21] and references thereof).

**Theorem 5.36** (Hadwin, 1998 [40]) *CH implies that any two maximal chains in $\mathcal{P}(C(H))$ are order-isomorphic.*

*Proof* One can show that $\mathcal{P}(C(H))$ has a similar saturation property and then use the same back-and-forth argument. □

**Conjecture 5.37** (Hadwin, 1998 [40]) *CH is equivalent to "any two maximal chains in $\mathcal{P}(C(H))$ are order-isomorphic".*

This conjecture seems unlikely to be true, and the analogous statement for $\mathcal{P}(\mathbb{N})/\mathrm{Fin}$ is not true.

**Theorem 5.38** (essentially Shelah–Steprāns) *There is a model of ¬CH in which all maximal chains in $\mathcal{P}(\mathbb{N})/\mathrm{Fin}$ are isomorphic.*

*Proof* Add $\aleph_2$ Cohen reals to a model of CH. We can then build up an isomorphism between any two maximal chains in the generic model in essentially the same way as a nontrivial automorphism of $\mathcal{P}(\mathbb{N})/\mathrm{Fin}$ is built up in [59]. □

The above proof cannot be straightforwardly adapted to the case of $\mathcal{P}(C(H))$.

By forcing towers in $\mathcal{P}(\mathbb{N})/\mathrm{Fin}$ of different cofinalities, one can construct maximal chains in $(\mathcal{P}(\mathbb{N})\setminus\{\mathbb{N}\})/\mathrm{Fin}$ of different cofinalities (in particular, they are non-isomorphic). The same thing works for $\mathcal{P}(C(H)) \setminus \{\pi(I)\}$.

**Theorem 5.39** (Wofsey, 2006 [65]) *There is a forcing extension in which there are maximal chains in $\mathcal{P}(C(H)) \setminus \{\pi(I)\}$ of different cofinalities (and $2^{\aleph_0} = \aleph_2$).*

*Idea of the proof* A standard forcing that adds maximal chains of different cofinalities to $\mathcal{P}(\mathbb{N})/\mathrm{Fin}$ works. □

## 6 More on pure states

Recall that a state of a C*-algebra is pure if it cannot be written as a nontrivial linear combination of two distinct nonzero states (§4.1). We now look at some set-theoretic problems concerning pure states on C*-algebras.

**Lemma 6.1** *If B is abelian and A is a unital subalgebra of B then any pure state of B restricts to a pure state of A*

*Proof* A state on either algebra is pure if and only if it is multiplicative. It follows that the restriction of a pure state is pure. □

However, in general the restriction of a pure state to a unital subalgebra need not be pure.

**Example 6.2** If $\omega_\xi$ is a vector state of $\mathcal{B}(H)$ and $\mathcal{A}$ is the atomic masa diagonalized by a basis $(e_n)$, then $\omega_\xi \restriction \mathcal{A}$ is pure if and only if $|(\xi \mid e_n)| = 1$ for some $n$. Indeed, $\mathcal{A}$ is isomorphic to $\ell^\infty$, which is in turn isomorphic to $C(\beta\mathbb{N})$. Therefore (cf. Exercise 4.16) a state of $\mathcal{A}$ is pure if and only if it is the evaluation functional at some point of $\beta\mathbb{N}$, or equivalently, if it is a limit of the vector states $\omega_{e_n}$ under an ultrafilter (such states reoccur in Example 6.31 below).

**Lemma 6.3** *If $A$ is an abelian C*-algebra generated by its projections than a state $\phi$ of $A$ is pure if and only if $\phi(p) \in \{0, 1\}$ for every projection $p$ in $A$.*

*Proof* Let us first consider the case when $A$ is unital. By the Gelfand–Naimark theorem we may assume $A$ is $C(X)$ for a compact Hausdorff space $X$. By Exercise 4.16 a state $\phi$ of $C(X)$ is pure if and only if there is $x \in X$ such that $\phi(f) = f(x)$ for all $f$. Such a state clearly satisfies $\phi(p) \in \{0, 1\}$ for each projection $p$ in $C(X)$.

If $\phi(p) \in \{0, 1\}$ for every projection $p$, then $\mathcal{F} = \{p : \phi(p) = 1\}$ is a filter such that for every $p$ either $p$ or $I - p$ is in $\mathcal{F}$. (Here $\mathcal{F}$ is a 'conventional' filter, not to be confused with quantum filters introduced after Lemma 6.41.) By our assumption, $X$ is zero-dimensional (cf. Exercise 5.11). Therefore $\mathcal{F}$ converges to a point $x$. We claim that $\phi(f) = f(x)$ for all $f \in C(X)$. Pick $f \in C(X)$ and $\epsilon > 0$. Let $U \subseteq X$ be a clopen neighborhood of $x$ such that $|f(y) - f(x)| < \epsilon$ for all $y \in U$, and let $p$ be the projection corresponding to the characteristic function of $U$. Then $\phi(p) = 1$ and by Lemma 4.8 we have $\phi(f) = \phi(pfp)$. On the other hand, with $\lambda = f(x)$ we have $\|pfp - \lambda p\| < \epsilon$, hence $|\phi(f) - \lambda| < \epsilon$. Since $\epsilon > 0$ was arbitrary we conclude that $\phi(f) = \lambda = f(x)$.

If $A$ is not unital, then $A$ is isomorphic to $C_0(X)$ for a locally compact Hausdorff space $X$. Consider it as a subalgebra of $C(\beta X)$ and use an argument similar to the above. □

**Proposition 6.4** *Let $B$ be a unital abelian C*-algebra and $A \subseteq B$ be a unital subalgebra. If every pure state of $A$ extends to a unique pure state of $B$, then $A = B$.*

*Proof* We have $B = C(X)$, where $X$ is the space of pure states on $B$. Since $B$ is abelian, every point of $X$ gives a pure state on $A$. We claim that $A$ separates points of $X$ (cf. Exercise 3.11). Assume the contrary and let $x \neq y$ be points of $X$ such that $f(x) = f(y)$ for all $f \in A$. Then $f \mapsto f(x)$ is a pure state of $A$ that has two distinct extensions to a pure state of $B$, contradicting our assumption. By Stone–Weierstrass we have $A = C(X)$. □

Without the assumption that $B$ is abelian the conclusion of Proposition 6.4 is no longer true. Let $B = M_{2^\infty}$ and let $A$ be its standard masa—the limit of algebras of diagonal matrices. Then $A$ is isomorphic to $C(2^{\mathbb{N}})$ and each pure state $\phi$ of $A$ is an evaluation function at some $x \in 2^{\mathbb{N}}$. Assume $\psi$ is a state extension of $\phi$ to $M_{2^\infty}$. In each $M_{2^n}(\mathbb{C})$ there is a 1-dimensional projection $p_n$ such that $\phi(p_n) = 1$, and therefore Lemma 4.8 implies that for all $a \in M_{2^n}(\mathbb{C})$ we have $\phi(a) = \phi(p_n a p_n) =$ the diagonal entry of the $2^n \times 2^n$ matrix $p_n a p_n$ determined by $p_n$. Since $\bigcup_n M_{2^n}(\mathbb{C})$ is dense in $M_{2^\infty}$, state $\psi$ is uniquely determined by $\phi$.

If $A \subseteq B$ are C*-algebras we say that $A$ *separates pure states of* $B$ if for all pure states $\psi \neq \phi$ of $B$ there is $a \in A$ such that $\phi(a) \neq \psi(a)$.

*Exercise* 6.5 Give an example of a C*-algebra $B$ and its unital subalgebra $A$ such that $A$ separates pure states of $B$ but every pure state of $A$ has a unique extension to a state of $B$. (Hint: See Exercise 6.25.)

**Problem 6.6** (Noncommutative Stone-Weierstrass problem) Assume $A$ is a unital subalgebra of $B$ and $A$ separates $\mathbb{P}(B) \cup \{0\}$. Does this necessarily imply $A = B$?

For more on this problem see e.g., [55].

*Exercise* 6.7 Prove that for an irreducible representation $\pi\colon A \to \mathcal{B}(H)$ we have $\pi[A] \supseteq \mathcal{K}(H)$ if and only if $\pi[A] \cap \mathcal{K}(H) \neq \{0\}$.

**Definition 6.8** (Kaplansky) A C*-algebra $A$ is of *type I* if for every irreducible representation $\pi : A \to \mathcal{B}(H)$ we have $\pi[A] \supseteq \mathcal{K}(H)$.

Type I C*-algebras are also known as GCR, postliminal, postliminary, or smooth. Here GCR stands for 'Generalized CCR' where CCR stands for 'completely continuous representation'; 'completely continuous operators' is an old-fashioned term for compact operators. See [50, §6.2.13] for an amusing explanation of the terminology (cf. footnote in §5.1). Type I C*-algebras should not be confused with type I von Neumann algebras: $\mathcal{B}(H)$ is a type I von Neumann algebra but is not a type I C*-algebra.

**Definition 6.9** A C*-algebra is *simple* if and only if it has no nontrivial (closed two-sided) ideals.

Recall that the pure states of a C*-algebra correspond to its irreducible representations (Lemma 4.14) and that pure states are equivalent if and only if the corresponding irreducible representations are equivalent (Proposition 4.23).

**Lemma 6.10** *If a type I C*-algebra has only one pure state up to equivalence then it is isomorphic to $\mathcal{K}(H)$ for some $H$.*

*Proof* Assume $A$ is of type I and all of its pure states are equivalent. It is not difficult to see that $A$ has to be simple. Therefore any irreducible representation is an isomorphism and therefore $\pi[A] = \mathcal{K}(H)$. □

The converse of Lemma 6.10 is a theorem of Naimark (Theorem 6.14). C*-algebras that are not type I are called *non-type I* or *antiliminary* (cf. discussion of this terminology in the introduction to [8]). Theorem 6.11 is the key part of Glimm's characterization of type I C*-algebras ([37], see also [50, Theorem 6.8.7]). Its proof contains a germ of what became known as the *Glimm-Effros Dichotomy* ([41]).

**Theorem 6.11** (Glimm) *If $A$ is a non-type-I* C*-*algebra then there is a subalgebra $B \subsetneq A$ that has a quotient isomorphic to $M_{2^\infty}$.*

*Proof* See [50, §6.8]. □

The following straightforward calculation will be used in Corollary 6.13.

**Lemma 6.12** *Assume $\varphi$ is a state of $A$ and $u$ and $v$ are unitaries in $A$ such that $\|u - v\| < \epsilon$. Then $\|\varphi \circ \operatorname{Ad} u - \varphi \circ \operatorname{Ad} v\| < 2\epsilon$.*

*Proof* It suffices to consider the case when $v = I$ and $\|u - I\| < \epsilon$. Then for $a \in A$ we have $\|a - uau^*\| = \|au - ua\| \leq \|au - a\| + \|a - ua\| < 2\epsilon\|a\|$. Therefore we have $\|\varphi(a) - \varphi(uau^*)\| \leq \|\varphi(a - uau^*)\| < 2\epsilon\|a\|$ for all $a \in A$ and $\|\varphi - \operatorname{Ad} u\varphi\| < 2\epsilon$ follows. □

**Corollary 6.13** (Akemann–Weaver, 2002 [2]) *If $A$ is non-type-I and has a dense subset of cardinality $< 2^{\aleph_0}$, then $A$ has nonequivalent pure states.*

*Proof* By Glimm's Theorem, a quotient of a subalgebra of $A$ is isomorphic to $M_{2^\infty}$, and the pure states $\varphi_f$ on $M_{2^\infty}$ then lift and extend to pure states $\psi_f$ of $A$. Furthermore, if $f \neq g$ then $\|\psi_f - \psi_g\| = 2$, since the same is true of $\varphi_f$ and $\varphi_g$. In particular, if $\psi$ is any pure state on $A$, then by Lemma 6.12 the unitaries that turn $\psi$ into $\psi_f$ must be far apart (distance $\geq 1$) from unitaries that turn $\psi$ into $\psi_g$. Since $A$ does not have a subset of cardinality $2^{\aleph_0}$ such that any two points are far apart from each other, $\psi$ cannot be equivalent to every $\psi_f$. □

## 6.1 Naimark's theorem and Naimark's problem

The starting point of this subsection is the following converse of Lemma 6.10.

**Theorem 6.14** (Naimark, 1948) *Any two pure states on $\mathcal{K}(H)$ are equivalent, for any (not necessarily separable) Hilbert space $H$.*

We shall sketch a proof of this theorem later on.

*Question* 6.15 (Naimark, 1951) If all pure states on a C*-algebra $A$ are equivalent, is $A$ isomorphic to $\mathcal{K}(H)$ for some Hilbert space $H$?

Note that by Lemma 6.10 and Corollary 6.13, any counterexample to this must be non-type I and have no dense subset of cardinality $< 2^{\aleph_0}$. A similar argument shows that a counterexample cannot be a subalgebra of $\mathcal{B}(H)$ for a Hilbert space with a dense subset of cardinality $< 2^{\aleph_0}$.

The proof of Naimark's theorem will require some terminology. Recall that a vector state on $\mathcal{B}(H)$ corresponding to a unit vector $\eta$ is defined by $\omega_\eta(a) = (a(\eta), \eta)$.

**Definition 6.16** An operator $a \in \mathcal{B}(H)$ is a *trace class operator* if for some orthogonal basis $E$ of $H$ we have $\sum_{e \in E}(|a|e, e) < \infty$. For a trace class operator $a$ define its *trace* as

$$\mathrm{tr}(a) = \textstyle\sum_{e \in E}(ae, e).$$

*Exercise* 6.17 Prove the following.

1. Trace class operators form an ideal in $\mathcal{B}(H)$ that is not norm-closed.
   (Hint: See [51].)
2. $\mathrm{tr}(ab) = \mathrm{tr}(ba)$ for any trace class operator $a$ and any operator $b$. In particular, this sum does not depend on the choice of the orthonormal basis.
   (Hint: This is similar to the finite-dimensional case.)
3. Every trace class operator is compact.
   (Hint: It can be approximated by finite rank operators.

For unit vectors $\eta_1$ and $\eta_2$ in $H$ define a rank one operator $b_{\eta_1,\eta_2} : H \to H$ by

$$b_{\eta_1,\eta_2}(\xi) = (\xi, \eta_2)\eta_1.$$

This is a composition of the projection to $\mathbb{C}\eta_2$ with the partial isometry sending $\eta_2$ to $\eta_1$.

**Lemma 6.18** *Given a functional $\phi$ in the dual of $\mathcal{K}(H)$ there is a trace class operator $u$ such that $\phi(a) = \mathrm{tr}(ua)$ for all $a \in \mathcal{K}(H)$. If $\phi \geq 0$ then $u \geq 0$.*

*Proof* For the existence, see e.g., [51, Theorem 3.4.13]. To see $u$ is positive, pick $\eta \in H$. Then $ub_{\eta,\eta}(\xi) = u((\xi,\eta)\eta) = (\xi,\eta)u(\eta) = b_{u(\eta),\eta}(\xi)$. Therefore

$$0 \le \phi(b_{\eta,\eta}) = \operatorname{tr}(ub_{\eta,\eta}) = \operatorname{tr}(b_{u(\eta),\eta}) = \sum_{e\in E}(b_{u(\eta),\eta}(e), e) = \sum_{e\in E}(ub_{\eta,\eta}e, e) = (u(\eta),\eta).$$

(In the last equality we change the basis to $E'$ so that $\eta \in E'$.) □

**Proposition 6.19** *Every pure state $\phi$ of $\mathcal{K}(H)$ is equal to the restriction of some vector state to $\mathcal{K}(H)$.*

*Proof* By Lemma 6.18 we have a trace class operator $u$ such that $\phi(a) = \operatorname{tr}(ua)$ for all $a \in \mathcal{K}(H)$. Since $u$ is a positive compact operator, it is by the Spectral Theorem diagonalizable so we can write $u = \sum_{e\in E} \lambda_e e^*$ with the appropriate choice of the basis $E$. Thus $\phi(a) = \operatorname{tr}(ua) = \operatorname{tr}(au) = \sum_{e\in E}(aue, e) = \sum_{e\in E} \lambda_e(ae, e) \ge \lambda_{e_0}(ae_0, e_0)$, for any $e_0 \in E$. Since $\phi$ is a pure state, for each $e \in E$ there is $t_e \in [0, 1]$ such that $t_e\phi(a) = \lambda_{e_0}(ae_0, e_0)$. Thus exactly one $t_e = t_{e_0}$ is nonzero, and $a \mapsto \lambda_{e_0}(ae_0, e_0)$. □

*Proof of Theorem 6.14* If $\xi$ and $\eta$ are unit vectors in $H$, then the corresponding vector states $\omega_\xi$ and $\omega_\eta$ are clearly equivalent, via any unitary that sends $\xi$ to $\eta$. Hence the conclusion follows from Proposition 6.19 below. □

## 6.2 A counterexample to Naimark's problem from $\diamondsuit$

We shall now sketch a recent result of Akemann and Weaver, giving a consistent counterexample to Naimark's problem. One of the most interesting set-theoretic problems about C*-algebras is whether a positive solution to Naimark's problem is consistent with ZFC. A positive answer would open a possibility of having an interesting representation theory for not necessarily separable C*-algebras (see the introduction to [2]). The following lemma is based on recent work of Kishimoto–Ozawa–Sakai and Futamura–Kataoka–Kishimoto.

**Lemma 6.20** (Akemann–Weaver, 2004 [2]) *Let $A$ be a simple separable unital C*-algebra and let $\varphi$ and $\psi$ be pure states on $A$. Then there is a simple separable unital $B \supseteq A$ such that*

1. *$\varphi$ and $\psi$ extend to states $\varphi'$, $\psi'$ on $B$ in a unique way.*

2. *$\varphi'$ and $\psi'$ are equivalent.*

*Proof* Omitted. □

It is not known whether this lemma remains true when the separability assumption is dropped. However, Kishimoto–Ozawa–Sakai proved that their result used in the proof of Lemma 6.20 fails for nonseparable algebras. A very simple example was given in [24].

We shall now briefly describe Jensen's $\diamondsuit$ principle. Set-theoretically informed readers may want to skip ahead to Theorem 6.22. Recall that a subset $C$ of $\omega_1$ is *closed* if for every countable $A \subsetneq C$ we have that $\sup A \in C$. It is *unbounded* if it $\sup C = \omega_1$. By $\diamondsuit$ we denote Jensen's diamond principle on $\omega_1$. One of its equivalent reformulations states that there are functions $h_\alpha : \alpha \to \omega_1$, for $\alpha < \omega_1$, such that for every $g : \omega_1 \to \omega_1$, the set $\{\alpha : g \restriction \alpha = h_\alpha\}$ is stationary.

There are several revealing reformulations of $\diamondsuit$ (see [47, Chapter II]), and the following one was suggested by Weaver.

*Exercise* 6.21 Consider $T = \omega_1^{<\omega_1}$ as a tree with respect to the end-extension ordering. Show that $\diamondsuit$ is equivalent to the assertion that there is $t_\alpha$ in $T$ of length $\alpha$ such that for every $\omega_1$-branch $b$ of $T$ the set of all $\alpha$ such that $b \restriction \alpha = t_\alpha$ is stationary.

**Theorem 6.22** (Akemann–Weaver, 2004 [2]) *Assume $\diamondsuit$. Then there is a* C*-*algebra $A$, all of whose pure states are equivalent, which is not isomorphic to $\mathcal{K}(H)$ for any $H$.*

*Proof* We construct an increasing chain of simple separable unital C*-algebras $A_\alpha$ ($\alpha \leq \omega_1$). We also construct pure states $\psi_\alpha$ on $A_\alpha$ such that for $\alpha < \beta$, $\psi_\beta \restriction A_\alpha = \psi_\alpha$. For each $A_\alpha$, let $\{\varphi_\alpha^\gamma\}_{\gamma<\omega_1}$ enumerate all of its pure states.

If $\alpha$ is limit, we let $A_\alpha = \varinjlim_{\beta\to\alpha} A_\beta$ and $\psi_\alpha = \varinjlim \psi_\beta$.

Let us consider the successor case, when $A_\alpha$ is defined and we want to define $A_{\alpha+1}$. Suppose there is $\varphi \in \mathbb{P}(A_\alpha)$ such that $\varphi \restriction A_\beta = \varphi_\beta^{h_\alpha(\beta)}$ for all $\beta < \alpha$ (if no such $\varphi$ exists, let $A_{\alpha+1} = A_\alpha$). Note that $\bigcup_{\beta<\alpha} A_\beta$ is dense in $A_\alpha$ since $\alpha$ is limit, so there is at most one such $\varphi$. By Lemma 6.20, let $A_{\alpha+1}$ be such that $\psi_\alpha$ and $\varphi$ have unique extensions to $A_{\alpha+1}$ that are equivalent, and let $\psi_{\alpha+1}$ be the unique extension of $\psi_\alpha$.

Let $A = A_{\omega_1}$ and $\psi = \psi_{\omega_1}$. Then $A$ is unital and infinite-dimensional, so $A$ is not isomorphic to any $\mathcal{K}(H)$. Let $\varphi$ be any pure state on $A$; we will show that $\varphi$ is equivalent to $\psi$, so that $A$ has only one pure state up to equivalence.

*Claim* 6.23 $S = \{\alpha : \varphi \restriction A_\alpha \text{ is pure on } A_\alpha\}$ contains a club.

*Proof* For $x \in A$ and $m \in \mathbb{N}$ the set

$$T_{m,x} = \Big\{\alpha \colon x \in A_\alpha \text{ and } (\exists \psi_1, \psi_2 \in \mathbb{S}(A_\alpha))$$
$$\varphi \restriction A_\alpha = \frac{\psi_1 + \psi_2}{2} \text{ and } |\varphi(x) - \psi_1(x)| \geq \frac{1}{m}\Big\}$$

is bounded in $\omega_1$. Indeed, if it were unbounded, we could take a limit of such $\psi_i$ (with respect to an ultrafilter) to obtain states $\psi_i$ on $A$ such that $\varphi = \frac{\psi_1+\psi_2}{2}$ but such that $|\varphi(x) - \psi_1(x)| \geq \frac{1}{m}$, contradicting purity of $\varphi$. Since each $A_\alpha$ is separable, we can take a suitable diagonal intersection of the $T_{m,x}$ over all $m$ and all $x$ in a dense subset of $A$ to obtain a club contained in $S$. □

Now let $h : S \to \omega_1$ be such that $\varphi \restriction A_\alpha = \varphi_\alpha^{h(\alpha)}$ for all $\alpha \in S$. Since $S$ contains a club, there is some limit ordinal $\alpha$ such that $h \restriction \alpha = h_\alpha$. Then by construction, $\varphi \restriction A_{\alpha+1}$ is equivalent to $\psi_{\alpha+1}$; say $\varphi \restriction A_{\alpha+1} = u\psi_{\alpha+1}u^*$ for a unitary $u$. For each $\beta \geq \alpha$, $\psi_\beta$ extends uniquely to $\psi_{\beta+1}$, so by induction we obtain that $\psi$ is the unique extension of $\psi_{\alpha+1}$ to $A$. Since $\varphi \restriction A_{\alpha+1}$ is equivalent to $\psi_{\alpha+1}$, it also has a unique extension to $A$, which must be $\varphi$. But $u\psi u^*$ is an extension of $\varphi \restriction A_{\alpha+1}$, so $\varphi = u\psi u^*$ and is equivalent to $\psi$. □

## 6.3 Extending pure states on masas

By Exercise 4.16, a state on an abelian C*-algebra is pure if and only if it is multiplicative, i.e., a *-homomorphism. If the algebra is generated by projections then this is equivalent to asserting that $\phi(p) \in \{0, 1\}$ for every projection $p$ (Lemma 6.3).

**Definition 6.24** A masa in a C*-algebra $A$ has the *extension property* (EP) if each of its pure states extends uniquely to a pure state on $A$.

If $\mathcal{A} \subseteq \mathcal{B}(H)$ is a masa and $\phi$ is a vector state on $\mathcal{A}$ then $\phi$ extends uniquely to a pure state of $\mathcal{B}(H)$. This is essentially an easy consequence of Lemma 4.8. By Theorem 4.21 all non-vector pure states are singular and thus define pure states on $C(H)$. These two observations together imply that a masa $\mathcal{A} \subset \mathcal{B}(H)$ has the EP if and only if $\pi[\mathcal{A}] \subset C(H)$ has the EP

*Exercise* 6.25 Let $\mathcal{A}$ be the CAR algebra and let $\mathcal{D}$ be the masa generated by diagonal matrices (cf. Exercise 5.22). Show that $\mathcal{D}$ has the extension property. (Hint: Do the finite-dimensional case first. That is, show that the masa consisting of diagonal matrices in $M_n(\mathbb{C})$ has the extension property. See also Exercise 4.11.)

**Theorem 6.26** (Kadison–Singer, 1959, [44]) *Atomless masas in $\mathcal{B}(H)$ do not have the EP.*

*Proof* Omitted. □

**Theorem 6.27** (Anderson, 1978 [4]) *CH implies there is a masa in $C(H)$ that has the EP.*

*Proof* Omitted. □

Note that Anderson's theorem does not give a masa on $\mathcal{B}(H)$ with the EP, since his masa on $C(H)$ does not lift to a masa on $\mathcal{B}(H)$. The following is a famous open problem (compare with Problem 6.6).

**Problem 6.28** (Kadison–Singer, 1959 [44]) Do atomic masas of $\mathcal{B}(H)$ have the EP?

This is known to be equivalent to an *arithmetic statement* (i.e., a statement all of whose quantifiers range over natural numbers). As such, it is absolute between transitive models of ZFC and its solution is thus highly unlikely to involve set theory. For more on this problem see [16] and [63]. However, there are related questions that seem more set-theoretic. For example, consider the following conjecture:

**Conjecture 6.29** (Kadison–Singer, 1959 [44]) *For every pure state $\varphi$ of $\mathcal{B}(H)$ there is a masa $\mathcal{A}$ such that $\varphi \restriction \mathcal{A}$ is multiplicative (i.e., pure).*

We could also make the following stronger conjecture:

**Conjecture 6.30** *For every pure state $\varphi$ of $\mathcal{B}(H)$ there is an* atomic *masa $\mathcal{A}$ such that $\varphi \restriction \mathcal{A}$ is multiplicative.*

**Example 6.31** Let $\mathcal{U}$ be an ultrafilter on $\mathbb{N}$ and $(e_n)$ be an orthonormal basis for $H$. Then

$$\varphi_{\mathcal{U}}^{(\vec{e})}(a) = \lim_{n \to \mathcal{U}} (ae_n \mid e_n)$$

is a state on $\mathcal{B}(H)$. It is singular if and only if $\mathcal{U}$ is nonprincipal (if $\{n\} \in \mathcal{U}$, then $\varphi_{\mathcal{U}}^{(\vec{e})} = \omega_{e_n}$).

We say a state of the form $\varphi_{\mathcal{U}}^{(\vec{e})}$ for some basis $(e_n)$ and some ultrafilter

$\mathcal{U}$ is *diagonalizable*. As noted in Example 6.2, the restriction of a diagonalizable state to the corresponding atomic masa is a pure state of the masa, and every pure state of an atomic masa is of this form.

**Theorem 6.32** (Anderson, 1979 [6]) *Diagonalizable states are pure.*

*Proof* Omitted. □

**Conjecture 6.33** (Anderson, 1981 [7]) *Every pure state on $\mathcal{B}(H)$ is diagonalizable.*

**Proposition 6.34** *If atomic masas do have the EP, then Anderson's conjecture is equivalent to Conjecture 6.30.*

*Proof* If atomic masas have the EP, a pure state on $\mathcal{B}(H)$ is determined by its restriction to any atomic masa on which it is multiplicative. Any multiplicative state on an atomic masa extends to a diagonalizable state, so this means that a pure state restricts to a multiplicative state if and only if it is diagonalizable. □

We now prove an affirmative answer for a special case of the Kadison-Singer problem. We say an ultrafilter $\mathcal{U}$ on $\mathbb{N}$ is a *Q-point* (sometimes called *rare ultrafilter*) if every partition of $\mathbb{N}$ into finite intervals has a transversal in $\mathcal{U}$. The existence of Q-points is known to be independent from ZFC, but what matters here is that many ultrafilters on $\mathbb{N}$ are not Q-points.

Fix a basis $(e_n)$ and let $\mathcal{A}$ denote the atomic masa of all operators diagonalized by it. In the following proof we write $P_X$ for $P_X^{(\vec{e})}$.

**Theorem 6.35** (Reid, 1971 [53]) *If $\mathcal{U}$ is a Q-point then the diagonal state $\varphi_{\mathcal{U}} \upharpoonright \mathcal{A}$ has a unique extension to a pure state of $\mathcal{B}(H)$.*

*Proof* Fix a pure state $\varphi$ on $\mathcal{B}(H)$ extending $\varphi_{\mathcal{U}} \upharpoonright \mathcal{A}$ and let $a \in \mathcal{B}(H)$. Without a loss of generality $\mathcal{U}$ is nonprincipal so $\varphi$ is singular.

Choose finite intervals $(J_i)$ such that $\mathbb{N} = \bigcup_n J_n$ and

$$\|P_{J_m} a P_{J_n}\| < 2^{-m-n}$$

whenever $|m - n| \geq 2$. This is possible by (2) and (3) of Example 3.1 since $aP_{J_m}$ and $P_{J_m}a$ are compact. (See [26, Lemma 1.2] for details.) Let $X \in \mathcal{U}$ be such that $X \cap (J_{2i} \cup J_{2i+1})$ has a unique element, $n(i)$, for all $i$. Then for $Q_i = P_{\{n(i)\}}$ and $f_i = e_{n(i)}$ we have $\varphi(\sum_i Q_i) = 1$ and

$$QaQ = \sum_i Q_i a \sum_i Q_i = \sum_i Q_i a Q_i + \sum_{i \neq j} Q_i a Q_j.$$

The second sum is compact by our choice of $(J_i)$, and

$$Q_i a Q_i = (af_i \mid f_i) Q_i.$$

Now as we make $X \in \mathcal{U}$ smaller and smaller, $\sum_{i \in X}(ae_i \mid e_i)P_{\{i\}}$ gets closer and closer to $(\lim_{i \to \mathcal{U}}(ae_i \mid e_i)) \sum P_i = \varphi_{\mathcal{U}}(a) \sum P_i$. Thus

$$\lim_{X \to \mathcal{U}} \pi(P_X a P_X - \varphi_{\mathcal{U}}(a) P_X) \to 0.$$

Since $\varphi$ is singular and $\varphi(P_X) = \varphi_{\mathcal{U}}(P_X) = 1$, by Lemma 4.8 $\varphi(a) = \varphi(P_X a P_X) = \varphi_{\mathcal{U}}(a)$. Since $a$ was arbitrary, $\varphi = \varphi_{\mathcal{U}}$. □

## 6.4 A pure state that is not multiplicative on any masa in $\mathcal{B}(H)$

The following result shows that Conjecture 6.30 is not true in all models of ZFC. The following theorem follows from a stronger result, Theorem 6.46, whose proof will be sketched below.

**Theorem 6.36** (Akemann–Weaver, 2005 [3]) *CH implies there is a pure state $\varphi$ on $\mathcal{B}(H)$ that is not multiplicative on any atomic masa.*

The basic idea of constructing such a pure state is to encode pure states as "quantum ultrafilters"; a pure state on the atomic masa $\ell^\infty \subset \mathcal{B}(H)$ is equivalent to an ultrafilter. By the following result, states on $\mathcal{B}(H)$ correspond to finitely additive maps from $\mathcal{P}(\mathcal{B}(H))$ into $[0, 1]$.

**Theorem 6.37** (Gleason) *Assume $\mu : \mathcal{P}(\mathcal{B}(H)) \to [0, 1]$ is such that $\mu(p + q) = \mu(p) + \mu(q)$ whenever $pq = 0$. Then there is a unique state on $\mathcal{B}(H)$ that extends $\mu$.*

*Proof* Omitted. □

We need to go a little further and associate certain 'filters' of projections to pure states of $\mathcal{B}(H)$.

**Definition 6.38** A family $\mathbb{F}$ of projections in a C*-algebra is a *filter* if

1. For any $p, q \in \mathbb{F}$ there is $r \in \mathbb{F}$ such that $r \leq p$ and $r \leq q$.
2. If $p \in \mathbb{F}$ and $r \geq p$ then $r \in \mathbb{F}$.

The filter generated by $\mathbb{X} \subseteq \mathcal{P}(A)$ is the intersection of all filters containing $\mathbb{X}$ (which may not actually be a filter in general if $\mathcal{P}(A)$ is not a lattice).

We say that a filter $\mathcal{F} \subset \mathcal{P}(C(H))$ *lifts* if there is a commuting family $\mathbb{X} \subseteq \mathcal{P}(\mathcal{B}(H))$ that generates a filter $\mathbb{F}$ such that $\pi[\mathbb{F}] = \mathcal{F}$. Note that, unlike the case of quotient Boolean algebras, $\pi^{-1}[\mathcal{F}]$ itself is not a filter because there exist projections $p, q \in \mathcal{B}(H)$ such that $\pi(p) = \pi(q)$ but $p \wedge q = 0$ (Lemma 5.25).

*Question* 6.39 Does every maximal filter $\mathcal{F}$ in $\mathcal{P}(C(H))$ lift?

Maximal filters in $\mathcal{P}(C(H))$ can have rather interesting properties, as the following result shows.

**Theorem 6.40** (Anderson, [5]) *There are a singular pure state $\varphi$ of $\mathcal{B}(H)$, an atomic masa $\mathcal{A}_1$, and an atomless masa $\mathcal{A}_2$ such that both $\varphi \restriction \mathcal{A}_1$ and $\varphi \restriction \mathcal{A}_2$ are multiplicative.*

*Proof* Omitted. □

**Lemma 6.41** (Weaver, 2007) *For $\mathcal{F}$ in $\mathcal{P}(\mathcal{B}(H))$ the following are equivalent:*

*(A)* $\|p_1 p_2 \cdots p_n\| = 1$ *for any $p_1, \cdots, p_n \in \mathcal{F}$ and $\mathcal{F}$ is maximal with respect to this property.*
*(B)* *For all $\epsilon > 0$ and for all finite $F \subseteq \mathcal{F}$ there is a unit vector $\xi$ such that $\|p\xi\| > 1 - \epsilon$ for all $p \in F$.*

*Proof* Since $\|p_1 p_2 \cdots p_n\| \leq \|p_1\| \cdot \|p_2\| \cdot \ldots \cdot \|p_n\| = 1$, clause (A) is equivalent to stating that for every $\epsilon > 0$ there is a unit vector $\xi$ such that $\|p_1 p_2 \cdots p_n \xi\| > 1 - \epsilon$. The remaining calculations are left as an exercise to the reader. Keep in mind that, for a projection $p$, the value of $\|p\xi\|$ is close to $\|\xi\|$ if and only if $\|\xi - p\xi\|$ is close to 0. □

We call an $\mathcal{F}$ satisfying the conditions of Lemma 6.41 a *quantum filter*. Such an $\mathcal{F}$ is a *maximal quantum filter* if it is not properly included in another quantum filter.

**Theorem 6.42** (Farah–Weaver, 2007) *Let $\mathcal{F} \subseteq \mathcal{P}(C(H))$. Then the following are equivalent:*

1. *$\mathcal{F}$ is a maximal quantum filter,*
2. *$\mathcal{F} = \mathcal{F}_\varphi = \{p : \varphi(p) = 1\}$ for some pure state $\varphi$.*

*Proof* (1⇒2): For a finite $F \subseteq \mathcal{F}$ and $\epsilon > 0$ let

$$X_{F,\epsilon} = \{\varphi \in \mathbb{S}(\mathcal{B}(H)) : \varphi(p) \geq 1 - \epsilon \text{ for all } p \in F\}.$$

If $\xi$ is as in (B) then $\omega_\xi \in X_{F,\epsilon}$.

Since $X_{F,\epsilon}$ is weak*-compact, $\bigcap_{(F,\epsilon)} X_{F,\epsilon} \neq \emptyset$, and any extreme point of the intersection is a pure state with the desired property.[4]

(2⇒1). If $\varphi(p_j) = 1$ for $j = 1, \dots, k$, then $\varphi(p_1 p_2 \dots p_k) = 1$ by Lemma 4.8, hence (A) holds. It is then not hard to show that $\mathcal{F}_\varphi$ also satisfies (B) and is maximal. □

**Lemma 6.43** *Let $\mathcal{F}$ be a maximal quantum filter, let $(\xi_n)$ be an orthonormal basis, and let $\mathbb{N} = \bigcup_{j=1}^n A_j$ be a finite partition. If there is a $q \in \mathcal{F}$ such that $\|P_{A_j}^{(\vec{\xi})} q\| < 1$ for all $j$, then $\mathcal{F}$ is not diagonalized by $(\xi_n)$. In other words, the corresponding pure state is not diagonalized by $(\xi_n)$.*

*Proof* Assume $\mathcal{F}$ is diagonalized by $(\xi_n)$ and let $\mathcal{U}$ be such that $\mathcal{F} = \varphi_{\mathcal{U}}^{(\vec{\xi})}$. Then $A_j \in \mathcal{U}$ for some $j$, but $\|P_{A_j}^{(\vec{\xi})} q\| < 1$ for $q \in \mathcal{F}$, contradicting the assumption that $\mathcal{F}$ is a filter. □

**Lemma 6.44** *Let $(e_n)$ and $(\xi_n)$ be orthonormal bases. Then there is a partition of $\mathbb{N}$ into finite intervals $(J_n)$ such that for all $k$,*

$$\xi_k \in \overline{\operatorname{span}}\{e_i : i \in J_n \cup J_{n+1}\}$$

*(modulo a small perturbation of $\xi_k$) for some $n = n(k)$.*

*Proof* Omitted. □

For $(J_n)$ as in Lemma 6.44 let

$$D_{\vec{J}} = \{q : \|P_{J_n \cup J_{n+1}}^{(\vec{e})} q\| < 1/2 \text{ for all } n\}$$

**Lemma 6.45** *Each $D_{\vec{J}}$ is dense in $\mathcal{P}(C(H))$, in the sense that for any noncompact $p \in \mathcal{P}(\mathcal{B}(H))$, there is a noncompact $q \leq p$ such that $q \in D_{\vec{J}}$.*

*Proof* Taking a basis for range($p$), we can thin out the basis and take appropriate linear combinations to find such a $q$. □

Recall that $\mathfrak{d}$ is the minimal cardinality of a cofinal subset of $\mathbb{N}^{\mathbb{N}}$ under the pointwise order, and we write $\mathfrak{t}^*$ for the minimal length of a maximal decreasing well-ordered chain in $\mathcal{P}(C(H)) \setminus \{0\}$. In particular, CH (or MA) implies that $\mathfrak{d} = \mathfrak{t}^* = 2^{\aleph_0}$.

**Theorem 6.46** (Farah–Weaver) *Assume $\mathfrak{d} \leq \mathfrak{t}^*$.*[5] *Then there exists a pure state on $\mathcal{B}(H)$ that is not diagonalized by any atomic masa.*

[4] It can be proved, using a version of Kadison's Transitivity Theorem ([38]), that this intersection is actually a singleton.

[5] The sharpest hypothesis would be a non-commutative analogue of the inequality $\mathfrak{d} <$"the Novák number of $\mathcal{P}(C(H))$."

*Proof* We construct a corresponding maximal quantum filter. By the density of $D_{\vec{J}}$ and $\mathfrak{d} \leq \mathfrak{t}^*$, it is possible to construct a maximal quantum filter $\mathcal{F}$ such that $\mathcal{F} \cap D_{\vec{J}} \neq \emptyset$ for all $\vec{J}$. Given a basis $(\xi_k)$, pick $(J_n)$ such that $\xi_k \in J_{n(k)} \cup J_{n(k)+1}$ (modulo a small perturbation) for all $k$. Let $A_i = \{k \mid n(k) = i \mod 4\}$ for $0 \leq i < 4$. Then if $q \in \mathcal{F} \cap D_{\vec{J}}$, $\|P^{(\vec{\xi})}_{A_i} q\| < 1$ for each $i$. By Lemma 6.43, $\mathcal{F}$ is not diagonalized by $(\xi_n)$. □

The theory of quantum filters was refined by Bice ([12]).

## References


[1] *AimPL: Set theory and C* algebras, available at http://aimpl.org/settheorycstar*, 2012, American Institute of Mathematics.

[2] C. Akemann and N. Weaver, *Consistency of a counterexample to Naimark's problem*, Proc. Natl. Acad. Sci. USA **101** (2004), no. 20, 7522–7525.

[3] ———, *$\mathcal{B}(H)$ has a pure state that is not multiplicative on any MASA*, Proc. Natl. Acad. Sci. USA **105** (2008), no. 14, 5313–5314.

[4] J. Anderson, *A maximal abelian subalgebra of the Calkin algebra with the extension property*, Math. Scand. **42** (1978), no. 1, 101–110.

[5] ———, *Extensions, restrictions, and representations of states on $C^*$-algebras*, Trans. Amer. Math. Soc. **249** (1979), no. 2, 303–329.

[6] ———, *Extreme points in sets of positive linear maps on $\mathcal{B}(\mathcal{H})$*, J. Funct. Anal. **31** (1979), no. 2, 195–217.

[7] ———, *A conjecture concerning the pure states of $B(H)$ and a related theorem*, Topics in modern operator theory (Timişoara/Herculane, 1980), Operator Theory: Adv. Appl., vol. 2, Birkhäuser, Basel, 1981, pp. 27–43. MR 672813 (83k:46050)

[8] William Arveson, *An invitation to $C^*$-algebras*, Springer-Verlag, New York, 1976, Graduate Texts in Mathematics, No. 39.

[9] ———, *A short course on spectral theory*, Graduate Texts in Mathematics, vol. 209, Springer-Verlag, New York, 2002.

[10] I. Ben Yaacov, A. Berenstein, C.W. Henson, and A. Usvyatsov, *Model theory for metric structures*, Model Theory with Applications to Algebra and Analysis, Vol. II (Z. Chatzidakis et al., eds.), London Math. Soc. Lecture Notes Series, no. 350, Cambridge University Press, 2008, pp. 315–427.

[11] T. Bice, *Set theory of Hilbert spaces and Hilbert space operators*, Master's thesis, Kobe University, February 2009.

[12] ———, *Filters in C*-algebras*, Canadian J. Math. (to appear).

[13] B. Blackadar, *Operator algebras*, Encyclopaedia of Mathematical Sciences, vol. 122, Springer-Verlag, Berlin, 2006, Theory of $C^*$-algebras and von Neumann algebras, Operator Algebras and Non-commutative Geometry, III.

[14] A. Blass, *Combinatorial cardinal characteristics of the continuum*, Handbook of set theory. Vols. 1, 2, 3, Springer, Dordrecht, 2010, pp. 395–489. MR 2768685

[15] L.G. Brown, R.G. Douglas, and P.A. Fillmore, *Extensions of C*-algebras and K-homology*, Annals of Math. **105** (1977), 265–324.

[16] P.G. Casazza and J.C. Tremain, *The Kadison-Singer problem in mathematics and engineering*, Proc. Natl. Acad. Sci. USA **103** (2006), no. 7, 2032–2039.

[17] John B. Conway, *A course in functional analysis*, second ed., Graduate Texts in Mathematics, vol. 96, Springer-Verlag, New York, 1990.

[18] M. J. Crabb, *A new prime $C^*$-algebra that is not primitive*, J. Funct. Anal. **236** (2006), no. 2, 630–633.

[19] Kenneth R. Davidson, *$C^*$-algebras by example*, Fields Institute Monographs, vol. 6, American Mathematical Society, Providence, RI, 1996.

[20] G. A. Elliott and A. S. Toms, *Regularity properties in the classification program for separable amenable $C^*$-algebras*, Bull. Amer. Math. Soc. (N.S.) **45** (2008), no. 2, 229–245.

[21] I. Farah, *How many Boolean algebras $\mathcal{P}(\mathbb{N})/I$ are there?*, Illinois J. Math. **46** (2003), 999–1033.

[22] ———, *A twist of projections in the Calkin algebra*, preprint, available at http://www.math.yorku.ca/~ifarah, 2006.

[23] ———, *The relative commutant of separable C*-algebras of real rank zero*, Jour. Funct. Analysis **256** (2009), 3841–3846.

[24] ———, *Graphs and CCR algebras*, Indiana Univ. Math. Journal **59** (2010), 10411056.

[25] ———, *All automorphisms of all Calkin algebras*, Math. Research Letters **18** (2011), 489–503.

[26] ———, *All automorphisms of the Calkin algebra are inner*, Annals of Mathematics **173** (2011), 619–661.

[27] ———, *A dichotomy for the Mackey Borel structure*, Proceedings of the 11th Asian Logic Conference In Honor of Professor Chong Chitat on His 60th Birthday (T. Arai et al., eds.), World Scientific, 2011, pp. 86–93.

[28] I. Farah and B. Hart, *Countable saturation of corona algebras*, preprint, arXiv:1112.3898v1, 2011.

[29] I. Farah, B. Hart, and D. Sherman, *Model theory of operator algebras II: Model theory*, preprint, arXiv:1004.0741, 2010.

[30] ———, *Model theory of operator algebras I: Stability*, Bull. London Math. Soc. (to appear), http://arxiv.org/abs/0908.2790.

[31] I. Farah and T. Katsura, *Nonseparable UHF algebras I: Dixmier's problem*, Adv. Math. **225** (2010), no. 3, 1399–1430.

[32] I. Farah, N.C. Phillips, and J. Steprāns, *The commutant of $L(H)$ in its ultrapower may or may not be trivial*, Math. Annalen **347** (2010), 839–857.

[33] I. Farah, A. Toms, and A. Törnquist, *The descriptive set theory of C*-algebra invariants*, Appendix with Caleb Eckhardt. Preprint, arXiv:1112.3576.

[34] ———, *Turbulence, orbit equivalence, and the classification of nuclear $C^*$-algebras*, J. Reine Angew. Math. (to appear).

[35] L. Ge and D. Hadwin, *Ultraproducts of $C^*$-algebras*, Recent advances in operator theory and related topics (Szeged, 1999), Oper. Theory Adv. Appl., vol. 127, Birkhäuser, Basel, 2001, pp. 305–326.

[36] S. Ghasemi, **-homomorphims from SAW*-algebras into tensor product of infinite-dimensional C*-algebras*, preprint, York University, 2012.

[37] J. Glimm, *Type I $C^*$-algebras*, Ann. of Math. (2) **73** (1961), 572–612.

[38] J. G. Glimm and R. V. Kadison, *Unitary operators in $C^*$-algebras*, Pacific J. Math. **10** (1960), 547–556.

[39] J.G. Glimm, *On a certain class of operator algebras*, Trans. Amer. Math. Soc. **95** (1960), 318–340.

[40] D. Hadwin, *Maximal nests in the Calkin algebra*, Proc. Amer. Math. Soc. **126** (1998), 1109–1113.

[41] L.A. Harrington, A.S. Kechris, and A. Louveau, *A Glimm–Effros dichotomy for Borel equivalence relations*, Journal of the Amer. Math. Soc. **4** (1990), 903–927.

[42] B. E. Johnson and S. K. Parrott, *Operators commuting with a von Neumann algebra modulo the set of compact operators*, J. Functional Analysis **11** (1972), 39–61.

[43] V.F.R. Jones, *Von Neumann algebras*, 2003, lecture notes, http://math.berkeley.edu/~vfr/VonNeumann.pdf.

[44] R.V. Kadison and I.M. Singer, *Extensions of pure states*, Amer. J. Math. **81** (1959), 383–400.

[45] T. Katsura, *Non-separable AF-algebras*, Operator Algebras: The Abel Symposium 2004, Abel Symp., vol. 1, Springer, Berlin, 2006, pp. 165–173.

[46] D. Kerr, H. Li, and M. Pichot, *Turbulence, representations, and trace-preserving actions*, Proc. London Math. Soc. (3) **100** (2010), no. 2, 459–484.

[47] K. Kunen, *An introduction to independence proofs*, North–Holland, 1980.

[48] Н. Лузин, *О частях натурального ряда*, Изв. АН СССР, серия мат. **11,** №5 (1947), 714–722.

[49] E. Odell and Th. Schlumprecht, *The distortion problem*, Acta Mathematica **173** (1994), 259–281.

[50] G.K. Pedersen, *$C^*$-algebras and their automorphism groups*, London Mathematical Society Monographs, vol. 14, Academic Press Inc. [Harcourt Brace Jovanovich Publishers], London, 1979.

[51] ———, *Analysis now*, Graduate Texts in Mathematics, vol. 118, Springer-Verlag, New York, 1989.

[52] N.C. Phillips and N. Weaver, *The Calkin algebra has outer automorphisms*, Duke Math. Journal **139** (2007), 185–202.

[53] G. A. Reid, *On the Calkin representations*, Proc. London Math. Soc. (3) **23** (1971), 547–564.

[54] M. Rørdam, *Classification of nuclear $C^*$-algebras.*, Encyclopaedia of Mathematical Sciences, vol. 126, Springer-Verlag, Berlin, 2002, Operator Algebras and Non-commutative Geometry, 7.

[55] Shôichirô Sakai, *$C^*$-algebras and $W^*$-algebras*, Classics in Mathematics, Springer-Verlag, Berlin, 1998, Reprint of the 1971 edition.

[56] R. Sasyk and A. Törnquist, *Borel reducibility and classification of von Neumann algebras*, Bull. Symbolic Logic **15** (2009), no. 2, 169–183.

[57] ———, *The classification problem for von Neumann factors*, Journal of Functional Analysis **256** (2009), 2710–2724.

[58] ———, *Turbulence and Araki-Woods factors*, J. Funct. Anal. **259** (2010), no. 9, 2238–2252. MR 2674113

[59] S. Shelah and J. Steprāns, *Non-trivial homeomorphisms of $\beta N \setminus N$ without the continuum hypothesis*, Fundamenta Mathematicae **132** (1989), 135–141.

[60] ———, *Masas in the Calkin algebra without the continuum hypothesis*, Preprint, 2009.

[61] N. Weaver, *Mathematical quantization*, Studies in Advanced Mathematics, Chapman & Hall/CRC, Boca Raton, FL, 2001.

[62] ———, *A prime $C^*$-algebra that is not primitive*, J. Funct. Anal. **203** (2003), no. 2, 356–361.

[63] ———, *The Kadison-Singer problem in discrepancy theory*, Discrete Math. **278** (2004), no. 1-3, 227–239.

[64] ———, *Set theory and $C^*$-algebras*, Bull. Symb. Logic **13** (2007), 1–20.

[65] E. Wofsey, $P(\omega)/\text{fin}$ *and projections in the Calkin algebra*, Proc. Amer. Math. Soc. **136** (2008), no. 2, 719–726.

[66] B. Zamora Aviles, *The structure of order ideals and gaps in the Calkin algebra*, Ph.D. thesis, York University, 2009.

[67] R. J. Zimmer, *Essential results of functional analysis*, Chicago Lectures in Mathematics, University of Chicago Press, Chicago, IL, 1990.

# 4

# Set Mapping Reflection

Justin Moore[a] and David Milovich

---

The sixth Appalachian Set Theory workshop was held at Pennsylvania State University in State College on May 31, 2008. The lecturer was Justin Moore. As a graduate student David Milovich assisted in writing this chapter, which is based on the workshop lectures.

---

## 1 Introduction

The goal of these lectures is to give an exposition of the concept of an *open stationary set*, an associated reflection principle (for lack of a better word), and a list of examples of how this sort of consideration arises naturally in the context of modern set theory. We will begin with a list of seemingly unrelated questions.

*Question* 1.1 Does PFA imply there is a well ordering of $\mathcal{P}(\omega_1)$ which is definable over $\langle H(\aleph_2), \in\rangle$ (with parameters)?

*Question* 1.2 Is it consistent that every Aronszajn line contains a Countryman suborder?

*Question* 1.3 Is it consistent that for all $c\colon [\omega_1]^2 \to 2$ there exist $A, B \in [\omega_1]^{\omega_1}$ such that $c$ is constant on $\{\{\alpha,\beta\} : \alpha < \beta \wedge \alpha \in A \wedge \beta \in B\}$?

Let us focus on the second question for a moment. Consider the following analogy. Recall that a forcing $Q$ satisfies the *countable chain*

[a] The research of Justin Moore was supported in part by NSF grant DMS-0757507. Any opinions, findings, and conclusions or recommendations expressed in this material are those of the authors and do not necessarily reflect the views of the National Science Foundation.

*condition (c.c.c.)* if every uncountable collection of conditions in $Q$ contains two compatible conditions. Similarly, $Q$ satisfies *Knaster's Condition (Property K)* if every uncountable collection of conditions contains an uncountable subcollection of pairwise compatible conditions. It is easily verified that the product of a c.c.c. forcing and one with Property K is c.c.c.. A consequence of this is that a Property K forcing cannot destroy a counterexample to Souslin's Hypothesis. Hence while the forcing axiom for c.c.c. forcings (a.k.a. $\mathrm{MA}_{\aleph_1}$) does imply Souslin's Hypothesis, the forcing axiom for Property K forcings is consistent with the failure of Souslin's Hypothesis.

What if the common and widely successful methods for building proper forcings inadvertently satisfied a stronger form of properness and that counterexamples to Question 1.2 were preserved by this stronger condition?

It turns out that this is indeed the case and we will now formulate a combinatorial obstruction of this sort. A $\mho$-*sequence* is a sequence $\langle f_\alpha : \alpha < \omega_1 \rangle$ of continuous functions $f_\alpha : \alpha \to \omega$ such that if $E \subseteq \omega_1$ is closed and unbounded, there is a $\delta$ such that $f_\delta$ takes all values in $\omega$ on $E \cap \delta$. Notice that if $f : \delta \to \omega$ is continuous and $\delta$ is a limit ordinal, then there is a cofinal $C \subseteq \delta$ of ordertype $\omega$ such that $f(\xi)$ depends only on $|C \cap \xi|$. That is $f$ is obtained by coloring the intervals in $\delta$ between points of $C$. Jensen's principle $\diamondsuit$ easily implies the existence of a $\mho$-sequence. Since only the club filter is quantified over in the definition of a $\mho$-sequence, $\mho$-sequences are preserved by c.c.c. forcing. This is because if $E$ is a club in a c.c.c. forcing extension, $E$ contains a club from the ground model (this appears as an exercise in [9]). In fact a much broader class of proper forcings preserve $\mho$-sequences; this will be discussed more later. In [12] it was shown that the existence of a $\mho$-sequence implies the existence of an Aronszajn line with no Countryman suborder.

While Question 1.3 has a negative answer [16], the construction in [12] served as a precursor to the ZFC construction of a coloring $c$ as in Question 1.3 (even though [12] was published considerably after [16]). We will see that a positive answer to Question 1.1 holds and that this is related to the existence of a weak analogue of a $\mho$-sequence which exists on $[\omega_2]^\omega$.

The focus of this note will be to examine a principle, MRP, which provides a general framework for eliminating combinatorial obstructions such as $\mho$-sequences and for tapping into additional strength of the Proper Forcing Axiom (PFA). After defining the principle, we will present a number of case studies of how this principle is applied.

The reader is assumed to have familiarity with set theory at the level of Kunen's [9]. Additional background can be found in [7]. In order to make the discussion of consistency less cumbersome, we will generally assume unless otherwise stated that the existence of a supercompact cardinal is consistent.

## 2 The club filter and stationary sets

Central to our discussion will be the "club filter" of countable sets on a given uncountable set $X$. Henceforth, our convention is that $X$ is an uncountable set, $\theta$ is an uncountable regular cardinal, and $[X]^\omega = \{A \subseteq X : |A| = \aleph_0\}$.

**Definition 2.1** The *Ellentuck topology* on $[X]^\omega$ is generated by the basic open sets

$$[a, N] = \{A \in [X]^\omega : a \subseteq A \subseteq N\}$$

where $a$ ranges over $[X]^{<\omega}$ and $N$ ranges over $[X]^\omega$.

It is not difficult to show that in fact the basic open sets in this topology are closed as well and hence that the topology is regular and Hausdorff.

**Definition 2.2** A *club*[1] in $[X]^\omega$ is a subset that is Ellentuck closed and cofinal in $([X]^\omega, \subseteq)$.

Observe that if $X = \omega_1$, then $\omega_1$ is club when viewed as a subset of $[\omega_1]^\omega$. Hence every closed unbounded subset of $\omega_1$ is club when viewed as a subset of $[\omega_1]^\omega$ and if $E \subseteq [\omega_1]^\omega$ is club, then $E \cap \omega_1$ is closed and unbounded.

The two other competing definitions of "club" which occur in the literature are (i) sets of the form $E_f$ and (ii) subsets $E$ of $[X]^\omega$ which are cofinal and closed under unions of countable chains. The following facts show that this is an intermediate notion. In particular, the definition of *stationary* does not depend on which definition is used.

**Definition 2.3** $S \subseteq [X]^\omega$ is *stationary* if $S \cap E \neq \emptyset$ for every club $E$.

[1] Note that "club" is a misnomer since it suggests the meaning of being "closed and unbounded." In fact it means closed and cofinal.

*Fact* 2.4 If $f \colon X^{<\omega} \to X$ and $E_f = \{M \in [X]^\omega : f``M^{<\omega} \subseteq M\}$, then $E_f$ is club. Moreover if $E$ is club, then there is a $f \colon X^{<\omega} \to X$ such that $E_f \subseteq E$.

If $f$ is as in the above definition and $f''M^{<\omega} \subseteq M$, then we say that $M$ is *closed under* $f$.

*Fact* 2.5 If $E \subseteq [X]^\omega$ is club and $\mathcal{N} \subseteq E$ is countable and linearly ordered by $\subseteq$, then $\bigcup\mathcal{N}$ is in $E$.

The next fact states the quintessential properties of clubs and stationary sets.

*Fact* 2.6 A countable intersection of clubs is a club. Equivalently, a partition of a stationary set into countably many pieces has a stationary piece.

## 3 Elementary submodels

Unless otherwise specified, $\theta$ will denote a regular uncountable cardinal.

**Definition 3.1** $H(\theta)$ is the collection of all sets of hereditary cardinality less than $\theta$. We will identify $H(\theta)$ with the structure $(H(\theta), \in)$.

The following observations are useful. Some require proof (which we leave to the reader).

1. $H(\theta)$ is a set (not a proper class) of cardinality $2^{<\theta}$.
2. $\langle H(\theta), \in\rangle$ satisfies ZFC except possibly the power set axiom.
3. $\mathrm{Ord}^{H(\theta)} = \theta$.
4. If $A, B \in H(\theta)$, then $A \times B \in H(\theta)$.
5. If $A \in H(\theta)$, then $\mathcal{P}(A) \subseteq H(\theta)$. In particular, if $A, B \in H(\theta)$, then $|A| = |B|$ if and only if $H(\theta) \models |A| = |B|$.
6. $H(\theta^+)$ is an element of $H(2^{\theta^+})$.

**Definition 3.2** We say $M$ is a *countable elementary submodel* of $H(\theta)$ and write $M \prec H(\theta)$ if $M \in [H(\theta)]^\omega$ and, for every logical formula $\varphi$ with parameters in $M$, $M \models \varphi$ if and only if $H(\theta) \models \varphi$.

Note our convention that $M \prec H(\theta)$ always implies $|M| = \aleph_0$. This is not standard, but it will considerably simplify writing at times.

*Fact* 3.3 There is a function $f \colon H(\theta)^{<\omega} \to H(\theta)$ such that if $M \in [H(\theta)]^\omega$ and $M$ is closed under $f$, then $M \prec H(\theta)$.

*Fact* 3.4 If $M \prec H(\theta)$ and $X \in H(\theta)$ and $X$ is definable from parameters in $M$, then $X \in M$.

*Fact* 3.5 If $M \prec H(\theta)$, then $M \cap \omega_1$ is a countable ordinal that is not in $M$.

*Fact* 3.6 If $X \in H(\theta)$ is uncountable and $A \in [H(\theta)]^{\leq\omega}$, then $\{M \cap X : A \subseteq M \prec H(\theta)\}$ contains a club.

*Fact* 3.7 If $A \in M \prec H(\theta)$ and $A \not\subseteq M$, then $A$ is uncountable. Also, if $M \prec H(\theta)$, then, for all $A \in M \cap \mathcal{P}(\omega_1)$, $A$ is uncountable if and only if $A \cap M$ is unbounded in $\omega_1 \cap M$.

*Fact* 3.8 If $X, S \in M \prec H(\theta)$ and $S \subseteq [X]^\omega$ and $M \cap X \in S$, then $S$ is stationary.

*Fact* 3.9 $\{M : M \prec H(\theta^+)\}$ is in $H(2^{\theta^+})$ but is not definable in $H(\theta^+)$.

For more of the basics of Stationary sets, see Chapter 8 of Jech [7]. From this point on it will be convenient to adopt the convention that, unless otherwise stated, $X$ is an uncountable set which is an element of $H(\theta)$.

## 4 The strong reflection principle

Before formulating MRP, we will first define the simpler *Strong Reflection Principle (*SRP*)* of Todorcevic [1]. We will then recall some conclusions and arguments which will serve as a starting point for the development of MRP. The material in this section is based on [1].

Recall that if $M, N \prec H(\theta)$ and $\lambda$ is a cardinal, we say that $N$ *$\lambda$-end extends* $M$ if $M \cap \lambda = N \cap \lambda$ and $M \subseteq N$. We will only be interested in $\omega_1$-end extensions and will refer to them simply as *end extensions*. The following fact will be used frequently.

*Fact* 4.1 If $\langle N_\xi : \xi < \omega_1 \rangle$ is a continuous $\in$-increasing sequence of countable elementary submodels of some $H(\theta)$, and $\overline{N}$ is a countable elementary submodel of $H(\theta)$ with $\langle N_\xi : \xi < \omega_1 \rangle$ in $\overline{N}$, then $\overline{N}$ end extends $N_\delta$ where $\delta = \overline{N} \cap \omega_1$.

*Proof* By Fact 3.7, $N_\xi$ is a subset of $\overline{N}$ for all $\xi < \delta$. By continuity of $\langle N_\xi : \xi < \omega_1 \rangle$, $N_\delta \subseteq \overline{N}$. Also, by continuity of $\langle N_\xi : \xi < \omega_1 \rangle$, the map $\xi \mapsto N_\xi \cap \omega_1$ is continuous. It follows from Fact 2.4 that $E = \{\xi < \omega_1 : N_\xi \cap \omega_1 = \xi\}$ is a club. Since this club is in $\overline{N}$, it contains $\delta$ as an element by Fact 3.8 and hence $N_\delta \cap \omega_1 = \delta = \overline{N} \cap \omega_1$. □

SRP asserts that if $X \in H(\theta)$ with $X$ uncountable and $S \subseteq [X]^\omega$, then there exists a continuous $\in$-chain $\langle N_\xi : \xi < \omega_1\rangle$ of elementary submodels of $H(\theta)$ such that, for all $\xi < \omega_1$, $X \in N_\xi$ and we have $N_\xi \cap X \in S$ if and only if there exists an end extension $M$ of $N_\xi$ such that $M \cap X \in S$. We say such an $\langle N_\xi : \xi < \omega_1\rangle$ is a *strong reflecting sequence* of $S$. The power of the continuity assumption lies in the ability to generate end extensions via Fact 4.1.

Recall that if $S \subseteq [X]^\omega$ is stationary, then we say $S$ *reflects* if there is a continuous $\in$-chain $\langle N_\xi : \xi < \omega_1\rangle$ of countable elementary submodels of $H(\theta)$ such that $\{\xi < \omega_1 : N_\xi \cap X \in S\}$ is stationary. The following proposition justifies the "strong" in Strong Reflection Principle.

**Proposition 4.2** *If $S \subseteq [X]^\omega$ is stationary and $\langle N_\xi : \xi < \omega_1\rangle$ strongly reflects $S$, then $\Xi = \{\xi < \omega_1 : N_\xi \cap X \in S\}$ is stationary.*

*Proof* Suppose not and let $E \subseteq \omega_1$ be a club disjoint from $\Xi$. Choose $M \prec H(\theta)$ such that $E, \langle N_\xi : \xi < \omega_1\rangle \in M$ and $M \cap X \in S$. By Fact 4.1, $M$ is an end extension of $N_\delta$ where $\delta = M \cap \omega_1$. Notice also that $\delta$ is in $E$ by Fact 3.8 and hence $N_\delta$ is not in $S$. But this is a contradiction to our assumption that $\langle N_\xi : \xi < \omega_1\rangle$ is a strong reflecting sequence for $S$. □

**Proposition 4.3** SRP *implies that if $S_\xi$ ($\xi < \omega_2$) are stationary subsets of $\omega_1$, then there are $\xi < \eta$ such that $S_\xi \cap S_\eta$ is stationary.*

*Proof* Let $\langle S_\xi : \xi < \omega_2\rangle$ be given. Define $\Gamma \subseteq [\omega_2]^\omega$ to be the collection of all $P$ such that $P \cap \omega_1$ is an ordinal and there is an $\alpha$ in $P$ such that $P \cap \omega_1$ is in $S_\alpha$.

Applying SRP, there is a continuous chain $N_\xi$ ($\xi < \omega_1$) of countable elementary submodels of $H(\aleph_3)$, each containing $\langle S_\xi : \xi < \omega_2\rangle$ as a member, and such that for all $\xi < \omega_1$, if $N_\xi$ has an end extension $\bar{N}$ with $\overline{N} \cap \omega_2$ in $\Gamma$, then $N_\xi \cap \omega_2$ is in $\Gamma$. Since $\bigcup_{\xi<\omega_1} N_\xi$ has cardinality $\omega_1$, it suffices to show that if $\beta < \omega_2$, then there is an $\alpha$ in some $N_\xi \cap \omega_2$ such that $S_\alpha \cap S_\beta$ is stationary. To this end, let $\beta$ be given and let $\overline{N}$ be a countable elementary submodel of $H(\aleph_3)$ such that $\langle N_\xi : \xi < \omega_1\rangle$ and $\beta$ are in $\overline{N}$ and $\delta = \overline{N} \cap \omega_1$ is in $S_\beta$. By Fact 4.1, $\overline{N}$ is an end extension of $N_\delta$ which is moreover in $\Gamma$. By assumption, $N_\delta$ is in $\Gamma$ and therefore there is an $\alpha$ in $N_\delta$ such that $\delta$ is in $S_\alpha$. Finally, by Fact 3.8, $S_\alpha \cap S_\beta$ is stationary since it contains $\delta = \overline{N} \cap \omega_1$. □

## 5 The Set Mapping Reflection Principle

Now we will turn to the Set Mapping Reflection Principle (MRP).

**Definition 5.1** Let $X, \theta$ be fixed. Suppose $\Sigma$ is a map such that $\mathrm{dom}(\Sigma)$ is a club subset of $\{M : M \prec H(\theta)\}$, and $\Sigma(M) \subseteq [X]^\omega$ for all $M$. We say $\Sigma$ is an *open set mapping* if $\Sigma(M)$ is open (in the Ellentuck topology) for all $M$.

Typically, $\Sigma(M)$ will actually be a subset of $[M \cap X]^\omega$.

**Definition 5.2** We say $S \subseteq [X]^\omega$ is *$M$-stationary* if, for all club $E \in M$, $S \cap M \cap E \neq \emptyset$. A set mapping $\Sigma$ is *open stationary* if $\Sigma(M)$ is open and $M$-stationary for all $M$.

Notice that open subsets of $[X]^\omega$ which are stationary are trivial in the sense that their complements are closed and not cofinal in $\langle [X]^\omega, \subseteq \rangle$. But it is not difficult to show that there are, for a given $M$, $\Sigma_0, \Sigma_1 \subseteq [M \cap X]^\omega$ which have empty intersection and which are each open and $M$-stationary.

**Example 5.3** Let $X = \omega_1$. For each $M \prec H(\theta)$, choose $\alpha < M \cap \omega_1$ and set $\Sigma(M) = \{\gamma \in [\omega_1]^\omega : \alpha \in \gamma \subseteq M \cap \omega_1\}$. Then $\Sigma$ is trivially open stationary.

**Definition 5.4** We say a sequence of sets indexed by ordinals is a *continuous $\in$-chain* if it is $\subseteq$-continuous and $\in$-increasing.

**Definition 5.5** An open stationary set mapping $\Sigma$ *reflects* if there exists a continuous $\in$-chain $\langle N_\xi : \xi < \omega_1 \rangle$ such that, for all $\nu < \omega_1$, $N_\nu \in \mathrm{dom}(\Sigma)$ and there exists $\nu_0 < \nu$ such that $N_\xi \cap X \in \Sigma(N_\nu)$ for all $\xi$ satisfying $\nu_0 < \xi < \nu$. We say that such an $\langle N_\xi : \xi < \omega_1 \rangle$ is a *reflecting sequence* for $\Sigma$.

**Definition 5.6** The *Set Mapping Reflection Principle (*MRP*)* is the assertion that all open stationary set mappings reflect.

One can view MRP as asserting that every open stationary $\Sigma$ contains a copy of Example 5.3.

**Theorem 5.7** MRP *implies the existence of a well ordering of $\mathcal{P}(\omega_1)/NS$ which is definable over $(H(\aleph_2), \in)$ with parameters.*

**Definition 5.8** Given $A, B \subseteq \omega_1$, define $A \equiv_{\mathrm{NS}} B$ to mean $A \triangle B$ is non stationary.

The following fact follows easily from the existence of a partition of $\omega_1$ into $\omega_1$ pairwise disjoint stationary sets.

*Fact* 5.9 There exist $2^{\aleph_1}$-many $\equiv_{\mathrm{NS}}$-equivalence classes.

*Proof* (of Theorem 5.7) Fix $\langle C_\delta : \delta \in \mathrm{Lim}(\omega_1)\rangle$ such that, for all $\delta$, $\mathrm{otp}(C_\delta) = \omega$ and $C_\delta$ is cofinal in $\delta$. If $A \subseteq B \in [\mathrm{Ord}]^{\leq\omega}$, $\sup A < \sup B$, and $B$ has no maximum, then set $w(A, B) = |\pi^{-1}(C_\delta) \cap \sup A|$ where $\delta = \mathrm{otp}(B)$ and $\pi$ is the unique order isomorphism from $B$ to $\delta$. Note that $w(A, B)$ is necessarily finite.

Set $X = \omega_2$ and $\theta = \left(2^{2^{\aleph_1}}\right)^+$. Given $M \prec H(\theta)$, define $\Sigma_<(M)$ to be the set of $A \in [\omega_2]^\omega$ for which the following conditions hold:

$$\sup(A \cap \omega_1) < \sup(M \cap \omega_1),$$

$$\sup A < \sup(M \cap \omega_2),$$

$$w(A \cap \omega_1, M \cap \omega_1) < w(A, M \cap \omega_2).$$

Analogously define $\Sigma_\geq(M)$ with $\geq$ replacing $<$ in the last inequality. Intuitively, $\Sigma_<(M)$ consists of those countable subsets of $M \cap \omega_2$ whose intersection with $\omega_2$ is "higher" in $M \cap \omega_2$ than its intersection with $\omega_1$ is in $M \cap \omega_1$.

Observe that $\Sigma_<(M)$ and $\Sigma_\geq(M)$ are open. To see this, suppose that $A \subseteq M \cap \omega_2$ with

$$\sup(A \cap \omega_1) < \sup(M \cap \omega_1),$$

$$\sup A < \sup(M \cap \omega_2),$$

and let $\alpha$ and $\beta$ be the least elements of $A \cap \omega_1$ and $A$, respectively, such that

$$w(\alpha, M \cap \omega_1) = w(A \cap \omega_1, M \cap \omega_1)$$

$$w(A \cap \beta, M \cap \omega_2) = w(A, M \cap \omega_2)$$

If $A$ is in $\Sigma_<(M)$, then $[\{\alpha, \beta\}, A] \subseteq \Sigma_<(M)$ and similarly for $\Sigma_\geq(M)$.

*Claim* 5.10 $\Sigma_<(M)$ is $M$-stationary.

*Proof* Let $E \in M$ be club subset of $[\omega_2]^\omega$. Observe that there exists $\alpha < \omega_1$ such that $\{\sup A : A \in E \wedge A \cap \omega_1 = \alpha\}$ is cofinal in $\omega_2$. Let $\alpha \in M$ be as above. Let $n = w(\alpha, M \cap \omega_1)$. Let $\beta \in M \cap \omega_2$ satisfy $w(\beta \cap M, \omega_2 \cap M) > n$. Pick $A \in E$ such that $A \cap \omega_1 = \alpha$ and $\sup A > \beta$. Since $E \in M$, we may assume $A \in M$. Thus, $A \in \Sigma_<(M) \cap M \cap E$. □

*Claim* 5.11 $\Sigma_{\geq}(M)$ is $M$-stationary.

*Proof* Let $E \in M$ be club subset of $[\omega_2]^{\omega}$. Let $N \in M$ satisfy $N$ be an elementary submodels of $H(2^{\aleph_1^+})$ with $|N| = \aleph_1$, and $\{E\} \cup \omega_1 \subseteq N$. Observe that by elementarity of $M$, $\sup(N \cap \omega_2)$ is an element of $M$ and hence $\sup(N \cap M \cap \omega_2) < \sup(M \cap \omega_2)$. Set

$$n = w(N \cap M \cap \omega_2, M \cap \omega_2)$$

and $E_0 = E \cap N$. Then $w(A \cap \omega_2, M \cap \omega_2) \leq n$ for all $A \in E_0 \cap M$. Moreover, $E_0$ is club in $[\omega_2 \cap N]^{\omega}$; hence, $\{\sup(A \cap \omega_1) : A \in E_0 \cap M\}$ is unbounded in $M \cap \omega_1$. Hence, there exists $A \in E_0 \cap M$ such that $w(A \cap \omega_1, M \cap \omega_1) \geq n$. Hence, $\Sigma_{\geq}(M) \cap M \cap E \neq \emptyset$. □

For each $A \subseteq \omega_1$, let $\Sigma_A(M) = \Sigma_{<}(M)$ if $M \cap \omega_1 \in A$ and $\Sigma_A(M) = \Sigma_{\geq}(M)$ if $M \cap \omega_1 \notin A$. Let $\langle N_\xi : \xi < \omega_1 \rangle$ reflect $\Sigma_A$. Set $\delta = \bigcup_{\xi<\omega_1} N_\xi \cap \omega_2$. Since $\langle N_\xi : \xi < \omega_1 \rangle$ is a continuous $\in$-chain of elementary submodels of $H(\theta)$, we have $\omega_1 \subseteq \bigcup_{\xi<\omega_1} N_\xi \prec H(\theta)$ and $\{N_\xi \cap \omega_2 : \xi < \omega_1\}$ a club subset of $[\delta]^{\omega}$. Hence, $\delta$ is an ordinal such that $\mathrm{cf}(\delta) = \omega_1 < \delta < \omega_2$. Moreover, $\delta$ satisfies the following property $\phi(A, \delta)$:

There is a club $\mathcal{M} \subseteq [\delta]^{\omega}$ which is well ordered by $\subseteq$ and is such that for all limit $\nu < \omega_1$ there is a $\nu_0 < \nu$ with $\nu$ is in $A$ if and only if

$$w(M_\xi \cap \omega_1, M_\nu \cap \omega_1) < w(M_\xi, M_\nu)$$

whenever $\nu_0 < \xi < \nu$.

Here $M_\xi$ is the $\xi^{\text{th}}$ element of $\mathcal{M}$ in its $\subseteq$-increasing enumeration. If we let $\delta_A$ be the least ordinal such that $\phi(A, \delta)$ holds, then $A \mapsto \delta_A$ is definable over $H(\aleph_2)$ with parameter $\langle C_\nu : \nu \in \mathrm{Lim}(\omega_1) \rangle$. Hence it suffices to prove the following claim.

*Claim* 5.12 If $A$ and $B$ are subsets of $\omega_1$ and $\phi(A, \delta) \wedge \phi(B, \delta)$ holds for some $\delta$, then $A \equiv_{\mathrm{NS}} B$.

*Proof* First observe that if $\mathcal{N}$ is a club witnessing $\phi(A, \delta)$ and $\mathcal{N}' \subseteq \mathcal{N}$ is also club, then $\mathcal{N}'$ also witnesses $\phi(A, \delta)$. Hence if $\phi(A, \delta) \wedge \phi(B, \delta)$, there is a single club $\langle N_\xi : \xi < \omega_1 \rangle$ in $[\delta]^{\omega}$ which witnesses both $\phi(A, \delta)$ and $\phi(B, \delta)$. Let $E = \{N_\xi \cap \omega_1 : \xi < \omega_1\}$. It suffices to show that no limit point of $E$ is in $A \triangle B$. To see this, let $\nu$ be a limit ordinal. $N_\nu \cap \omega_1$ is in $A$ iff there are arbitrarily large $\xi < \nu$ such that

$$w(N_\xi \cap \omega_1, N_\nu \cap \omega_1) < w(N_\xi \cap \omega_2, N_\nu \cap \omega_2)$$

iff $N_\nu \cap \omega_1$ is in $B$. □

This also completes the proof of the theorem. □

*Remark* 5.13 The main result of [3] shows that the coding in the proof of Theorem 5.7 above necessarily yields $2^{\aleph_0} = 2^{\aleph_1}$. Notice, however, that the forcing in Theorem 6.8 used to reflect an open stationary set mapping does not introduce new reals.

There is an analogous proof that SRP implies $2^{\aleph_1} = \aleph_2$ which can be described as follows. Suppose $S$ is a stationary co-stationary subset of $\omega_1$. For each $A \subseteq \omega_1$, set

$$\Gamma_A = \{X \in [\omega_2]^\omega : X \cap \omega_1 \in A \leftrightarrow \text{otp}(X) \in S\}.$$

Woodin's statement $\psi_{\text{AC}}$ is the assertion that for every $A \subseteq \omega_1$ and for every stationary co-stationary $S \subseteq \omega_1$, there is a $\delta < \omega_2$ of cofinality $\omega_1$ and a club $E$ in $[\delta]^\omega$ which is contained in $\Gamma_A$. Notice that for a given $\delta$ of cofinality $\omega_1$, if $\Gamma_A$ and $\Gamma_B$ both contains a club in $[\delta]^\omega$ for $A, B \subseteq \omega_1$, then $A$ and $B$ differ by a non stationary set. Hence $\psi_{\text{AC}}$ implies $2^{\aleph_1} = \aleph_2$.

## 6 PFA **implies** MRP

Let $Q$ be a forcing (*i.e.*, a poset with a maximum element). For us, $p \leq q$ means $p$ is stronger than $q$. The smallest $\theta$ for which "$G \subseteq Q$ is generic over $H(\theta)$" makes sense is $\theta = |2^Q|^+$, assuming the underlying set of $Q$ is $|Q|$.

**Definition 6.1** $Q$ is *proper* if, for all sets $X$, forcing with $Q$ preserves all stationary subsets of $[X]^\omega$.

The following characterization, which is due to Jech, provides the standard method for verifying a forcing $Q$ is proper. In order to state this characterization in a concise manner, it is helpful to make the following additional definition.

**Definition 6.2** Suppose that $Q$ is a forcing and $M \prec H(|2^Q|^+)$ with $Q$ in $M$. A condition $\bar{q}$ in $Q$ is *$(M, Q)$-generic* if whenever $D \subseteq Q$ is a dense open set in $M$ and $r \leq \bar{q}$, there is an element of $D \cap M$ compatible with $r$. Equivalently, $\bar{q}$ forces that $\dot{G} \cap M$ is generic over $M$ (here $\dot{G}$ is the $Q$-name for the generic filter).

**Proposition 6.3** *A forcing $Q$ is proper if and only if whenever $\mathcal{P}(Q) \in M \prec H(\theta)$ and $q \in Q \cap M$, there exists $\bar{q} \leq q$ which is* $(M, Q)$-generic.

*Remark* 6.4 The assumption that $\mathcal{P}(Q)$ is in $M$ is natural since then the collection of all dense open subsets of $Q$ is an element of $M$. The meaning of the statement "$\mathcal{P}(Q)$ in $M$" should be clear but is somewhat subtle: we want the powerset of $Q$'s underlying set (which we also denote by $Q$) to be in $M$, as well as the order on $Q$. In the above proposition we can also fix $\theta$ to be minimal with the property that $\mathcal{P}(Q)$ is in $H(\theta)$.

**Definition 6.5** The Proper Forcing Axiom (PFA) is the assertion that if $Q$ is proper and $D_\xi$ is predense in $Q$ for all $\xi < \omega_1$, then there exists $G \subseteq Q$ such that $G$ is a filter and $G \cap D_\xi \neq \emptyset$ for all $\xi < \omega_1$.

The following two classes of forcings and their iterations (which are also proper by a well known theorem of Shelah [17]) already are sufficient to yield many of the consequences of PFA (including the failure of $\square(\theta)$ and $2^{\aleph_0} = \aleph_2$). For more about proper forcing, see [6], [7], [17], [21].

**Example 6.6** Every c.c.c. forcing is proper. To see this, note that the definition of $(M, Q)$-generic remains unchanged if one replaces "dense open" with "maximal antichain." It then follows from Fact 3.7, that every condition in a c.c.c. forcing $Q$ is $(M, Q)$-generic for any relevant $M$.

**Example 6.7** Every $\sigma$-closed forcing is proper. To see this, start with $q_0 = q \in M$. Let $\{A_n\}_{n<\omega}$ enumerate all predense sets in $M$. Construct $\langle q_n : n < \omega\rangle$ decreasing in $Q \cap M$ such that $q_{n+1}$ is below an element of $A_n \cap M$. Any lower bound for $\{q_n\}_{n<\omega}$ is an $(M, Q)$-generic condition and such a lower bound exists by assumption that $Q$ is $\sigma$-closed.

The proof of the following theorem is a standard verification of properness.

**Theorem 6.8** PFA *implies* MRP.

*Proof* Let $\Sigma$ be an open stationary set mapping (with $X$ and $\theta$ as before). Let $Q$ be the set of all continuous $\in$-chains $\langle N_\xi : \xi \leq \alpha\rangle$ in $\mathrm{dom}(\Sigma)$ for which $\alpha < \omega_1$ and, for all limit $\nu \leq \alpha$, there exists $\nu_0 < \nu$ such that $N_\xi \cap X \in \Sigma(N_\nu)$ for all $\xi \in \nu \setminus \nu_0$. That is $Q$ consists of all countable partial reflecting sequences for $\Sigma$. Order $Q$ by extension. The following claim implies that it suffices to prove that $Q$ is proper.

*Claim* 6.9 Given that $Q$ is proper, $\{q \in Q : \alpha \in \mathrm{dom}(q)\}$ is dense for all $\alpha < \omega_1$.

*Proof* The set $\{q \in Q : x \in \bigcup \mathrm{ran}(q)\}$ is dense for all $x \in H(\theta)$; hence, $\mathbb{1} \Vdash \check{H}(\theta) = \bigcup_{q \in \dot{G}} \bigcup \mathrm{ran}(q)$. Since $Q$ is proper, it does not collapse $\omega_1$, so, since $H(\theta)$ is uncountable, $\mathbb{1} \Vdash \alpha \in \mathrm{dom}(\bigcup \dot{G})$ for all $\alpha < \omega_1$. □

Set $\lambda = 2^{<\theta}$ and let $\Sigma, Q \in M \prec H(2^{\lambda^+})$. Fix $q_0 \in Q \cap M$. Observe that if $\langle q_n : n < \omega \rangle$ is a decreasing sequence in $Q \cap M$ such that each $q_{n+1}$ is below some element of $D_n$, the nth dense subset of $Q$ in $M$, and $q_\omega = \bigcup_{n<\omega} q_n = \langle N_\xi : \xi < \alpha \rangle$, then $\bigcup_{\xi<\alpha} N_\xi = M \cap H(\theta)$ because, for all $x \in M \cap H(\theta)$, the set of $p \in Q$ such that $x \in \bigcup \mathrm{ran}(p)$ is dense in $Q$. So, to prove $Q$ is proper, it suffices to show that $N_\xi \cap X \in \Sigma(M \cap H(\theta))$ for all $\xi \in \alpha \setminus \mathrm{dom}(q_0)$, for then $q_\omega \cup \{\langle \alpha, M \cap H(\theta) \rangle\}$ will be an $(M, Q)$-generic element of $Q$ below $q_0$. Therefore, it suffices to show that, given $q_n \in Q \cap M$, there exists $q_{n+1} \in D_n \cap M$ such that $q_{n+1} \leq q_n$ and $q_{n+1}(\xi) \cap X \in \Sigma(M \cap H(\theta))$ for all $\xi \in \mathrm{dom}(q_{n+1}) \setminus \mathrm{dom}(q_n)$.

Using $M$-stationarity of $\Sigma(M \cap H(\theta))$, let $N \in M$ satisfy $D_n, q_n, Q, \Sigma \in N \prec H(\lambda^+)$ and $N \cap X \in \Sigma(M \cap H(\theta))$. Using openness of $\Sigma(M \cap H(\theta))$, let $a \in [N \cap X]^{<\omega}$ satisfy $[a, N \cap X] \subseteq \Sigma(M \cap H(\theta))$. Set $q = q_n \cup \{\langle \mathrm{dom}(q_n), P \rangle\}$ where $P \in N \cap \mathrm{dom}(\Sigma)$ and $q_n(\max(\mathrm{dom}(q_n))) \cup a \subseteq P$. Since $q, D_n, Q \in N$, there exists $q_{n+1} \in D_n \cap N$ such that $q_{n+1} \leq q$. Since $D_n, q, Q, N \in M$, we may assume $q_{n+1} \in M$. Since $q_{n+1} \in N$ and $a \subseteq P$, every $\xi \in \mathrm{dom}(q_{n+1}) \setminus \mathrm{dom}(q_n)$ satisfies $q_{n+1}(\xi) \in [a, N \cap X]$. □

Now consider the following strengthening of properness.

**Definition 6.10** A forcing $Q$ is $\omega$-proper if, given $q \in Q$ and an $\in$-chain $\langle N_i : i < \omega \rangle$ of elementary submodel of $H(|2^Q|^+)$ such that $q, Q \in N_0$, there exists $\bar{q} \leq q$ such that $\bar{q}$ is $(N_i, Q)$-generic for all $i < \omega$.

To see the relevance of $\omega$-proper, consider the club of those $M \prec H(\theta^+)$ (for some fixed $\theta$) which is a union of an increasing chain of $M_i \prec M$ of elements of $M$. If $N$ is an elementary submodel of $H(2^{\theta^+})$ and we define

$$\Sigma(N) = [H(\theta^+)]^\omega \setminus \{N_i : i < \omega\}$$

where $N_i$ is the increasing sequence chosen for $N \cap H(\theta^+)$, then any forcing which reflects $\Sigma$, can not be $\omega$-proper. This is so even if we weaken $\omega$-properness to *weak $\omega$-properness* where we require only the existence of a $\bar{q}$ which forces that

$$\{i < \omega : \dot{G} \cap N_i \text{ is } N_i\text{-generic}\}$$

is infinite (this definition is due to Eisworth and Nyikos [4]). This observation can be cast into a theorem (due to Shelah) as follows.

**Theorem 6.11** *(Weakly) $\omega$-proper forcings preserve (weak) club guessing sequences on $\omega_1$. Moreover weakly $\omega$-proper forcings preserve $\mho$-sequences.*

*Proof* We will only prove the theorem for club guessing sequences. Let $\langle C_\alpha : \alpha < \omega_1\rangle$ be club guessing. Let $Q$ be $\omega$-proper, $\dot{D}$ be a $Q$-name for a club subset of $\omega_1$ and $q$ be in $Q$. It suffices to find an extension of $q$ which forces that $\dot{D}$ contains some $C_\alpha$. Let $\langle M_\xi : \xi < \omega_1\rangle$ be a continuous $\in$-chain in $H(|2^Q|^+)$ such that $Q \in M_0$. Set

$$E = \{\xi < \omega_1 : \omega_1 \cap M_\xi = \xi\}.$$

Hence, $C_\delta \subseteq E$ for some $\delta < \omega_1$. Let $N_i = M_{\xi_i}$ where $\langle \xi_i : i < \omega\rangle$ is an increasing enumeration of $C_\delta$. Then every $\bar{q}$ that is $(N_i, Q)$-generic for $i < \omega$ forces $C_\delta \subseteq \dot{D}$ as desired. □

It is worth noting that the $\epsilon$-collapse forcing of Baumgartner (which forms the cornerstone of the "models as side conditions" method due to Todorcevic [19], [21]), can also be shown to be weakly $\omega$-proper. Very few applications of PFA prior to [14] required more than the forcing axiom for weakly $\omega$-proper forcings.

## 7 Influence of MRP on the club filter

In this section we will consider how the assumption MRP influences the combinatorics of the club filter. The first is hardly more than an observation.

**Definition 7.1** If $X$ and $Y$ are countable subsets of $\omega_1$ which are closed in their supremum, then we say $X$ *measures* $Y$ if there is a $\xi < \sup X$ such that $X \setminus \xi$ is either contained in or disjoint from $Y$.

We define *measuring* to be the assertion that for every sequence $\langle D_\alpha : \alpha < \omega_1\rangle$ with $D_\alpha \subseteq \alpha$ closed for all $\alpha < \omega_1$, there is a club $E \subseteq \omega_1$ such that $E \cap \alpha$ measures $D_\alpha$ whenever $\alpha$ is a limit point of $E$. Notice that measuring implies the non existence of $\mho$-sequences: if $\langle f_\alpha : \alpha < \omega_1\rangle$ is a $\mho$-sequence, then for any $i$, $\langle f_\alpha^{-1}(i) : \alpha < \omega_1\rangle$ is not measured by any club.

**Theorem 7.2** MRP *implies measuring.*

*Proof* This is not hard to verify; we will prove a more general statement in the next section. □

We will justify the name "measuring" momentarily. First it will be helpful make a definition and prove a few claims.

**Definition 7.3** If $M \prec H(\theta)$, we say that a club $E \subseteq \omega_1$ *diagonalizes M's club filter* if $\delta = M \cap \omega_1$ is a limit point of $E$ and whenever $D \subseteq \omega_1$ is a club in $M$, there is a $\delta_0 < \delta$ such that $E \cap (\delta_0, \delta) \subseteq D$.

*Claim* 7.4 If $\langle N_\xi : \xi < \omega_1 \rangle$ is a continuous $\in$-chain of countable elementary submodels of some $H(\theta)$ for $\theta \geq \omega_2$, then $E = \{N_\xi \cap \omega_1 : \xi < \omega_1\}$ diagonalizes the club filter of $N_\nu$ whenever $\nu < \omega_1$ is a limit ordinal.

*Proof* If $\nu$ is a limit ordinal and $D \subseteq \omega_1$ is a club in $N_\nu$, then by continuity of the sequence there is a $\xi < \nu$ such that $D$ is in $N_\xi$. By Fact 3.8, $N_\eta \cap \omega_1$ is in $D$ whenever $\xi < \eta$. It follows that if $\delta = N_\nu \cap \omega_1$ and $\delta_0 = N_\xi \cap \omega_1$, then

$$E \cap (\delta_0, \delta) = \{N_\eta \cap \omega_1 : \xi < \eta < \nu\}$$

is contained in $D$. □

*Claim* 7.5 If $E$ diagonalizes the club filter of $M$, then $[\omega_1]^\omega \setminus E$ is $M$-stationary.

*Proof* Suppose $D \subseteq \omega_1$ is a club in $M$. Since the limit points $D'$ of $D$ is also a club in $M$, $(D \setminus D') \cap E$ is bounded in $M \cap \omega_1$ and hence $D \setminus (D' \cap E)$ is non-empty. □

**Proposition 7.6** *The following are equivalent:*

1. *Measuring holds.*
2. *If $M$ is a countable elementary submodel of $H(\aleph_2)$ and $Y \subseteq M \cap \omega_1$ is closed and in* Hull$(M \cup \{M \cap \omega_1\})$*, then there is a club $E \subseteq \omega_1$ in $M$ such that $E \cap M$ is either contained in or disjoint from $Y$ (i.e. $Y$ is measured by the club filter of $M$).*

*Proof* To see the forward implication, suppose measuring holds and let $M$ and $Y$ be given. Since $Y$ is in the Skolem hull of $M \cup \{M \cap \omega_1\}$, there is a function $f$ in $M$ defined on $\omega_1$ such that $f(M \cap \omega_1) = Y$. Without loss of generality, $f(\alpha)$ is a closed subset of $\alpha$ for each $\alpha < \omega_1$. Applying measuring in $M$, there is a club $E \subseteq \omega_1$ such that $E \cap \alpha$ measures $f(\alpha)$ for each $\alpha$ which is a limit point of $E$. By elementarity, $M \cap \omega_1$ is a limit point of $E$. By removing an initial part of $E$ if necessary, we may assume that $E \cap M$ is either contained in or disjoint from $Y$.

To see the reverse implication, suppose $\langle Y_\alpha : \alpha < \omega_1 \rangle$ is a sequence such that for all $\alpha < \omega_1$, $Y_\alpha$ is a closed subset of $\alpha$.

Let $\langle N_\xi : \xi < \omega_1 \rangle$ be a continuous $\in$-chain of countable elementary submodels such that $\langle Y_\alpha : \alpha < \omega_1 \rangle$ is in $N_0$. Let $E = \{N_\xi \cap \omega_1 : \xi < \omega_1\}$. By Claim 7.4, $E$ diagonalizes the club filter of $N_\nu$ whenever $\nu$ is a limit ordinal. Also, if $\delta$ is a limit point of $E$, then there is a limit ordinal $\nu$ such that $N_\nu \cap \omega_1 = \delta$. By our assumption (2.), there is a club $D$ in $N_\nu$ which is either contained in or disjoint from $Y_\delta$. It follows that $E \cap \delta$ measures $Y_\delta$. □

Measuring is arguably the simplest consequence of PFA which is not known to be (in)consistent with CH.

**Problem 7.7** Is measuring consistent with CH?

Eisworth and Nyikos have shown that measuring for sequences of clopen $Y_\alpha \subseteq \alpha$ is consistent with CH [4]. It is similarly known by [17] that measuring for sequences $\langle Y_\alpha : \alpha < \omega_1 \rangle$ on which the ordertype function is regressive is consistent with CH.

Now we consider a coherence property of the club filter considered by Larson [11].

**Definition 7.8** Let (+) denote the statement that there exists a stationary $S \subseteq [H(\aleph_2)]^\omega$ such that, for all $M, M' \in S$, if $M \cap \omega_1 = M' \cap \omega_1$, then, for every $E \in M$ and $E' \in M'$ such that $E$ and $E'$ are club subsets of $\omega_1$, the set $E \cap E' \cap M \cap \omega_1$ is cofinal in $M \cap \omega_1$. Let (−) denote ¬(+).

Larson[11] showed that (+) follows from club guessing on $\omega_1$. Very recently Tetsuya Ishiu has shown that (+) is consistent with the failure of club guessing (even in the presence of CH).

**Theorem 7.9** MRP *implies* (−).

*Proof* Fix a stationary set $S$. We will show that $S$ does not witness (+). Suppose that $M$ is such that $S \in M \prec H(2^{\aleph_1^+})$. Ask:

(∗) Is there an end extension $\widetilde{M}$ of $M$ such that $\widetilde{M} \cap H(\aleph_2) \in S$ and there exists a club $E_M \subseteq \omega_1$ such that $E_M \in \widetilde{M}$ and $E_M$ diagonalizes the club filter of $M$?

If "no," then set $\Sigma(M) = [\omega_1]^\omega$. If "yes," then set $\Sigma(M) = [\omega_1]^\omega \setminus E_M$ for some such $E_M$. In either case, $\Sigma(M)$ is open. Moreover, $\Sigma(M)$ is $M$-stationary by the above observation.

Suppose $\langle N_\xi : \xi < \omega_1 \rangle$ reflects $\Sigma$. Let $\langle N_\xi : \xi < \omega_1 \rangle \in \overline{M} \prec H(2^{\aleph_1^+})$ and $\overline{M} \cap H(\aleph_2) \in S$. By Fact 4.1, $\overline{M}$ end extends $N_\delta$ where $\delta = \overline{M} \cap \omega_1$. Also $E = \{N_\xi \cap \omega_1 : \xi < \omega_1\}$ is in $\overline{M}$ and by Claim 7.4 diagonalizes the club filter of $N_\delta$. Hence the answer to $(*)$ is "yes" for $N_\delta$. Therefore there exist an end extension $\tilde{M} \in S$ of $N_\delta$ and a club $E' \in \tilde{M}$ such that $E'$ diagonalizes the club filter of $N_\delta$ and $\Sigma(N_\delta) = [\omega_1]^\omega \setminus E'$. Hence, for some $\xi < \delta$, we have $N_\nu \cap \omega_1 \notin E'$ for all $\nu \in \delta \setminus \xi$; hence, $E \cap E' \cap \delta$ is bounded in $\delta$. The sets $\overline{M} \cap H(\aleph_2)$ and $\tilde{M} \cap H(\aleph_2)$ now demonstrate that $S$ does not witness (+). □

## 8 The influence of MRP beyond $H(\aleph_2)$

In this section we will study the influence of MRP on sets higher up in the cumulative hierarchy. It is based on work of Viale.

Let $\kappa$ be an uncountable regular cardinal. Suppose $\mathcal{I}$ is an ideal of closed subsets of $\kappa$ (in the order topology). *I.e.*, for all $I_0, I_1 \in \mathcal{I}$, we have $I_0$ closed, every closed subset of $I_0$ in $\mathcal{I}$, and $I_0 \cup I_1 \in \mathcal{I}$. Consider the following three conditions on $\mathcal{I}$:

(I1) If $\beta < \kappa$, then there exists $\mathcal{J} \in [\mathcal{I}]^{\leq\omega}$ such that $\beta + 1 = \bigcup \mathcal{J}$.
(I2) If $X \in [\kappa]^\omega$, then $\mathcal{I} \restriction X$ is countably generated. *I.e.*, there exists $\mathcal{J} \in [\mathcal{I}]^\omega$ such that, for all $I \in \mathcal{I}$, there exists $J \in \mathcal{J}$ such that $I \cap X \subseteq J \cap X$.
(I3) If $Z \subseteq \kappa$ is unbounded, then there exists $Y \in [Z]^\omega$ such that $Y \nsubseteq I$ for all $I \in \mathcal{I}$.

Recall that the Singular Cardinals Hypothesis (SCH) is the assertion that if $\lambda$ is a singular strong limit cardinal, then $2^\lambda = \lambda^+$. By a theorem of Silver [18], if SCH fails at $\lambda$ and $\mathrm{cf}(\lambda)$ is uncountable, then there is a stationary set of singular $\mu < \lambda$ such that SCH fails at $\mu$. In particular, if $\lambda$ is the least singular cardinal such that $2^\lambda > \lambda^+$, then $\mathrm{cf}(\lambda) = \omega$.

**Theorem 8.1** *If* SCH *fails, then there exist $\kappa$ and $\mathcal{I}$ satisfying (I1), (I2), and (I3).*

*Proof* Assume SCH fails. Then there exists $\mu$ such that $\omega = \mathrm{cf}(\mu) < \mu = 2^{<\mu}$ and $\mu^+ < 2^\mu$. Set $\kappa = \mu^+$ and let $\mu = \sup_{n<\omega} \mu_n$. For all $\beta < \kappa$, let $\beta + 1 = \bigcup_{n<\omega} E^\beta_n$ where $\langle E^\beta_n : n < \omega \rangle$ is an ascending sequence of closed subsets of $\kappa$ such that $|E^\beta_n| \leq \mu_n$ for all $n$. By modifying the sequence, arrange that $E^\beta_{n+1} \supseteq \bigcup_{\alpha \in E^\beta_n \cap \beta} E^\alpha_n$ for all $n$. Let $\mathcal{I}$ be the ideal generated by $\{E^\beta_n : n < \omega \wedge \beta < \kappa\}$. Then (I1) is trivially true. For (I2), fix $X \in [\kappa]^\omega$. Then, for all $\beta < \kappa$, $\{E^\beta_n \cap X : n < \omega\}$ generates an ideal $\mathcal{I}^\beta_X$ on $\mathcal{P}(X)$. Moreover, if $\beta < \beta'$, then $\mathcal{I}^\beta_X \subseteq \mathcal{I}^{\beta'}_X$. Since $\mathrm{cf}(\kappa) > 2^{2^{\aleph_0}}$, as $\beta$

increases, this restricted ideal eventually stabilizes to $\mathcal{I} \restriction X$. Therefore, $\{E_n^\beta \cap X\}_{n<\omega}$ generates $\mathcal{I} \restriction X$ if $\beta$ is large enough. Finally, (I3) holds simply because $|\mathcal{I}| = 2^{<\mu} \cdot \mu^+ < 2^\mu = \mu^\omega = \mathrm{cf}(\langle [\mu^+]^\omega, \subseteq \rangle)$. □

**Theorem 8.2** MRP *implies that there does not exist a regular cardinal* $\kappa$ *and an ideal* $\mathcal{I}$ *on* $\kappa$ *which satisfies* (*I*1), (*I*2), *and* (*I*3).

*Proof* Seeking a contradiction, suppose $\kappa$ and $\mathcal{I}$ satisfy (I1), (I2), and (I3). For each $\alpha \in \mathrm{Lim}(\omega_1)$, fix a cofinal $C_\alpha \subseteq \alpha$ such that $\mathrm{otp}(C_\alpha) = \omega$. Fix, for each $M \prec H(2^{\kappa^+})$, a $\subseteq$-increasing sequence $\langle I_n^M : n < \omega \rangle$ of elements of $\mathcal{I}$ which generates $\mathcal{I} \restriction M$. Let $\Sigma(M)$ be the set of $N \in [\kappa]^\omega$ for which $\sup(N \cap \omega_1) < M \cap \omega_1$ and $\sup N \notin I_n^M$ where $n = |C_{M\cap\omega_1} \cap N|$.

*Claim* 8.3 $\Sigma(M)$ is open and $M$-stationary.

*Proof* To see that $\Sigma(M)$ is open, suppose that $N$ is in $\Sigma(M)$ and let $\alpha$ be an element of $N \cap \omega_1$ which is an upper bound for $N \cap C_\delta$ where $\delta = M \cap \omega_1$. Since $\beta = \sup N$ is not in $I_n^M$, where $n = |C_\delta \cap N|$, there is a $\beta_0$ in $N$ such that $(\beta_0, \beta) \cap I_n^M = \emptyset$. Then $[\{\alpha, \beta_0\}, N] \subseteq \Sigma(M)$.

To see that $\Sigma(M)$ is $M$-stationary, suppose $E \in M$ is a club subset of $[\kappa]^\omega$. Fix $\alpha < M \cap \omega_1$ such that $\Gamma = \{\sup A : A \in E \wedge A \cap \omega_1 = \alpha\}$ is cofinal in $\kappa$. Set $E' = \{A \in E : A \cap \omega_1 = \alpha\}$. Using (I3), pick $X \in [\Gamma]^\omega \cap M$ such that $X$ is not contained in any $I \in \mathcal{I}$. Pick $\beta \in X \setminus I_n^M$ where $n = |C_\delta \cap \alpha|$ where $\delta = M \cap \omega_1$. Then $\beta \in M$ because $X \in M$ and $X$ is countable. Pick $A \in E' \cap M$ such that $\sup A = \beta$. Then $A \in \Sigma(M) \cap M \cap E$. Thus, $\Sigma(M)$ is $M$-stationary. □

Suppose $\langle N_\xi : \xi < \omega_1 \rangle$ is a reflecting sequence and let $E = \{\sup(N_\xi \cap \kappa) : \xi < \omega_1\}$. Using (I1), pick $I \in \mathcal{I}$ such that $I \cap E$ is cofinal in $E$. Let $\delta$ be such that $\Xi = \{\xi < \delta : \sup(N_\xi \cap \kappa) \in I\}$ is cofinal in $\delta$. Then there exists $n$ such that $I \cap N_\delta \subseteq I_n^{N_\delta}$. Pick $\xi \in \Xi$ such that $|C_{N_\delta \cap \omega_1} \cap N_\xi| \geq n$ and $N_\xi \cap \kappa \in \Sigma(N_\delta)$. Then $\sup(N_\xi \cap \kappa) \notin I^{N_\delta}_{|N_\xi \cap C_\delta|}$ and therefore not in $I_n^{N_\delta}$ by our assumption that $\langle I_k^{N_\delta} : k < \omega \rangle$ is $\subseteq$-increasing. Hence $\sup(N_\xi \cap \kappa) \notin I$, in contradiction with how we chose $\xi$. □

**Corollary 8.4** MRP *implies* SCH.

Now recall the definition of $\square(\kappa)$.[2]

**Definition 8.5** $\square(\kappa)$ asserts that there exists a sequence $\langle C_\alpha : \alpha \in \mathrm{Lim}(\kappa) \rangle$ satisfying the following conditions:

[2] $\square_\kappa$ and $\square(\kappa)$ are different principles. They are related in the sense that $\square(\kappa^+)$ is a formal weakening of $\square_\kappa$.

1. $C_\alpha$ is closed and cofinal in $\alpha$.
2. If $\alpha \in \mathrm{Lim}(C_\beta)$, then $C_\alpha = C_\beta \cap \alpha$.
3. There is no closed and cofinal $E \subseteq \kappa$ such that $C_\beta = E \cap \beta$ for all $\beta \in \mathrm{Lim}(E)$.

We can similarly use this route to show that MRP implies $\square(\kappa)$ fails.

**Theorem 8.6** *[22] If $\square(\kappa)$ holds, then there exists $\mathcal{I}$ satisfying (I1), (I2), and (I3).*

*Proof* Fix a $\square(\kappa)$ sequence $\langle C_\alpha : \alpha < \kappa\rangle$ and, following [20], define

$$\varrho_2(\alpha,\alpha) = 0$$

$$\varrho_2(\alpha,\beta) = 1 + \varrho_2(\alpha, \min C_\beta \setminus \alpha)$$

whenever $\alpha < \beta < \kappa$. Notice that $\varrho_2(\alpha,\beta) \leq 1$ if and only if $\alpha$ is in $C_\beta$. It is sufficient to show that the ideal $\mathcal{I}$ generated by the sets

$$I_{\beta,n} = \{\alpha \in \beta : \varrho_2(\alpha,\beta) \leq n\}$$

satisfies (I1)-(I3). The following identity clarifies the relationship between $\mathcal{I}$ and the ideal generated by $\langle C_\alpha : \alpha < \kappa\rangle$:

$$I_{\beta,n+1} = I_{\beta,n} \cup \bigcup_{\gamma \in I_{\beta,n}} C_\gamma \setminus \sup(I_{\beta,n} \cap \gamma).$$

That $\mathcal{I}$ satisfies (I1) is trivial, since for every $\alpha < \beta$, $\varrho_2(\alpha,\beta) < \omega$. The following are standard properties of $\varrho_2$ [20]:

1. for all $\alpha < \beta < \kappa$, the set $\{|\rho_2(\xi,\alpha) - \rho_2(\xi,\beta)| : \xi < \alpha\}$ is bounded in $\omega$.
2. $\varrho_2[Z]^2$ is unbounded in $\omega$ for every $Z$ unbounded in $\kappa$;

To see that $\mathcal{I}$ satisfies (I2), notice that, for all $\alpha < \beta < \kappa$ and $n < \omega$, we have $I_{\beta,n} \cap \alpha \subseteq I_{\alpha,k}$ for some $k < \omega$, and $I_{\alpha,k} \subseteq I_{\beta,l}$ for some $l < \omega$. Hence, if $X \in [\kappa]^\omega$ and $\beta \geq \sup(X)$, then $\{I_{\beta,n} \cap X : n < \omega\}$ generates $\mathcal{I} \upharpoonright X$. For (I3), suppose $Z \subseteq \kappa$ is unbounded. Then $\varrho_2[Z]^2$ is unbounded in $\omega$ and in particular there is a $\beta < \kappa$ such that $\varrho_2[Z \cap \beta]^2$ is unbounded in $\omega$. It follows that $Z \cap \beta$ is not contained in an element of $\mathcal{I}$. □

For the sake of demonstration, we will now give a direct proof that MRP implies the failure of $\square$.

**Theorem 8.7** *For all regular $\kappa$,* MRP *implies $\neg\square(\kappa)$.*

*Proof* Suppose not; let $\langle C_\alpha : \alpha < \kappa\rangle$ witness $\square(\kappa)$. For all $M \prec H(2^{\kappa^+})$ such that $\langle C_\alpha : \alpha < \kappa\rangle \in M$, set $\Sigma(M)$ equal to the set of $N \in [\kappa]^\omega$ for which $\sup N < \sup(M \cap \kappa)$ and $\sup N \notin C_{\sup(M\cap\kappa)}$. Suppose $\langle N_\xi : \xi < \omega_1\rangle$ reflects $\Sigma$. Set $E_0 = \{\sup(N_\xi \cap \kappa)\}_{\xi<\omega_1}$. Then $E_0$ is closed and cofinal in some $\delta$ with $\mathrm{cf}(\delta) = \omega_1$. Let $E = C_\delta \cap E_0$. Suppose $\nu \in \mathrm{Lim}(E)$. Let $\eta < \omega_1$ be such that $\nu = \sup(N_\eta \cap \kappa)$. Then $C_\nu = C_\delta \cap \nu$ and $\eta \in \mathrm{Lim}(\omega_1)$. Let $\eta_0 < \eta$ be such that $N_\xi \cap \kappa \in \Sigma(N_\eta)$ for all $\xi \in \eta \setminus \eta_0$. Let $\nu' \in E$ be such that $\sup(N_{\eta_0} \cap \kappa) \leq \nu' < \sup(N_\nu \cap \kappa)$. Then there exists $\xi \in \eta \setminus \eta_0$ such that $\nu' = \sup(N_\xi \cap \kappa)$. Since $N_\xi \cap \kappa \in \Sigma(N_\nu)$, we have $\sup(N_\xi \cap \kappa) \notin C_\nu$. But $\sup(N_\xi \cap \kappa) \in E \cap \nu \subseteq C_\delta \cap \nu = C_\nu$, which is absurd. Thus, it suffices to show that $\Sigma$ is open stationary.

Fix $M \in \mathrm{dom}(\Sigma)$ and $N \in \Sigma(M)$. Then $\sup N \notin C_{\sup(M\cap\kappa)}$; hence, there exists $\alpha \in N$ such that $(\alpha, \sup N) \cap C_{\sup(M\cap\kappa)} = \emptyset$; hence, $[\{\alpha\}, N] \subseteq \Sigma(M)$. Thus, $\Sigma(M)$ is open. Next, let $E \in M$ be a club subset of $[\kappa]^\omega$. Set $\Gamma = \{\sup A : A \in E\}$, which is closed and cofinal in $\kappa$. Then it suffices to show that $\Gamma \cap M \not\subseteq C_{\sup(M\cap\kappa)}$. Seeking a contradiction, suppose $\Gamma \cap M \subseteq C_{\sup(M\cap\kappa)}$. Then, for all $\{\alpha < \beta\} \in [\mathrm{Lim}(\Gamma \cap M)]^2$, we have $C_\alpha = C_{\sup(M\cap\kappa)} \cap \alpha$ and $C_\beta = C_{\sup(M\cap\kappa)} \cap \beta$; whence, $C_\alpha = C_\beta \cap \alpha$. By elementarity, we have $C_\alpha = C_\beta \cap \alpha$ for all $\{\alpha < \beta\} \in [\mathrm{Lim}(\Gamma)]^2$. Therefore, $C = \bigcup_{\alpha\in\mathrm{Lim}(\Gamma)} C_\alpha$ is closed and cofinal in $\kappa$ and $C_\alpha = C \cap \alpha$ for all $\alpha \in \mathrm{Lim}(C)$, in contradiction with (3.) of the definition of $\square(\kappa)$. □

## 9 The 0-1 law for open set mappings

Let $\Sigma$ be an open set mapping (for some $X, \theta$). Are there some conditions that, in the presence of MRP or some stronger assumption, ensure that every $\Sigma(M)$ is trivial, *i.e.*, either $\Sigma(M)$ is not $M$-stationary or contains $E \cap M$ for some club $E \subseteq [X]^\omega$? Note that $\Sigma_<$ and $\Sigma_\geq$ are an example of nontriviality since they are everywhere disjoint open stationary set mappings.

**Definition 9.1** The *0-1 law* for open set mappings asserts that, if $\Sigma$ is an open set mapping defined on a club in $H(\theta)$ and for all $M \in \mathrm{dom}(\Sigma)$,

1. for all $P \in \Sigma(M)$ and all end extensions $\overline{P}$ of $P$, we have $\overline{P} \in \Sigma(M)$, and,
2. for all end extensions $\overline{M}$ of $M$ in $\mathrm{dom}(\Sigma)$, we have $\Sigma(\overline{M}) \cap M = \Sigma(M) \cap M$,

then there exists a club $E^* \subseteq \mathrm{dom}(\Sigma)$ such that, for all $M \in E^*$, there exists a club $E \subseteq [X]^\omega$ such that $E \in M$ and either $E \cap M \subseteq \Sigma(M)$ or $E \cap M \cap \Sigma(M) = \emptyset$.

**Example 9.2** [8] Let $T$ be an $\omega_1$-tree and $B$ be the set of uncountable maximal chains in $T$. Suppose $\{t_{\delta,n}\}_{n<\omega}$ is the $\delta^{\text{th}}$ level of $T$. Define $\Sigma_n(M)$ to be the set of $N \in [H(\aleph_2)]^\omega$ for which either $N$ is not an elementary submodel of $H(\aleph_2)$, or there exists $b \in N \cap B$ such that $t_{\delta,n} \restriction (N \cap \omega_1) \in b$ where $\delta = M \cap \omega_1$. If each $\Sigma_n$ satisfies the 0-1 law, then $T$ has at most $\aleph_1$-many branches. In fact more generally, if $\mathcal{B}$ is a collection of uncountable downward $<_T$-closed subsets of $T$ which have pairwise countable intersection, then if each $\Sigma_n$ satisfies the 0-1 law, $\mathcal{B}$ has at most $\aleph_1$ many elements. This latter statement is known as *Aronszajn tree saturation*. It is equivalent to the corresponding statement in which $T$ is Aronszajn.

In order to prove the *0-1 law*, we will need a strengthening of MRP.

**Definition 9.3** Let SMRP (Strong Mapping Reflection Principle) assert that if $\Sigma$ is an open stationary set mapping on some domain $S \subseteq [H(\theta)]^\omega$ that is not necessarily club, then there is a strong reflecting sequence $\langle N_\xi : \xi < \omega_1 \rangle$ for $S$ such that, for all $\nu \in \operatorname{Lim}(\omega_1)$, if $N_\nu \in S$, then there exists $\nu_0 < \nu$ such that $N_\xi \cap X \in \Sigma(N_\nu)$ for all $\xi \in \nu \setminus \nu_0$.

MM implies SMRP. If $\operatorname{dom}(\Sigma)$ is a club, then SMRP is just MRP. If $\Sigma$ is always trivial, then SMRP is just SRP.

**Theorem 9.4** SMRP *implies the 0-1 law for open set mappings.*

*Proof* Let $\Sigma$ satisfying (1.) and (2.) be given. Let $S$ denote the set of $M \in \operatorname{dom}(\Sigma)$ for which $\Sigma(M)$ is $M$-stationary. Let $\langle N_\xi : \xi < \omega_1 \rangle$ strongly reflect $S$ and reflect $\Sigma$ in the sense of SMRP. Set

$$E^* = \{M \in \operatorname{dom}(\Sigma) : \langle N_\xi : \xi < \omega_1 \rangle \in M\}.$$

Let $M \in E^*$ be arbitrary. If $\Sigma(M)$ is not $M$-stationary, then there is trivially a club $E \subseteq [X]^\omega$ such that $E \in M$ and $E \cap M \cap \Sigma(M) = \emptyset$. Therefore, we may assume $\Sigma(M)$ is $M$-stationary. Then $N_\delta \in S$ where $\delta = M \cap \omega_1$ because $M \cap \omega_1 = N_\delta \cap \omega_1$. Let $\delta_0 < \delta$ be such that $N_\xi \cap X \in \Sigma(N_\delta)$ for all $\xi \in \delta \setminus \delta_0$. Define $E_0$ to be the set of all $N$ in $[H(\theta)]^\omega$ such that $N \cap \omega_1$ is a limit ordinal and for all $\xi < N \cap \omega_1$, $N_\xi \subseteq N$ and $\delta_0 \in N$. If $N \in E_0 \cap M$, then $N$ end extends $N_\xi$ for some $\xi \in \delta \setminus \delta_0$. Since $\delta_0 < \xi$, $N_\xi \cap X \in \Sigma(N_\delta)$ and therefore $N_\xi \cap X \in \Sigma(M)$ by (2.). Hence $N \cap X \in \Sigma(M)$ by (1.). If we set $E = \{N \cap X : N \in E_0\}$, then $E$ is a club subset of $[X]^\omega$, $E \in M$, and $E \cap M \subseteq \Sigma(M)$. □

*Remark* 9.5    It is often the case that an open set mapping $\Sigma$ satisfies the following additional hypothesis: if $\overline{N}$ is an end extension of $N$ and $\Sigma(\overline{N})$ is $\overline{N}$-stationary, then $\Sigma(N)$ is $N$-stationary. This is satisfied, for instance, in the open set mappings defined in Example 9.2 and in [15]. It follows from the above argument that MRP implies the 0-1 law for open set mappings which satisfy this additional condition.

# 10 Open problems

Here is a collection of open problems which are related to MRP.

While MRP implies $2^{\aleph_0} = 2^{\aleph_1} = \aleph_2$, the forcing to reflect a single open stationary set mapping does not introduce new reals. Many of these forcings are moreover *completely proper* (see Eisworth and Moore's lecture notes in this volume). It is not known whether the forcing axiom for completely proper forcings is consistent with CH and the corresponding fragments of MRP already seem to capture much of the generality of this question. An important special case of this is the following.

**Problem 10.1**    Is *measuring* consistent with CH?

Similarly, we would like to know how much of MRP is consistent with $2^{\aleph_0} > \aleph_2$. Aspero and Mota have announced that *measuring* is consistent with $2^{\aleph_0} > \aleph_2$.

A common theme with many of SRP's consequences is that they are typical examples of what MM implies but PFA was not known to imply. Most of the remaining consequences of SRP can not follow from PFA because, in general, they can not be forced with a proper forcing notion (in contrast to PFA). Some, however, follow from MRP via a proof similar to that used in the case of SRP. The following is an open problem in this vein.

**Problem 10.2**    Does MRP imply that there is a $Q$ and a generic elementary embedding $j : V \to M \subseteq V^Q$ such that $\mathrm{crit}(j) = \omega_1$ and $M^{<\omega_1} \subseteq M$?

If $\mathrm{NS}_{\omega_1}$ is saturated, then there is such a generic embedding if we take $Q = \mathcal{P}(\omega_1)/\mathrm{NS}_{\omega_1}$. In particular, the problem has a positive answer if one replaces MRP with SRP. Also, if there is a Woodin cardinal, then the "countable tower" $Q_{<\delta}$ is such a forcing (see [10]). This latter example makes it very difficult to prove a negative answer to this question.

While this would not serve to give a better lower bound on the consistency strength of PFA, it would be interesting if there are other ways to establish lower bounds on the consistency strength of PFA which do not involve the failure of $\square$ (and are at least at the level of a Mahlo cardinal).

**Problem 10.3** What is the consistency strength of the *0-1 law for open set mappings*?

The *0-1 law* implies that there are no Kurepa trees and hence that $\omega_2$ is inaccessible to subsets of $\omega_1$. This seems to be the best known lower bound.

Let $Q_0$ be a forcing of size $\aleph_1$ and $\mathcal{A}$ be a collection of $\aleph_1$ many maximal antichains of $Q_0$. We say that $Q_0$ *$\mathcal{A}$-embeds* into $Q$ if there is a injection $f : Q_0 \to Q$ which preserves order, compatibility, incompatibility, and the maximality of elements of $\mathcal{A}$. Many problems concerning applications of forcing axioms reduce to the question of when a given pair $(Q_0, \mathcal{A})$ as above can be $\mathcal{A}$-embedded into a proper forcing $Q$. In [15], MRP was useful in providing an answer to this in a special case. It seems natural (though ambitious) to ask if it is possible to use MRP to prove a general result of this form.

**Problem 10.4** Assume MRP and let $Q_0$ be a partial order and $\mathcal{A}$ a collection of maximal antichains of $Q_0$ such that $|Q_0|, |\mathcal{A}| \leq \aleph_1$. Is there an informative necessary and sufficient condition for when $Q_0$ can be $\mathcal{A}$-embedded into a proper forcing $Q$?

For instance, is there an upper bound on the cardinality of such $Q$ which is expressible in terms of the $\beth$ function? For those $Q_0$ which can be $\mathcal{A}$-embedded into a proper $Q$, is there a canonical form that $Q$ can be assumed to take? Does the answer to these questions change if "proper" is replaced by "preserves $NS_{\omega_1}$?"

## 11 Further reading

The notion of set mapping reflection was introduced in [14] in order to prove that BPFA implies that there is a well ordering of $\mathbb{R}$ which is definable over $H(\aleph_2)$ (and consequently that $L(\mathcal{P}(\omega_1))$ satisfies the Axiom of Choice). This paper also establishes that PFA implies MRP and that MRP implies the failure of $\square(\kappa)$ at all regular $\kappa > \omega_1$. Caicedo and Veličković [2] built on these ideas to show that BPFA implies there is a well ordering of $\mathbb{R}$ which is $\Delta_1$-definable with parameters in $H(\aleph_2)$

(the complexity of the well ordering presented in Theorem 5.7 is $\Delta_2$). Given the parameters, this complexity is optimal.

In [15], Moore used MRP in conjunction with BPFA to prove that every Aronszajn line contains a Countryman suborder. The 0-1 law was isolated from that proof and serves as the sole use of MRP in that paper. In [8], König, Larson, Moore, and Veličković further analyzed the role of the 0-1 law [15]. While the end goal was to reduce the consistency strength of the results of [15], much of [8] concerns a study of Aronszajn tree saturation. In particular, it is shown that MRP implies A-tree saturation and that the conjunction of A-tree saturation and BPFA implies that every Aronszajn line contains a Countryman suborder. It is not known whether the existence of a non-saturated Aronszajn tree implies that there is an Aronszajn line with no Countryman suborder.

The analysis of Aronszajn tree saturation was used explicitly in [13] to establish the consistent existence of a universal Aronszajn line. Similar combinatorial arguments were used by Ishiu and Moore [5] to characterize (assuming $\mathrm{PFA}^+$) when a linear order contains an Aronszajn suborder. In [12], Moore showed that some non-trivial application of MRP is needed for the results of [15]. (In particular, if every Aronszajn line contains a Countryman suborder, then club guessing fails). Viale [23] proved that MRP implies the Singular Cardinals Hypothesis.

## References


[1] M. Bekkali. *Topics in set theory*. Springer-Verlag, Berlin, 1991. Lebesgue measurability, large cardinals, forcing axioms, $\rho$-functions, Notes on lectures by Stevo Todorčević.

[2] Andrés E. Caicedo and Boban Veličković. The bounded proper forcing axiom and well orderings of the reals. *Mathematical Research Letters*, 13(3):393–408, 2006.

[3] Keith Devlin and Saharon Shelah. A weak version of $\diamondsuit$ which follows from $2^{\aleph_0} < 2^{\aleph_1}$. *Israel Journal of Math*, 29(2–3):239–247, 1978.

[4] Todd Eisworth and Peter Nyikos. First countable, countably compact spaces and the continuum hypothesis. *Trans. Amer. Math. Soc.*, 357(11):4269–4299, 2005.

[5] Tetsuya Ishiu and Justin Tatch Moore. Minimality of non $\sigma$-scattered orders. submitted, April 2007.

[6] T. Jech. *Multiple forcing*, volume 88 of *Cambridge Tracts in Mathematics*. Cambridge University Press, Cambridge, 1986.

[7] Thomas Jech. *Set theory*. Springer Monographs in Mathematics. Springer-Verlag, Berlin, 2003. The third millennium edition, revised and expanded.

[8] Bernhard König, Paul Larson, Justin Tatch Moore, and Boban Veličković. Bounding the consistency strength of a five element linear basis. *Israel Journal of Math*, 164, 2008.

[9] Kenneth Kunen. *An introduction to independence proofs*, volume 102 of *Studies in Logic and the Foundations of Mathematics*. North-Holland, 1983.

[10] Paul B. Larson. *The stationary tower*, volume 32 of *University Lecture Series*. American Mathematical Society, Providence, RI, 2004. Notes on a course by W. Hugh Woodin.

[11] Paul B. Larson. The canonical function game. *Arch. Math. Logic*, 44(7):817–827, 2005.

[12] Justin Tatch Moore. Aronszajn lines and the club filter. preprint 9/11/07.

[13] Justin Tatch Moore. A universal Aronszajn line. submitted 2/26/08.

[14] Justin Tatch Moore. Set mapping reflection. *J. Math. Log.*, 5(1):87–97, 2005.

[15] Justin Tatch Moore. A five element basis for the uncountable linear orders. *Annals of Mathematics (2)*, 163(2):669–688, 2006.

[16] Justin Tatch Moore. A solution to the L space problem. *Jour. Amer. Math. Soc.*, 19(3):717–736, 2006.

[17] Saharon Shelah. *Proper and improper forcing*. Springer-Verlag, Berlin, second edition, 1998.

[18] Jack Silver. On the singular cardinals problem. In *Proceedings of the International Congress of Mathematicians (Vancouver, B. C., 1974), Vol. 1*, pages 265–268. Canad. Math. Congress, Montreal, Que., 1975.

[19] Stevo Todorcevic. A note on the proper forcing axiom. In *Axiomatic set theory (Boulder, Colo., 1983)*, volume 31 of *Contemp. Math.*, pages 209–218. Amer. Math. Soc., Providence, RI, 1984.

[20] Stevo Todorcevic. Partitioning pairs of countable ordinals. *Acta Math.*, 159(3-4):261–294, 1987.

[21] Stevo Todorcevic. *Partition Problems In Topology*. Amer. Math. Soc., 1989.

[22] Stevo Todorcevic. A dichotomy for P-ideals of countable sets. *Fund. Math.*, 166(3):251–267, 2000.

[23] Matteo Viale. The proper forcing axiom and the singular cardinal hypothesis. *Jour. Symb. Logic*, 71(2):473–479, 2006.

# 5

# An introduction to hyperlinear and sofic groups

Vladimir G. Pestov and Aleksandra Kwiatkowska

---

The seventh Appalachian Set Theory workshop was held at Cornell University in Ithaca on November 22, 2008. The lecturer was Vladimir Pestov. As a graduate student Aleksandra Kwiatkowska assisted in writing this chapter, which is based on the workshop lectures.

---

## Abstract

This is an edited write-up of lecture notes of the 7-th Appalachian set theory workshop of the same title led by the first named author at the Cornell University on November 22, 2008. A draft version of the notes was prepared by the second named author. This presentation is largely complementary to the earlier survey by the first-named author (*Hyperlinear and sofic groups: a brief guide*, Bull. Symb. Logic **14** (2008), pp. 449-480).

## 1 Motivation: group matrix models in the sense of classical first-order logic

In these lectures, we will deal with a class of groups called *hyperlinear groups,* as well as its (possibly proper) subclass, that of *sofic groups.* One natural way to get into this line of research is through the theory of operator algebras. Here, the hyperlinear groups are sometimes referred to as "groups admitting matrix models". This can be indeed interpreted as a genuine model-theoretic statement, within a suitable version of logic. Namely, a group $G$ is said to admit matrix models if every existential sentence of the first-order theory of $G$ is satisfied in matrix groups.

What makes the concept interesting — and difficult to work with — is that at the matrix group end it is not the classical first-order logic that one has in mind, but rather a version of continuous logic with truth values in the unit interval $[0,1]$. By way of motivation, let us try to understand first what we get by considering a class of groups admitting matrix models in the sense of the traditional binary logic.

The language of group theory, which we will denote $L$, is a first-order predicate calculus with equality, having one ternary predicate letter $S$ and a constant symbol $e$. The group operation is coded as follows: $S(x,y,z)$ if $xy = z$. In addition, we have variables, the equality symbol $=$, logical connectives, and quantifiers. Now consider a (countable) group $G$. A formula is said to be a *sentence* if all its variables are bound within quantifiers. These formulas say something definite about the structure of a group, hence the following definition. The *theory* of $G$, denoted $\mathrm{Th}(G)$, is defined to be the set of all sentences of the predicate calculus $L$ which are valid in $G$. The fact that the language is first-order implies that the variables only range over $G$ (and not, for instance, over families of subsets of $G$). We will further denote by $\mathrm{Th}^{\exists}(G)$ the subset of $\mathrm{Th}(G)$ consisting of all existential first-order sentences, that is, those of the form $\exists x_1 \exists x_2 \ldots \exists x_n\, \phi(x_1,\ldots,x_n)$.

Let us introduce an *ad hoc* notion. Say that $G$ *admits matrix models "in the classical sense,"* if every quantifier-free (open) formula in first order logic can be satisfied in $GL(n,\mathbb{K})$ for some $n$ and some field $\mathbb{K}$. Even more precisely: whenever $G \models \phi(g_1, g_2 \ldots, g_k)$, where $\phi$ is an open formula and $g_1, g_2, \ldots, g_k \in G$, then, for a suitable natural number $n$ and some $g'_1, g'_2, \ldots, g'_k \in GL(n,\mathbb{K})$,

$$GL(n,\mathbb{K}) \models \phi(g'_1, g'_2, \ldots, g'_k).$$

Here we take matrix groups to be as general as possible: all groups of the form $GL(n,\mathbb{K})$, where $n \in \mathbb{N}$ and $\mathbb{K}$ is an arbitrary field, are allowed.

What groups admit matrix models "in the classical sense"? It turns out this class can be described in a very transparent way, and does not in fact depend on the choice of a field $\mathbb{K}$. We obtain this description in the rest of the present Section. First, a preliminary observation.

*Observation* A group $G$ admits matrix models "in the classical sense" if and only if $G$ locally embeds into matrix groups, that is, for every finite $F \subseteq G$ there is a natural number $n$, a field $\mathbb{K}$, and an injective map $i\colon F \to GL(n, \mathbb{K})$ so that, whenever $x, y \in F$ and $xy \in F$, one has

$$i(xy) = i(x)i(y).$$

Such a mapping $i$ as above is called a *local monomorphism,* or a *partially defined monomorphism.*

*Proof* The *necessity* follows from the fact that the conjunction of all the formulas of the form $\neg(g_i = g_j)$, $i \neq j$, as well as $S(g_i, g_j, g_k)$, where $g_i, g_j, g_k \in F$ and $g_i g_j = g_k$, $1 \leq i, j, k \leq n$, is satisfied in $G$ and so in a suitable linear group. Here we denote $F = \{g_1, g_2, \dots, g_n\}$ and $i(g_i) = g_i'$, where $g_i'$ are chosen as in the paragraph preceding the Observation.

To prove *sufficiency,* let $\phi = \phi(g_1, g_2 \dots, g_k)$ be an open formula satisfied in $G$. Write $\phi$ in a disjunctive normal form. The atomic formulas are of the form $\neg(x = y)$, $S(x, y, z)$, or $\neg S(x, y, z)$. All occurrences of atoms of the type $\neg S(x, y, z)$ can be replaced with formulas $S(x, y, w) \wedge \neg(w = z)$, where $w$ is a new variable suitably interpreted in $G$. Denote $\phi'$ the resulting open formula, having probably more variables, which is satisfied in $G$. This $\phi'$ is written in a disjunctive normal form, with atomic formulas of the kind either $\neg(x = y)$ or $S(x, y, z)$. As a consequence of our assumptions, a conjunctive clause of such atoms is satisfied in some $GL(n, \mathbb{K})$, and the same of course applies to the disjunction of a set of conjunctive clauses. Since $GL(n, \mathbb{K}) \models \phi'$, the formula $\phi$ is satisfied in $GL(n, \mathbb{K})$ as well. □

To take the next step, we introduce the following notion.

**Definition 1.1** A group $G$ is *residually finite* if it satisfies one of the following equivalent conditions:

1. for every $g \in G$, $g \neq e$ there exists a normal subgroup $N$ of finite index such that $g \notin N$,
2. for every finite subset $F \subseteq G$ there is a homomorphism $h$ from $G$ to a finite group with $h \restriction F$ being an injection,
3. $G$ is a subgroup of a direct product of a family of finite groups.

*Equivalence of the conditions*
(1)⇒(3): For each $g \in G \setminus \{e\}$, choose a normal subgroup $N_g$ of finite index not containing $g$, and let $\pi_g\colon G \to G/N_g$ denote the corresponding

quotient homomorphism. One has $\pi_g(g) \neq e$. Consequently, the diagonal product of all $\pi_g$, sending each $x \in G$ to the element $(\pi_g(x))_{g\in G\setminus\{e\}}$ of the direct product of quotient groups, is a monomorphism.
(3)⇒(2): here $h$ is a projection on the product of a suitable finite subfamily of groups.
(2)⇒(1): take $F = \{e, g\}$. □

**Example 1.2**

1. Finite groups are residually finite;
2. finitely generated abelian groups are residually finite;
3. free groups are residually finite.

*Proof of (3.) in Example 1.2* The free group of countably many generators $F_\infty$ can be embedded into the free group on two generators $F_2$. (Namely, if $a, b$ are free generators of $F_2$, then the conjugates $b$, $aba^{-1}$, $a^2ba^{-2}, \ldots$ are free generators of a subgroup they generate.) Therefore it it enough to do the proof for $F_2$. First we show that

$$F_2 < SL(2, \mathbb{Z}),$$

where $SL(2, \mathbb{Z})$ denotes the group of $2\times2$ matrices of determinant equal to 1. We will prove that

$$A = \begin{pmatrix} 1 & 2 \\ 0 & 1 \end{pmatrix}$$

and

$$B = \begin{pmatrix} 1 & 0 \\ 2 & 1 \end{pmatrix}$$

are free generators.

Consider subspaces of $\mathbb{R}^2$, $X = \{(x, y)\colon |x| > |y|\}$ and $Y = \{(x, y)\colon |x| < |y|\}$. Note that for every $n \in \mathbb{Z}$, $A^n$ maps $Y$ into $X$, and $B^n$ maps $X$ into $Y$.

Although it is obvious that $AB \neq \mathbb{I}$, another way to see this is to observe that the conjugate $A^2BA^{-1}$ maps $Y$ into $X$. Hence $A^2BA^{-1} \neq \mathbb{I}$, and therefore $AB \neq \mathbb{I}$. The argument easily generalizes to show that an arbitrary word $A^{n_1}B^{m_1} \ldots A^{n_k}B^{m_k} \neq \mathbb{I}$.

To finish the proof, we have to show that $SL(2, \mathbb{Z})$ is residually finite. For every prime $p$ the quotient homomorphism $h_p\colon SL(2, \mathbb{Z}) \to SL(2, \mathbb{Z}_p)$ sends a matrix over $\mathbb{Z}$ to one over $\mathbb{Z}_p$ by taking its entries mod $p$. The family of homomorphisms $h_p$ is easily seen to separate points in $SL(2, \mathbb{Z})$. □

Now comes a classical result.

**Theorem 1.3** (Malcev) *Every finitely generated subgroup $G$ of the linear group $GL(n,\mathbb{K})$ is residually finite.*

*Sketch of a proof* Let $A_1, A_2, \ldots, A_n$ be any finite set of matrices generating $G$. Without loss in generality, assume that the identity matrix is among them. Let $X$ denote the set of elements of $\mathbb{K}$ formed as follows: whenever $k$ is an entry of $A_i A_j^{-1}$, then we put $k, k^{-1}, k-1$ and $(k-1)^{-1}$ into $X$ whenever they are defined. Let $R \subseteq \mathbb{K}$ be the ring generated by $X$. It is an integral domain, and so for every maximal ideal $I$ of $R$ the quotient ring $R/I$ is a field.

Consider a natural homomorphism $\phi\colon G \to GL(n, R/I)$ induced by the quotient modulo $I$. Note that when an element has its inverse in $R$, it cannot be in $I$, and so by the choice of generators of $R$, none of matrices $A_i A_j^{-1}$ is equal to the identity in $GL(n, \mathbb{R}/I)$. Thus, $\phi(A_i) \neq \phi(A_j)$, and since $A_i$ were arbitrary, we conclude that homomorphisms $\phi$ as above separate points of $G$.

It remains to notice that the field $R/I$ is finite, because a finitely generated ring that is a field is finite. This part of the proof requires most effort, and the details can be found e.g. in [6], Theorem 6.4.12. □

Our purpose is served by the following concept, which is more general than residual finiteness.

**Definition 1.4** (Vershik and Gordon [28]) A group $G$ is said to be *locally embeddable into finite groups* (an *LEF group*, for short) if for every finite subset $F \subseteq G$ there is a partially defined monomorphism $i$ of $F$ into a finite group.

*Remark* 1.5 Every residually finite group is LEF, which is immediate from (2.).

Another source of LEF groups is given by the following notion.

**Definition 1.6** A group is *locally finite* if every finite set is contained in a finite subgroup (i.e. if every finitely generated subgroup is finite).

**Example 1.7** $S_\infty^{fin}$ = the group of finitely supported bijections of $\mathbb{N}$ is locally finite, but not residually finite. Indeed, the only normal subgroup of $S_\infty^{fin}$ is the group of finitely supported bijections of $\mathbb{N}$ of even sign.

The following result is folk knowledge.

**Theorem 1.8** *For a group $G$ the following are equivalent:*

1. *$G$ admits matrix models "in the classical sense,"*

2. *G is LEF.*

*Proof* (i) ⇒ (ii): Let $F$ be a finite subset of $G$. Then there is a local monomorphism $i: F \to GL(n, \mathbb{K})$. The group $\langle i(F) \rangle$ is residually finite by Theorem 1.3, hence there is a homomorphism $j: \langle i(F) \rangle \to H$, where $H$ is a finite group, whose restriction to $i(F)$ is injective. The composition $j \circ i: F \to H$ is the required partial monomorphism.

(i) ⇐ (ii): Let $F$ be a finite subset of $G$. There is a partial monomorphism $i: F \to H$, where $H$ is a finite group. Let $j: H \to S_n$ be an embedding into a finite permutation group (every finite group embeds into some $S_n$). To finish the proof we notice that $S_n$ embeds into $GL(n, \mathbb{K})$ for an arbitrary field $\mathbb{K}$: to a permutation $\sigma$ we assign the matrix $A = (a_{ij})_{i \le n, j \le n}$ by letting $a_{ij} = 1$ if $\sigma(i) = j$, and $a_{ij} = 0$ otherwise. □

Notice that the above proof, the choice of a field $\mathbb{K}$ does not matter.

To finish this introductory section, we will show that not every group is LEF. Letting $N(r_1, r_2, \ldots, r_m)$ denote the normal subgroup generated by elements $r_1, r_2, \ldots, r_m$, a group $G$ is *finitely presented* when $G \cong F_n / N(r_1, r_2, \ldots, r_m)$, where $r_1, r_2, \ldots, r_m$ is a finite collection of relators and $n \in \mathbb{N}$.

**Proposition 1.9** *Suppose G is an infinite simple finitely presented group. Then G is not LEF.*

*Proof* Represent $G$ as $F_n/N$, where, for short, $N = N(r_1, r_2, \ldots, r_m)$. Let $X$ be the set of free generators of $F_n$. Denote by $d$ the word metric on $F_n$ with respect to $X$, given by

$$d(x, y) = \min\{i : y = b_1 b_2 \ldots b_i x;\ b_1, b_2, \ldots, b_i \in X \cup X^{-1}\}.$$

Let $R$ be so large that the $R$-ball $B_R$ around identity in $F_n$ contains the relators $r_1, r_2, \ldots, r_m$. Denote $\pi: F_n \to F_n/N$ the canonical homomorphism and put $\widetilde{B}_R = \pi(B_R)$.

Suppose that $G$ is LEF. Let $j: \widetilde{B}_R \to H$ be an injection into a finite group $H$ preserving partial multiplication. Define a homomorphism $h: F_n \to H$ by the condition $h(x) = j \circ \pi(x)$, $x \in X$. The kernel $N' = \ker(h)$ is a proper subset of $F_n$ and a proper superset of $N$ (as $F_n/N$ is infinite). This contradicts the simplicity of $G$. □

**Example 1.10** *Thompson's groups:*

$F$ = all orientation-preserving piecewise linear homeomorphisms of $[0, 1]$ with finitely many non-smooth points which are all contained in the set of dyadic rationals, and the slopes being integer powers of 2.

$T$ = all orientation-preserving piecewise linear homeomorphisms of $\mathbb{T} = \mathbb{R}/\mathbb{Z}$ with finitely many non-smooth points which are contained in dyadic rationals, and slopes being integer powers of 2.

$V$ = all orientation-preserving piecewise linear bijections of $[0, 1]$ (not necessarily continuous), with finitely many points of discontinuity, all contained in the set of dyadic rationals, the slopes being integer powers of 2.

Every group $F, T, V$ is finitely presented. Moreover $T$ and $V$, and the commutator of $F$ are simple. A standard reference to Thompson's groups is the survey by Cannon, Floyd and Parry [7].

By Proposition 1.9, Thompson's groups are not LEF.

## 2 Algebraic ultraproducts

An important role played by ultraproducts in logic and model theory is well known. Hyperlinear/sofic groups are no exception, and in the subsequent sections ultraproducts of metric groups will have a significant impact. In this section we will discuss algebraic ultraproducts of groups, and show how to reformulate in their language the existence of matrix models "in the classical sense". In particular, the ultraproduct technique allows for a simpler proof of Theorem 1.8, bypassing Malcev's theorem 1.3.

Recall that, given a family $G_\alpha$, $\alpha \in A$,s of groups and an ultrafilter $\mathcal{U}$ on the index set $A$, the (*algebraic*) *ultraproduct* of the family $(G_\alpha)$ is defined as follows:

$$\left(\prod_{\alpha\in A} G_\alpha\right)_{\mathcal{U}} = \left(\prod_{\alpha\in A} G_\alpha\right)/N_{\mathcal{U}},$$

where

$$N_{\mathcal{U}} = \{x \colon x \sim_{\mathcal{U}} e\}$$

and

$$x \sim_{\mathcal{U}} y \iff \{\alpha \in A \colon x_\alpha = y_\alpha\} \in \mathcal{U}.$$

Notice that $N_{\mathcal{U}}$ is a normal subgroup of the direct product of groups $G_\alpha$.

In a similar way, one can define an algebraic ultraproduct of a family of any algebraic structures of the same signature. In particular, if $\mathbb{K}_\alpha$, $\alpha \in A$ are fields, then the subset $\mathcal{I}_{\mathcal{U}} = \{x \colon x \sim_{\mathcal{U}} 0\}$ is a maximal ideal of the direct product ring $\prod_{\alpha\in A} \mathbb{K}_\alpha$, and the corresponding quotient field

$(\prod_{\alpha\in A}\mathbb{K}_\alpha)_{\mathcal{U}}$ is called the ultraproduct of the fields $\mathbb{K}_\alpha$ modulo $\mathcal{U}$. (It is useful to notice that the underlying set of the algebraic ultraproduct is independent of the algebraic structure, because only $=$ is used in the definition of the equivalence relation $\sim_{\mathcal{U}}$.)

Now it is easy to make the following observations, going back to Jerzy Łoś [21].

1. Let $\mathcal{U}$ be a nonprincipal ultrafilter on the natural numbers, and let $A = \cup_n A_n$ be the union of an increasing chain of some algebraic structures (e.g. groups, fields, ...). Then $A$ canonically embeds in the ultraproduct $(\prod_n A_n)_{\mathcal{U}}$. (To every $a \in A$ one associates an equivalence class containing any eventually constant sequence stabilizing at $a$.)
2. Let $n \in \mathbb{N}$ and let $\mathbb{K}_\alpha$, $\alpha \in A$ be fields. Then for every ultrafilter $\mathcal{U}$ on $A$ the groups $(\prod_{\alpha\in A} GL(n,\mathbb{K}_\alpha))_{\mathcal{U}}$ and $GL\left(n,(\prod_{\alpha\in A}\mathbb{K}_\alpha)_{\mathcal{U}}\right)$ are isomorphic.

   (The canonical ring isomorphism between the rings $M_n(\prod\mathbb{K}_\alpha)$ and $\prod_\alpha M_n(\mathbb{K}_\alpha)$ factors through the relation $\sim_{\mathcal{U}}$ to a ring isomorphism between the rings $(\prod_{\alpha\in A} M_n(\mathbb{K}_\alpha))_{\mathcal{U}}$ and $M_n\left((\prod_{\alpha\in A}\mathbb{K}_\alpha)_{\mathcal{U}}\right)$. The ultraproduct of the general linear groups of $K_\alpha$ sits inside the former ring as the group of all invertible elements, while the general linear group of the ultraproduct of $\mathbb{K}_\alpha$ is by its very definition the group of invertible elements of the latter ring.)
3. The ultraproduct of a family of ultraproducts is again an ultraproduct.
4. The ultraproduct of a family of algebraically closed fields is algebraically closed.
5. Let $X_n$ be non-empty finite sets and let $\mathcal{U}$ be an ultrafilter on the set of natural numbers. If for every $N \in \mathbb{N}$ $\{n \in \mathbb{N}\colon |X_n| < N\} \notin \mathcal{U}$, then the cardinality of the ultraproduct of $X_n$ mod $\mathcal{U}$ equals $\mathfrak{c}$.
6. An algebraic ultraproduct of a family of LEF groups is again an LEF group.

The following result is weaker than Malcev's theorem (of which it is a corollary thanks to observation (2) above), but is nonetheless strong enough for our purposes.

*Observation* Every field $\mathbb{K}$ embeds, as a subfield, into a suitable ultraproduct of a family of finite fields.

*Proof* We will only give an argument in the case where the cardinality of $\mathbb{K}$ does not exceed that of the continuum, leaving an extension to the general case to the reader. Let at first $p > 0$ be a positive characteristic. Select an increasing chain of finite algebraic extensions of $\mathbb{F}_p$ whose union is the algebraic closure, $\overline{\mathbb{F}_p}$, of $\mathbb{F}_p$ (e.g. $(\mathbb{F}_{p^k})$), and fix a free ultrafilter on $\mathbb{N}$. The ultraproduct modulo $\mathcal{U}$ of finite fields forming this chain contains $\overline{\mathbb{F}_p}$ by (1.). By (3.), there is an ultraproduct of the family $(\mathbb{F}_p)$ containing a non-trivial ultrapower of $\overline{\mathbb{F}_p}$ as a subfield. This ultrapower, denote it $\mathbb{K}_p$, is an algebraically closed field by (4.), and its transcendence degree is $\mathfrak{c}$ by force of (5.). Since in a given characteristic two algebraically closed fields of the same transcendence degree are isomorphic (Steinitz' theorem), our result now follows in the case of prime characteristic. To settle the case of characteristic zero, notice that the ultraproduct of all fields $\mathbb{K}_p$ modulo a nonprincipal ultrafilter over the prime numbers is an algebraically closed field of characteristic zero and of transcendence degree continuum. □

In the following strengthening of Theorem 1.8, the equivalence (1.) $\iff$ (2.) is an immediate consequence of a well-known general result in logic, see Lemma 3.8 in Chapter 9 [2], but this is not the main point here.

**Theorem 2.1** *For a group G the following are equivalent:*

1. *G admits matrix models "in the classical sense", that is, every existential sentence from the first-order theory of G is valid in some matrix group.*
2. *$G < (\prod_i GL(n_i, \mathbb{K}_i))_{\mathcal{U}}$ for some family of fields $\mathbb{K}_i$, natural numbers $n_i$, and an ultrafilter $\mathcal{U}$.*
3. *G is LEF.*
4. *G embeds into the algebraic ultraproduct of a family of permutation groups of finite rank.*
5. *For every field $\mathbb{K}$, $G < (\prod_i GL(n_i, \mathbb{K}))_{\mathcal{U}}$ for a suitably large index set and a suitable ultrafilter $\mathcal{U}$.*

*Proof* (1.)$\Rightarrow$(2.): On $\mathscr{P}_{fin}(G)$, the family of all finite subsets of $G$ ordered by inclusion, take an ultrafilter containing all upper cones, that is, the sets

$$\{\Phi \in \mathscr{P}_{fin}(G) \colon \Phi \supseteq F\},$$

where $F \in \mathscr{P}_{fin}(G)$. For each $\Phi \in \mathscr{P}_{fin}(G)$ choose a field $\mathbb{K}_\Phi$, a natural number $n_\Phi$, and an injection $j_\Phi \colon \Phi \to GL(n_\Phi, \mathbb{K}_\Phi)$ preserving partial multiplication (Observation 1). Define

$$j\colon G \to \left( \prod_{\Phi \in \mathscr{P}_{fin}(G)} GL(n_\Phi, \mathbb{K}_\Phi) \right)_{\mathcal{U}}$$

by

$$j(g) = [j_\Phi(g)]_{\mathcal{U}}$$

(when $g \notin \Phi$, $j_\Phi(g)$ denotes an arbitrary element of $GL(n_\Phi, \mathbb{K})$). This $j$ is an embedding of groups.

(2.) ⇒ (3.). By embedding every field $\mathbb{K}_i$ into an ultraproduct of finite fields (Obs. 2), and using observations (2.) and (3.), we can assume without loss in generality that all $\mathbb{K}_i$ are finite fields. Let $F \subseteq G$ be finite. For every $g \in F$, pick a representative $(j(g)_i) \in \prod_i GL(n_i, \mathbb{K}_i)$ of the equivalence class $[j(g)]_{\mathcal{U}}$. For every index $i$, there is now a well-defined mapping $F \ni g \mapsto j(g)_i \in GL(n_i, \mathbb{K}_i)$. The set of all indices $i$ for which $j(g)_i$ is a local monomorphism must belong to the ultrafilter and so is non-empty. Choose an index $i$ from this set and notice that the group $GL(n_i, \mathbb{K}_i)$ is finite.

(3.) ⇒ (4.): A similar argument to the proof of implication (1.)⇒(2.), only take as $j_\Phi$ a local monomorphism from $\Phi$ into a suitable finite group of permutations (which exists since $G$ is assumed LEF).

(4.) ⇒ (5.): Here use the fact that $S_n$ sits inside of the group $GL(n, \mathbb{K})$ as a subgroup for every $\mathbb{K}$ and $n$.

(5.)⇒(1.): Let $j\colon G < \prod_i (GL(n_i, \mathbb{K}))_{\mathcal{U}}$ be an embedding, and let $F \subseteq G$ be finite. Then $j \restriction F$ is a partial monomorphism, and so $\{i\colon j_i \restriction F \text{ is a partial monomorphism}\} \in \mathcal{U}$ (so in particular is nonempty). The result now follows by Observation 1. □

Overall, we can see that theory of groups admitting matrix models "in the classical sense" is more or less fully understood. This approach can be seen as a "toy example" (to borrow another expression from theoretical physics) of more interesting and mysterious theories of group matrix models, to which we proceed now.

## 3 Ultraproducts of metric structures

The concept of a matrix model adequate for the needs of operator algebraists is less strict than the one "in classical sense". We do not aim to ascertain that two elements of a matrix group, $x$ and $y$, are equal. Instead, given an $\varepsilon > 0$, we are allowed to interpret a formula $x = y$ in a matrix group in such a way that the "truth value" of the equality is $> 1 - \varepsilon$. This is understood in the sense

$$d(x, y) < \varepsilon,$$

where $d$ is a distance on a matrix group in question and $x, y$ are elements of the group. Accordingly, instead of the algebraic ultraproduct of matrix groups, we will form the *metric ultraproduct,* factoring out pairs of infinitesimally close elements.

The aim of this section is to formulate an adequate version of an ultraproduct of a family of metric groups, and to give some examples.

To make a good choice of a distance $d$ as above, let us first examine the notion of the Banach space ultraproduct, which is well established.

## 3.1 Ultraproducts of normed spaces

Let $(E_\alpha)_{\alpha\in A}$ be a family of normed spaces and let $\mathcal{U}$ be an ultrafilter on the index set $A$. Define the $\ell^\infty$*-type sum* of the spaces $E_\alpha$,

$$\mathscr{E} = \oplus^{\ell^\infty} E_\alpha = \left\{ x \in \prod_\alpha E_\alpha \colon \sup_\alpha \|x_\alpha\| < \infty \right\}.$$

This $\mathscr{E}$ is a normed linear space containing every $E_\alpha$ as a normed subspace. The norm on $\mathscr{E}$ is given by:

$$\|x\| = \sup_{\alpha\in A} \|x_\alpha\|_\alpha .$$

Consider

$$\mathscr{N}_\mathcal{U} = \left\{ x\colon \lim_{\alpha\to\mathcal{U}} \|x_\alpha\| = 0 \right\},$$

where we let $\lim_{\alpha\to\mathcal{U}} y_\alpha = y$ if for every $\varepsilon > 0$, $\{\alpha\colon |y_\alpha - y| < \varepsilon\} \in \mathcal{U}$. If the $y_\alpha$ are uniformly bounded, then $\lim_{\alpha\to\mathcal{U}} y_\alpha$ exists and is unique.

This $\mathscr{N}_\mathcal{U}$ is a closed linear subspace of $\mathscr{E}$. Now we define the metric ultraproduct of the family $(E_\alpha)_{\alpha\in A}$ modulo the ultrafilter $\mathcal{U}$ as the normed quotient space

$$\left(\prod E_\alpha\right)_\mathcal{U} = \mathscr{E}/\mathscr{N}_\mathcal{U}.$$

It is a linear space equipped with the norm

$$\|[x]_\mathcal{U}\| = \lim_{\alpha\to\mathcal{U}} \|x_\alpha\| .$$

A version of the diagonal argument shows that when the ultrafilter $\mathcal{U}$ is not countably complete (in particular, is non-principal), then the ultraproduct $E = (\prod E_\alpha)_{\mathcal{U}}$ is a Banach space. To see this, let $(x_k)$ be a Cauchy sequence of elements of $E$. For each $i \in \mathbb{N}$, fix $N(i)$ so that

$$\forall N', N \geq N(i), \quad \|x_{N'} - x_N\| < 2^{-i}.$$

For every $k$, select a representative $(x_k^\alpha)_{\alpha \in A} \in \mathscr{E}$ of the equivalence class $x_k$. Given an $i \in \mathbb{N}_+$, define

$$I_i = \{\alpha \in A \colon \left\|x^\alpha_{N(i)} - x^\alpha_{N(i+1)}\right\| < 2^{-i}\}.$$

Every $I_i \in \mathcal{U}$, and without loss in generality, we may assume that $I_1 \supseteq I_2 \supseteq \ldots$ and $\bigcap_{i=1}^\infty I_i = \emptyset$ (countable incompleteness of $\mathcal{U}$). Now define an element $x \in \mathscr{E}$ by

$$x|_{I_i \setminus I_{i+1}} = x_{N(i)}|_{I_i \setminus I_{i+1}}.$$

Then the equivalence class $[x]_{\mathcal{U}}$ is the limit of our Cauchy sequence $(x_k)$.

If for some natural number $n$ the set of indices $\alpha$ with $\dim E_\alpha = n$ is in $\mathcal{U}$, then the ultraproduct is a normed linear space of dimension $n$. If it is not the case for any $n$, then yet another variation of Cantor's argument establishes that the ultraproduct is a non-separable Banach space.

## 3.2 Ultraproducts of metric groups

We would like to have a similar construction for metric groups as we had for normed spaces. First we show that if we just equip the groups with left-invariant metrics (and every metrizable group admits a compatible left-invariant metric by the result of Kakutani below), some problems arise. Hence, we will have to assume that metrics are bi-invariant. Not every metrizable group has a compatible bi-invariant metric.

**Theorem 3.1** (Kakutani) *Every metrizable topological group admits a compatible left-invariant metric, i.e. a metric $d$ such that for every $g \in G$*

$$d(gx, gy) = d(x, y).$$

□

Let $(G_\alpha, d_\alpha)_{\alpha \in A}$ be a family of topological groups equipped with compatible left-invariant metrics, and let $\mathcal{U}$ be an ultrafilter on $A$. We can form an ultraproduct of the family $(G_\alpha, d_\alpha)$ following the same steps as for normed spaces, but the resulting object will not, in general, be a metric group, only a homogeneous metric space, as the following example shows.

**Example 3.2** Let $S_\infty$ denote the infinite symmetric group consisting of all self-bijections of a countably infinite set $\omega$. The *standard Polish topology* on $S_\infty$ is the topology of pointwise convergence on the discrete topological space $\omega$. In other words, it is induced from the product topology on $\omega^\omega$. As shown by Kechris and Rosendal [20], the standard Polish topology is the only non-trivial separable group topology on $S_\infty$. This topology admits the following compatible left-invariant metric:

$$d(\sigma,\tau) = \sum_{i=1}^{\infty}\{2^{-i} : \sigma(i) \neq \tau(i)\}.$$

Let us try to form an ultrapower of the metric group $(S_\infty, d)$ with regard to a nonprincipal ultrafilter $\mathcal{U}$ on the natural numbers. Every sequence $x \in (S_\infty)^{\mathbb{N}}$ is "bounded" in the sense that $\sup_n d(e_n, x_n) < \infty$, and so the analogue of the space $\mathscr{E}$ is the full Cartesian product group $\mathscr{G} = (S_\infty)^{\mathbb{N}}$ itself. Define

$$\mathscr{N} = \left\{x\colon \lim_{\alpha\to\mathcal{U}} d(x_\alpha, e) = 0\right\}.$$

The estimate

$$\begin{aligned} d(xy,e) &= d(y,x^{-1}) \\ &\leq d(y,e) + d(x^{-1},e) \\ &= d(y,e) + d(e,x) \end{aligned}$$

shows that $\mathscr{N}$ is a subgroup of $\mathscr{G}$. (Notice the use of left-invariance of $d$.) However, it is not a *normal* subgroup. To see this, consider two sequences of transpositions of $\omega$, $x = (x_i) = ((i, i+1))_{i\in\omega}$ and $y = (y_i) = ((1, i))$. Then it is easily seen that $x \in \mathscr{N}$, and yet $y^{-1}xy \notin \mathscr{N}$. Thus, although the homogeneous factor-space $\mathscr{G}/\mathscr{N}$ admits a $\mathscr{G}$-invariant metric

$$d(x,y) = \lim_{n\to\mathcal{U}} d_n(x_n, y_n),$$

it is not a group.

If we want to get a *metric group* as a result of an ultraproduct construction, we must use *bi-invariant* metrics:

$$d(gx, gy) = d(x,y) = d(xg, yg).$$

If $(G_\alpha, d_\alpha)$, $\alpha \in A$, is a family of groups equipped with bi-invariant metrics and $\mathcal{U}$ is an ultrafilter on the index set $A$, then the subgroup

$$\mathcal{N} = \left\{x\colon \lim_{\alpha\to\mathcal{U}} d(x_\alpha, e) = 0\right\}$$

is easily seen to be a normal subgroup of

$$\mathcal{G} = \oplus^{\ell^\infty} G_\alpha = \left\{x \in \prod_\alpha G_\alpha\colon \sup_\alpha d(x_\alpha, e) < \infty\right\}, \tag{5.1}$$

and the quotient group

$$\left(\prod_{\alpha\in A} G_\alpha\right)_{\mathcal{U}} = \mathcal{G}/\mathcal{N}$$

is well-defined. It is a metric group equipped with the bi-invariant metric

$$d(x\mathcal{N}, y\mathcal{N}) = \lim_{\alpha\to\mathcal{U}} d_\alpha(x_\alpha, y_\alpha)$$

and the corresponding group topology. It will be referred to as the *metric ultraproduct* of the family $(G_\alpha, d_\alpha)_{\alpha\in A}$ modulo $\mathcal{U}$.

Just as in the case of normed spaces, the ultraproduct of a family of groups with bi-invariant metrics is a complete topological group, which is either non-separable or locally compact (assuming $\mathcal{U}$ to be non countably complete). Moreover, in all the examples we will be considering below, the domain of the ultraproduct coincides with the full cartesian product, because all the metrics are uniformly bounded from above. (In fact, one can always replace a bi-invariant metric $d$ on a group with the bounded bi-invariant metric $\min\{d, 1\}$).

Here are a few of the most important examples of groups equipped with natural bi-invariant metrics.

**Example 3.3** The symmetric group $S_n$ of finite rank $n$ equipped with the *normalized Hamming distance*:

$$d_{hamm}(\sigma, \tau) = \frac{1}{n}\sharp\{i\colon \sigma(i) \neq \tau(i)\}.$$

**Example 3.4** The unitary group of rank $n$,

$$U(n) = \{u \in M_n(\mathbb{C})\colon u^* u = uu^* = \mathrm{Id}\},$$

equipped with the *normalized Hilbert-Schmidt metric*:

$$d_{HS}(u,v) = \|u-v\|_2 = \sqrt{\frac{1}{n}\sum_{i,j=1}^{n}|u_{ij}-v_{ij}|^2}.$$

This is the standard $\ell^2$ distance between matrices viewed as elements of an $n^2$-dimensional Hermitian space $\mathbb{C}^{n^2}$, which is normalized so as to make the identity matrix have norm one. The metric is easily checked to be bi-invariant, by rewriting the definition of the distance,

$$\begin{aligned} d_{HS}(u,v) &= \frac{1}{\sqrt{n}}\sqrt{\operatorname{tr}((u-v)^*(u-v))} \qquad (5.2)\\ &= \sqrt{2-\widetilde{\operatorname{tr}}_n(u^*v)-\widetilde{\operatorname{tr}}_n(v^*u)}, \end{aligned}$$

where $\widetilde{\operatorname{tr}}_n = n^{-1/2}\operatorname{tr}$ is the normalized trace on $U(n)$, and using the characteristic property of trace:

$$\operatorname{tr}(AB) = \operatorname{tr}(BA).$$

We will use the notation $U(n)_2$ for the group $U(n)$ equipped with the normalized Hilbert-Schmidt distance.

**Example 3.5** The group $U(n)$ equipped with the *uniform operator metric:*

$$d_{unif}(u,v) = \|u-v\| = \sup_{\|x\|\le 1}\|(u-v)(x)\|.$$

Larger matrix groups, such as $GL(n,\mathbb{K})$ and their closed non-compact subgroups, typically do not possess any compatible bi-invariant metrics whatsoever. For instance, the following is a well-known observation.

**Example 3.6** The group of invertible matrices $GL(n,\mathbb{R})$, as well as the special linear group $SL(n,\mathbb{R})$, do not admit bi-invariant metrics compatible with their standard locally euclidean topology (induced from $M_n(\mathbb{R}) \cong \mathbb{R}^{n^2}$). (Hint of a proof: if such a metric existed, then the group in question would possess *small invariant neighbourhoods,* that is, conjugation-invariant open sets would form a basis at identity. But this is not the case. The details can be found in [18].)

## 4 Groups admitting matrix models (hyperlinear groups)

In this Section, we define the central concept of a hyperlinear group, and outline a version of model theory for metric structures which provides a rigorous framework for treating hyperlinear groups as groups admitting matrix models which are unitary groups with the Hilbert–Schmidt distance.

**Definition 4.1** A countable discrete group $G$ is *hyperlinear* (or: *admits matrix models*) if it is isomorphic to a subgroup of a metric ultraproduct of a suitable family of unitary groups of finite rank, with their normalized Hilbert-Schmidt distances.

More exactly, $G$ is hyperlinear if there are a set $A$, an ultrafilter $\mathcal{U}$ on $A$, a mapping $\alpha \mapsto n(\alpha)$ and an imbedding

$$G < \left(\prod_{\alpha}(U(n(\alpha)), d_{HS})\right)_{\mathcal{U}}.$$

The model theory of metric structures as developed by Ben-Yaacov, Berenstein, Ward Henson and Usvyatsov [3] allows to see the above definition as a genuine statement about a possibility to interpret every sentence of the theory of $G$ in some matrix group $U(n)_2$. We will not attempt to develop this viewpoint systematically, limiting ourselves to a few indicative remarks.

The space of truth values in this version of continuous logic is the unit interval $\mathbb{I} = [0, 1]$. The truth value is interpreted as a measure of closeness, and in particular the truth value of the formula $x = y$ is $d(x, y)$.

The two quantifiers are inf (continuous analogue of $\exists$) and sup (analogue of $\forall$). Predicates are (bounded, uniformly continuous) functions $M^n \to [0, 1]$, e.g. the counterpart of the equality relation = is the distance function $d\colon M^2 \to [0, 1]$.

Similarly, functions $M^n \to M$ are subject to the uniform continuity restriction. Connectives are all continuous functions $[0, 1]^n \to [0, 1]$. If $d$ is a trivial ($\{0, 1\}$-valued) metric, one recovers the usual predicate logic with truth values 0 and 1, which have swapped their places.

Every sentence in the classical theory can be thus interpreted as a sentence formed in the continuous logic, but not vice-versa.

Formulas are defined inductively, just like in the classical logic. All variables and constants are terms, and whenever $f$ is a function symbol and $t_1, \ldots, t_n$ are terms, then $f(t_1, t_2, \ldots, t_n)$ is a term. An atomic formula is an expression of the form either $P(t_1, \ldots, t_n)$ or $d(t_1, t_2)$, where

$P$ is an $n$-ary predicate symbol and $t_i$ are terms. Formulas are build from atomic formulas, using two rules: if $u\colon [0,1]^n \to [0,1]$ is a continuous function (that is, a connective) and $\varphi_1, \ldots, \varphi_n$ are formulas, then $u(\varphi_1, \ldots, \varphi_n)$ is a formula; if $\varphi$ is a formula and $x$ is a variable, then $\sup_x \varphi$ and $\inf_x \varphi$ are formulas.

A *metric structure*, $\mathcal{M}$, is a complete bounded metric space equipped with a family of predicates and functions. For instance, if $(G, d)$ is a complete bounded metric group and $d$ is bi-invariant, then $G$ can be treated as a metric structure equipped with the predicate $S(g, h, k) = d(gh, k)$, the inversion function $i\colon G \to G$, and the identity, given by the function $e\colon \{*\} \to G$ (a homomorphism from the trivial group to $G$). Notice that $S$ and $i$ are uniformly continuous due to the bi-invariance of the metric $d$.

The *value* of a sentence $\sigma$ in a metric structure $\mathcal{M}$ is a number $\sigma^{\mathcal{M}} \in [0,1]$ defined by induction on (variable-free) formulas, beginning with the convention that $d(t_1, t_2)^{\mathcal{M}}$ is just the value of the distance between $t_1$ and $t_2$. The value of $P(t_1, t_2, \ldots, t_n)$ and $u(\sigma_1, \sigma_2, \ldots, \sigma_n)$, where $P$ is an $n$-ary predicate symbol, $t_i$ are terms, $u$ is a continuous function $[0,1]^n \to [0,1]$, and $\sigma_j$ are sentences, is defined in a natural way. Finally,

$$\left(\sup_x \varphi(x)\right)^{\mathcal{M}} = \sup_x \varphi(x)^{\mathcal{M}},$$

and similarly

$$\left(\inf_x \varphi(x)\right)^{\mathcal{M}} = \inf_x \varphi(x)^{\mathcal{M}}.$$

A sentence of the form $\inf_x \varphi(x)$, where $x = (x_1, x_2, \ldots, x_n)$, is called an inf-*sentence,* and serves as an analogue of an existential sentence in the classical binary logic.

Notice that the normalized Hilbert-Schmidt distance $d_{HS}$ takes values in the interval $[0,2]$, so if we want the values of sentences to belong to $[0,1]$, we may wish to use the distance $d = \min\{d_{HS}, 1\}$ instead.

Every formula of the first-order theory of groups admits a "translation" into a formula of the continuous logic theory of groups equipped with a bi-invariant metric. Namely, the symbols $S$, $i$ and $e$ are replaced with the corresponding predicate and function symbols described above, the logical connectives $\wedge$ and $\vee$ become, respectively, continuous functions max and min from $[0,1]^2$ to $[0,1]$, while $\neg$ is replaced with the function $t \mapsto 1 - t$, and the quantifiers $\exists_x$ and $\forall_x$ are turned into $\inf_x$ and $\sup_x$, accordingly. Under this translation, sentences go to sentences, existential sentences go to inf-sentences, and so on. Intuitively, under this "translation," exact statements become approximate. It is in this sense that we treat sentences of Th($G$) as sentences of the continuous logic theory of unitary groups $U(n)_2$ in the statement of the next result.

In connection with item (2.) below, remember that $U(n)_2$ embeds isometrically into the (renormalized) Euclidean space $\ell^2(n^2)$, and so the ultraproduct of unitary groups isometrically embeds into the corresponding Hilbert space ultraproduct. In this sense, one can talk about orthogonality.

**Theorem 4.2** *For a group G, the following are equivalent.*

1. *G is hyperlinear,*
2. *G embeds into a metric ultraproduct* $(\prod_i U(n_i)_2)_{\mathcal{U}}$ *of a family of unitary groups as an orthonormal system of vectors,*
3. *For every finite* $F \subseteq G$ *and every* $\varepsilon > 0$, *there are* $n \in \mathbb{N}$ *and an* $(F, \varepsilon)$-almost homomorphism $j\colon F \to U(n)_2$, *that is, a map with the property*

   *(a) if* $g, h \in F$ *and* $gh \in F$, *then* $d(j(g)j(h), j(gh)) < \varepsilon$,

   *which is in addition* uniformly injective *on F in the sense that:*

   *(b) if* $g, h \in F$ *and* $g \neq h$, *then* $d(g, h) > \sqrt{2} - \varepsilon$.
4. *G admits matrix models in the sense of continuous logic: for every existential first-order sentence* $\sigma \in \mathrm{Th}^{\exists}(G)$ *and each* $\varepsilon > 0$ *there is n such that*
$$\sigma^{U(n)_2} < \varepsilon.$$
5. *The same conditions (a) and (b) as in item (3.), but with* $\sqrt{2} - \varepsilon$ *in (b) replaced by a fixed positive value, e.g.* $10^{-10}$.

*Proof* (1.)⇒(2.): This is the key to the entire result, whence the rest follows easily. Given a monomorphism

$$i\colon G \hookrightarrow \left(\prod_{\alpha} U(n_\alpha)_2\right)_{\mathcal{U}}, \tag{5.3}$$

and any two distinct elements $g, h \in G$, we can of course guarantee that the images $i(g)$ and $i(h)$ are at a strictly positive distance from each other, but no more than that: something like $d(i(g), i(h)) = 10^{-10}$ is definitely a possibility, and in fact the image $i(G)$ in the induced topology may even happen to be a non-discrete group. We will now construct a re-embedding, $j$, of $G$ into another metric ultraproduct of unitary groups, where the distance between $j(g)$ and $j(h)$ will be always equal to $\sqrt{2}$.

The Hermitian space $M_n(\mathbb{C})$, which we identify with $\mathbb{C}^{n^2}$, admits a natural action of the unitary group $U(n)$ by conjugations:

$$u \cdot M = u^* M u, \quad u \in U(n), \quad M \in M_n(\mathbb{C}).$$

This action is by linear operators and preserves the Hermitian inner product, for instance, since Formula (5.2), without the scalar factor in front, gives the Hilbert distance on the space $M_n(\mathbb{C})$. Thus, we obtain a unitary representation $U(n) \to U(n^2)$. Denote it $i_n^{(2)}$ (in fact, the more precise symbol would be $\bar{i}_n \otimes i_n$).

We want to compute the distance induced on $U(n)$ by the embedding $i_n^{(2)}\colon U(n) \hookrightarrow U(n^2)_2$ as above. For this purpose, again according to Equation (5.2), it suffices to know the restriction of the trace $\mathrm{tr}_{n^2}$ to $U(n)$, that is, the composition $\mathrm{tr}_{n^2} \circ i_n^{(2)}$. The matrices $E_{ij}$ whose $(i, j)$-th position is one and the rest are zeros form an orthonormal basis of $M_n(\mathbb{C})$, and so for every linear operator $T$ on $M_n(\mathbb{C})$,

$$\mathrm{tr}(T) = \sum_{ij} \langle T(E_{ij}), E_{ij} \rangle = \sum_{ij} \left( T(E_{ij}) \right)_{ij}.$$

Since $(u \cdot E_{ij})_{ij} = (u^* E_{ij} u)_{ij} = \overline{u_{ji}} u_{ji}$, we conclude: for every $u \in U(n)$,

$$\mathrm{tr}_{n^2}(i_n^{(2)}(u)) = \overline{\mathrm{tr}_n(u)}\, \mathrm{tr}_n(u) = |\mathrm{tr}_n(u)|^2.$$

The same clearly holds with regard to the normalized traces on both unitary groups. Since at the same time the trace is a linear functional, we deduce from Equation (5.2):

$$\begin{aligned} d_{HS,n^2}\left(i_n^{(2)}(u), i_n^{(2)}(v)\right) &= \frac{1}{\sqrt{n^2}} \sqrt{\mathrm{tr}_{n^2}((i_n^{(2)}(u) - i_n^{(2)}(v))^* (i_n^{(2)}(u) - i_n^{(2)}(v)))} \\ &= \frac{1}{n} \sqrt{\mathrm{tr}_{n^2}(2\mathbb{I} - \mathrm{tr}_{n^2}(i_n^{(2)}(u^* v)) - \mathrm{tr}_{n^2}(i_n^{(2)}(v^* u)))} \\ &= \sqrt{2 - 2\left|\widetilde{\mathrm{tr}}_n(u^* v)\right|^2}, \end{aligned}$$

where $\widetilde{\mathrm{tr}}_n$ denotes the normalized trace on $U(n)$.

Compare this to:

$$d_{HS,n}(u, v) = \sqrt{2 - \widetilde{\mathrm{tr}}_n(u^* v) - \widetilde{\mathrm{tr}}_n(v^* u)}.$$

Since $\widetilde{\mathrm{tr}}_n(u^*v) + \widetilde{\mathrm{tr}}_n(v^*u) = 2\left|\widetilde{\mathrm{tr}}_n(u^*v)\right|$, the last two equations imply:

$$d_{HS,n^2}\left(i_n^{(2)}(u), i_n^{(2)}(v)\right) = d_{HS,n}(u,v)\sqrt{2 - \frac{d_{HS,n}(u,v)^2}{2}}. \tag{5.4}$$

If we now define recurrently group embeddings $i_n^{(2^k)}: U(n) \hookrightarrow U(n^{2^k})$, $k = 2, 3, \ldots$, it follows that for any two elements $u, v \in U(n)$ satisfying $0 < d_{HS,n}(u,v) < 2$ the iterated distances inside of the groups $U(n^{2^k})_2$ converge to $\sqrt{2}$ in the limit $k \to \infty$.

Now let us get back to the initial group embedding from Equation (5.3). First, we want to assure that the pairwise distances within the image $i(G)$ are strictly less than 2. This is achieved by throwing inside the ultraproduct a pile of rubbish, as follows: embed every $U(n_\alpha)$ into a unitary group of twice the rank using block-diagonal matrices:

$$U(n_\alpha) \ni u \mapsto \left(\begin{array}{c|c} u & 0 \\ \hline 0 & \mathbb{I}_n \end{array}\right) \in U(2n_\alpha).$$

The resulting composition mapping

$$i' : G \to \left(\prod_\alpha U(n_\alpha)_2\right)_{\mathcal{U}} \to \left(\prod_\alpha U(2n_\alpha)_2\right)_{\mathcal{U}}$$

is still a group monomorphism, but all the distances between elements of $G$ are now cut by half and so the diameter of $i'(G)$ is $\leq 1$. So we can assume without loss in generality that the original embedding $i$ has this property.

On the new index set $B = A \times \mathbb{N}_+$ choose an ultrafilter $\mathcal{V}$ satisfying two properties:

1. the projection of $\mathcal{V}$ along the first coordinate is the initial ultrafilter $\mathcal{U}$, and
2. if the intersection of a subset $X \subseteq A \times \mathbb{N}_+$ with every fiber $\{a\} \times \mathbb{N}_+$ is cofinite, then $X \in \mathcal{V}$.

Lift the monomorphism $i$ in an arbitrary way to a map

$$\bar{i} : G \to \prod_{\alpha \in A} U(n_\alpha)$$

and define a map

$$\bar{j} : G \to \prod_{(\alpha,k) \in B} U\left(n_\alpha^{2^k}\right)$$

by letting $\bar{j}_{\alpha,k}(g) = \bar{i}^{(2^k)}_{n_\alpha}(i_\alpha(g))$. This $\bar{j}$ determines a map

$$j\colon G \to \left(\prod_{(\alpha,k)\in B} U\left(n_\alpha^{2^k}\right)\right)_{\mathcal{V}},$$

and it is not hard to see that $j$ is a group monomorphism with the property that the images of every two distinct elements of $G$ are at a distance exactly $\sqrt{2}$ from each other.

(2.)⇒(3.): Given a group embedding $i$ as in Equation (5.3) with the property that $i(g)$ and $i(h)$ are orthogonal whenever $g \neq h$, let $F \subseteq G$ be finite and let $\varepsilon > 0$. Let $\bar{i}$ denote any lifting of $i$ to a map from $G$ to the direct product of $U(n_\alpha)$. Denote $C$ the set of indices $\alpha$ for which $\bar{i}_\alpha$ is an $(F, \varepsilon)$-almost monomorphism which in addition satisfies

$$\sqrt{2} - \varepsilon < d(\bar{i}_\alpha(g), \bar{i}_\alpha(h)) < \sqrt{2} + \varepsilon$$

for all $g, h \in F$, $g \neq h$. Then $C \in \mathcal{U}$ and in particular $C$ is non-empty.

(3.)⇒(4.): again, as in the proof of sufficiency in Observation 1, it is enough to consider the case of a conjunction of atomic formulas of the form $\neg(x = y)$ or $S(x, y, z)$. When dealing with negation, remember that we replace the metric $d_{HS}$ with $\min\{1, d_{HS}\}$, and so the condition $d_{HS}(x, y) > \sqrt{2} - \varepsilon$ implies $\neg(x = y)^{U(n)_2} < \varepsilon$.

(4.)⇒(5.): quite obvious.

(5.)⇒(1.): the argument is just a slight variation of the proof of the implication "(1.)⇒(2.)" in Theorem 2.1, so we leave the details to the reader. They can be found in the proof of Th. 3.5 in [23], see also Corollary 5.10 in [3]. □

The meat of the above theorem (the equivalence of conditions 1, 2, 3, and 5) is variously attributed either to Radulescu [24] or to an earlier work of Kirchberg.

A *closed L-condition* is an expression of the form $\varphi^{\mathcal{M}} = 0$, where $\varphi$ is a sentence of the language of continuous logic. Notice that this means $\mathcal{M} \models \varphi$. A *theory* is a set of closed $L$-conditions. It may be slightly unsettling to observe that the characterization of hyperlinear groups in Theorem 4.2, item (4.), is not, strictly speaking, stated in terms of the *theory of unitary groups*. However, this is most naturally fixed, as follows. In the statement of the following result, the *ultrapower* of $G$, as usual, means the ultraproduct of a family of metric groups metrically isomorphic to $G$.

**Corollary 4.3** *Let $U = (U, d)$ be a group equipped with a bi-invariant metric and satisfying two conditions:*

*(a) For every n, the group $U(n)_2$ embeds into U as a metric subgroup, and*
*(b) U embeds into an ultraproduct of groups $U(n)_2$ as a metric subgroup.*

*Then the following are equivalent for an arbitrary group G:*

1. *G is hyperlinear,*
2. *every existential sentence $\sigma$ of the first-order theory of G satisfies $\sigma^U = 0$, that is, belongs to the continuous theory of U:*

$$\mathrm{Th}^{\exists}(G) \subseteq \mathrm{Th}_c(U).$$

3. *G embeds into a metric ultrapower of U.*

(Here $\mathrm{Th}_c(U)$ denotes of course the first-order continuous logic theory of $U$.)

*Proof* (1.)⇒(2.): Thanks to Theorem 4.2, for every $\varepsilon > 0$ we have $\sigma^U < \varepsilon$, whence the conclusion follows.

(2.)⇒(1.): Denote, for simplicity, by $P$ a metric ultraproduct of the unitary groups of finite rank containing $U$ as a metric subgroup. If $\sigma \in \mathrm{Th}^{\exists}(G)$ and $\varepsilon > 0$ is any, then we have $\sigma^P < \varepsilon$ and a by now standard argument using a lift of the monomorphism $G \hookrightarrow U \hookrightarrow P$ to the direct product of unitary groups implies the existence of $n$ with $\sigma^{U(n)_2} < \varepsilon$. Now we conclude by Theorem 4.2.

(1.) ⟺ (3.): It is enough to notice that every metric ultrapower of $U$ is contained in some metric ultraproduct of unitary groups of finite rank, via a rather straightforward reindexing procedure, and vice versa. □

Here is just one example of a group $U$ as above, and the most economical one.

**Example 4.4** The group monomorphism

$$U(n)_2 \ni u \mapsto \begin{pmatrix} u & 0 \\ 0 & u \end{pmatrix} \in U(2n)_2$$

is an isometry with regard to the normalized Hilbert-Schmidt distances on both groups. It generates an increasing chain of unitary groups

$$U(1)_2 < U(2)_2 < \ldots < U(2^n)_2 < U(2^{n+1})_2 < \ldots.$$

The union of the chain, $\cup_{n=1}^{\infty} U(2^n)$, is a group which supports a naturally defined bi-invariant Hilbert-Schmidt metric. The completion of this group is a Polish group, denoted $U(R)$ and called, in full, the "unitary group of the hyperfinite factor $R$ of type $II_1$ equipped with the ultraweak topology." Regarded as a metric group, $U(R)$ clearly satisfies the hypothesis of Corollary 4.3.

It remains unknown whether every group is hyperlinear, and this is presently one of the main open questions of the theory.

The origin of the concept of a hyperlinear group is described in the survey [23], §7, whose duplication we try to avoid inasmuch as possible. In brief, it is motivated by Connes' Embedding Conjecture [8], which states that every von Neumann factor of type $II_1$ embeds into an ultrapower of $R$, the (unique) hyperfinite factor of type $II_1$, traditionally denoted $R^{\omega}$. Existence of a non-hyperlinear group would imply a negative answer to Connes' Embedding Conjecture, and send far-reaching ripples.

In a highly interesting historical remark at the beginning of a recent preprint [25], David Sherman brings attention to the 1954 article [31] by Fred Wright, which essentially contained a construction of the ultraproduct of von Neumann factors of type $II_1$ (the so-called *tracial ultraproduct*). It was done in the language of maximal ideals rather than ultrafilters, but the two approaches are equivalent. Sherman notes: "An amusing consequence is that the tracial ultraproduct is older than the "classical" ultraproduct from model theory (Łoś [21] in 1955)." Since the metric ultraproduct of unitary groups $(\prod U(n)_2)_{\mathcal{U}}$ is isomorphic, as a metric subgroup, to the unitary group of the tracial ultraproduct of finite-dimensional matrix algebras (considered by Wright as an example, *loco citato*), it means that the metric ultraproduct of groups considered in these notes historically made its appearance — albeit an implicit one — before the algebraic ultraproduct of groups, as described in Section 2.

## 5 Sofic groups

Sofic groups are those groups admitting models which are finite symmetric groups with the normalized Hamming distance — that is, matrix models of a more restrictive kind, meaning that every sofic group is hyperlinear. This Section largely mirrors the preceding Section 4, and we show first examples of sofic groups towards the end.

**Definition 5.1** A discrete group $G$ is *sofic* if it is isomorphic to a subgroup of a metric ultraproduct of a suitable family of symmetric groups of finite rank with their normalized Hamming distances.

In other words, there is a set $A$, a nonprincipal ultrafilter $\mathcal{U}$ on $A$, and a mapping $\alpha \mapsto n(\alpha)$ so that

$$G < \left( \prod_{\alpha} (S_{n(\alpha)}, d_{hamm}) \right)_{\mathcal{U}} .$$

Again, one can reformulate the concept in the language of the existence of models which are finite symmetric groups with their normalized Hamming distances.

**Theorem 5.2** *For a group $G$, the following conditions are equivalent.*

1. *$G$ is sofic.*
2. *$G$ embeds into an ultraproduct of symmetric groups of finite rank in such a way that every two distinct elements in the image are at a distance* 1 *from each other.*
3. *For every finite $F \subseteq G$ and every $\varepsilon > 0$, there are $n$ and an $(F, \varepsilon)$-almost homomorphism $j \colon F \to S_n$ which is uniformly injective:*

$$d_{hamm}(j(g), j(h)) \geq 10^{-10}$$

*whenever $g, h \in F$ and $g \neq h$.*

4. *For every existential sentence $\sigma$ of the first-order theory of $G$ and each $\varepsilon > 0$, there exists $n$ so that*

$$\sigma^{S_n} < \varepsilon.$$

The proof is very similar to, but quite a bit easier than, that of Theorem 4.2, with the implication (1.)⇒(2.) again being central. We have chosen not to duplicate the proof which can be found in the survey [23] of the first-named author (see Theorem 3.5). Theorem 5.2 (save condition (4)) was established by Elek and Szabó [13], who were, it seems, already aware of Radulescu's result for hyperlinear groups [24] (that is, our Theorem 4.2).

Every finite symmetric group $S_n$ canonically embeds into the unitary group $U(n)$, and their distances are easily seen to satisfy:

$$d_{hamm}(\sigma, \tau) = \frac{1}{2} \left( d_{HS}(A_\sigma, A_\tau) \right)^2 .$$

Now Condition (3.) in Theorem 5.2, jointly with Condition (3.) in Theorem 4.2, imply:

**Corollary 5.3** (Elek and Szabó [13]) *Every sofic group is hyperlinear.* □

Again, it is unknown whether every group is sofic, or whether every hyperlinear group is sofic.

One can also state a close analogue of Corollary 4.3.

**Corollary 5.4** *Let $S = (S, d)$ be a group equipped with a bi-invariant metric and satisfying two conditions:*

*(a) For every $n$, the group $S_n$, with its normalized Hamming distance, embeds into $S$ as a metric subgroup, and*
*(b) $S$ embeds into an ultraproduct of groups $S_n$ as a metric subgroup.*

*Then the following are equivalent for an arbitrary group $G$:*

1. *$G$ is sofic,*
2. *every existential sentence $\sigma$ of the first-order theory of $G$ satisfies $\sigma^S = 0$:*

$$\mathrm{Th}^{\exists}(G) \subseteq \mathrm{Th}_c(S).$$

3. *$G$ embeds into a metric ultrapower of $S$.*

There exist natural examples of groups $S$ satisfying the above, and the following is, in a sense, the simplest among them.

**Example 5.5** Let $\lambda$ denote the Lebesgue measure on the unit interval $[0, 1]$. Equip the group $\mathrm{Aut}([0, 1], \lambda)$ of measure-preserving transformations with the uniform metric

$$d_{unif}(\sigma, \tau) = \lambda\{t \in [0, 1] : \sigma(t) \neq \tau(t)\}.$$

This metric is bi-invariant, complete, and makes $\mathrm{Aut}([0, 1], \lambda)$ into a non-separable group.

For every $n$, realize $S_{2^n}$ as the group of measure preserving transformations of the interval $[0, 1]$ whose restriction to every interval $[i2^{-n}, (i+1)2^{-n}]$, $i = 0, 1, \ldots, 2^{n-1}$, is a translation. The restriction of the uniform distance $d_{unif}$ to $S_{2^n}$ equals the normalized Hamming distance, and for every $n$

$$S_{2^n} < S_{2^{n+1}}.$$

The uniform closure of the union of the chain of subgroups $\bigcup_n S_{2^n}$ in the group Aut([0, 1], $\lambda$) is denoted $[E_0]$. This is a Polish group equipped with a bi-invariant metric. The name for this object is somewhat long: *"the full group of the hyperfinite aperiodic ergodic measure-preserving equivalence relation"*. It is easy to verify that the group $[E_0]$ satisfies the assumptions of Corollary 5.4. In addition, it sits naturally as a closed topological subgroup of the group $U(R)$.

Here are just a few words of explanation of where this group and its name come from; we refer to [19] for details and references. Let $\mathscr{R}$ be a Borel equivalence relation on a standard Borel space $X$ equipped with a finite measure $\mu$. The *full group* of $\mathscr{R}$ in the sense of Dye, denoted $[\mathscr{R}]$, is the subgroup of all non-singular transformations $\sigma$ of $(X, \mu)$ with the property $(x, \sigma(x)) \in \mathscr{R}$ for $\mu$-a.e. $x$. (A transformation is non-singular if it takes null sets to null sets.) If equipped with the uniform metric, $[\mathscr{R}]$ is a Polish group.

The relation $\mathscr{R}$ is *hyperfinite* if it is the union of an increasing chain of relations each having finite equivalence classes, and it is *aperiodic* if $\mu$-almost all $\mathscr{R}$-equivalence classes are infinite. The relation $\mathscr{R}$ is ergodic if every $\mathscr{R}$-saturated measurable subset of $X$ is either a null set or has full measure. Finally, $\mathscr{R}$ is *measure-preserving* if the full group $[\mathscr{R}]$ consists of measure-preserving transformations. An example of such a relation is the *tail equivalence relation* on the compact space $\{0, 1\}^\omega$ equipped with the product of uniform measures: $xE_0y$ if and only if there is $N$ such that for all $n \geq N$, $x_n = y_n$. It can be proved that the group $[\mathscr{R}]$ of every hyperfinite ergodic ergodic measure-preserving equivalence relation is isometrically isomorphic to $[E_0]$ as defined above.

Historically the first ever example of a hyperlinear group which is not obviously such belongs to Connes [8] and, independently, Simon Wassermann [29]: the free non-abelian group. Recall that every non-abelian free group is LEF.

**Theorem 5.6** *Every LEF group is sofic (hence hyperlinear).*

*Proof* It is easily seen that the LEF property of a group $G$ is equivalent to the following: for every sentence $\sigma \in \mathrm{Th}^{\exists}(G)$ there is $n$ with $\sigma^{S_n} = 0$. Now Condition (3.) of Theorem 5.2 applies. □

In fact, it appears that all the presently known particular examples of hyperlinear groups are at the same time known to be sofic. However, there is an interesting class of groups potentially able to distinguish between soficity and hyperlinearity and pointed out by Ozawa in [22]. These are wreath products $\mathbb{Z}_2 \wr G$ where $G$ is a sofic group, that is, the semi-direct products $G \ltimes \mathbb{Z}_2^G$ with regard to the natural action of $G$ by permutations. That such groups are hyperlinear, follows from Theorem 2 in Elek and Lippner [12].

The second basic class of sofic groups is that of *amenable* groups, and we proceed to examine it in the next section.

## 6 Amenability

Amenability has its origins in the Banach–Tarski paradox which says that we can partition a solid unit ball in the three-dimensional Euclidean space $\mathbb{R}^3$ into a finite number of pieces (five is enough) such that by rearranging them via isometries of $\mathbb{R}^3$, we can obtain two unit solid balls in $\mathbb{R}^3$.

Expanding on this idea, we say that a group $G$ admits a *paradoxical decomposition* if there are pairwise disjoint subsets $A_1, A_2, \ldots, A_n$, $B_1, B_2, \ldots, B_m$ of $G$ and elements $g_1, g_2, \ldots, g_n, h_1, h_2, \ldots, h_m \in G$ such that

$$G = \bigcup_i g_i A_i = \bigcup_j h_j B_j.$$

**Example 6.1** The group $F_2$ admits a paradoxical decomposition. Indeed, let $a, b$ be free generators of $F_2$. Let $w(a)$ be the set of all reduced words in $F_2$ with the first letter equal to $a$. Similarly define $w(a^{-1})$, $w(b)$ and $w(b^{-1})$.

Note that

$$\begin{aligned} F_2 &= \{e\} \cup w(a) \cup w(a^{-1}) \cup w(b) \cup w(b^{-1}) \\ &= w(a) \cup a\left(w(a^{-1})\right) \\ &= w(b) \cup b\left(w(b^{-1})\right). \end{aligned}$$

**Theorem 6.2** *For a discrete group G the following are equivalent:*

1. *G does* not *admit a paradoxical decomposition,*
2. *G admits a finitely additive probability measure $\mu$, defined on the power set $\mathcal{P}(G)$, which is invariant under left translations,*
3. *G admits a* (left) invariant mean, *i.e. a positive linear functional $\phi\colon \ell^\infty(G) \to \mathbb{C}$ satisfying $\phi(1) = 1$ and invariant under left translations,*

4. *There is an invariant regular probability measure on the Stone-Čech compactification $\beta G$,*
5. (Følner's condition)*: for every finite $F \subseteq G$ and $\varepsilon > 0$, there is a finite $\Phi \subseteq G$ (a* Følner set *for $F$ and $\varepsilon$) such that for each $g \in F$,*

$$|g\Phi \bigtriangleup \Phi| < \varepsilon|\Phi|,$$

6. (Reiter's condition (P1))*: for every finite $F \subseteq G$ and $\varepsilon > 0$, there is $f \in \ell^1(G)$ with $\|f\|_{\ell^1(G)} = 1$ and such that for each $g \in F$, $\|f - {}^g f\|_{\ell^1} < \varepsilon$.* (Here ${}^g f(x) = f(g^{-1}x)$).

A countable group is called *amenable* if it satisfies one of the equivalent conditions of Theorem 6.2.

*Proof* (2)⇒(1): It is clear that the presence of finitely additive measure invariant under left translations precludes the possibility of a paradoxical decomposition.

(3)⇒(2): Put $\mu(A) = \phi(\chi_A)$.

(3)⇔(4): Banach algebras $\ell^\infty(G)$ and $C(\beta G)$ are canonically isomorphic, and the isomorphism preserves the action of $G$ by isometries. Positive linear functionals on $C(\beta G)$ correspond to regular measures on $\beta G$ via Riesz representation theorem. Left-invariant means on $\ell^\infty(G)$ correspond to invariant regular probability measures on $C(\beta G)$.

(5)⇒(6): For a given $\varepsilon > 0$ and $F \subseteq G$ take $\Phi$ as in (5.). Now the function

$$f = \frac{\chi_\Phi}{|\Phi|} \in \ell^1(G)$$

has the required property.

(6)⇒(3): The closed unit ball $B$ of the dual Banach space to $\ell^\infty(G)$ is compact in the weakest topology making all evaluation mappings $\phi \mapsto \phi(x)$ continuous, $x \in \ell^\infty(G)$ (the *Banach–Alaoglu theorem*). The set of all means on $\ell^\infty(G)$ (that is, positive linear functionals $\phi$ satisfying $\phi(1) = 1$) is a weak* closed subset of $B$, and so is weak* compact as well. Every element $f \in \ell^1(G)$ can be viewed as a bounded linear functional on $\ell^\infty(G)$, and so by (6), there is a net $(f_\alpha)$ of means on $\ell^\infty(G)$ such that for every $g \in G$ and $h \in \ell^\infty(G)$,

$$\lim_\alpha \langle ({}^g f_\alpha - f_\alpha), h\rangle = 0. \qquad (+)$$

(As they say, the net $(f_\alpha)$ weak* converges to invariance.) Let $f$ be a weak* cluster point of this net, that is, a limit of a convergent subnet (which exists by weak* compactness). By (+), $f$ is invariant.

(1)⇒(5): We need a version of classical Hall's matching theorem (for a proof, see e.g. [4], Corollary III.3.11, or below).

**Theorem 6.3** (Hall's (2, 1)-matching theorem) *Let*

$$\Gamma = (V, E) = (A, B, E)$$

*be a bipartite graph, where $V$ denotes vertices, $E$ denotes edges, $V = A \sqcup B$. Assume the degree of every vertex in $A$ is finite. Suppose further that for every finite $X \subseteq A$, $|\Gamma(X)| \geq 2|X|$, where $\Gamma(X)$ denotes the set of edges having a vertex in $X$. Then there are two injections $i$ and $j$ with domains equal to $A$, disjoint images in $B$, and such that $(a, i(a)), (a, j(a)) \in E$, whenever $a \in A$.*

Suppose that the Følner's condition fails. Fix a finite $F \subseteq G$, $\varepsilon > 0$ so that for every finite $\Phi \subseteq G$ there is $g \in F$ such that

$$|g\Phi \bigtriangleup \Phi| \geq \varepsilon|\Phi|.$$

Consider a graph $\Gamma = (V, E)$ with $V = G \cup G$ and

$$(g, h) \in E \Leftrightarrow \exists_{x \in F^k}\ h = xg,$$

where $k$ is large so that for every finite $X \subseteq A$ the condition $|\Gamma(X)| \geq 2|X|$ holds.

Apply Hall's theorem and get injections $i$ and $j$. For $s, t \in F^k$ define

$$\Omega_{s,t} = \{g \in G \colon i(g) = sg \text{ and } j(g) = tg\}.$$

Then $\{s\Omega_{s,t} \colon s \in F^k\} \cup \{t\Omega_{s,t} \colon t \in F^k\}$ is a family of sets such that each two are either pairwise disjoint or equal. Note that

$$\bigcup_{s \in F^k} s^{-1}(s\Omega_{s,t}) = G = \bigcup_{t \in F^k} t^{-1}(t\Omega_{s,t}).$$

This contradicts (1). □

The present elegant proof of the implication (1)⇒(5) is relatively recent, see Deuber, Simonovits and Sós [11]. In its present form, it only applies to discrete groups, and it would be interesting to know whether the idea can be made to work for locally compact groups as well, where the classical argument remains rather more complicated.

For a detailed treatment of amenability and related topics in the same spirit, see the survey article [10], containing in particular a proof of the (2, 1)-matching theorem (§35). Here is a different argument.

*Proof of Hall's* $(2,1)$*-matching theorem* It is enough to establish the result for finite graphs and use the standard compactness argument. (In the spirit of these notes: choose injections $i_{A'}$, $j_{A'}$ for every induced subgraph on vertices $A' \sqcup \Gamma(A')$, where $A' \subseteq A$ is finite. Choose a suitable ultrafilter $\mathcal{U}$ on the family of all finite subsets of $A$. The ultralimit $i(a) = \lim_{A' \to \mathcal{U}} i(a)$ is well-defined, similarly for $j$, and the pair of injections $i, j$ is as desired.)

We use induction on $n = |A|$. For $n = 1$ the result is obvious. Let now $|A| = n + 1$. Assume without loss in generality that $E$ is a minimal set of edges satisfying the assumptions of Theorem. It suffices now to verify that for every $a \in A$, $|E(a)| = 2$.

Suppose towards a contradiction that there is an $a \in A$ with $E(a) \subseteq B$ containing at least three distinct points, $b, c, d$. The minimality of $E$ means none of the edges $(a, b)$, $(a, c)$ and $(a, d)$ can be removed, as witnessed by finite sets $X_b, X_c, X_d \subseteq A \setminus \{a\}$ with $\Gamma(X_b \cup \{a\}) \setminus \{b\}$ containing $\leq 2|X_b| + 1$ points, and so on. This in particular implies $|\Gamma(X_b \cup \{a\})| = 2|X_b| + 2$, and similarly for $c$ and $d$. Since every set $X_z$, $z \in \{b, c, d\}$ must contain a point adjacent to a point in $\{b, c, d\} \setminus \{z\}$, at least one of these sets is a proper subset of $A \setminus \{a\}$. Fix a $z \in \{b, c, d\}$ with this property and denote $S = X_z \cup \{a\}$. One has $1 \leq |S| \leq n$ and $|\Gamma(S)| = 2|S|$.

Let $\Gamma_1$ be the induced subgraph on vertices $S \sqcup \Gamma(S)$, and let $\Gamma_2$ be the induced subgraph on the remaining vertices of $\Gamma$, that is, $(A \setminus S) \sqcup (B \setminus \Gamma(S))$. Both $\Gamma_1$ and $\Gamma_2$ satisfy the assumptions of Theorem. For $\Gamma_1$ this is obvious: if $Y \subseteq S$, then $\Gamma_1(Y) = \Gamma(Y)$. For $\Gamma_2$, if we assume that $Z \subseteq A \setminus S$ is such that $|\Gamma_2(Z)| < 2|Z|$, we get a contradiction:

$$|\Gamma(S \sqcup Z)| = |\Gamma(S) \cup \Gamma(Z)| = |\Gamma(S) \sqcup \Gamma_2(Z)| = 2|S| + |\Gamma_2(Z)| < 2(|S| + |Z|).$$

Since the cardinality of $S$ and of $A \setminus S$ is less than $n + 1$, the graphs $\Gamma_i$, $i = 1, 2$ admit $(2, 1)$-matchings, say $i_1, j_1$ and $i_2, j_2$ respectively. The images of four mappings are all pairwise disjoint. A concatenation of $i$'s and $j$'s gives a $(2, 1)$-matching of $\Gamma$, whose set of edges is strictly contained in $E$, contradicting the minimality of the latter. □

**Theorem 6.4** *Every amenable group is sofic (hence hyperlinear).*

*Proof* Let $F \subseteq G$ be finite, and let $\varepsilon > 0$. Choose a Følner set, $\Phi$, for the pair $(F, \varepsilon)$.

The map $x \mapsto gx$ is well-defined on a subset of $\Phi$ containing $> (1 - \varepsilon)|\Phi|$ points, and by extending it to a self-bijection of $\Phi$ one gets a

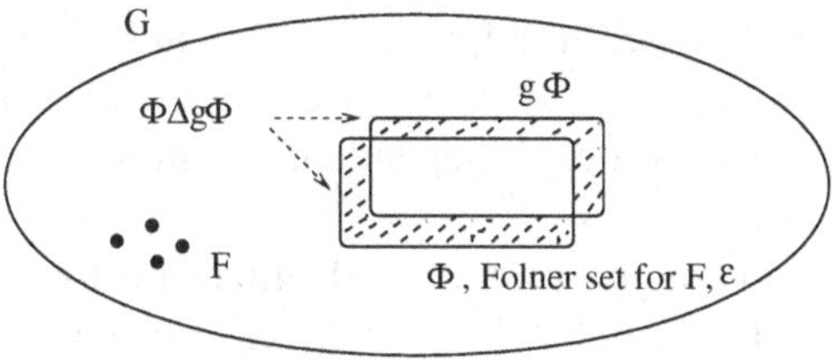


Figure 5.1 A Følner set.

$(F, 2\varepsilon)$-homomorphism to the symmetric group $S_{|\Phi|}$ satisfying condition (3.) of Theorem 5.2. □

The two results (Theorem 5.6 and Theorem 6.4) can be combined as follows. A group $G$ is *initially subamenable* (Gromov) if every finite subset $F \subseteq G$ admits an $(F, 0)$-almost monomorphism into an amenable group $\Gamma$, that is, $F$ embeds into $\Gamma$ with the partial multiplication preserved.

**Corollary 6.5** (Gromov) *Every initially subamenable group is sofic.*

As observed (independently) by Simon Thomas and Denis Osin, no finitely presented simple non-amenable group (for instance Thompson's groups $V$ and $T$) is initially subamenable. (The argument is simple, and, as pointed out to me by the author of [16], of the same kind as that used in the paper to show that the Grigorchuk groups are not finitely presented.) Apparently, it remains unknown whether the three Thompson's groups are sofic. Examples of sofic groups which are not initially subamenable have been given by Cornulier [9]. (Earlier an example of a non-initially subamenable group which is hyperlinear was presented by Thom [26].)

## 7 Universal hyperlinear and sofic groups without ultraproducts

Metric ultraproducts of groups $U(n)_2$ can be considered as universal hyperlinear groups, and those of groups $S_n$ as universal sofic groups. They are studied from this viewpoint in [27], where it is shown that if the Continuum Hypothesis fails, then there exist $2^{2^{\aleph_0}}$ pairwise-nonisomorphic metric ultraproducts of groups $S_n$ over the index set of integers.

As seen from Theorems 4.2 and 5.2, the definitions of hyperlinear and sofic groups can be restated in a form independent from ultraproducts. In view of this, one would expect the existence of "canonical" universal hyperlinear/sofic groups, independent of ultraproducts. Such a construction indeed exists.

Let $(G_\alpha, d_\alpha)$ be a family of groups equipped with bi-invariant metrics. As before (Equation (5.1)), form the $\ell^\infty$-type direct sum $\mathcal{G} = \oplus^{\ell^\infty}_{\alpha\in A} G_\alpha$ of the groups $G_\alpha$; when the diameters of $G_\alpha$ are uniformly bounded from above, $\mathcal{G}$ is just the direct product of the groups in question, equipped with the supremum distance.

Now define the $c_0$*-type sum* of groups $G_\alpha$ by letting

$$\oplus^{c_0}_{\alpha\in A} G_\alpha = \{x \in \mathcal{G}\colon \lim_\alpha d(e_\alpha, x_\alpha) = 0\}.$$

In other words, $x \in \oplus^{c_0}_{\alpha\in A} G_\alpha$ if and only if

$$\forall\varepsilon > 0,\ \ \{\alpha \in A\colon d(e_\alpha, x_\alpha) > \varepsilon\} \text{ is finite.}$$

It is easily seen that $\oplus^{c_0}_{\alpha\in A} G_\alpha$ forms a closed normal subgroup of $\mathcal{G}$, and so the quotient group $\oplus^{\ell^\infty}_{\alpha\in A} G_\alpha / \oplus^{c_0}_{\alpha\in A} G_\alpha$ is equipped with a complete bi-invariant metric.

To simplify the notation, we will write

$$\oplus^{\ell^\infty/c_0}_{\alpha\in A} G_\alpha = \oplus^{\ell^\infty}_{\alpha\in A} G_\alpha / \oplus^{c_0}_{\alpha\in A} G_\alpha,$$

and call the resulting complete metric group the $\ell^\infty/c_0$*-type product* of the groups $G_\alpha$, $\alpha \in A$.

**Example 7.1** In the case where all $G_\alpha$ are equal to the additive group of the scalar field (e.g. $\mathbb{R}$ or $\mathbb{C}$), the resulting metric group is just the additive group of the well-known Banach space $\ell^\infty/c_0$, which is isometric to the space of continuous functions on the remainder $\beta\mathbb{N} \setminus \mathbb{N}$ of the Stone-Čech compactification of $\mathbb{N}$. This motivates our terminology and notation.

**Theorem 7.2** *Let G be a countable group.*

1. *G is hyperlinear if and only if G is isomorphic to a subgroup of* $\oplus^{\ell^\infty/c_0}_{n\in\mathbb{N}} U(n)_2$.
2. *G is sofic if and only if G is isomorphic to a subgroup of* $\oplus^{\ell^\infty/c_0}_{n\in\mathbb{N}} S_n$.

*Proof* We treat only the hyperlinear case. Write $G$ as $F_\infty/N$, where $N$ is a normal subgroup in the free group of countably many generators. Let $\pi\colon F_\infty \to G$ be the corresponding quotient homomorphism. For every $n$, denote $B_n$ the set of reduced words in $F_\infty$ of length $\leq n$ on the first $n$ generators of $F_\infty$, and let $\widetilde{B_n} = \pi(B_n)$ denote the image of $B_n$ in $G$.

*Necessity.* Suppose $G$ is hyperlinear. For every $n$, fix a $(\widetilde{B_n}, 1/n^2)$-almost monomorphism $\tilde{j}_n$ to some unitary group $U(k_n)_2$ with images of every two distinct elements being at a distance at least $10^{-10}$. The composition $\pi \circ \tilde{j}_n$ is defined on the first $n$ generators of $F_\infty$. Extend it by the constant map $e$ over the rest of them. Denote by $j_n \colon F_\infty \to \prod_n U(n)_2$ the unique homomorphism on the free group assuming the given values on free generators. An induction on the word length using bi-invariance of the Hilbert-Schmidt metric shows that for all $x \in B_n$, one has $d(\tilde{j}_n(\pi(x)), j_n(x)) < 1/n$. In particular, for every $x \in B_n \cap N$, one has $d_{HS}(j_n(x), e) < 1/n$, and if $x, y \in B_n$ and $x \neq y$, then $d(j_n(x), j_n(y)) \geq 10^{-10} - 2/n$.

One can surely assume without loss of generality the unitary groups $U(k_n)$ to have distinct dimensions. Define a homomorphism

$$h \colon F_\infty \to \prod_n U(n)_2$$

by setting

$$h(x)_n = \begin{cases} j_m(x), & \text{if } n = k_m \text{ for some } m, \\ e, & \text{otherwise.} \end{cases}$$

This $h$ has two properties:

1. if $x \in N$, then $d(j_n(x), e) \to 0$ as $n \to \infty$, and therefore $h(x) \in \oplus^{c_0}_{n \in \mathbb{N}} U(n)_2$,
2. if $x \neq y \mod N$, then $d(h(x), h(y)) = \sup_n d(j_n(x), j_n(y)) \geq 10^{-10}$.

This means that the homomorphism $h$ taken modulo $\oplus^{c_0}_{n \in \mathbb{N}} U(n)_2$ factors through $N$ to determine a group monomorphism from $G$ into $\oplus^{\ell^\infty/c_0}_{n \in \mathbb{N}} U(n)_2$.

*Sufficiency.* Now suppose $G$ embeds as a subgroup into the $\ell^\infty/c_0$-type product of unitary groups. Every non-principal ultrafilter $\mathcal{U}$ on the natural numbers gives rise to a quotient homomorphism from $\prod_n U(n)_2$ to the metric ultraproduct $(\prod_n U(n)_2)_\mathcal{U}$, and since the normal subgroup $\oplus^{c_0} U(n)$ always maps to identity, we get a family of homomorphisms

$$h_\mathcal{U} \colon \oplus^{\ell^\infty/c_0}_{n \in \mathbb{N}} U(n)_2 \to \left(\prod_n U(n)_2\right)_\mathcal{U}.$$

Let $x \in \prod_n U(n)_2 \setminus \oplus^{c_0} U(n)_2$. There are $\varepsilon > 0$ and an infinite set

$A \subseteq \mathbb{N}$ so that for all $n \in A$ one has $d_{HS,n}(x_n, e) > \varepsilon$. If now $\mathcal{U}$ is an ultrafilter on $\mathbb{N}$ containing the set $A$, then $d(h_{\mathcal{U}}(x), e) \geq \varepsilon$.

We have shown that the homomorphisms $h_{\mathcal{U}}$ separate points, and consequently the $\ell^\infty/c_0$ type sum of the unitary groups is isomorphic with a subgroup of the product of a family of hyperlinear groups. In order to conclude that $G$ is hyperlinear, it remains to notice that the class of hyperlinear groups is closed under passing to subgroups and direct products. The first is obvious, and for the second embed the group $U(n) \times U(m)$ into $U(n+m)$ via block-diagonal matrices, make an adjustment for the distances, and use Theorem 4.2.

The sofic case is completely similar. □

(The above result is very close in spirit to Prop. 11.1.4 in [6].)

The above theorem can be generalized to groups $G$ of any cardinality, in which case the universal group will be the $\ell^\infty/c_0$-type product of the family of all groups $U(n)_2$ (respectively $S_n$), $n \in \mathbb{N}$, each one taken $|G|$ times.

In view of this result, the $\ell^\infty/c_0$-type products of metric groups deserve further attention, including from the viewpoint of the model theory of metric structures.

In our view, it would also be interesting to know what can be said about subgroups of $\ell^\infty/c_0$-type sums of other groups, especially the unitary groups $U(n)$, $n \in \mathbb{N}$, with the uniform operator metric.

## 8 Sofic groups as defined by Gromov, and Gottschalk's conjecture

Sofic groups were first defined by Gromov [17], under the descriptive name of *groups with initially subamenable Cayley graphs* (the name "sofic groups" belongs to Benjy Weiss [30]). In this Section we will finally state Gromov's original definition, and explain a motivation: the Gottschalk's Surjunctivity Conjecture.

A directed graph $\Gamma$ is *edge-coloured* if there are a set $C$ of *colours* and a mapping $E(\Gamma) \to C$. We will also say that $\Gamma$ is *edge $C$-coloured.*

Here is a natural example how the edge-colouring comes about. Let $G$ be a finitely-generated group. Fix a finite symmetric set $V$ of generators of $G$ not containing the identity $e$. The *Cayley graph* of $G$ (defined by $V$) is a non-directed graph having elements of $G$ as vertices, with $(g, h)$ being adjacent if and only if $g^{-1}h \in V$, that is, there is an edge from

$g$ to $h$ iff one can get to $h$ by multiplying $g$ with a generator $v \in V$ on the right. Since the generator $v$ associated to a given edge is unique, the Cayley graph is edge $V$-coloured.

The *word distance* in the Cayley graph of $G$ is the length of the shortest path between two vertices. It is a left-invariant metric. Notice that as finitely coloured graphs, every two closed balls of a given radius $N$ are naturally isomorphic to one another (by means of a uniquely defined left translation), which is why we will simply write $B_N$ without indicating the centre.

**Definition 8.1** (Gromov) The Cayley graph of a finitely generated group $G$ is *initially subamenable* if for every natural $N$ and $\varepsilon > 0$ there is a finite edge $V$-coloured graph $\Gamma$ with the property that for the fraction of at least $(1 - \varepsilon)|\Gamma|$ of vertices $x$ of $\Gamma$ the $N$-ball $B_N$ around $x$ is isomorphic, as an edge $V$-coloured graph, to the $N$-ball in $G$.

It can be seen directly that the above definition does not depend on the choice of a particular finite set of generators, but this will also follow from the next theorem. The equivalence of the original Gromov's concept of a group whose Cayley graph is initially subamenable with most other definitions of soficity mentioned in these notes belongs to Elek and Szabó [13]. Notice that the restriction to finitely generated groups in Gromov's definition is inessential — as it is should now be obvious to the reader, soficity is a local property in the sense that a group $G$ is sofic if and only if every finitely generated subgroup of $G$ is sofic.

**Theorem 8.2** *Let $G$ be a finitely generated group with a finite symmetric generating set $V$. Then $G$ is sofic if and only if the Cayley graph of $(G, V)$ is initially subamenable.*

*Proof* *Necessity.* Assume $G$ is sofic. Let $\varepsilon > 0$ and $N \in \mathbb{N}$ be given. Choose a $(B_N, \varepsilon)$-almost monomorphism, $j$, to a permutation group $S_n$, with the property that the images of every two distinct elements of $B_N$ are at a distance $> (1 - \varepsilon/|B_N|^2)$ from each other. Define a directed graph $\Gamma$ whose vertices are integers $1, 2, \dots, n$ (i.e., elements of the set $[n]$ upon which $S_n$ acts by permutations), and $(m, k)$ is an edge coloured with a $v \in V$ if and only if $j(v)m = k$. If $v, u \in B_N$ and $v \neq u$, then for at least $(1 - \varepsilon/|B_N|^2)n$ vertices $m$ one has $v(m) \neq u(m)$. It follows that the map $v \mapsto v(m)$ is one-to-one on $B_N$, and consequently an isomorphism of $V$-coloured graphs, for at least $(1 - \varepsilon)n$ vertices $m$.

$\Leftarrow$: In the presence of an edge-colouring, every element $w \in B_N$ determines a unique translation of $\Gamma$ that is well-defined at all but $< \varepsilon|\Gamma|$ of its vertices (just follow, inside $\Gamma$, any particular string of colours leading up to $w$ in the original ball). This defines a $(B_N, \varepsilon)$-almost homomorphism into the permutation group on the vertices of $\Gamma$, which is uniformly $(1-\varepsilon)$-injective. □

The graphs $\Gamma$ as above can be considered as finite clones of $G$, grown artificially using some sort of genetic engineering.

Here is an open problem in topological dynamics that motivated Gromov to introduce sofic groups. Let $G$ be a countable group, $A$ a finite set equipped with a discrete topology. The Tychonoff power $A^G$ is a Cantor space (i.e., a compact metrizable zero-dimensional space without isolated points), upon which $G$ acts by translations:

$$(g \cdot x)(h) = x(g^{-1}h).$$

Equipped with this action of $G$ by homeomorphisms, $A^G$ is a *symbolic dynamical system,* or a *shift.* An *isomorphism* between two compact $G$-spaces $X$ and $Y$ is a homeomorphism $f\colon X \to Y$ which commutes with the action of $G$:

$$f(g \cdot x) = g \cdot f(x) \text{ for all } g \in G,\ x \in A^G.$$

**Conjecture 8.3** (Gottschalk's Surjunctivity Conjecture, 1973, [14]) *For every countable group $G$ and every finite set $A$, the shift system $A^G$ contains no proper closed $G$-invariant subspace $X$ isomorphic to $A^G$ itself.*

The Conjecture remains open as of time of writing these notes, and the following is the strongest result to date.

**Theorem 8.4** (Gromov [17]) *Gottschalk's Surjunctivity Conjecture holds for sofic groups.*

*Sketch of the proof* Let $\Phi\colon A^G \to A^G$ be an endomorphism of $G$-spaces, that is, such a continuous mapping that $\Phi(gf) = g(\Phi(f))$ for every $g \in G$. Consider the mapping $A^G \to A$ which sends $f \mapsto \Phi(f)(e)$. Since $A$ is finite, the preimage of every $a \in A$ is a clopen set. Hence there is a finite $F' \subseteq G$ and $\Phi_0\colon A^{F'} \to A$ such that for every $g \in G$,

$$\Phi(f)(g) = \Phi(g^{-1}f(e)) = \Phi_0(g^{-1}f \restriction F') = \Phi_0(f \restriction gF').$$

Assume that $\Phi$ is injective, then there is an inverse $\Psi$: image $(\Phi) \to A^G$. The map $\Psi$ is determined by a certain finite $F'' \subseteq G$, and by $\Psi_0$: image $(\Phi) \restriction A^{F''} \to A$. That means that both $\Phi$ and its inverse are encoded locally.

Now assume in addition that $\Phi$ is not onto. Choose a finite symmetric subset $B_1 \subseteq G$ which is big enough both to store complete information about $\Phi$ and its inverse, and so that the restriction of $\Phi(A^G)$ to $B_1$ is not onto. From now on, without loss in generality, we can replace $G$ with a subgroup generated by $B_1$.

Grow a finite $B_1$-coloured graph $\Gamma$ whose number of vertices we will denote $N = N(\varepsilon) = |V(\Gamma)|$, which locally looks like $B_5$ around at least $(1-\varepsilon)N$ vertices.

Using the local representations $\Phi_0$ and $\Psi_0$, we construct maps

$$\widetilde{\Phi}, \widetilde{\Psi} \colon A^\Gamma \to A^\Gamma .$$

It follows that the size of the image of $\widetilde{\Phi}$ is at least $|A|^{(1-\varepsilon)N}$.

Now choose in $\Gamma$ a maximal system of disjoint balls of radius 1. The number of vertices in the union of those balls is at least $cN(1-\varepsilon)$ for some $0 < c < 1$ (which only depends on $G$ and $B_1$). It follows that *image*$(\widetilde{\Phi})$ has size at most $|A|^{N-cN(1-\varepsilon)}(|A|^{|B_1|}-1)^{cN(1-\varepsilon)/|B_1|}$. By combining the two observations, we get:

$$|A|^{(1-\varepsilon)N} \leq |A|^{N-cN(1-\varepsilon)}(|A|^{|B_1|}-1)^{cN(1-\varepsilon)/|B_1|},$$

that is,

$$|A|^{(1-\varepsilon)} \leq |A|^{(1-c+c\varepsilon)}(|A|^{|B_r|}-1)^{c/|B_1|}$$

for every $\varepsilon > 0$. We get a contradiction by sending $\varepsilon \downarrow 0$. □

The above proof belongs to Benjy Weiss and is worked out in great detail in [30]. The original proof of Gromov [17] was different.

It would be interesting to know whether Gromov's theorem can be extended to hyperlinear groups.

## 9 Near actions and another criterion of soficity by Elek and Szabó

Let $(X, \mu)$ be a measure space, where the measure $\mu$ is at least finitely additive. A *near-action* of a group $G$ on $(X, \mu)$ is an assignment to $g \in G$ of a measure-preserving mapping $\tau_g \colon X \to X$ defined $\mu$-a.e. in such a way that

$$\tau_{gh} = \tau_g \circ \tau_h$$

in the common domain of definition of both sides. A near-action is *essentially free* if for every $g \neq e$ and for $\mu$-a.e. $x \in X$, $\tau_g x \neq x$.

**Theorem 9.1** (Elek and Szabó [13]) *A group $G$ is sofic if and only if it admits an essentially free near-action on a set $X$ equipped with a finitely-additive probability measure $\mu$ defined on the family $\mathscr{P}(X)$ of all subsets of $X$.*

*Sketch of the proof* *Necessity* ($\Rightarrow$): Let $G$ be a sofic group. For every finite $F \subseteq G$ and each $k \in \mathbb{N}_+$ select a $(F, 1/k)$-almost homomorphism $j_{(F,k)}$ with values in some permutation group $S_{n(F,k)}$, which is uniformly $(1-\varepsilon)$-injective. As usual, we think of $S_{n(F,k)}$ as the group of self-bijections of the set $[n(F,k)] = \{1, 2, \ldots, n(F,k)\}$. Form a disjoint union

$$X = \bigsqcup_{F,k} [n(F,k)].$$

Let $\mathcal{U}$ be any ultrafilter on the directed set of all pairs $(F,k)$ containing every upper cone $\{(F,k)\colon\ F \supseteq F_0,\ k \geq k_0\}$. For every $A \subseteq X$ the formula

$$\mu(A) = \lim_{(F,k)\to\mathcal{U}} \frac{|A \cap [n(F,k)]|}{n(F,k)}$$

defines a finitely-additive probability measure, $\mu$, on the power set of $X$. Given $g \in G$, the rule

$$\tau_g(x) = j_{(F,k)}(x), \text{ if } x \in [n(F,k)] \text{ and } g \in F$$

defines $\mu$-a.e. a measure-preserving transformation of $X$. It is easy to verify that $\tau$ is an essentially free near-action of $G$.

*Sufficiency* ($\Leftarrow$): Here the proof follows rather closely the arguments used to establish the implications (2) $\Rightarrow$ (1) $\Rightarrow$ (5) in Theorem 6.2. □

The above criterion stresses yet again that soficity is a weaker version of amenability. To the best of authors' knowledge, no analogous criterion for hyperlinear groups is known yet.

## 10 Discussion and further reading

It is still hard to point to any obvious concrete candidates for examples of non-sofic or non-hyperlinear groups.

One class of groups rather allergic to amenability and its variations is formed by Kazhdan groups, or groups with property $(T)$. Let $G$ be a group, and let $\pi\colon G \to U(\mathcal{H})$ be a unitary representation. Then $\pi$ *admits almost invariant vectors* if for every finite $F \subseteq G$ there are $\varepsilon > 0$

and $x \in \mathscr{H}$ such that $||x|| = 1$, and for every $g \in F$, $||x - \pi(g)(x)|| < \varepsilon$. A group $G$ has *Kazhdan's property*, or *property* $(T)$, if whenever a unitary representation $\pi$ of $G$ admits almost invariant vectors, it has a fixed non-zero vector. For an introduction into this vast subject, see [1]. And is the simplest example of an "allergy" mentioned above.

**Theorem 10.1** *If a group is amenable and has property $(T)$, then it is finite.*

The proof follows at once from the definition combined with the following equivalent characterization of amenability:

*Reiter's condition (P2):* for every $F \subseteq G$ and $\varepsilon > 0$, there is $f \in \ell^2(G)$ with $||f||_{\ell^2(G)} = 1$ and such that for each $g \in F$, $||f - gf||_{\ell^2(G)} < \varepsilon$.

Here is a much more difficult result in the same vein:

**Theorem 10.2** (Kirchberg, Valette) *If a group with property (T) embeds into the group $U(R)$ (in particular, into its subgroup $[\mathscr{R}]$), then it is residually finite.*

It is in view of such results that Ozawa asked whether every finitely generated Kazhdan group that is sofic is residually finite. A negative answer was announced by Thom [26]. Consequently, a hope to use property $(T)$ in order to construct non-hyperlinear groups is a bit diminished now, but surely not gone, as it remains in particular unknown whether finitely generated simple Kazhdan groups can be hyperlinear/sofic.

The present notes have been organized so as to minimize an overlap with the survey [23] by the first-named author. We recommend the survey as a useful complementary source for a number of topics which were not mentioned in the workshop lectures because of lack of time, including the origin and significance of the class of hyperlinear groups (Connes' Embedding Conjecture [8, 6]), links of the present problematics with solving equations in groups, and more, as well as a longer bibliography and a number of (overwhelmingly still open) questions. Among interesting recent developments are sofic measure-preserving equivalence relations [12] and a theory of entropy for measure-preserving actions of sofic groups [5].

## Acknowledgements

The first-named author thanks the Organizers of the 7th Appalachian set theory workshop, especially Ernest Schimmerling and Justin Moore, for their hospitality and patience. Thanks go to David Sherman for the illuminating historical remark at the end of Section 4, and to Peter Mester for correcting an oversight in the earlier version of the notes. The authors are grateful to a team of anonymous referees who have produced a most helpful report of an astonishing size (12 typed pages long).

## References

[1] M.B. Bekka, P. de la Harpe, and A. Valette, *Kazhdan's Property (T),* New Mathematical Monographs **11**, Cambridge University Press, 2008.

[2] J.L. Bell and A.B. Slomson, *Models and Ultraproducts. An introduction,* Dover Publications, Inc., Mineola, NY, 2006 reprint of the 1974 3rd revised edition.

[3] I. Ben Yaacov, A. Berenstein, C.W. Henson, and A. Usvyatsov, *Model Theory for Metric Structures,* in: Model theory with applications to algebra and analysis. Vol. 2, 315–427, London Math. Soc. Lecture Note Ser., **350**, Cambridge Univ. Press, Cambridge, 2008.

[4] B. Bollobás, *Modern Graph Theory,* Graduate Texts in Mathematics, **184**, Springer-Verlag, New York, 1998.

[5] L. Bowen, *Measure conjugacy invariants for actions of countable sofic groups,* J. Amer. Math. Soc. **23** (2010), 217–245.

[6] N.P. Brown and N. Ozawa, *$C^*$-Algebras and Finite-Dimensional Approximations,* Graduate Studies in Mathematics **88**, American Mathematical Society, Providence, R.I., 2008.

[7] J.W. Cannon, W.J. Floyd, and W.R. Parry, *Introductory notes on Richard Thompson's groups,* Enseign. Math. (2) **42** (1996), 215–256.

[8] A. Connes, *Classification of injective factors,* Ann. of Math. **104** (1976), 73–115.

[9] Y. Cornulier, *A sofic group away from amenable groups,* Math. Ann. **350** (2011), 269–275.

[10] P. de la Harpe, R.I. Grigorchuk, and T. Ceccherini-Silberstein, *Amenability and paradoxical decompositions for pseudogroups and discrete metric spaces,* Tr. Mat. Inst. Steklova **224** (1999), Algebra. Topol. Differ. Uravn. i ikh Prilozh., 68–111 (in Russian); English translation in Proc. Steklov Inst. Math. **224** (1999), 57–97.

[11] W.A. Deuber, M. Simonovits, and V.T. Sós, *A note on paradoxical metric spaces,* Studia Sci. Math. Hungar. **30** (1995), 17–23. An annotated 2004 version is available at:
`http://novell.math-inst.hu/~miki/walter07.pdf`

[12] G. Elek and G. Lippner, *Sofic equivalence relations,* J. Funct. Anal. **258** (2010), 1692–1708.

[13] G. Elek and E. Szabó, *Hyperlinearity, essentially free actions and $L^2$-invariants. The sofic property,* Math. Ann. 332 (2005), no. 2, 421–441.

[14] W. Gottschalk, *Some general dynamical notions,* in: Recent Advances in Topological Dynamics, Lecture Notes Math. **318**, Springer-Verlag, Berlin, 1973, pp. 120–125.

[15] F.P. Greenleaf, *Invariant Means on Topological Groups,* Van Nostrand Mathematical Studies **16**, Van Nostrand – Reinhold Co., NY–Toronto–London–Melbourne, 1969.

[16] R.I. Grigorchuk, *Degrees of growth of finitely generated groups and the theory of invariant means,* Math. USSR-Izv. **25** (1985), no. 2, 259300.

[17] M. Gromov, *Endomorphisms of symbolic algebraic varieties,* J. Eur. Math. Soc. (JEMS) 1 (1999), no. 2, 109–197.

[18] E. Hewitt and K.A. Ross, *Abstract Harmonic Analysis. Vol. 1 (2nd ed.),* Springer–Verlag, NY a.o., 1979.

[19] A.S. Kechris and B.D. Miller, *Topics in orbit equivalence,* Lecture Notes Math. **1852**, Springer-Verlag, Berlin, 2004.

[20] A.S. Kechris and C. Rosendal, *Turbulence, amalgamation and generic automorphisms of homogeneous structures,* Proc. Lond. Math. Soc. (3) **94** (2007), 302–350.

[21] J. Łoś, *Quelques remarques, théorèmes et problèmes sur les classes définissables d'algèbres,* in: Mathematical interpretation of formal systems, North-Holland Publishing Co., Amsterdam, 1955, pp. 98–113.

[22] N. Ozawa, *Hyperlinearity, sofic groups and applications to group theory,* handwritten note, 14 pp., August 2009, available at `http://www.ms.u-tokyo.ac.jp/~narutaka/NoteSofic.pdf` (accessed on April 24, 2012).

[23] V.G. Pestov, *Hyperlinear and sofic groups: a brief guide,* Bull. Symb. Logic **14** (2008), 449–480.

[24] F. Radulescu, *The von Neumann algebra of the non-residually finite Baumslag group* $\langle a, b | ab^3a^{-1} = b^2 \rangle$ *embeds into* $R^\omega$, in: Hot topics in operator theory, 173–185, Theta Ser. Adv. Math., 9, Theta, Bucharest, 2008 (prepublished as arXiv:math/0004172v3, 2000, 16 pp.)

[25] David Sherman, *Notes on automorphisms of ultrapowers of* $II_1$ *factors,* Studia Math. **195** (2009), 201–217.

[26] A. Thom, *Examples of hyperlinear groups without factorization property,* Groups Geom. Dyn. **4** (2010), 195–208.

[27] S. Thomas, *On the number of universal sofic groups,* Proc. Amer. Math. Soc. **138** (2010), 2585–2590.

[28] A.M. Vershik and E.I. Gordon, *Groups that are locally embeddable in the class of finite groups,* St. Petersburg Math. J. **9** (1998), no. 1, 49–67.

[29] S. Wassermann, *On tensor products of certain group* $C^*$*-algebras,* J. Functional Analysis **23** (1976), 239–254.

[30] B. Weiss, *Sofic groups and dynamical systems,* Sankhyā Ser. A 62 (2000), no. 3, 350–359.
Available at: `http://202.54.54.147/search/62a3/eh06fnl.pdf`

[31] F.B. Wright, *A reduction for algebras of finite type,* Ann. of Math. (2) **60** (1954), 560–570.

# 6

# Aronszajn Trees and the SCH

Itay Neeman[a] and Spencer Unger

The eighth Appalachian Set Theory workshop was held at Carnegie Mellon University in Pittsburgh on February 28, 2009. The lecturer was Itay Neeman. As a graduate student Spencer Unger assisted in writing this chapter, which is based on the workshop lectures.

## 1 Introduction

The purpose of the workshop was to present a recent theorem due to Neeman [16].

**Theorem 1.1** *From large cardinals, it is consistent that there is a singular strong limit cardinal $\kappa$ of cofinality $\omega$ such that the Singular Cardinal Hypothesis fails at $\kappa$ and the tree property holds at $\kappa^+$.*

The notes are intended to give the reader the flavor of the argument without going into the complexities of the full proof in [16]. Having read these notes, the motivated reader should be prepared to understand the full argument. We begin with a discussion of trees, which are natural objects in infinite combinatorics. One topic of interest is whether a tree has a cofinal branch. For completeness we recall some definitions.

**Definition 1.2** Let $\lambda$ be a regular cardinal and $\kappa$ be a cardinal.

1. A *$\lambda$-tree* is a tree of height $\lambda$ with levels of size less than $\lambda$.

[a] This material is based upon work supported by the National Science Foundation under Grant No. DMS-055622.

2. A *cofinal branch* through a tree of height $\lambda$ is a linearly ordered subset of order type $\lambda$.
3. A *$\lambda$-Aronszajn tree* is a $\lambda$-tree with no cofinal branch.
4. A $\kappa^+$-tree is *special* if there is a function $f : T \to \kappa$ such that for all $x, y \in T$, if $x\ T\ y$ then $f(x) \neq f(y)$.

*Remark* 1.3 It is easy to see that a special $\kappa^+$-tree is in fact a $\kappa^+$-Aronszajn tree. Moreover, a special $\kappa^+$-tree remains special after cardinal preserving forcing.

For a regular cardinal $\lambda$ we can ask whether all $\lambda$-trees have a cofinal branch. This leads to the definition of the tree property.

**Definition 1.4** For a regular cardinal $\lambda$, $\lambda$ has the *tree property* if every $\lambda$-tree has a cofinal branch. Equivalently, $\lambda$ has the tree property if and only if there are no $\lambda$-Aronszajn trees.

We list a few classical results:

- (König [11]) $\aleph_0$ has the tree property.
- (Aronszajn, see [12]) $\aleph_1$ does not have the tree property.
- (Specker [19]) If $\kappa^{<\kappa} = \kappa$, then there is a special $\kappa^+$-Aronszajn tree.
- (Keisler and Tarski [10]) If $\kappa$ is strongly inaccessible, then $\kappa$ has the tree property if and only if $\kappa$ is weakly compact.
- (Jensen [9]) There is special $\kappa^+$-Aronszajn tree if and only if the weak square property $\square_\kappa^*$ holds.

Forcing is required to answer questions about small cardinals and the tree property. Again we list some results.

- (Mitchell and Silver [15]) Con(ZFC + "There is a weakly compact cardinal") implies Con(ZFC + "$\aleph_2$ has the tree property").
- (Magidor and Shelah [14]) From large cardinals it is consistent with ZFC that $\aleph_{\omega+1}$ has the tree property.

We turn to discussion of the Singular Cardinal Hypothesis.

**Definition 1.5** The *Singular Cardinal Hypothesis* at a singular cardinal $\kappa$ (SCH$_\kappa$) is the assertion, "If $2^{<\kappa} = \kappa$, then $2^\kappa = \kappa^+$." The Singular Cardinal Hypothesis (SCH) is the assertion, "For all singular cardinals $\kappa$, SCH$_\kappa$".

Easton [4] established that it is consistent that the continuum function ($\kappa \mapsto 2^\kappa$) have any reasonable behavior on regular cardinals. However, the possible behavior of the continuum function on singular cardinals is unclear. In particular, results of inner model theory show that large cardinals are required to construct a model with the failure of SCH. The first such construction started from a supercompact cardinal $\kappa$ and proceeded in two steps. The first step was forcing to make $2^\kappa > \kappa^+$ while maintaining the measurability of $\kappa$. This forcing is due to Silver and an account can be found in [1]. The second step was to use Prikry forcing [17] to make $\kappa$ singular of cofinality $\omega$ without adding bounded subsets of $\kappa$ or collapsing cardinals. Gitik determined the exact strength of the failure of SCH.

**Theorem 1.6** *(Gitik [6])* $Con(\neg \mathrm{SCH}) \Leftrightarrow Con(\exists\kappa\; o(\kappa) = \kappa^{++})$.

We will now argue that there is a $\kappa^+$-Aronszajn tree in the Prikry-Silver model for the failure of SCH. By the result of Specker, we have a special Aronszajn tree at the successor of any inaccessible cardinal. In particular there is a special $\kappa^+$-tree in the ground model for the construction. By Remark 1.3 above, special trees are preserved by cardinal preserving forcing. Both steps above are cardinal preserving. So the original model for the failure of $\mathrm{SCH}_\kappa$ has a special $\kappa^+$-Aronszajn tree. These ideas can be found in work of Ben-David and Magidor [2].

This brings up the following question, which Woodin asked in the early 80s for $\kappa = \aleph_\omega$, and Woodin and others asked in the late 80s in general [5]. For $\kappa$ singular of cofinality $\omega$, does the failure of SCH imply that there exists a $\kappa^+$-Aronszajn tree? The question was intended to test whether the original way to obtain the failure of SCH was the only way. Gitik and Magidor [7] showed that there is a different way to get the failure of SCH. They proved that one can add $\kappa^{++}$ Prikry sequences without adding bounded subsets of $\kappa$. However, Woodin's question remained open. It turns out that there are other ways to make SCH fail, but they all still gave Aronszajn trees.

Many people had hope for a positive answer. The line of thought was that the failure of SCH at $\kappa$ would imply some intermediate combinatorial principle, like approachability or weak square, and then from this one could construct a $\kappa^+$-Aronszajn tree. It turns out that the first step fails. In particular, Gitik and Sharon [8] showed that the failure of SCH does not imply approachability. Cummings and Foreman [3] showed that there is a PCF theoretic object called a bad scale in the Gitik-Sharon model; this implies the failure of approachability.

The purpose of the workshop is to present a proof that the answer to the general question is *no*. By using a variation of the forcing from [8], we construct a model in which we have the failure of $SCH_\kappa$ at a singular cardinal of cofinality $\omega$ and there are no $\kappa^+$-Aronszajn trees.

## 2 The tree property

In order to motivate the use of large cardinals in our main result, we outline an application of inner model theory which shows that the tree property at the successor of a singular cardinal has high consistency strength. Let $\nu$ be a singular cardinal. As we mentioned in the introduction if the weak square principle $\square^*_\nu$ holds then there is a special $\nu^+$-Aronszajn tree. By inner model theory, if there are no inner models with large cardinals then there is a model $K \subseteq V$ such that

1. $\nu^+ = (\nu^+)^K$, and
2. the weak square principle $\square^*_\nu$ (in fact $\square_\nu$) holds in $K$.

It follows that in $V$ there is a special $\nu^+$-Aronszajn tree. So large cardinals are required in order to obtain the tree property at the successor of a singular cardinal.

In [14], Magidor and Shelah show that the successor of a singular limit of strongly compact cardinals has the tree property. This result is important because we can view our proof as using the same idea, but with generic large cardinal properties in place of real large cardinal properties. For our context we state Magidor and Shelah's theorem using supercompact cardinals.

**Theorem 2.1** *Let $\nu$ be a singular limit of supercompact cardinals, then $\nu^+$ has the tree property.*

We will prove the theorem for a cofinality $\omega$ limit of supercompact cardinals, because this reflects the proof of our main theorem. For the proof of this result we will have two lemmas, which we call the Spine Lemma and the Traction Lemma. For the main result of these notes we will have to prove new versions of both lemmas.

*Proof of Theorem 2.1* Let $T$ be a $\nu^+$-tree. Without loss of generality, level $\alpha$ of $T$ is $\{\langle\alpha,\xi\rangle \mid \xi < \nu\}$. Let $\langle\kappa_n \mid n < \omega\rangle$ be an increasing sequence of supercompact cardinals cofinal in $\nu$.

**Spine Lemma 1** *There exist an $n < \omega$ and a cofinal set $C \subseteq \nu^+$ such that for all $\alpha < \beta$ both in $C$, there are $\xi, \zeta < \kappa_n$ such that $\langle\alpha,\xi\rangle\ T\ \langle\beta,\zeta\rangle$.*

We call this Lemma the Spine Lemma, because we can view it as picking out a narrow essential component of the tree. The set $\{\langle \alpha, \xi \rangle \mid \alpha \in C, \xi < \kappa_n\}$ has the property that any two levels from $C$ contain nodes in the spine which are related in the tree. Spine Lemma 1 shows that a set with this spine-like property exists.

To clarify the difference between members of the domain and members of the codomain of our elementary embeddings, we will write ordinals of the domain as the usual Greek letters and ordinals of the codomain as Greek letters with a superscript *.

*Proof* For convenience we let $\kappa =_{def} \kappa_0$. Let $\pi : V \to M$ be a $\nu^+$-supercompactness embedding with critical point $\kappa$; that is $\mathrm{crit}(\pi) = \kappa$, $\pi(\kappa) > \nu^+$ and ${}^{\nu^+}M \subseteq M$. Let $\gamma^* =_{def} \sup \pi``\nu^+$. We claim that $\gamma^* < \pi(\nu^+)$. By elementarity and since $\nu^+$ is regular in $V$, we have that $\pi(\nu^+)$ is regular in $M$. So it suffices to show that the cofinality of $\gamma^*$ is less than $\pi(\nu^+)$ in $M$. The closure of $M$ under $\nu^+$-sequences implies that $\pi \restriction \nu^+ \in M$. Moreover, $\pi \restriction \nu^+$ is an order preserving bijection from $\nu^+$ to $\pi``\nu^+$. Therefore, the cofinality of $\gamma^*$ is $\nu^+$ when computed in $M$. This finishes the claim that $\gamma^* < \pi(\nu^+)$.

We fix some node $\langle \gamma^*, \eta^* \rangle$ on level $\gamma^*$ of $\pi(T)$. For each $\alpha < \nu^+$, there is a unique $\xi^*_\alpha < \pi(\nu)$ such that

$$\langle \pi(\alpha), \xi^*_\alpha \rangle \; \pi(T) \; \langle \gamma^*, \eta^* \rangle.$$

As $\langle \pi(\kappa_n) \mid n < \omega \rangle$ is cofinal in $\pi(\nu)$, there is $n_\alpha < \omega$ such that $\xi^*_\alpha < \pi(\kappa_{n_\alpha})$. As $\nu^+$ is regular, there exist $n < \omega$ and $C \subseteq \nu^+$ cofinal such that $n_\alpha = n$ for all $\alpha \in C$. We check that this $n$ and $C$ are as required for Spine Lemma 1. Fix $\alpha < \beta$ both in $C$. By the choice of $C$ we have

$$\langle \pi(\alpha), \xi^*_\alpha \rangle \; \pi(T) \; \langle \gamma^*, \eta^* \rangle,$$
$$\langle \pi(\beta), \xi^*_\beta \rangle \; \pi(T) \; \langle \gamma^*, \eta^* \rangle.$$

Since $\pi(T)$ is a tree, it follows that $\langle \pi(\alpha), \xi^*_\alpha \rangle \; \pi(T) \; \langle \pi(\beta), \xi^*_\beta \rangle$. We can collect this information in $M$ and then use elementarity to bring the conclusion back to $V$.

$$M \models \exists \xi^*, \zeta^* < \pi(\kappa_n) \, \langle \pi(\alpha), \xi^* \rangle \; \pi(T) \; \langle \pi(\beta), \zeta^* \rangle.$$

Thus by elementarity

$$V \models \exists \, \xi, \zeta < \kappa_n \, \langle \alpha, \xi \rangle \; T \; \langle \beta, \zeta \rangle.$$

□

**Traction Lemma 1** *There exist a cofinal set $J \subseteq C$ and a map $\alpha \mapsto \xi_\alpha$ such that for all $\alpha < \beta$ both in $J$, $\langle \alpha, \xi_\alpha \rangle \ T \ \langle \beta, \xi_\beta \rangle$.*

We have thinned the tree to the spine $\{\langle \alpha, \xi \rangle \mid \alpha \in C, \xi < \kappa_n\}$, which satisfies the conclusion of Spine Lemma 1. Now we take an embedding with a sufficiently high critical point and this embedding will only stretch the spine vertically but not horizontally. In other words we put the spine in traction.

*Proof* Let $\pi : V \to M$ be a $\nu^+$-supercompactness embedding with critical point $\kappa_{n+1}$. The key point about the new embedding is that $\pi(\kappa_n) = \kappa_n$ and for all sets $A$ of size $\kappa_n$, $\pi(A) = \pi``A$. This will be important when we work with the level sets of the spine. As before we can argue that $\sup \pi``\nu^+ < \pi(\nu^+)$. By elementarity, $\pi(C)$ is unbounded in $\pi(\nu^+)$. Working in $M$, let $\gamma^*$ be the least element of $\pi(C)$ above $\sup \pi``\nu^+$. By applying elementarity to the conclusion of Spine Lemma 1, for each $\alpha \in C$ there exist $\xi^*_\alpha, \eta^*_\alpha < \pi(\kappa_n)$ such that $\langle \pi(\alpha), \xi^*_\alpha \rangle \ \pi(T) \ \langle \gamma^*, \eta^*_\alpha \rangle$

We claim that there are $\eta^* < \kappa_n$ and $J \subseteq C$ cofinal, such that if $\alpha \in J$, then $\eta^*_\alpha = \eta^*$. In $M$ level $\gamma^*$ of the spine has size $\pi(\kappa_n) = \kappa_n$. So $\{\eta^*_\alpha \mid \alpha \in C\}$ has at most size $\kappa_n$. Therefore by the regularity of $\nu^+$, the map taking $\alpha \in C$ to $\eta^*_\alpha$ must be constant with some value $\eta^*$ on an unbounded set $J \subseteq C$. Let $\xi_\alpha = \xi^*_\alpha$ for $\alpha \in J$. Each $\xi_\alpha < \kappa_n$, so $\pi(\xi_\alpha) = \xi^*_\alpha$. We claim that $J$ and $\alpha \mapsto \xi_\alpha$ satisfy the conclusion of Traction Lemma 1. Fix $\alpha < \beta$ both in $J$. Then

$$\langle \pi(\alpha), \pi(\xi_\alpha) \rangle \ \pi(T) \ \langle \gamma^*, \eta^* \rangle,$$
$$\langle \pi(\beta), \pi(\xi_\beta) \rangle \ \pi(T) \ \langle \gamma^*, \eta^* \rangle.$$

Again using the fact that $\pi(T)$ is a tree, it follows that

$$M \models \langle \pi(\alpha), \pi(\xi_\alpha) \rangle \ \pi(T) \ \langle \pi(\beta), \pi(\xi_\beta) \rangle.$$

Lastly by elementarity $V \models \langle \alpha, \xi_\alpha \rangle \ T \ \langle \beta, \xi_\beta \rangle$. □

To finish the proof, we notice that the map $\alpha \mapsto \xi_\alpha$ for $\alpha \in J$ enumerates a cofinal branch through $T$. This completes the proof of Theorem 2.1. □

## 3 Diagonal Prikry forcing

In this section we define a forcing due to Gitik and Sharon [8] and prove of some of its properties. The forcing that we define in this section is different from Prikry's diagonal Prikry forcing. Prikry's diagonal

Prikry forcing uses an increasing $\omega$-sequence of measurable cardinals $\langle \kappa_n \mid n < \omega \rangle$ where we use one measure on each $\kappa_n$. Note that the measures involved become more complete as $n$ increases. The result is that the the forcing adds no bounded subsets of $\sup_{n<\omega} \kappa_n$. The forcing from [8] is quite different. Again we use an $\omega$-sequence of measures but the completeness of each measure is the same. The result is that some cardinals are collapsed, but this is by design. We begin with $\kappa$ supercompact and $\nu > \kappa$, $\mathrm{cof}(\nu) = \omega$. We define a diagonal Prikry forcing to make $\mathrm{cof}(\kappa) = \omega$ and $|\nu| = \kappa$ while preserving $\nu^+$ and cardinals less than or equal to $\kappa$.

To begin we fix $\langle \kappa_n \mid n < \omega \rangle$ increasing and cofinal in $\nu$ with each $\kappa_n$ regular. Then for each $n$ we let $\mathcal{U}_n$ be a supercompactness measure on $\mathcal{P}_\kappa(\kappa_n)$. The supercompactness measure $\mathcal{U}_n$ can be derived from an embedding $j$ which witnesses that $\kappa$ is $\kappa_n$-supercompact. We define $\mathcal{U}_n =_{def} \{X \subseteq \mathcal{P}_\kappa(\kappa_n) \mid j``\kappa_n \in j(X)\}$. Note that $\mathcal{U}_n$ concentrates on $K_n =_{def} \{a \subseteq \kappa_n \mid |a| < \kappa,\ a \cap \kappa \text{ inaccessible}\}$. Let $K = \bigcup_{n<\omega} = K_n$. We define an ordering on $K$ as follows. Let $a, b \in K$, $a \prec b$ if and only if $b \supseteq a$ and $b \cap \kappa > |a|$. For $a \in K$, define $\mathrm{Cone}(a) = \{b \in K \mid a \prec b\}$. Note that if $a \in K_n$, then for all $i \geq n$, $\{b \in \mathcal{P}_\kappa(\kappa_i) \mid b \in \mathrm{Cone}(a)\} \in \mathcal{U}_i$. We use this ordering in the definition of the forcing and it allows us to formulate a notion of diagonal intersection which is crucial in proving that $\kappa$ is a cardinal in the extension.

We are ready to define our poset $\mathbb{P}$. Conditions are of the form $p = \langle g_p, A_p \rangle$, with

$$\begin{aligned} g_p &= \langle g_p(0), \ldots\ g_p(k-1) \rangle, \\ A_p &= \langle A_p(k), A_p(k+1), \ldots \rangle, \end{aligned}$$

where for all $n < k$ $g_p(n) \in K_n$ and for all $n \geq k$ $A_p(n) \subseteq K_n$ and has $\mathcal{U}_n$-measure 1. Lastly, we require that $g_p = \langle g_p(0), \ldots\ g_p(k-1) \rangle$ satisfy for all $n < m < k$ $g_p(m) \in \mathrm{Cone}(g_p(n))$. In the above condition we call the natural number $k$ the *length of p* and denote it $\ell(p)$. The ordering is defined as follows. $q \leq p$ if and only if

1. $g_q$ extends $g_p$, that is $g_q \restriction \mathrm{dom}\, g_p = g_p$,
2. $A_q(n) \subseteq A_p(n)$ for all $n \geq \ell(g_q)$,
3. $g_q(n) \in A_p(n)$ for all $n$ such that $\ell(g_p) \leq n < \ell(g_q)$.

Let $G$ be $\mathbb{P}$-generic over $V$. Define $g = \bigcup\{g_p \mid p \in G\}$. Then $g =_{def} \langle g(n) \mid n < \omega \rangle$ is a sequence with $g(n) \in K_n$ for all $n$. The intuition is that conditions in $\mathbb{P}$ are a description of $g$. A condition $p$ specifies a finite initial segment of $g$, and gives a restriction on later terms of $g$, namely $g(i) = g_p(i)$ for $i < \ell(p)$ and $g(i) \in A_p(i)$ for $i \geq \ell(p)$.

By genericity $\bigcup_{n<\omega} g(n) = \nu$. So in $V[G]$, $|\nu| = \sum_{n<\omega} |g(n)| \leq \kappa$, because $|g(n)| < \kappa$ for all $n$. In fact for any $\tau$ such that $\kappa \leq \tau \leq \nu$, $\tau = \bigcup_{n<\omega} g(n) \cap \tau$ and $|g(n) \cap \tau| < \tau$. It follows that if $\tau$ as above is regular in $V$, we have $cf(\tau) = \omega$ in $V[G]$.

We show that $\nu^+$ is preserved by an argument that is typical of Prikry forcings.

**Lemma 3.1** *$\mathbb{P}$ is $\nu^+$-cc.*

*Proof* For $A = \langle A(n) \mid k \leq n < \omega \rangle$ and $B = \langle B(n) \mid k \leq n < \omega \rangle$, define

$$A \cap B = \langle A(n) \cap B(n) \mid k \leq n < \omega \rangle.$$

If $p, q$ are conditions and $g_p = g_q$, then $p, q$ are compatible, because $\langle g_p, A_p \cap A_q \rangle$ is stronger than both. This shows that members of an antichain must have different stems. To determine the chain condition we count the number of stems. To count the number of stems we need a result of Solovay [18].

**Theorem 3.2** *If $\kappa$ is strongly compact and $\mu > \kappa$ is regular, then $\mu^{<\kappa} = \mu$.*

Therefore we have

$$|\{g_p \mid p \in \mathbb{P}\}| = \sum_{n<\omega} |\mathcal{P}_\kappa(\kappa_n)| = \nu.$$

Assume that there is $\langle p_\alpha \mid \alpha < \nu^+ \rangle$ an antichain in $\mathbb{P}$. Then there are conditions in the antichain with the same stem. This is a contradiction, because we just showed that such conditions are compatible. □

We would like to see that cardinals below $\kappa$ are preserved. In fact, we show that no bounded subsets of $\kappa$ are added. To do this we prove that $\mathbb{P}$ has the *Prikry property*, which we will define below. First, we introduce some notation. Let $\varphi = \varphi(\dot{x}_1, \ldots, \dot{x}_n)$ be a statement in the forcing language. We write $h \Vdash \varphi$ (for a finite stem $h$) if there is some $A$ such that $\langle h, A \rangle$ is a condition forcing $\varphi$.

*Note* $h \Vdash \varphi$ and $h \Vdash \neg\varphi$ is impossible, as we would have $\langle h, A \rangle \Vdash \varphi$ and $\langle h, B \rangle \Vdash \neg\varphi$, but $\langle h, A \rangle$ and $\langle h, B \rangle$ are compatible.

We write $h$ decides $\varphi$ ($h \parallel \varphi$), if $h \Vdash \varphi$ or $h \Vdash \neg\varphi$. Next we define *diagonal intersection* which is an essential concept in the proof of the Prikry property.

**Definition 3.3** Let $\langle A_s \mid s \text{ is a stem}\rangle$ be a sequence such that for each $s$, $A_s$ is a sequence of measure one sets such that $\langle s, A_s\rangle$ is a condition in $\mathbb{P}$. Then the *diagonal intersection* of the above sequence, $\triangle_s A_s$, is a sequence of sets whose $n^{th}$ coordinate is the set $\{x \in \mathcal{P}_\kappa(\kappa_n) \mid \text{for all } h \text{ if } h^\frown x \text{ is a stem, then } x \in A_h(n)\}$.

*Fact* 3.4 The $n^{th}$ coordinate of the diagonal intersection is measure one for $\mathcal{U}_n$.

*Fact* 3.5 Let $\langle A_s \mid s \text{ is a stem}\rangle$ be a sequence such that $\langle s, A_s\rangle \in \mathbb{P}$ and let $A^*$ be their diagonal intersection. If $\langle s, A_s\rangle$ decides $\varphi$, then $\langle s, A^*(\ell(s)), A^*(\ell(s)+1)\dots\rangle$ decides $\varphi$ in the same way.

*Proof* Without loss of generality assume that $\langle s, A_s\rangle \Vdash \varphi$. Using the definition of diagonal intersection, any extension of

$$\langle s, A^*(\ell(s)), A^*(\ell(s)+1)\dots\rangle$$

is compatible with $\langle s, A_s\rangle$. Hence there is a dense set of conditions below $\langle s, A^*(\ell(s)), A^*(\ell(s)+1)\dots\rangle$ which force $\varphi$. □

**Prikry Lemma** *For every stem h, and for every formula $\varphi$ in the forcing language, $h \parallel \varphi$. Equivalently, for every condition p there is a $q \leq p$ such that q decides $\varphi$ and $g_p = g_q$.*

We begin with a claim that provides the induction step in the proof of the Prikry Lemma.

*Claim* 3.6 If $h \nparallel \varphi$ with $\ell(h) = k$ implies for $\mathcal{U}_k$ almost every $a$, $h^\frown a \nparallel \varphi$.

*Proof of Claim* We prove that if $B =_{def} \{a \in \mathcal{P}_\kappa(\kappa_k) \mid h^\frown a \text{ decides } \varphi\} \in \mathcal{U}_k$, then $h$ decides $\varphi$. Assume that $B \in \mathcal{U}_k$. For each $b \in B$ there is a sequence of measure one sets $A_b$ such that $\langle h^\frown b, A_b\rangle$ decides $\varphi$. We can partition $B$ into the set of those $b$ such that $\langle h^\frown b, A_b\rangle \Vdash \varphi$ and the set of $b$ such that $\langle h^\frown b, A_b\rangle \Vdash \neg\varphi$. Exactly one of these sets must be measure one for $\mathcal{U}_k$. Without loss of generality, we let $B' \in \mathcal{U}_k$ such that for all $b \in B'$, $\langle h^\frown b, A_b\rangle \Vdash \varphi$.

Consider the collection of stems $f$ such that $f \upharpoonright k = h$ and $f(k) \in B'$. If there is a sequence of measure one sets $C$ such that $\langle f, C\rangle$ decides $\varphi$, then let $A_f$ be one such sequence. Otherwise let $A_f = \langle \mathcal{P}_\kappa(\kappa_{\ell(f)}), \dots\rangle$. Let $A$ be the diagonal intersection of the sequence $\langle A_f \mid f \text{ extends } h \text{ and } f(k) \in B'\rangle$.

We claim that $\langle h, \langle B'\rangle^\frown A\rangle$ decides $\varphi$ and hence $h$ decides $\varphi$. Suppose

$\langle f, C\rangle \le \langle h, \langle B'\rangle^\frown A\rangle$. By an easy induction using the definition of diagonal intersection, for all $m$ with $k \le m < \ell(f)$, $f(m) \in A_{f\restriction m}(m)$. Using the fact that $f \restriction (k+1)$ forces $\varphi$ and another easy induction, we see that $f$ forces $\varphi$. So $A_f$ was chosen so that $\langle f, A_f\rangle \Vdash \varphi$. Hence $\langle f, A_f \cap C\rangle \le \langle f, C\rangle \le \langle h, \langle B'\rangle^\frown A\rangle$. So there is a dense set of conditions below $\langle h, \langle B'\rangle^\frown A\rangle$ that decides $\varphi$. This finishes the claim. □

*Proof of the Prikry Lemma* We assume for a contradiction that there is a stem, $h$, and a statement $\varphi$ such that $h \not\parallel \varphi$. Suppose $f$ is a stem of length $n \ge k$ extending $h$ such that $f$ does not decide $\varphi$. Then by the claim there is a $\mathcal{U}_n$ measure one set of extensions of $f$ that do not decide $\varphi$. Let $A_f(n)$ be this set and let $A_f(m) = \mathcal{P}_\kappa(\kappa_m)$ for all $m > n$. Let $A_f = \langle A_f(n), A_f(n+1)\ldots\rangle$. Let $A$ be the diagonal intersection of the sequence $\langle A_f \mid f \text{ extends } h\rangle$.

We claim that no extension of $\langle h, A\rangle$ decides $\varphi$. This will be our contradiction. Let $\langle f, B\rangle \le \langle h, A\rangle$ with $\ell(f) = n$. An easy inductive argument using the definition of diagonal intersection shows that for all $m$ with $k \le m < n$, $f(m) \in A_{f\restriction m}(m)$. So by the choice of $A_{f\restriction m}(m)$ for each $m$ as above, $f$ does not decide $\varphi$. So no condition extending $\langle h, A\rangle$ decides $\varphi$. However it is a general forcing fact that we can always extend to decide a statement. This is a contradiction. □

**Lemma 3.7** *$\mathbb{P}$ adds no bounded subsets of $\kappa$.*

*Proof* Let $\tau$ and $\theta$ be cardinals with $\tau \le \theta < \kappa$. Suppose $\dot{f}$ is a $\mathbb{P}$-name for a function from $\tau$ to $\theta$. We assume that $\mathbb{1}_{\mathbb{P}} \Vdash \dot{f} : \check{\tau} \to \check{\theta}$. By the Prikry Lemma, for each $\alpha < \tau$ and $\beta < \theta$ there is $A_{\alpha,\beta}$ such that $\langle \emptyset, A_{\alpha,\beta}\rangle \parallel \dot{f}(\check{\alpha}) = \check{\beta}$. Let

$$A =_{def} \bigcap_{\substack{\alpha<\tau \\ \beta<\theta}} A_{\alpha,\beta}.$$

Then $\langle \emptyset, A\rangle$ is a condition by the $\kappa$-completeness of each $\mathcal{U}_n$. Moreover, $\langle \emptyset, A\rangle$ decides $\dot{f}$ completely. So $\dot{f}[G] \in V$. □

It follows easily that cardinals less than $\kappa$ are preserved. So we have defined a version of diagonal Prikry forcing $\mathbb{P}$. We showed that it singularizes $\kappa$ while preserving $\kappa$ as a cardinal, and that it collapses $\nu$ to have size $\kappa$ while preserving $\nu^+$. The main difference in our use of this forcing will be a different choice for the cardinal $\nu$.

### 3.1 Gitik-Sharon

What follows is a short summary of Gitik and Sharon's [8] use of diagonal Prikry forcing. To begin we start with the statement of a theorem of Laver [13], which is used in their work and which we will use as well.

**Theorem 3.8** *Assuming there is a supercompact cardinal $\kappa$, then there is a forcing extension in which $\kappa$ is still supercompact and remains supercompact under any $\kappa$-directed closed forcing. We say $\kappa$ is* indestructibly supercompact.

For the Gitik and Sharon model, we start from $\kappa$ indestructibly supercompact, with GCH holding above $\kappa$. Take $\nu = \kappa^{+\omega}$, $\kappa_n = \kappa^{+n}$. First, let $\mathbb{A} = Add(\kappa, \nu^{++})$, and take $E$ to be $\mathbb{A}$-generic over $V$. Note that $\kappa$ is still supercompact in $V[E]$ by Theorem 3.8. Now take $\mathbb{P}$ to be diagonal Prikry forcing for $\kappa$ and $G$ $\mathbb{P}$-generic over $V[E]$. In $V[E][G]$, $\kappa$ is singular of cofinality $\omega$, SCH fails at $\kappa$ and the approachability property fails at $\kappa$. With more work than we have outlined above, the Gitik-Sharon paper showed that there is a very good scale on $\kappa$. Cummings and Foreman extended this to show that there is a bad scale on $\kappa$.

## 4 The proof of Theorem 1

Recall the question, 'Does the tree property at $\kappa^+$ imply $\text{SCH}_\kappa$?' We will show that the answer is no. We will start with $\nu$ as a limit of $\omega$ many supercompact cardinals, $\kappa_n$ for $n < \omega$. Let $\kappa = \kappa_0$ be indestructibly supercompact. Let $\mathbb{A}$ be $\text{Add}(\kappa, \nu^{++})$. Let $E$ be $\mathbb{A}$-generic over $V$. Let $G$ be $\mathbb{P}$-generic over $V[E]$. We will start by showing that $\nu^+$ has the tree property in $V[E]$ and then we will show that it still has the tree property in $V[E][G]$.

### 4.1 The tree property in $V[E]$

**Theorem 4.1** *$\nu^+$ has the tree property in $V[E]$.*

*Proof* Let $T \in V[E]$ be a $\nu^+$-tree. Without loss of generality level $\alpha$ of $T$ is $\{\langle\alpha, \xi\rangle \mid \xi < \nu\}$. Fortunately, the Spine Lemma is exactly as before.

**Spine Lemma 2** *In $V[E]$, there is a $C \subseteq \nu^+$ cofinal and $n < \omega$, such that for all $\alpha < \beta$ both in $C$, there are $\xi, \zeta < \kappa_n$ such that $\langle\alpha, \xi\rangle\ T\ \langle\beta, \zeta\rangle$*

Using the indestructibility of $\kappa$, we have that $\kappa$ is still supercompact in $V[E]$. Therefore the argument from the proof of Spine Lemma 1 works. Next we reformulate the Traction Lemma. Before we used a supercompactness embedding $\pi : V \to M$ with critical point point $\kappa_{n+1}$. This time we will lift the elementary embedding $\pi$ to the universe of $V[E]$ by passing to a further extension $V[E][F]$. The use of the embedding is the same, but additional work is needed to show that the forcing we used to add the embedding did not add the branch through $T$.

**Traction Lemma 2** *In $V[E]$ there is a $J \subseteq C$ cofinal and a map $\alpha \mapsto \xi_\alpha$ such that for all $\alpha < \beta$ both in $J$, $\langle \alpha, \xi_\alpha \rangle \; T \; \langle \beta, \xi_\beta \rangle$.*

*Proof* Let $F$ be $\mathrm{Add}(\kappa, \pi(\nu^{++}))$-generic over $V[E]$. In $V[E][F]$, we claim that $\pi$ can be extended to an elementary embedding $\pi^* : V[E] \to M[E^*]$ for some $E^* \in V[E][F]$ which is $M$-generic for $\pi(\mathbb{A})$. By work of Silver, it suffices to arrange that $\pi``E \subseteq E^*$. We can do this by interleaving the generic $F$ with the pointwise image of $E$ under $\pi$ to create the generic $E^*$. We do this working in $V[E][F]$. Enumerate $F$ as $\langle f_\alpha : \alpha < \pi(\nu^{++}) \rangle$ where each $f_\alpha$ is a function from $\kappa$ to 2. Let $\langle g_\alpha : \alpha < \nu^{++} \rangle$ be a similar enumeration of $E$. Now let $E^* =_{def} \langle h_\beta : \beta < \pi(\nu^{++}) \rangle$, where $h_{\pi(\alpha)} = g_\alpha$ for each $\alpha < \nu^{++}$ and $h_\beta = f_\alpha$ where $\beta$ is the $\alpha^{th}$ member of $\pi(\nu^{++}) \setminus \pi``\nu^{++}$. It is easy to see that $E^*$ is $M$-generic for $\pi(\mathbb{A})$ and that $\pi``E \subseteq E^*$. Hence we can lift the embedding to the generic extension $V[E]$.

Now we repeat the proof of Traction Lemma 1 using $\pi^*$. So in $V[E][F]$ we get that there is a $J \subseteq C$ cofinal and $\alpha \mapsto \xi_\alpha$ for $\alpha \in J$ such that for $\alpha < \beta$ both in $J$, $\langle \alpha, \xi_\alpha \rangle \; T \; \langle \beta, \xi_\beta \rangle$. So we got a branch through $T$ in $V[E][F]$, call it $b$. We want to show that $b \in V[E]$. This will follow from the next lemma.

*Note* In the statement of the lemma below we use the notation $\mathbb{B}^\lambda$ for a power of the poset $\mathbb{B}$. This notation is ambiguous. However the support of the power that is used is not important so long as we have the hypotheses.

**Lemma S** *Let $\theta$ be a cardinal. Let $S$ be a tree of height $\theta$. Let $\mathbb{B}$ be a poset. Assume that*

1. *$\mathbb{B} \times \mathbb{B}$ is* $\mathrm{cof}(\theta)$*-cc,*
2. *$\mathbb{B}^{|S|^+}$ does not collapse $|S|^+$.*

*Then $\mathbb{B}$ does not add cofinal branches through $S$. More precisely, if $F$ is $\mathbb{B}$-generic over $V$ and $b \in V[F]$ is a branch through $S$, then $b \in V$.*

*Note* In the above formulation $S$ need not be a $\theta$-tree.

We will use this lemma in $V[E]$ with $S = T$, $\theta = \nu^+$ and $\mathbb{B} = Add(\kappa, \pi(\nu^{++}))$. We need to check that the hypotheses hold. $\mathbb{B} \times \mathbb{B}$ is $\kappa^+$-cc, because it is $Add(\kappa, \pi(\nu^{++}) + \pi(\nu^{++}))$. Also $\mathbb{B}^{\nu^{++}}$ is $Add(\kappa, \nu^{++} \cdot \pi(\nu^{++}))$ if we use supports of size $< \kappa$. Hence $\mathbb{B}^{\nu^{++}}$ is $\kappa^+$-cc and does not collapse $\nu^{++}$. So we have finished with Traction Lemma 2 except for the proof of Lemma S.

*Proof of Lemma S* We can assume that $\theta$ is regular. If $\theta$ were not regular, then we could replace it with cof($\theta$) and $S$ by its restriction to cof($\theta$) many levels cofinal in $\theta$. Let $b \in V[F]$ be a cofinal branch through $S$. Fix a name $\dot{b}$ such that $\dot{b}[F] = b$. Suppose for a contradiction that $\Vdash_{\mathbb{B}} \dot{b} \notin \check{V}$

We force with $\mathbb{B}^* = \mathbb{B}^{|S|^+}$, letting $F^*$ be $\mathbb{B}^*$-generic over $V$. We write $F^*$ as a product of generics $\prod_{\delta < |S|^+} F_\delta$. Let $b_\delta = \dot{b}[F_\delta]$. Using the assumption that $\Vdash_{\mathbb{B}} \dot{b} \notin V$, it follows that for all $\delta_1, \delta_2 < |S|^+$, $b_{\delta_1} \neq b_{\delta_2}$. To show this we consider $\mathbb{B} \times \mathbb{B}$. Let $\dot{b}_{left}$ and $\dot{b}_{right}$ be the $\mathbb{B} \times \mathbb{B}$-names for the interpretations of $\dot{b}$ by the left and right generics. The assumption $\Vdash_{\mathbb{B}} \dot{b} \notin \check{V}$ implies $\Vdash_{\mathbb{B} \times \mathbb{B}} \dot{b}_{left} \neq \dot{b}_{right}$. Since $F_{\delta_1} \times F_{\delta_2}$ is generic for $\mathbb{B} \times \mathbb{B}$, we have $b_{\delta_1} \neq b_{\delta_2}$.

Let $H \prec V_\rho$ for a sufficiently large regular cardinal $\rho$. Since $\theta$ is regular, we can arrange that $\{\theta, S, \dot{b}, \mathbb{B}, \mathbb{B}^*\} \subseteq H$, $H \cap \theta$ is an ordinal and $|H| < \theta$. Since $\mathbb{B}$ is $\theta$-cc, each antichain of $\mathbb{B}$ in $H$, is contained in $H$. So for all $\delta$, $F_\delta$ is $\mathbb{B}$-generic over $H$, $H[F_\delta] \prec V_\rho[F_\delta]$ and $H[F_\delta] \cap V = H$. We can argue similarly for $\mathbb{B} \times \mathbb{B}$ and $F_{\delta_1} \times F_{\delta_2}$ for any $\delta_1, \delta_2$.

*Note* We are essentially arguing that $\mathbb{B} \times \mathbb{B}$ satisfies a version of properness for an arbitrary regular cardinal $\theta$.

Let $\eta = H \cap \theta$. Working in $V[F^*]$, for each $\delta$ let $\beta_\delta$ be the node of $b_\delta$ on level $\eta$. There are $|S|$ possibilities for $\beta_\delta$ and by assumption $|S|^+$ is a cardinal in $V[F^*]$. So there are $\delta_1$ and $\delta_2$ such that $\beta_{\delta_1} = \beta_{\delta_2}$. We will work with $F_{\delta_1} \times F_{\delta_2}$ as a generic for $\mathbb{B} \times \mathbb{B}$. Recall $\Vdash_{\mathbb{B} \times \mathbb{B}} \dot{b}_{left} \neq \dot{b}_{right}$ So by elementarity of $H[F_{\delta_1} \times F_{\delta_2}]$ in $V_\rho[F_{\delta_1} \times F_{\delta_2}]$, there is a condition $\langle p_1, p_2 \rangle \in (F_{\delta_1} \times F_{\delta_2}) \cap H$ forcing this. Again using elementarity we can extend $\langle p_1, p_2 \rangle$ to $\langle p_1', p_2' \rangle \in F_{\delta_1} \times F_{\delta_2} \cap H$ such that there is $\gamma \in H$ with

$$\langle p_1', p_2' \rangle \Vdash \dot{b}_{left}(\gamma) \neq \dot{b}_{right}(\gamma).$$

This implies that $b_{\delta_1}(\gamma) \neq b_{\delta_2}(\gamma)$. As $S$ is a tree and $\gamma < \eta$ we get $b_{\delta_1}(\eta) \neq b_{\delta_2}(\eta)$, a contradiction. □

This finishes the proofs of both Traction Lemma 2 and Theorem 4.1.

By Lemma S, the branch that we found above is in $V[E]$ and thus the tree property holds in $V[E]$ at $\nu^+$. □

□

## 4.2 The tree property in $V[E][G]$

In this section the proof becomes more difficult. We show that the tree property holds at $\nu^+ = \kappa^+$ in $V[E][G]$. Let $T \in V[E][G]$. Without lost of generality, level $\alpha$ of $T$ is $\{\langle\alpha,\xi\rangle \mid \xi < \kappa\}$. Let $\dot{T} \in V[E]$ such that $\dot{T}[G] = T$. We assume that $\mathbb{1}_{\mathbb{P}} \Vdash$ "$\dot{T}$ is a tree with the above form". Working in $V[E]$ we formulate the Spine Lemma.

**Spine Lemma 3** *In $V[E]$ there are $n < \omega$ and $C \subseteq \nu^+$ cofinal so that for all $\alpha < \beta$ both in $C$, there are $\xi, \zeta < \kappa$ and a stem $h$ of length $n$ such that $h \Vdash \langle\alpha,\xi\rangle\ \dot{T}\ \langle\beta,\zeta\rangle$.*

*Proof* As in the proof of Spine Lemma 2, we fix $\pi : V[E] \to M$, a $\nu^+$-supercompactness embedding with critical point $\kappa$ in $V[E]$. Let $G^*$ be $\pi(\mathbb{P})$-generic over $M$. Let $T^* = \pi(\dot{T})[G^*]$. Let $\gamma^* = \sup \pi``\nu^+$ and fix $\eta^*$ such that $\langle\gamma^*,\eta^*\rangle$ is a node of $T^*$ on level $\gamma^*$.

Working in $M[G^*]$, for each $\alpha < \nu^+$ we fix $\xi^*_\alpha$ such that $\langle\pi(\alpha),\xi^*_\alpha\rangle\ T^*\ \langle\gamma^*,\eta^*\rangle$. There is a condition $p_\alpha \in G^*$ forcing this and we let $n_\alpha$ be the length of the stem of $p_\alpha$. The sequences $\langle\xi^*_\alpha \mid \alpha < \nu^+\rangle$ and $\langle n_\alpha \mid \alpha < \nu^+\rangle$ are in $M[G^*]$, since $\pi \restriction \nu^+$ belongs to $M$. Note that $\nu^+$ is a cardinal in $M[G^*]$ and $\nu^+ < \pi(\kappa)$. So we can find an unbounded set of $\alpha < \nu^+$ and a fixed $n$ such that $n_\alpha = n$. Let $h^*$ be the unique stem of length $n$ of some condition in $G^*$. Then define $C$ to be the set of $\alpha < \nu^+$ such that there are a condition in $r \in \pi(\mathbb{P})$ with stem $h^*$ and an ordinal $\xi^*_\alpha$ such that $r \Vdash \langle\pi(\alpha),\xi^*_\alpha\rangle\ \pi(\dot{T})\ \langle\gamma^*,\eta^*\rangle$. Our definition of $C$ does not make use of $G^*$ and hence $C \in M$. Moreover, $C$ is unbounded by the choice of $h^*$.

*Claim* 4.2 This $C, n$ satisfy the requirements of Spine Lemma 3

Note that in this situation we have $M \subseteq V[E]$ and hence $C \in V[E]$. If $\alpha < \beta$ are both in $C$, then

$$h^* \Vdash \langle\pi(\alpha),\xi^*_\alpha\rangle\ \pi(\dot{T})\ \langle\gamma^*,\eta^*\rangle \text{ and } \langle\pi(\beta),\xi^*_\beta\rangle\ \pi(\dot{T})\ \langle\gamma^*,\eta^*\rangle.$$

Since $\Vdash \pi(\dot{T})$ is a tree, $h^* \Vdash \langle\pi(\alpha),\xi^*_\alpha\rangle\ \pi(\dot{T})\ \langle\pi(\beta),\xi^*_\beta\rangle$.

So $M \models$ "There are a stem $h$ of length $n$, and $\xi, \zeta < \kappa$ such that $h \Vdash \langle\pi(\alpha),\xi\rangle\ \pi(\dot{T})\ \langle\pi(\beta),\zeta\rangle$." By elementarity, there is a stem $h \in V[E]$ of length $n$ such that $h \Vdash \langle\alpha,\xi\rangle\ \dot{T}\ \langle\beta,\zeta\rangle$. □

**Traction Lemma 3** *In $V[E]$ there are $J \subseteq C$ cofinal, a map $\alpha \mapsto \xi_\alpha (\alpha \in J)$ and a stem $\bar{h}$ such that for all $\alpha < \beta$ both in $J$, $\bar{h} \Vdash \langle \alpha, \xi_\alpha \rangle \dot{T} \langle \beta, \xi_\beta \rangle$.*

*Remark* 4.3 Our notation is deceptive. This will not finish the proof. We could have $\bar{h}^\frown a \Vdash \neg\langle \alpha, \xi_\alpha \rangle \dot{T} \langle \beta, \xi_\beta \rangle$!

*Proof* Let $\pi : V \to M$ be a $\nu^+$-supercompactness embedding with critical point $\kappa_{n+1}$, where $n$ is given by Spine Lemma 3, and let $F$ be $Add(\kappa, \pi(\nu^{++}))$-generic over $V[E]$. Again in $V[E][F]$, $\pi$ extends to an embedding $\pi^* : V[E] \to M[E^*]$. As before let $\gamma^*$ be least in $\pi^*(C)$ greater than $\sup \pi^{*\text{``}}\nu^+$. We apply Spine Lemma 3 in $M[E^*]$. For each $\alpha \in C$, we have $\xi^*_\alpha, \eta^*_\alpha$ and $h^*_\alpha$ such that

$$h^*_\alpha \Vdash \langle \pi^*(\alpha), \xi^*_\alpha \rangle\ \pi^*(\dot{T})\ \langle \gamma^*, \eta^*_\alpha \rangle$$

with $\ell(h^*_\alpha) = n$.

We are going to stabilize $h^*_\alpha$ and $\eta^*_\alpha$. In $V[E][F]$, there is $J \subseteq C$ cofinal and a fixed stem $\bar{h}$ of length $n$ and an $\eta < \kappa$ such that $\alpha \in J$ implies $\eta^*_\alpha = \eta$ and $h^*_\alpha = \bar{h}$. In the above we used that $\mathrm{crit}(\pi^*) = \kappa_{n+1}$, to obtain the fact that stems of length $n$ are the same in $\pi^*(\mathbb{P})$ and $\mathbb{P}$. So $\alpha < \beta$ both in $J$ implies

$$\bar{h} \Vdash \langle \pi^*(\alpha), \xi^*_\alpha \rangle\ \pi^*(\dot{T})\ \langle \gamma^*, \eta \rangle,$$
$$\bar{h} \Vdash \langle \pi^*(\beta), \xi^*_\beta \rangle\ \pi^*(\dot{T})\ \langle \gamma^*, \eta \rangle.$$

So $\bar{h} \Vdash \langle \pi^*(\alpha), \xi^*_\alpha \rangle\ \pi^*(\dot{T})\ \langle \pi^*(\beta), \xi^*_\beta \rangle$.

Set $\xi_\alpha = \xi^*_\alpha$. Note $\pi^*(\xi_\alpha) = \xi^*_\alpha$ and $\pi^*(\bar{h}) = \bar{h}$. So $\bar{h} \Vdash_{\mathbb{P}} \langle \alpha, \xi_\alpha \rangle\ \dot{T}\ \langle \beta, \xi_\beta \rangle$ by elementarity. This almost finishes the proof. In $V[E][F]$ we have the map $\alpha \mapsto \xi_\alpha$ for $\alpha \in J$, but we need to pull this back to $V[E]$.

We apply Lemma S. We do this by viewing the above map as a branch through a particular tree, a tree of attempts to create such a map. Without loss of generality, we may assume that $J$ is maximal, by which we mean if $\beta \in J$ and $\alpha < \beta$ such that there is a $\xi$ with $\bar{h} \Vdash \langle \alpha, \xi \rangle\ \dot{T}\ \langle \beta, \xi_\beta \rangle$ then $\alpha \in J$ and $\xi = \xi_\alpha$. Again we are using that $\Vdash \dot{T}$ is a tree. The fact that $J$ is maximal will allow us to code $J$ as a branch through a tree of height $\nu^+$. Let $f$ be the function $i \mapsto (\alpha_i, \xi_{\alpha_i})$ where $i \mapsto \alpha_i$ enumerates $J$ in increasing order. Note that for $i < \nu^+$, $f \restriction i \in V[E]$, because by maximality $f \restriction i$ is determined in $V[E]$ from $(\alpha_i, \xi_{\alpha_i})$ and $\bar{h}$. So $f$ is a branch through a tree in $V[E]$ of length $\nu^+$, namely the tree of attempts to construct such a function. We have already checked that $Add(\kappa, \pi(\nu^{++}))$

satisfies the hypotheses of the poset in Lemma S. Hence $f \in V[E]$ as required. □

Are we done? *No!* Let $\bar{h}$ witness the lemma and let $\ell(\bar{h}) = \bar{k}$. We can assume that $g = \bigcup\{g_p \mid p \in G\}$ extends $\bar{h}$. However $\{\langle\alpha, \xi_\alpha\rangle \mid \alpha \in J\}$ is not necessarily a branch. We know that for all $\alpha < \beta$ both in $J$, $\bar{h} \Vdash \langle\alpha, \xi_\alpha\rangle\ \dot{T}\ \langle\beta, \xi_\beta\rangle$. However there might be an $a \in \mathcal{P}_\kappa(\kappa_{\bar{k}})$ such that $\bar{h}^\frown a \Vdash \neg\langle\alpha, \xi_\alpha\rangle\ \dot{T}\ \langle\beta, \xi_\beta\rangle$. The set of such $a$ has measure zero, but need not be empty. For all we know $g(\bar{k})$ is such an $a$, then we would have $\neg\langle\alpha, \xi_\alpha\rangle\ \dot{T}[G]\ \langle\beta, \xi_\beta\rangle$ in the extension.

## The next step

The final step in the proof is to get the following.

*Fact* 4.4 In $V[E]$ there are $\rho < \nu^+$ and a sequence $\langle A_\alpha \mid \alpha \in J \setminus \rho\rangle$ with each $A_\alpha$ an $\omega$-sequence of measure one sets, such that for all $\alpha < \beta$ both in $J \setminus \rho$, $\langle\bar{h}, A_\alpha \cap A_\beta\rangle \Vdash \langle\alpha, \xi_\alpha\rangle\ \dot{T}\ \langle\beta, \xi_\beta\rangle$.

Recall that $A_\alpha$ and $A_\beta$ are sequences of measure one sets and that we intersect them pointwise. We will show that Fact 4.4 is enough to finish the proof of Theorem 1.

*Claim* 4.5 If $G$ is $\mathbb{P}$-generic over $V[E]$, then $G$ contains cofinally many of the conditions $\langle\bar{h}, A_\alpha\rangle$.

*Proof* Assume that $G$ only contains boundedly many of the above conditions and fix a condition $q_0$ forcing this. Note that any condition can be extended to one that satisfies the conclusion of Spine Lemma 3 and Traction Lemma 3. We can assume that $q_0$ satisfies the conclusions of the lemmas and we let $g_{q_0} = \bar{h}$, $J$, $\rho$ and $\langle A_\alpha \mid \alpha \in J \setminus \rho\rangle$ witness this. By the $\nu^+$-cc of our forcing, there is an $\alpha_0 < \nu^+$ such that $q_0$ forces for all $\alpha > \alpha_0$ in $J \setminus \rho$, $\langle\bar{h}, A_\alpha\rangle \notin G$. We take $\alpha \in J$ above $\alpha_0$. By our choice of $q_0$, $q_0$ and $\langle\bar{h}, A_\alpha\rangle$ are compatible. Let $r$ be their common extension. Obviously, $r \Vdash \langle\bar{h}, A_\alpha\rangle \in G$. However, we also have $r \Vdash \langle\bar{h}, A_\alpha\rangle \notin G$, because $r \leq q_0$. This is a contradiction. □

To finish the proof of Theorem 1, we note that if $G$ meets cofinally many of the conditions $\langle\bar{h}, A_\alpha\rangle$, then there is a branch through $T$ in $V[E][G]$. This is easy by the choice of the $A_\alpha$. So it remains to construct the sets $A_\alpha$. The idea is to construct $A_\alpha(n)$ by recursion on $n \geq \bar{k}$. In these notes, we will show how to do the construction for $n = \bar{k}$. For the full construction we refer the reader to [16]. The next lemma is the appropriate weakening of Fact 4.4.

**Final Lemma** *In $V[E]$ there are $\rho < \nu^+$ and sets $Z_\alpha$ for $\alpha \in J \smallsetminus \rho$, such that each $Z_\alpha$ has $\mathcal{U}_{\bar{k}}$-measure one and for all $\alpha < \beta$ both in $J \smallsetminus \rho$, for all $a \in Z_\alpha \cap Z_\beta$, $\bar{h}^\frown a \Vdash \langle \alpha, \xi_\alpha \rangle \dot{T} \langle \beta, \xi_\beta \rangle$.*

Before proceeding with the proof, we will explain a failed attempt which is quite instructive. Fix $\pi : V \to M$ a supercompactness embedding with critical point $\kappa_{\bar{k}+1}$. As before we can lift $\pi$ to the universe $V[E]$ to get an elementary embedding $\pi^* : V[E] \to M[E^*]$ where $E^*$ is generic for $\pi(Add(\kappa, \nu^{++}))$. As before we can take $\gamma^* > \sup \pi^* {}^{\prime\prime} \nu^+$, with $\gamma^* \in \pi^*(J)$. Then there is $\eta^*$ such that for each $\alpha$, we get $A_\alpha^*$ and $\xi_\alpha^*$ such that for all $x \in A_\alpha^*$, $\bar{h}^\frown x \Vdash \langle \pi^*(\alpha), \xi_\alpha^* \rangle \pi^*(\dot{T}) \langle \gamma^*, \eta^* \rangle$. Here $\eta^*$ is just $\pi^*(\alpha \mapsto \xi_\alpha)(\gamma^*)$. $A_\alpha^*$ has $\pi^*(\mathcal{U}_{\bar{k}})$-measure one. Note that $\mathcal{P}_\kappa(\kappa_{\bar{k}})$ is the same computed in both $V[E]$ and $V[E][F]$, since $Add(\kappa, \pi(\nu^{++}))$ is $\kappa$-closed. However, the powerset of $\mathcal{P}_\kappa(\kappa_{\bar{k}})$ is larger, so $\pi^*(\mathcal{U}_{\bar{k}})$ measures more sets. Also, elements of the powerset are fixed by $\pi^*$, so we have $\mathcal{U}_{\bar{k}} \subseteq \pi^*(\mathcal{U}_{\bar{k}})$. It follows that $A_\alpha^*$ need not be in $V[E]$ let alone measure one for $\mathcal{U}_{\bar{k}}$. So this will not work.

*Proof of the Final Lemma* The key idea is to work "vertically" instead of "horizontally". A vertical segment will use a version of $J$ for $h^\frown x$. So fix $\pi^*$, $\gamma^*$ and $\eta^*$ as in the last paragraph. For each $x \in \mathcal{P}_\kappa(\kappa_{\bar{k}})$, let $J_x = \{\alpha \in J \mid \bar{h}^\frown x \Vdash \langle \pi(\alpha), \xi_\alpha \rangle \pi^*(\dot{T}) \langle \gamma^*, \eta^* \rangle\}$. The "horizontal" sets are $\{x \mid \alpha \in J_x\}$. They have $\pi(\mathcal{U}_{\bar{k}})$-measure one, but they need not be in $V[E]$. So we are going to look at the "vertical" segments $J_x$.

*Claim* 4.6 *If $J_x$ is unbounded in $\nu^+$, then $J_x \in V[E]$*

*Proof* We apply Lemma S in the same way that we did for $J$. If $J_x$ is unbounded, then it is coded by a branch through a tree of height $\nu^+$ as follows. Again we assume that $J_x$ is maximal in the sense that for all $\alpha \in J_x$ and all $\beta < \alpha$ with $\beta \in J$, if $h^\frown x \Vdash \langle \beta, \xi_\beta \rangle \dot{T} \langle \alpha, \xi_\alpha \rangle$, then $\beta \in J_x$. We let $f_x \in V[E][F]$ be the map $i \mapsto (\alpha_i, \xi_{\alpha_i})$ that enumerates $J_x$ on the first coordinate in increasing order. Then for all $i < \nu^+$, $f_x \upharpoonright i \in V[E]$, since it is determined from $(\alpha_i, \xi_{\alpha_i})$, $\bar{h}$ and $x$. Hence $f_x$ is a branch through a tree of attempts to find it in $V[E]$. The other parameters in the application of Lemma S are the same as before. $\square$

Let $\dot{J}_x \in V[E]$ be a name for $J_x$ in $Add(\kappa, \pi(\nu^{++}))$. (Recall that $F$ was the generic object for this poset.) Since $Add(\kappa, \pi(\nu^{++}))$ is $\kappa^+$-cc, there is a set $K_x \in V[E]$ of size $\leq \kappa$ such that $\Vdash_{Add(\kappa, \pi(\nu^{++}))} "\dot{J}_x \in K_x$, if $\dot{J}_x$ is unbounded in $\nu^+$". By shrinking $K_x$, we may assume that for each $I \in K_x$

1. $I$ is unbounded in $\nu^+$
2. $(\beta \in I \wedge \alpha < \beta \wedge \alpha \in J) \Rightarrow (\alpha \in I \Leftrightarrow h^\frown x \Vdash \langle \alpha, \xi_\alpha \rangle \; \dot{T} \; \langle \beta, \xi_\beta \rangle)$

We call condition 2 *maximality*. Note that maximality is essentially the same condition we used in the applications of Lemma S. We can assume that each member of $K_x$ is maximal since any unbounded $I$ as above has a unique extension to an unbounded set which is maximal. This follows from the fact that $\dot{T}$ is forced to be a tree. In $V[E]$ we have the map $x \mapsto K_x$ where $K_x$ is the collection of candidates for $J_x$. We work to refine our knowledge of each $K_x$ and its members.

*Claim* 4.7 If $I, I' \in K_x$ are distinct, then they are disjoint on a tail.

*Proof* If there is a place, $\beta$, where $I, I'$ agree then by maximality they agree below $\beta$. So after the first place where $I, I'$ differ, they are disjoint. □

**Corollary 4.8** *For each x, there is $\rho_x < \nu^+$ such that if $I, I' \in K_x$ are distinct, then they are disjoint above $\rho_x$.*

*Proof* Recall that $|K_x| \leq \kappa$. So to find $\rho_x$ we take a supremum over the least place where any pair from $K_x$ differ. $\rho_x$ is a supremum over just $|K_x|^2$ many ordinals less than $\nu^+$ and hence it is less than $\nu^+$. □

Let $\rho = \sup_{x \in \mathcal{P}_\kappa(\kappa_{\bar{k}})} \rho_x$. Then $\rho < \nu^+$, since $|\mathcal{P}_\kappa(\kappa_{\bar{k}})| = \kappa_{\bar{k}} < \nu^+$. So for any $x$ and for any $I, I' \in K_x$ which are distinct, $I, I'$ are disjoint above $\rho$. We are going to work above $\rho$. Define a function $f$ on $\mathcal{P}_\kappa(\kappa_{\bar{k}}) \times (J \setminus \rho)$ by $f(x, \alpha) =$ the unique $I \in K_x$ such that $\alpha \in I$, if such $I$ exists and $f(x, \alpha)$ is undefined otherwise.

*Claim* 4.9 For $\alpha \in J \setminus \rho$, $\{x \mid f(x, \alpha)$ is defined$\}$ has $\mathcal{U}_{\bar{k}}$-measure one.

*Proof* Fix $\alpha \in J \setminus \rho$. First note that $f \in V[E]$ and hence $\{x \mid f(x, \alpha)$ is defined $\} \in V[E]$. Let $Y$ be its complement. Suppose for a contradiction that $Y$ has $\mathcal{U}_{\bar{k}}$-measure one.

Here we actually need the sets from our failed attempt at a proof of the Final Lemma. Recall, $A^*_\alpha$ was measure one for $\pi^*(\mathcal{U}_{\bar{k}})$. Furthermore, we had the property that there is $\eta^*$ such that for every $\alpha$, there is $\xi^*_\alpha$ such that for all $x \in A^*_\alpha$, $\bar{h}^\frown x \Vdash \langle \pi^*(\alpha), \xi^*_\alpha \rangle \; \pi^*(\dot{T}) \; \langle \gamma^*, \eta^* \rangle$. As $\mathrm{crit}(\pi^*) = \kappa_{\bar{k}+1}$, $\pi^*(Y) = Y$. By elementarity $Y$ has $\pi^*(\mathcal{U}_{\bar{k}})$-measure one. For every $\beta \in J$, the intersection of measure one sets $A^*_\alpha \cap A^*_\beta \cap Y$ is nonempty. For each $\beta$, let $x_\beta \in A^*_\alpha \cap A^*_\beta \cap Y$.

As $J$ is unbounded in $\nu^+$ and $|\mathcal{P}_\kappa(\kappa_{\bar{k}})| = \kappa_{\bar{k}}$, there are a fixed $x \in \mathcal{P}_\kappa(\kappa_{\bar{k}})$ and $U \subseteq J$ unbounded, such that $x = x_\beta$ for all $\beta \in U$. By the construction of the $A^*_\beta$s, we have $\bar{h}^\frown x \Vdash \langle\pi^*(\alpha), \xi^*_\alpha\rangle\ \pi^*(\dot{T})\ \langle\gamma^*, \eta^*\rangle$ and $\langle\beta, \xi^*\beta\rangle\ \pi^*(\dot{T})\ \langle\gamma^*, \eta^*\rangle$ for all $\beta \in U$. So by the definition of $J_x$, $\alpha \in J_x$ and $U \subseteq J_x$. But this means that $f(x, \alpha)$ was defined and equal to $J_x$, a contradiction. □

*Claim* 4.10 For $\alpha, \alpha'$ both in $J \setminus \rho$, the set $\{x \mid f(x, \alpha) = f(x, \alpha')\}$ has $\mathcal{U}_{\bar{k}}$-measure one.

*Proof* By the previous claim there is a measure one set where both are defined. Fix $x$ and suppose that $f(x, \alpha)$ and $f(x, \alpha')$ are both defined. Without loss of generality $\alpha < \alpha'$. We claim that $\alpha \in f(x, \alpha')$. Using maximality and the fact that $\alpha, \alpha' > \rho$, it suffices to check that $\alpha' \in f(x, \alpha')$, $\alpha < \alpha'$, $\alpha \in f(x, \alpha)$ and $\bar{h}^\frown x \Vdash \langle\alpha, \xi_\alpha\rangle\ \dot{T}\ \langle\alpha', \xi_{\alpha'}\rangle$. The first three are obvious, and the last one follows from the fact that $f(x, \alpha), f(x, \alpha')$ are candidates for $J_x$ and that $\Vdash \dot{T}$ is a tree. This finishes the proof as we have shown that everywhere $f(x, \alpha), f(x, \alpha')$ are defined they are equal. □

We are ready to define the measure one sets $Z_\alpha$. Let $\alpha_0$ be the least element of $J \setminus \rho$. Define $Z_\alpha = \{x \mid f(x, \alpha) = f(x, \alpha_0)$ where both are defined$\}$. By the previous claims $Z_\alpha$ has $\mathcal{U}_{\bar{k}}$-measure one. If $x \in Z_\alpha \cap Z_\beta$ then let $I = f(x, \alpha) = f(x, \beta) = f(x, \alpha_0)$ Recall, $I$ is maximal so $\bar{h}^\frown x \Vdash \langle\alpha, \xi_\alpha\rangle\ \dot{T}\ \langle\beta, \xi_\beta\rangle$, as required. □

The complete construction of the sequences of measure one sets $A_\alpha$ mentioned above works by recursion on the length of a stem $h$ extending $\bar{h}$. Using suitable inductive hypotheses, Neeman constructs $J^h, \rho_h$, which are analogs of the $J, \rho$ that we constructed above. From $J^h$ and $\rho_h$, Neeman obtains measure one sets $A^h_\alpha$ which are analogs of the $Z_\alpha$ that we constructed. To finish the proof, each $A_\alpha$ is essentially the diagonal intersection of the sets $A^h_\alpha$ for $h$ extending $\bar{h}$.

## 5 Open problems

We proved that the tree property at $\kappa^+$ does not imply $\mathrm{SCH}_\kappa$.

1. Does $\nu^+$ still have the tree property after cardinal preserving forcing?
2. Can we make $\nu$ of the result into $\aleph_\omega$ or some other small cardinal?

# References


[1] James E. Baumgartner, *Iterated forcing*, Surveys in set theory, London Math. Soc. Lecture Note Ser., vol. 87, Cambridge Univ. Press, Cambridge, 1983, pp. 1–59.

[2] Shai Ben-David and Menachem Magidor, *The weak $\square^*$ is really weaker than the full $\square$*, J. Symbolic Logic **51** (1986), no. 4, 1029–1033.

[3] James Cummings and Matthew Foreman, *Diagonal Prikry extensions*, J. Symbolic Logic **75** (2010), no. 4, 1382–1402.

[4] William B. Easton, *Powers of regular cardinals*, Ann. Math. Logic **1** (1970), 139–178.

[5] Matthew Foreman, *Some problems in singular cardinals combinatorics*, Notre Dame J. Formal Logic **46** (2005), no. 3, 309–322.

[6] Moti Gitik, *The strength of the failure of the singular cardinal hypothesis*, Ann. Pure Appl. Logic **51** (1991), no. 3, 215–240.

[7] Moti Gitik and Menachem Magidor, *The singular cardinal hypothesis revisited*, Set theory of the continuum (Berkeley, CA, 1989), Math. Sci. Res. Inst. Publ., vol. 26, Springer, New York, 1992, pp. 243–279.

[8] Moti Gitik and Assaf Sharon, *On SCH and the approachability property*, Proc. Amer. Math. Soc. **136** (2008), no. 1, 311–320 (electronic).

[9] R. Björn Jensen, *The fine structure of the constructible hierarchy*, Ann. Math. Logic **4** (1972), 229–308; erratum, ibid. 4 (1972), 443, With a section by Jack Silver.

[10] H. J. Keisler and A. Tarski, *From accessible to inaccessible cardinals. Results holding for all accessible cardinal numbers and the problem of their extension to inaccessible ones*, Fund. Math. **53** (1963/1964), 225–308.

[11] D. König, *Sur les correspondence multivoques des ensembles*, Fund. Math. **8** (1926), 114–134.

[12] D. Kurepa, *Ensembles ordonnés et ramifiés*, Publ. Math. Univ. Belgrade **4** (1935), 1–138.

[13] Richard Laver, *Making the supercompactness of $\kappa$ indestructible under $\kappa$-directed closed forcing*, Israel J. Math. **29** (1978), no. 4, 385–388.

[14] Menachem Magidor and Saharon Shelah, *The tree property at successors of singular cardinals*, Arch. Math. Logic **35** (1996), no. 5-6, 385–404.

[15] William Mitchell, *Aronszajn trees and the independence of the transfer property*, Ann. Math. Logic **5** (1972/73), 21–46.

[16] Itay Neeman, *Aronszajn trees and failure of the singular cardinal hypothesis*, J. Math. Log. **9** (2009), no. 1, 139–157.

[17] K. L. Prikry, *Changing measurable into accessible cardinals*, Dissertationes Math. Rozprawy Mat. **68** (1970), 55.

[18] Robert M. Solovay, *Strongly compact cardinals and the GCH*, Proceedings of the Tarski Symposium (Proc. Sympos. Pure Math., Vol. XXV, Univ. California, Berkeley, Calif., 1971) (Providence, R.I.), Amer. Math. Soc., 1974, pp. 365–372.

[19] E. Specker, *Sur un problème de Sikorski*, Colloquium Math. **2** (1949), 9–12.

# 7

# Iterated Forcing and the Continuum Hypothesis

Todd Eisworth, Justin Tatch Moore[a] and David Milovich

The ninth Appalachian Set Theory workshop was held at the Fields Institute in Toronto on May 29-30, 2009. The lecturers were Todd Eisworth and Justin Moore. As a graduate student David Milovich assisted in writing this chapter, which is based on the workshop lectures.

The notes which follow reflect the content of a two day tutorial which took place at the Fields Institute on 5/29 and 5/30 in 2009. Most of the content has existed in the literature for some time (primarily in the original edition of [10]) but has proved difficult to read and digest for various reasons. The only new material contained in these lectures concerns the notion of a *fusion scheme* presented in Sections 6 and 7 and even this has more to do with style than with mathematics. Our presentation of the iteration theorems follows [4]. The *k-iterability condition* is a natural extrapolation of what appears in [4] and [5], where the iteration theorem for the $\aleph_0$-iterability condition is presented (with a weakening of $< \omega_1$-properness). The formulation of complete properness is taken from [8]. We stress, however, these definitions and theorems are really technical and/or stylistic modifications of the theorems and definitions of Shelah presented in [10]. Those interested in further reading on the topic of the workshop should consult: [1], [4], [5], [8], [10], and [12]. We would like to thank Ilijas Farah, Miguel Angel Mota, Paul Shafer, and the anonymous referee for their careful reading and suggesting a number of improvements.

[a] The research of Justin Moore was supported in part by NSF grant DMS-0757507. Any opinions, findings, and conclusions or recommendations expressed in this material are those of the authors and do not necessarily reflect the views of the National Science Foundation.

# 1 Introduction

The focus of the following lectures is on forcing axioms in the presence of the Continuum Hypothesis. Not long after Solovay and Tennenbaum's proof that Souslin's Hypothesis was relatively consistent [11], Jensen showed that Souslin's Hypothesis is relatively consistent with CH (see [3]). While Martin's Maximum provides a provably optimal consistent forcing axiom [6], it is still not clear whether there is an optimal forcing axiom which is consistent with CH.[1] Over the last three decades, Shelah and others developed a number of sufficient conditions for establishing that consequences of forcing axioms are consistent with CH. The purpose of these lectures is to present these conditions in a form which strikes some balance between utility and ease of understanding.

We will begin by stating an open problem which seems to require new ideas and at the same time serves to illustrate what can be accomplished through existing methods. If $X$ and $Y$ are countable subsets of $\omega_1$ which are closed in their suprema, then we say that $X$ *measures* $Y$ if there is an $\alpha_0 < \alpha = \sup X$ such that $X \cap (\alpha_0, \alpha)$ is contained in or disjoint from $Y$. *Measuring* is the assertion that whenever $\langle D_\alpha : \alpha \in \omega_1 \rangle$ is a sequence with $D_\alpha$ a closed subset of $\alpha$ for each $\alpha \in \omega_1$, there is a club $E \subseteq \omega_1$ such that $E \cap \alpha$ measures $D_\alpha$ whenever $\alpha$ is a limit point of $E$.

*Question* 1.1 Is *measuring* consistent with CH?

It is easy to show that $\diamondsuit$ implies that measuring fails. We will also see that there is a canonical partial order for forcing an instance of measuring without adding reals. By the book keeping arguments of [11], the question reduces to showing that an iteration of these partial orders does not add new reals. Dealing with this difficulty — determining when an iteration of forcings does not add reals — will be the central theme throughout these lectures.

In order to demonstrate the type of problem which arises here, let us consider another combinatorial principle. Recall that a *ladder system* is a sequence $\langle C_\alpha : \alpha \in \lim(\omega_1) \rangle$ such that for each $\alpha \in \lim(\omega_1)$, $C_\alpha \subseteq \alpha$ is cofinal and has ordertype $\omega$. Let (U) be the assertion that if $\vec{C}$ is a ladder system and $g : \omega_1 \to 2$, then there is an $f : \omega_1 \to 2$ such that for each

[1] In September 2009 Aspero, Larson, and Moore announced that there are two $\Pi_2$ sentences $\psi_1$ and $\psi_2$ in the language of $(H(\omega_2), \in, \omega_1)$ such that it is forcible that $(H(\omega_2), \in, \omega_1)$ satisfy $\psi_i \wedge \mathrm{CH}$ for $i = 1$ or 2, but such that $\psi_1 \wedge \psi_2$ implies $\neg\mathrm{CH}$. This essentially rules out the possibility of a provably optimal forcing axiom which is consistent with CH.

$\alpha \in \lim(\omega_1)$,

$$f \upharpoonright C_\alpha \equiv_* g(\alpha).$$

Here $\equiv_* g(\alpha)$ means "takes the constant value $g(\alpha)$ except on a finite set." We will see that there is a partial order which forces an instance of (U) and which does not add new reals. Still, Devlin and Shelah [2] have shown that (U) implies $2^\omega = 2^{\omega_1}$ and in particular that CH fails. To see this, fix a bijection $h : \omega \to \omega \times \omega$ such that $i, j \leq n$ whenever $h(n+1) = (i, j)$. For each $g : \omega_1 \to 2$ construct a sequence of functions $f_n : \omega_1 \to 2$ such that $f_0 = g$ and

$$f_{n+1} \upharpoonright C_\alpha \equiv_* f_i(\alpha + j)$$

whenever $\alpha \geq \omega$ is a limit and $h(n+1) = (i, j)$. Given $f_k$ $(k \leq n)$, $f_{n+1}$ exists by applying (U) to the coloring $\alpha \mapsto f_i(\alpha + j)$ where $h(n+1) = (i, j)$. Now observe that for each limit $\alpha \geq \omega$, $\langle f_n \upharpoonright \alpha : n \in \omega \rangle$ uniquely determines $\langle f_n \upharpoonright \alpha + \omega : n \in \omega \rangle$. Hence, by the transfinite recursion theorem, $\langle f_n \upharpoonright \omega : n \in \omega \rangle$ uniquely determines $\langle f_n : n \in \omega \rangle$ and in particular uniquely determines $g = f_0$. Since $|(2^\omega)^\omega| = |2^\omega|$, it follows that $2^{\aleph_0} = 2^{\aleph_1}$.

In fact Devlin and Shelah showed that the following weak form of $\diamondsuit$ is equivalent to $2^{\aleph_0} < 2^{\aleph_1}$ [2]:

For all $F \colon 2^{<\omega_1} \to 2$ there exists a $g \colon \omega_1 \to 2$ such that for all $f \colon \omega_1 \to 2$ the set $\{\delta < \omega_1 : F(f \upharpoonright \delta) \neq g(\delta)\}$ is stationary.

Here $F$ can be viewed as a method for coding and *weak diamond* can be viewed as asserting that each of these coding methods fails to code at least some element of $2^{\omega_1}$. In the example just discussed, $F(f \upharpoonright \delta) = i$ if $f \upharpoonright C_\delta \equiv_* i$ (with $F(f \upharpoonright \delta)$ defined arbitrarily if $f \upharpoonright C_\delta$ is not eventually constant).

## 2 Proper forcing

Before proceeding it will be useful to review some terminology associated to proper forcing. First, let us agree to the following conventions about forcing. A forcing $Q$ is a partial order with the following additional properties:

- $Q$ is separative, *i.e.*, if $q \nleq p$, then there is an $r \leq q$ such that $r$ is incompatible with $p$.
- $Q$ has a maximal element.

The first condition is out of convenience for our general discussion of forcing and iterations. There is no loss of generality since we may always replace a given partial order (or even quasi-order) with its separative quotient and the underlying set with its cardinality. Similarly, we may always adjoin a maximal element to the partial order if it is not present. This will frequently be done without further mention. We will also use the following notation:

- If $P$ is a forcing, $G \subseteq P$ is $V$-generic, and $x$ is in $V[G]$, then $\dot{x}$ denotes a name such that $\dot{x}[G] = x$.
- For each $x$ in $V$, $\check{x}$ is the canonical name for $x$.

The choice of names in the first convention is not canonical but will be taken to be when possible (for instance the generic filter does have a canonical name, as do new sets which are explicitly constructed from it).

Recall that if $P$ is a forcing and $\dot{Q}$ is a $P$-name for a forcing, then the *two-step iteration* is defined by $P * \dot{Q} = \{p * \dot{q} : p \in P, \Vdash \dot{q} \in \dot{Q}\}$ and declaring that $p * \dot{q} \leq r * \dot{s}$ if and only if $p \leq r$ and $p \Vdash \dot{q} \leq \dot{s}$.

Regarding elementary submodels, we will adopt the following conventions:

1. $\chi$ always denotes a regular cardinal sufficiently large for the argument at hand;
2. $H(\chi)$ denotes the sets hereditarily of size less than $\chi$;
3. $<_\chi$ is a well-ordering of $H(\chi)$.

When discussing forcing at an abstract level, we all always assume that the underlying set of a given forcing is a cardinal. If $P$ is a notion of forcing and $\chi \geq \left(2^{|P|}\right)^+$, then all statements of interest about $P$ are absolute between $V$ and $H(\chi)$ (the reason for our assumption on the underlying set is that then $P$ is necessarily in $H(\chi)$). Recall that if $X$ is an uncountable set, then a *club* in $[X]^\omega$ is a set $E$ of the form $\{Z \in [X]^\omega : f''Z^{<\omega} \subseteq Z\}$ for some $f : X^{<\omega} \to X$. A subset $S \subseteq [X]^\omega$ is *stationary* if it intersects every club in $[X]^\omega$. The countable elementary submodels of $H(\chi)$ form a closed unbounded subset of $[H(\chi)]^\omega$.

Our usual situation is that $P$ is a notion of forcing, $P \in N \prec H(\chi)$, and $N$ is countable. We won't explicitly mention all the parameters included in $N$ (*e.g.*, $P, \leq_P, \mathbb{1}_P \in N \prec \langle H(\chi), \in, <_\chi\rangle$). Rather than say all this, we will just say, "*let $P \in N$ as usual*" or "*let $N$ be a suitable model for $P$.*" Unless explicitly stated otherwise, all elementary submodels are countable. It will be helpful to define some notation in this context.

**Definition 2.1** Given $P$ totally proper and $P \in N$ as usual, define

$$Gen(N, P) = \{G^* \subseteq N \cap P : G^* \text{ is an } N\text{-generic filter of } N \cap P\},$$

$$Gen^+(N, P) = \{G^* \in Gen(N, P) : G^* \text{ has a lower bound}\}, \text{ and}$$

$$Gen(N, P, p) = \{G^* \in Gen(N, P) : p \in G^*\}.$$

**Definition 2.2** Given $P \in N$ as usual, set $N^P = \{\dot{\tau} : \dot{\tau} \in N \wedge \dot{\tau}$ is a $P$-name$\}$. If $G \subseteq P$ is generic, then $N[G] \prec H(\chi)[G]$ where $N[G] = \{\dot{\tau}[G] : \dot{\tau} \in N^P\}$.

Is $N[G]$ as above a generic extension of $N$? This is where properness comes in.

- A condition $q \in P$ is *$(N, P)$-generic* if $q \Vdash N \cap \dot{G}_P \cap D \neq \emptyset$ for every dense $D \subseteq P$ for which $D \in N$.
- $P$ is *proper* if whenever $P \in N$ as usual and $p \in N \cap P$, there is an $(N, P)$-generic $q \leq p$.

*Remark* 2.3 This definition is robust with respect to demanding that $\chi = \left(2^{|P|}\right)^+$, that $\chi$ be arbitrarily large, that additional parameters be added to $H(\chi)$, and so forth.

**Definition 2.4** "$N \cap D$ is predense below $q$" means that every extension of $q$ has an extension below an element of $N \cap D$.

**Proposition 2.5** *Suppose $P \in N$ as usual. The following are equivalent.*

1. *$q$ is $(N, P)$-generic.*
2. *$N \cap D$ is predense below $q$ for all dense $D \subseteq P$ from $N$.*
3. *If $\mathcal{A}$ is a maximal antichain and $\mathcal{A} \in N$, then $q \Vdash \mathcal{A} \cap N \cap \dot{G}_P \neq \emptyset$.*

The following is a key property of proper forcing. In fact, together with the preservation of properness under countable support iterations, it gives the essential properties of properness.

**Proposition 2.6** *If $P$ is proper and $G \subseteq P$ is generic, then every countable set of ordinals in $V[G]$ is covered by a countable set from $V$. In particular:*

- *$P$ does not collapse $\omega_1$;*
- *$P$ adds a countable sequence of elements of $V$ if and only if it adds a new real.*

*Proof* Suppose $p$ is in $P$ and forces $\dot{A}$ is a countable set of ordinals. Let $\dot{\alpha}_n$ be forced by $p$ to be the $n^{\text{th}}$ element of $\dot{A}$, $N \prec H(\chi)$ be a suitable model with $p, P, \{\dot{\alpha}_n : n \in \omega\} \in N$, and $q \leq p$ be $(N, P)$-generic. We claim that $q \Vdash \dot{\alpha}_n \in \check{N}$ for all $n$. To see this, observe that $D_n = \{r \in P : r \text{ decides a value for } \dot{\alpha}_n\}$ is an element of $M$ (it is definable from parameters in $M$) and is a dense subset of $P$. It follows that $q \Vdash N \cap D_n \cap \dot{G}_P \neq \emptyset$. Let $G \subseteq P$ be generic with $q \in G$. Working in $V[G]$, we have that $\dot{\alpha}_n[G]$ is an ordinal and some $r \in N \cap D_n \cap G$ decides the value of $\alpha_n$. We recover the ordinal in $N$ from $r$ and $\dot{\alpha}_n$ by elementarity. □

**Theorem 2.7** *P is proper if and only if forcing with P preserves stationary subsets of* $[X]^\omega$ *for any uncountable set X.*

The following are easy observations which give important examples of proper forcings. Recall that a forcing $P$ has the *c.c.c.* if every antichain in $P$ is countable. A forcing $P$ is countably closed if every countable descending sequence in $P$ has a lower bound.

**Proposition 2.8** *Every partial order with the c.c.c. is proper.*

*Proof* Let $P$ be a c.c.c. forcing and let $P \in N$ be as usual. Every condition is $(N, P)$-generic because if $\mathcal{A}$ is a maximal antichain from $N$, then $\mathcal{A} \subseteq N$. □

**Proposition 2.9** *Every countably closed forcing is proper.*

*Proof* Let $P$ be a countably closed forcing and let $P \in N$ be as usual with $p \in N \cap P$. Build $\langle p_n \rangle_{n<\omega}$ such that $p_0 = p$, $p_{n+1} \leq p_n$, and $p_{n+1} \in D_n \cap N$. Let $q \leq p_n$ for all $n$. Then $q$ is $(N, P)$-generic. □

As simple as it is, this last construction provides the template for the constructions to come. In general if a proper forcing does not add new reals, it need not be the case that an arbitrary sequence of conditions has a lower bound. If some additional care is taken in constructing the sequence, however, one can often arrange that the resulting sequence is bounded.

Let $P$ be proper and not add reals. Let $N$ be as usual. Let $p$ be $(N, P)$-generic. Let $\langle D_n \rangle_{n<\omega}$ enumerate the dense subsets of $P$ from $N$. Let $\langle d^n_m \rangle_{m<\omega}$ enumerate $D_n \cap N$. We have that $p \Vdash \forall n\ \exists m\ d^n_m \in \dot{G}_P$ because $p$ is $(N, P)$-generic. Let $\dot{f}$ be a $P$-name such that $p \Vdash \forall n\ d^n_{\dot{f}(n)} \in \dot{G}_P$. Since $P$ doesn't add reals, we can find $g \in \omega^\omega$ and $q \leq p$ such that $q \Vdash \check{g} = \dot{f}$. For each $n$, $q \Vdash d^n_{g(n)} \in \dot{G}_P$; hence, $q \leq d^n_{g(n)}$. If this were not the case, $q$ would have an extension $r \perp d^n_{g(n)}$, which would imply that every generic filter $G$ with $r \in G$ would have incompatible elements, which is absurd. So, $q$ is $(N, P)$-generic in a strong sense: whenever $D \in N$ is dense in $P$, there is a $d \in N \cap D$ such that $q \leq d$. This motivates the following definition.

- $q$ is *totally* $(N, P)$*-generic* if $q$ extends an element of $N \cap D$ for any dense $D \subseteq P$ with $D \in N$.
- $P$ is *totally proper* if whenever $N, P$ are as usual, any $p \in N \cap P$ has a totally $(N, P)$-generic extension.

Of course being totally $(N, P)$-generic is equivalent to being a lower bound for a $(N, P)$-generic filter.

**Proposition 2.10** *$P$ is totally proper if and only if $P$ is proper and adds no new reals.*

An important point which we will come to momentarily is that, even in a totally proper forcing $P$, conditions which are $(N, P)$-generic need not be totally $(N, P)$-generic. It is true, however, that every $(N, P)$-generic condition in a totally proper forcing can be extended to a totally $(N, P)$-generic condition.

## 3 Two-step iterations

If $P$ is proper and $\Vdash_P \dot{Q}$ is proper, then $P * \dot{Q}$ is proper: $P$ preserves stationary subsets of $[\lambda]^\omega$ and then $Q$ preserves them, so $P * \dot{Q}$ preserves them. Similarly, if $P$ is totally proper and $\Vdash_P \dot{Q}$ is totally proper, then $P * \dot{Q}$ is totally proper.

Understanding preservation of properties such as properness and total properness in transfinite iterations is more subtle and ultimately requires a finer and more localized analysis of two step iterations. To illustrate this, suppose that $P$ is proper and that $P$ forces $\dot{Q}$ is proper. Let $N$ be as usual with $P * \dot{Q} \in N$. It can be shown that $p * \dot{q}$ is $(N, P * \dot{Q})$-generic if and only if $p$ is $(N, P)$-generic and $p \Vdash \dot{q}$ is $(N[\dot{G}_P], \dot{Q})$-generic.

This refinement fails for total properness and this is ultimately the source of all of the difficulties which we will encounter in these lectures. There are $N, P, \dot{Q}, p, \dot{q}$ such that $P$ is totally proper, $\Vdash_P \dot{Q}$ is totally proper, $p$ is totally $(N, P)$-generic, and $p \Vdash \dot{q}$ is totally $(N[\dot{G}_P], \dot{Q})$-generic, but $p * \dot{q}$ is *not* totally $(N, P * \dot{Q})$-generic.

This is best illustrated in an example.

**Example 3.1** Let $\vec{C} = \langle C_s : s \in \lim(\omega_1)\rangle$ be a ladder system. Let $g : \omega_1 \to \{0, 1\}$. Does there exist $f : \omega_1 \to \{0, 1\}$ such that for all $\delta \in \lim(\omega_1)$, $f \restriction C_\delta$ is eventually constant with value $g(\delta)$? Generally the answer is 'no' if, for instance, $\diamondsuit$ holds. That is, for a given ladder system one can use a $\diamondsuit$-sequence to predict possible uniformizing functions $f$ and build the desired coloring $g$. We'll force the answer to be yes, for a given $g$, without adding new reals. Define $P_g$ to be the collection of all countable approximations to the desired uniformizing function $f$. Specifically, $\mathrm{dom}(p) = \delta$ for some $\delta < \omega_1$, and if $\alpha \leq \delta$ is a limit ordinal, then $p \restriction C_\alpha$ is eventually constant with value $g(\alpha)$.

**Proposition 3.2** *$P_g$ is totally proper and forces the existence of a uniformizing function $f$ for the coloring $g$.*

The following three lemmas constitute the essence of the proof. Moreover, the role of each is quite typical in arguments of this sort.

**Lemma 3.3** *For each $\alpha < \omega_1$, the set of conditions $p$ for which $\alpha \in \mathrm{dom}(p)$ is dense.*

*Proof* Suppose that the lemma holds for all $\beta < \alpha$. Suppose $p \in P_g$ and $\alpha \notin \mathrm{dom}(p)$. Let $\mathrm{dom}(p) = \beta + 1$ for some $\beta < \alpha$. If $\alpha = \gamma + 1$, then extend $p$ to $q$ with $\gamma \in \mathrm{dom}(q)$; extend $q$ to $q \cup \{\langle\alpha, 0\rangle\}$. So, we may assume $\alpha$ is a limit ordinal. Let $\langle\gamma_n\rangle_{n<\omega}$ be strictly increasing with limit $\alpha$ and satisfy $\gamma_0 = \beta$. Build $\langle p_n\rangle_{n<\omega}$ such that $p_0 = p$, $p_{n+1} \leq p_n$, $\gamma_n \in \mathrm{dom}(p_n)$, and $p_{n+1}(\delta) = F(\alpha)$ for all $\delta \in \mathrm{dom}(p_{n+1} \setminus p_n) \cap C_\alpha$. The last requirement can be met because $P_g$ is closed with respect to finite modification. Finally, let $q = \bigcup_{n<\omega} p_n \cup \{\langle\alpha, 0\rangle\}$. □

**Lemma 3.4** *Suppose that $p_n$ $(n \in \omega)$ is a strictly descending sequence in $P_g$. The following are equivalent:*

1. *$p_n$ $(n \in \omega)$ has a lower bound in $P_g$.*
2. *$\bigcup p_n$ is in $P_g$.*
3. *if $\alpha = \mathrm{dom}(\bigcup_n p_n)$, then there is an $\alpha_0 < \alpha$ such that $p_n(\xi) = g(\alpha)$ whenever $\xi$ is in $C_\alpha \cap \mathrm{dom}(p_n)$ with $\alpha_0 < \xi$.*

*Proof* Follows from the definitions. □

*Proof* (of Proposition 3.2) Let $N$ be as usual with $P_g, g,$ etc. $\in N$. Let $p \in P_g \cap N$. Let $N_k$ $(k \in \omega)$ be an $\in$-chain of countable elementary submodels of $H(\omega_2)$ such that $g$ and $p$ are in $N_0$ and $N \cap H(\omega_2) = \bigcup_k N_k$. Let $\langle D_k\rangle_{k<\omega}$ enumerate the dense subsets of $P$ from $N$ such that $D_k$ is in $N_k$. We build $\langle p_k\rangle_{k<\omega}$ such that $p = p_0$, $p_k \geq p_{k+1} \in N_k \cap D_k$, and, for any $\alpha \in C_\delta \cap \mathrm{dom}(p_{k+1} \setminus p_k)$ where $\delta = N \cap \omega_1$, we have $p_{k+1}(\alpha) = g(\delta)$ if $\delta \in S$.

To see that this can be done, suppose we have $p_k$ and look at $N_k \cap C_\delta \setminus \mathrm{dom}(p_k)$. It's finite and hence an element of $N_k$. Inside $N_k$, extend $p_k$ to a condition $r$ such that $C_\delta \cap N_k \subseteq \mathrm{dom}(r)$. Modify $r$ on $C_\delta \cap N_k \setminus \mathrm{dom}(p_n)$ to agree with $g(\delta)$. This modification is finite and hence $r$ remains both in $N_k$ and $P_g$. Now extend $r$ to $p_{k+1} \in N_k \cap D_k$. $p_{k+1}$ is as desired. By Lemma 3.4, $p_k$ $(k \in \omega)$ has a lower bound as desired. □

**Example 3.5** Set $P = \langle 2^{<\omega_1}, \supseteq \rangle$ and let $\dot{g}$ be the generic function coded by $P$. Define $\dot{Q} = P_{\dot{g}}$ and let $N$ be as usual with $P, \dot{Q} \in N$. Let $p$ be $(N, P)$-generic with $\mathrm{dom}(p) = N \cap \omega_1$. We claim there is no $\dot{q}$ such that $p * \dot{q}$ is totally $(N, P * \dot{Q})$-generic. To see this, let $\delta = N \cap \omega_1$ and let $\langle \alpha_n \rangle_{n<\omega}$ enumerate $C_\delta$. Let

$$D_n = \{r * \dot{s} \in P * \dot{Q} : r * \dot{s} \text{ decides the value of } \dot{f}(\alpha_n)\}$$

where $\dot{f}$ is the function added by $\dot{Q}$. If some $p * \dot{q}$ could decide every $\dot{f}(\alpha_n)$, then it would also decide $\dot{g}(\delta)$ to be $\epsilon$, which is absurd, for we can extend $p$ to force $\dot{g}(\delta)$ to be $1 - \epsilon$.

This problem can be remedied by requiring $p$ to be totally generic over models above $N$ as well.

**Proposition 3.6** *Suppose that $P$ is totally proper and $\dot{Q}$ is a $P$-name for a totally proper forcing. Let $N_0 \in N_1$ be as usual with $P * \dot{Q} \in N_0$. If $p$ is totally $(N_i, P)$-generic for $i = 0, 1$, then there is a $\dot{q}$ such that $p * \dot{q}$ is totally $(N_0, P * \dot{Q})$-generic.*

*Proof* Set $G_i = \{r \in N_i \cap P : p \leq r\}$. We have the following facts.

1. $G_0 = N_0 \cap G_1$.
2. $G_0 \in N_1$.
3. $G_0$ has a lower bound $p' \in G_1$.

To see that this last fact is true, observe that the set of conditions which decide a particular value for $N_0 \cap \dot{G}_P$ is dense in $P$ and is in $N_1$. Choose $p' \in N_1 \cap G_1$ which decides $N_0 \cap \dot{G}_P$—it is decided to be $G_0$. Since $P$ is separative, $p'$ must extend every element of $G_0$.

Now fix $\dot{q} \in N_1 \cap \dot{Q}$ such that $\Vdash_P \dot{q}$ is totally $(N_0[\dot{G}_P], Q)$-generic and fix a dense $D \subseteq P * \dot{Q}$ such that $D \in N_0$. We need to produce an element $r * \dot{s}$ of $D \cap N_0$ such that $p \leq r$ and $p \Vdash \dot{q} \leq \dot{s}$. For this it is sufficient to prove that $p * \dot{q}$ decides $(\dot{G}_P * \dot{G}_Q) \cap D \cap N_0$. Let $D/\dot{G}_P$ be the $P$-name for the set of $\dot{s}[G_P]$ such that for some $r \in \dot{G}_P$, $r * \dot{s}$ is in $D$. Observe that

$D/\dot{G}_P$ is forced to be dense in $\dot{Q}$. Since $p$ is totally $(N_1, P)$-generic, it forces that $\dot{q}$ decides $D/\dot{G}_P \cap N_0$. Also, $p'$ decides $\dot{G}_P \cap N_0$. Therefore, $p * \dot{q}$ decides $(\dot{G}_P * \dot{G}_Q) \cap D \cap N_0$ as desired. □

It is important to note, however, that $p * \dot{q}$ is not $(N_1, P * \dot{Q})$-generic. For longer iterations, we expect to need more models above $N_0$ than just $N_1$ and for transfinite iterations we expect to need an infinite tower of models above $N_0$. This creates a new challenge. In order to describe it, we will need a definition.

**Definition 3.7** If $\chi$ is a regular uncountable cardinal, a *suitable tower of models* is a $\subseteq$-continuous sequence $\mathcal{N} = \langle N_\xi : \xi < \alpha \rangle$ of countable elementary submodels of $H(\chi)$ such that if $\xi < \alpha$, $\langle N_\eta : \eta \leq \xi \rangle$ is in $N_{\xi+1}$. We will say that $\mathcal{N}$ is suitable for $P$ if $N_0$ is suitable for $P$. We will abuse notation and write $P \in \mathcal{N}$ to mean $P \in N_0$. In what follows, $\mathcal{N}$ will always refer to a suitable tower of models.

If $P \in \mathcal{N}$ are as usual, then a condition $p$ is $(\mathcal{N}, P)$-generic if it is $(N, P)$-generic for each $N$ in $\mathcal{N}$. For finite towers of models $\mathcal{N}$, the existence of $(\mathcal{N}, P)$-generic conditions below elements of $N_0 \cap P$ is already guaranteed by the properness of $P$. For infinite towers of models, however, this is no longer the case and, stated for towers of height $\alpha$, this yields the definition of $\alpha$-properness. We will write *$(< \omega_1)$-proper* to mean *$\alpha$-proper for every $\alpha < \omega_1$*.

Another important point in the formulation of Proposition 3.6 is that $p$ is required to be *totally* $(N_1, P)$-generic. If $Q$ satisfies a strengthening of properness, which we will term *complete properness*, then this requirement can be relaxed to $p$ being $(N_1, P)$-generic. A precise definition of *complete properness* will be given in Section 7.

**Proposition 3.8** *Suppose P is totally proper and*

$$\Vdash_P \dot{Q} \textit{ is } \aleph_1\textit{-completely proper.}$$

*Given $N_0 \in N_1 \in N_2$ as usual and p that is $(N_2, P)$-generic, $(N_1, P)$-generic, and totally $(N_0, P)$-generic, there is a $\dot{q}$ such that $p * \dot{q}$ is totally $(N_0, P * \dot{Q})$-generic.*

To illustrate the significance of this change, let us return to Example 3.5. Replace $2^{<\omega_1}$ in the definition of $P$ with its regular open algebra. This forcing generates the same generic extensions as $2^{<\omega_1}$, but also has a well defined join operation. If $N_0 \in N_1 \in N_2$ are suitable models for $P$ as usual and $p \in P$ is totally $(N_0, P)$-generic and $(N_i, P)$-generic for

$i = 1, 2$, it is possible that $p$ does not decide $\dot{g}(\delta)$ where $\delta = N_0 \cap \omega_1$. To see this, let $p_0$ and $p_1$ be totally $(N_i, P)$-generic for $i = 0, 1, 2$ with $p_0 \restriction \delta = p_1 \restriction \delta$ and $p_0(\delta) \neq p_1(\delta)$. Then $p = p_0 \vee p_1$ is as desired.

The point is that the previous arguments still show there is no $\dot{q}$ such that $p * \dot{q}$ is totally $(N_0, P * \dot{Q})$-generic. Complete properness is in fact designed to avoid this sort of situation which we know from the demonstration in the introduction represents a fundamental obstruction to an iteration theorem for total properness.

We will see that transfinite iterations of forcings which are completely proper and $\alpha$-proper for every $\alpha < \omega_1$ do not add new reals. We will also see that the degree to which $\alpha$-properness is needed in this theorem is somewhat of a mystery.

## 4 Countable support iterations

**Definition 4.1** A *countable support (c.s.) iteration* $\mathbb{P} = \langle P_\alpha, \dot{Q}_\alpha \rangle_{\alpha < \varepsilon}$ satisfies the following conditions (which also define $P_\varepsilon$).

- $P_0 = \emptyset$.
- For all $\alpha \leq \varepsilon$, elements of $P_\alpha$ are functions with domain $\alpha$.
- For all $\alpha < \varepsilon$ $P_{\alpha+1}$ is forcing equivalent to $P_\alpha * \dot{Q}_\alpha$ as witnessed by a coordinate preserving function.
- If $\alpha \leq \varepsilon$ is a limit ordinal, then $P_\alpha$ is the set of countably supported functions $f$ for which $f \restriction \beta \in P_\beta$ for all $\beta < \alpha$.

If $\varepsilon$ is a limit ordinal, then for any generic $G \subseteq P_\varepsilon$, $p \in G$ if and only if $p \restriction \alpha \in G_\alpha$ for all $\alpha < \varepsilon$ where $G_\alpha = \{f \restriction \alpha : f \in G\}$.

**Theorem 4.2** *Let $\mathbb{P} = \langle P_\alpha, \dot{Q}_\alpha \rangle_{\alpha < \varepsilon} \in N$ be as usual, $\mathbb{P}$ be a countable support iteration of proper forcings, $\alpha \in N \cap \varepsilon$, $q$ be $(N, P_\alpha)$-generic, $q \Vdash \dot{p} \in N \cap P_\varepsilon$, and $q \Vdash \dot{p} \restriction \alpha \in \dot{G}_\alpha$. Then there is a $q^\dagger \in P_\varepsilon$ such that $q^\dagger$ is $(N, P_\varepsilon)$-generic, $q^\dagger \restriction \alpha = q$, and $q^\dagger \Vdash \dot{p} \in \dot{G}_\varepsilon$.*

**Lemma 4.3** *Suppose $P$ is proper, $\Vdash_P \dot{Q}$ is proper, $N$ is as usual, $P * \dot{Q} \in N$, $p$ is $(N, P)$-generic, $\dot{r}$ is a $P$-name for a condition in $P * \dot{Q}$ whose first component is forced by $p$ to be in $\dot{G}_P$, and $\dot{r} \in N$. There is a $\dot{q}$ such that $p * \dot{q}$ is $(N, P * \dot{Q})$-generic and $p * \dot{q} \Vdash \dot{r} \in \dot{G}_{P * \dot{Q}}$.*

*Proof of theorem* Proceed by induction on $\varepsilon$. The lemma is trivially true if $\varepsilon = 0$ and follows from Lemma 4.3 if $\varepsilon$ is a successor ordinal. Now assume $\varepsilon$ is a limit ordinal and let $\langle \alpha_n \rangle_{n < \omega}$ be strictly increasing, cofinal in $N \cap \varepsilon$, and such that $\alpha_0 = \alpha$. Let $\langle D_n \rangle_{n<\omega}$ enumerate the dense subsets of $P_\varepsilon$ from $N$. Define $\langle q_n, \dot{p}_n \rangle_{n<\omega}$ such that:

1. $\dot{p}_0 = \dot{p}$ and $q_0 = q$
2. $q_n$ is $(N, P_{\alpha_n})$-generic
3. $q_n \Vdash \dot{p}_{n+1}$ is a $P_{\alpha_n}$-name for a condition in $P_\varepsilon \cap N$
4. $q_n \Vdash \dot{p}_n \geq \dot{p}_{n+1} \in D_n$
5. $m < n$ implies $q_n \restriction \alpha_m = q_m$
6. $q_n \Vdash \dot{p}_n \restriction \alpha_n \in \dot{G}_{\alpha_n}$

Given $\dot{p}_n, q_n$, let $G_n \subseteq P_{\alpha_n}$ be generic with $q_n \in G_n$, so that $\dot{p}_n$ is interpreted as $p_n$ with $p_n \restriction \alpha_n \in G_n$. The set of restrictions to $\alpha_n$ of conditions in $D_n$ which extend $p_n$ is dense below $p_n \restriction \alpha_n$ in $P_{\alpha_n}$. This set is also in $N[G_n]$, so there exists $p_{n+1} \leq p_n$ such that $p_{n+1} \in N \cap D_n$ and $p_{n+1} \restriction \alpha_n \in G_n$. Let $\dot{p}_{n+1}$ be a $P_{\alpha_n}$-name forced by $q_n$ to have the above properties of $p_{n+1}$. Now apply our induction hypothesis to $\mathbb{P} \restriction \alpha_{n+1}, \alpha_n, q_n, \dot{p}_{n+1} \restriction \alpha_{n+1}$ to get $q_{n+1} \in P_{\alpha_{n+1}}$ such that (2.), (5.), and (6.) hold.

Let $q^\dagger = \bigcup_{n<\omega} q_n$ inside of $P_\varepsilon$. Strictly speaking, this union has domain $\sup(N \cap \varepsilon)$, which may be less than $\varepsilon$, but we extend the domain to all of $\varepsilon$ without increasing the support. It suffices to show that $q^\dagger \Vdash \dot{p}_n \in \dot{G}_{P_\varepsilon}$ for all $n$. We do this by proving that $q^\dagger \Vdash \dot{p}_n \restriction \alpha_m \in \dot{G}_{P_{\alpha_m}}$ for all $m$. Let $G \subseteq P_\varepsilon$ be generic with $q^\dagger \in G$. It follows that $q_m \in G \restriction \alpha_m$ for each $m$. Given $n$, we know that $p_n \restriction \alpha_n \in G \restriction \alpha_n$ by our construction. If $m > n$, we know that $p_m \restriction \alpha_m \in G \restriction \alpha_m$. Since $p_m \leq p_n$, we have $p_m \restriction \alpha_m \leq p_n \restriction \alpha_m$. Hence, $p_n \restriction \alpha_m \in G \restriction \alpha_m$; hence, $p_n \in G_{P_\varepsilon}$. Thus, $q^\dagger \Vdash \dot{p}_n \in \dot{G}_{P_\varepsilon}$. □

Recall the following definition:

**Definition 4.4** $Q$ is $\alpha$-proper if every $p \in Q$ extends to an $(\mathcal{N}, Q)$-generic condition, for any tower of models $\mathcal{N}$ of length $\alpha$ where $p \in N_0 \cap Q$ and $Q \in N_0$.

Note that $Q$ is totally proper and $\alpha$-proper if and only if $Q$ is "totally $\alpha$-proper" (i.e. suitable towers of models admit sufficiently many conditions which are totally generic for all of their models). The above discussion suggests that, at least at the level of proving things, $< \omega_1$-properness is a natural hypothesis. While its role in iteration theorems is not fully understood, a construction of Shelah presented in Section 9 shows that it can not be removed entirely as a hypothesis. As this conditions is satisfied by the forcing to uniformize a ladder system by countable approximations, we should expect an additional hypothesis is necessary in order to obtain an iteration theorem for not introducing new reals. The following theorem of Shelah is the prototypical example of such an iteration theorem. As we will work with a slightly different notion of "completeness," we will refer the interested reader to [10] for all undefined terminology.

**Theorem 4.5** *[10] A countable support iteration of forcings which are $\alpha$-proper for all $\alpha < \omega_1$ and $\mathbb{D}$-complete for a simple 2-completeness system $\mathbb{D}$ is totally proper.*

The following theorems are the analogues which we will develop in these lectures. The proof of the Main Theorem, along with the definition of the *iterability condition* can be found in Section 8; the proof of Proposition 4.6 and the definition of *complete properness* can be found in Section 7.

**Main Theorem** *If $\mathbb{P} = \langle P_\alpha, \dot{Q}_\alpha \rangle_{\alpha<\varepsilon}$ is a CS iteration of totally proper forcings, then $P_\varepsilon$ is totally proper provided that:*

1. *$\Vdash_{P_\alpha} \dot{Q}_\alpha$ is $<\omega_1$-proper.*
2. *$\dot{Q}_\alpha$ satisfies the iterability condition over $P_\alpha$.*

**Proposition 4.6** *If $P$ is totally proper and $P$ forces $\dot{Q}$ is a completely proper forcing, then $P * \dot{Q}$ satisfies the iterability condition.*

## 5 Limitations of $(< \omega_1)$-proper forcing

$(< \omega_1)$-properness is not merely a technical assumption satisfied by all forcings of interest. Recall the following definition.

**Definition 5.1** [9] ♣ asserts that there exists a ladder system $\vec{C}$ such that for all $X \in [\omega_1]^{\aleph_1}$ there exists $\delta$ such that $C_\delta \subseteq X$. *Club guessing (on $\omega_1$)*, denoted by $♣_C$, asserts the same as ♣ except that $X$ is required to be a club.

♣ ($♣_C$) is equivalent to ♣ (respectively $♣_C$) with the demand that there exist stationarily many $\delta$ as above. $\diamondsuit$ is equivalent to the conjunction of ♣ and CH and it is well known and easily demonstrated that $\mathrm{MA}_{\aleph_1}$ implies that ♣ fails. On the other hand, $♣_C$ is a much weaker assumption. It is well known that if $Q$ is a c.c.c. forcing, then every club in a generic extension by $Q$ contains a club from the ground model. In particular, $\clubsuit_C$ is preserved by c.c.c. forcing. Since it is possible to force MA + ¬CH with a c.c.c. forcing, $\clubsuit_C$ is consistent with $\mathrm{MA}_{\aleph_1}$.

The following proposition of Shelah connects this with the property of $(<\omega_1)$-properness.

**Proposition 5.2** *Suppose that* $\langle C_\alpha : \alpha \in \lim(\omega_1)\rangle$ *is a* $\clubsuit_C$*-sequence and* $Q$ *is an* $\omega$*-proper forcing. Forcing with* $Q$ *preserves that* $\langle C_\alpha : \alpha \in \lim(\omega_1)\rangle$ *is a* $\clubsuit_C$*-sequence.*

*Proof* Suppose that $\dot{E}$ is a $Q$-name for a club, $q$ is in $Q$, and let $\langle N_\xi : \xi < \omega_1\rangle$ be a suitable tower of models with $q$, $\dot{E}$, and $Q$ in $N_0$. Define $D = \{\xi : N_\xi \cap \omega_1 = \xi\}$ and let $\alpha$ be such that $C_\alpha \subseteq D$. Since $Q$ is $\omega$-proper, there is a $\bar{q} \leq q$ which is $(N_\xi, Q)$-generic for all $\xi$ in $C_\alpha$. This implies that $\bar{q}$ forces $\xi$ is in $\dot{E}$ for all $\xi$ in $C_\alpha$ and hence that $\check{C}_\alpha \subseteq \dot{E}$. □

A similar argument shows that if measuring fails and $Q$ is an $\omega$-proper forcing, then measuring fails after forcing with $Q$. Notice that *measuring* can be viewed as the ultimate failure of club guessing. If $\vec{D}$ is a ladder system and $E$ is a club, then for club-many $\delta \in E$ we have $\mathrm{otp}(\delta \cap E) = \delta$, so $E \cap (\delta_0, \delta) \subseteq D_\delta$ is impossible for all $\delta_0 < \delta$, and hence measuring implies that a tail of $E$ misses $D_\delta$. Thus, measuring implies that there is a club $F \subseteq E$ not guessed by $\vec{D}$. This remains true even for sequences where the map $\alpha \mapsto \mathrm{otp}(D_\alpha)$ is regressive.

## 6 Methods for verifying $(<\omega_1)$-properness

We will now turn to a framework for proving that forcings satisfy conditions such as total $(<\omega_1)$-properness and complete properness. The purpose is not to reformulate these conditions or simplify their statement. Rather it is to provide a sufficient criteria — analogous to the existence of an Axiom A structure — which is both easy to verify and sufficiently general to accommodate the important classes of examples of totally proper forcings.

**Definition 6.1** Let $Q$ be a fixed forcing notion with order $\leq$. A *fusion scheme* on $Q$ is an indexed family of partial orders $\leq_\sigma$ $(\sigma \in X^{<\omega})$ such that the following conditions are satisfied:

1. $\leq_\emptyset$ is $\leq$ and if $\sigma \subseteq \tau$, $q \leq_\tau p$ implies $q \leq_\sigma p$;
2. Player II has a winning strategy in $G(Q, \vec{\leq}, M)$ whenever $M$ is a suitable model for $Q$ and $\vec{\leq}$.

The game $G(Q, \vec{\lesssim}, M)$ is defined as follows. In the $n^{\text{th}}$ inning, Player I plays $q_n$ in $Q \cap M$ and Player II responds by playing $\sigma_n$ in $X^{<\omega}$. The players are required to play so that $q_{n+1} \leq_{\sigma_n} q_n$ and $\sigma_n \subseteq \sigma_{n+1}$; the first player to break one of these rules loses. If $q_n$ $(n \in \omega)$ is the result of a play of the game in which the players followed the rules, then Player II wins if either $\{q_n : n \in \omega\}$ does not generate an $(M, Q)$-generic filter or else there is a $\bar{q}$ with $\bar{q} \leq q_n$ for all $n \in \omega$.

This definition becomes of interest only when additional conditions are placed on the scheme. Before proceeding, let us see how Example 3.1 fits into this framework.

Let $\langle C_\alpha : \alpha \in \lim(\omega_1)\rangle$ be a ladder system and let $g : \omega_1 \to 2$ be a function. Let $P_g$ be the collection of all countable partial uniformizing functions with the order $\leq$ of extension. Set $X = \lim(\omega_1)$ and if $\sigma$ is in $X^{<\omega}$, define $q \leq_\sigma p$ iff whenever $\xi$ is in $\text{dom}(q) \setminus \text{dom}(p)$ and $i < |\sigma|$ is minimal such that $\xi$ is in $C_{\sigma(i)}$, $q(\xi) = g(\sigma(i))$. Player II's strategy is to play $\langle M \cap \omega_1\rangle$ in every round of $G(Q, \vec{\lesssim}, M)$. It should be clear that this defines a strategy for Player II which is winning from every initial position.

Fix a fusion scheme $\leq_\sigma$ $(\sigma \in X^{<\omega})$ on a forcing $Q$.

**Definition 6.2** The fusion scheme satisfies (TP) if whenever $M$ is a suitable model for the fusion scheme, $p$ is in $Q \cap M$, $D \subseteq Q$ is dense and in $M$, and $\sigma$ is in $X^{<\omega}$, there is a $q$ in $D \cap M$ such that $q \leq_\sigma p$.

We have already shown that the fusion scheme defined above on $P_g$ satisfies (TP).

**Theorem 6.3** *If a forcing $Q$ admits a fusion scheme satisfying (TP), then $Q$ is totally proper.*

*Proof* Let $\leq_\sigma$ $(\sigma \in X^{<\omega})$ be a fusion scheme on $Q$ which satisfies (TP). Suppose that $M$ is a suitable model for the fusion scheme and let $q$ be in $M$. Let $D_n$ $(n \in \omega)$ enumerate the dense subsets of $Q$ which are in $M$. Player I plays by the following strategy. To begin the game, Player I plays $q_0$. If in the $n^{\text{th}}$ inning $q_n$ and $\sigma_n$ were played, Player I picks a $q_{n+1}$ in $D_n \cap M$ which satisfies $q_{n+1} \leq_{\sigma_n} q_n$. This is possible by our assumption that the scheme satisfies (TP). Now let $q_n$ $(n \in \omega)$ be the result of a play of this strategy against Player II's winning strategy. Player I has arranged that $\{q_n : n \in \omega\}$ generates a $(M, Q)$-generic filter and hence there must be a $\bar{q}$ in $Q$ such that $\bar{q} \leq q_n$ for all $n$. Such a $\bar{q}$ is totally $(M, Q)$-generic. □

Next we will consider a condition on a fusion scheme which can be used to verify $\alpha$-properness for each $\alpha < \omega_1$. First we will need a preliminary definition.

**Definition 6.4** If $Q$ is a forcing equipped with a fusion scheme $\leq_\sigma$ ($\sigma \in X^{<\omega}$) and $\langle q_n : n < \omega \rangle$ is a $\leq$-descending sequence in $Q$, then we say that $\bar{q}$ is a *conservative lower bound* for $\langle q_n : n < \omega \rangle$ if whenever $\sigma \in X^{<\omega}$ is such that $\langle q_n : n < \omega \rangle$ is eventually $\leq_\sigma$-descending, $\bar{q} \leq_\sigma q_n$ for all but finitely many $n$. Here $\langle q_n : n < \omega \rangle$ is *eventually* $\leq$*-descending* if there is an $m$ such that $\langle q_n : m < n < \omega \rangle$ is $\leq$-descending.

Returning to Example 3.1, $\bar{q} = \bigcup_n q_n$ is a conservative lower bound for $\langle q_n : n \in \omega \rangle$ provided that $\langle q_n : n < \omega \rangle$ has a lower bound in $P_g$. This is in fact often the case in practice.

**Definition 6.5** A fusion scheme satisfies (A) if the following conditions are met:

1. for any countable $Q_0 \subseteq Q$, $\{\leq_\sigma \restriction Q_0 : \sigma \in X^{<\omega}\}$ is countable;
2. every bounded $\leq$-descending sequence in $Q$ has a conservative lower bound;
3. Player II has a winning strategy in $G(Q, \vec{\leq}, M)$ starting from any initial position.[2]

Notice that it is in fact trivial to verify that $P_g$ with its fusion scheme satisfies (A). The point of this definition and the following theorem is that (A) is usually trivial to verify for a given fusion scheme which satisfies it. This should be contrasted by the often tedious direct verification of $\alpha$-properness by induction on $\alpha$ (see the proof of Lemma 5.11 of [8]).

**Theorem 6.6** *If a forcing $Q$ admits a fusion scheme which satisfies (TP) and (A), then $Q$ is totally $\alpha$-proper for all $\alpha < \omega_1$.*

*Proof* Fix a forcing $Q$ and fusion scheme $\leq_\sigma$ ($\sigma \in X^{<\omega}$) on $Q$ which satisfies (TP) and (A). Let $\langle M_\xi : \xi < \alpha \rangle$ be a tower of models of ordertype $\alpha$ which is suitable for $Q$ and $\leq_\sigma$ ($\sigma \in X^{<\omega}$). As usual, we may assume that $\alpha$ is a limit ordinal. It will be convenient to adopt the convention that $M_\alpha = \bigcup_{\xi<\alpha} M_\xi$ and $M_{\alpha+1}$ is $H(\theta)$. We will verify the following by induction on $\zeta \leq \alpha$:

[2] I.e. given any partial play of the game, there is a strategy for Player II such that any completion of game play in which II follows this strategy in the remainder of the game results Player II winning.

If $\xi \leq \zeta$, $q$ is in $Q \cap M_\xi$ and is totally $(M_\eta, Q)$-generic for all $\eta < \xi$, and $\sigma$ is in $X^{<\omega}$, then there is a $\bar{q} \leq_\sigma q$ such that $\bar{q}$ is in $M_{\zeta+1}$ such that $\bar{q}$ is totally $(M_\eta, Q)$-generic for all $\eta \leq \zeta$.

Let $\xi < \zeta$ be given and assume the induction hypothesis holds for all smaller values of $\zeta$. Since our fusion scheme satisfies (A), $\{\leq_\tau \restriction M_\zeta : \tau \in X^{<\omega}\}$ is countable and hence contained in $M_{\zeta+1}$. Therefore we can choose a $\bar{\sigma}$ in $M_{\zeta+1} \cap X^{<\omega}$ such that $\leq_\sigma \restriction M_\zeta = \leq_{\bar{\sigma}} \restriction M_\zeta$.

If $\zeta = \zeta_0 + 1$ for some $\zeta_0$, then we can first apply the induction hypothesis to find a $q' \leq_\sigma q$ in $M_{\zeta_0+1}$ which is totally $(M_\eta, Q)$-generic for all $\eta \leq \zeta_0$. We now need to find a $\bar{q}$ in $M_{\zeta+1}$ such that $\bar{q} \leq_\sigma q'$ and $\bar{q}$ is totally-$(M_\zeta, Q)$-generic. To do this, we run the proof of Theorem 6.3 for $q'$ inside $M_{\zeta+1}$ except that we begin with the following position in $G(Q, \vec{\lesssim}, M_\zeta)$: $\langle q', \bar{\sigma}, q' \rangle$. What results will be a sequence $\langle q_n : n \in \omega \rangle$ in $M_{\zeta+1}$ which is $\leq_{\bar{\sigma}}$-descending, $\leq$-bounded, and which generates a $(M_\zeta, Q)$-generic filter. By our assumption and elementarity of $M_{\zeta+1}$, there is a $\bar{q}$ which a conservative lower bound for $\langle q_n : n \in \omega \rangle$. Consequently $\bar{q} \leq_\sigma q'$ and $\bar{q}$ is totally $(M_\zeta, Q)$-generic.

Next suppose that $\zeta$ is a limit and let $\langle \zeta_n : n < \omega \rangle$ be a sequence converging to $\zeta$ which is in $M_{\zeta+1}$ and such that $\zeta_0 = \xi$. By our assumption of suitability, $\langle M_{\zeta_n} : n < \omega \rangle$ is in $M_{\zeta+1}$. We will play $G(Q, \vec{\lesssim}, M_\zeta)$, this time starting from the position $\langle q, \bar{\sigma}, q \rangle$. Let $\sigma_0 = \bar{\sigma}$ and $q_0 = q$. We now describe a strategy for Player I to use in the remainder of the game. If $q_n$ and $\sigma_n$ were played in the previous round, Player I plays $q_{n+1}$ in $M_{\zeta_{n+1}+1}$ such that $q_{n+1} \leq_{\sigma_n} q_n$ and $q_{n+1}$ is totally $(M_\eta, Q)$-generic for all $\eta \leq \zeta_{n+1}$. In the end, Player I has arranged that $\langle q_n : n \in \omega \rangle$ will generate a $(M_\zeta, Q)$-generic filter since any dense subset of $Q$ in $M_\zeta$ will be in some $M_{\zeta_n}$. Let $\langle q_n : n < \omega \rangle$ be the result of a play of the game which is in $M_{\zeta+1}$ and in which Player I played this strategy but Player II won. Such a play exists by our assumptions and by elementarity of $M_{\zeta+1}$. Let $\bar{q}$ be a conservative lower bound for $\langle q_n : n < \omega \rangle$ which is in $M_{\zeta+1}$. It follows that $\bar{q} \leq_\sigma q$ and $\bar{q}$ is $(M_\eta, Q)$-generic for all $\eta \leq \zeta$. □

## 7 Complete properness

Now we will turn to the definition of *complete properness* and some analogous approaches to verifying it. First we will need some preliminary definitions. If $M$ and $N$ are sets, then $M \to N$ will symbolize an elementary embedding $\epsilon$ of $(M, \in)$ into $(N, \in)$ such that $\epsilon$ is in $N$ and $N$

satisfies that $M$ is countable (i.e. $N$ contains an injection of $M$ into $\omega$). If $X$ is an element of $M$, we will use $X^N$ to denote $\epsilon(X)$. If $G$ is a subset of $M$ which is not an element, we will use $G^N$ to denote the point-wise image of $G$ under $\epsilon$. If there is no cause for confusion, the superscript will sometimes be omitted to simplify notation.

**Definition 7.1** If $Q$ is a forcing, then a *$Q$-diagram* is a collection of the form $M \to N_i$ $(i \in \lambda)$ where $M$ is a suitable model for $Q$.

**Definition 7.2** If $Q$ is a forcing and $M$ is a suitable model and $M \to N$, then we say that a $(M, Q)$-generic filter $G \subseteq Q \cap M$ is *$\overrightarrow{MN}$-prebounded* if whenever $N \to N^*$ and $G$ is in $N^*$, then $N^* \models G^{N^*}$ is bounded.

**Definition 7.3** $Q$ is *$\lambda$-completely proper* if whenever $M \to N_i$ $(i < \kappa)$ is a $Q$-diagram for some $\kappa < 1 + \lambda$ and $q$ is in $Q^M$, there is a $(M, Q)$-generic filter $G$ which contains $q$ and is $\overrightarrow{MN_i}$-prebounded for all $i < \kappa$. We will write *completely proper* to mean 2-completely proper.

To see the relevance of this definition, let us return to our example of the uniformizing forcing $P_g$ for a ladder system coloring $g$ of $\vec{C}$. If $M \to N_i$ $(i < 2)$ is a $P_g$-diagram, then in general it is possible that

$$g^{N_0}(M \cap \omega_1) \neq g^{N_1}(M \cap \omega_1).$$

If this occurs and $C_\delta^{N_0} \cap C_\delta^{N_1}$ is infinite, then there is no $(M, P_g)$-generic $G$ which is $\overrightarrow{MN_i}$-prebounded for both $i$.

In general it seems difficult to prove that a specific forcing such as $P_g$ fails to be completely proper. For instance it is not clear at all whether every forcing $P_g$ fails to be completely proper. Usually there is a hypothetical scenario suggesting the forcing is not completely proper — such as what we have just demonstrated for $P_g$ — and there is a corresponding proof (via [2]) that iterations of forcings of this type can add new reals. We only know that some member of the given class of forcings (which shows up in a forcing extension) fails to be completely proper. This is not well understood. A (seemingly) isolated exception to this will be discussed in Section 9 below.

The connection of this notion to the $\lambda$-iterability condition is made clear by the following proposition.

**Proposition 7.4** *Suppose that $P$ is totally proper and $P$ forces that $\dot{Q}$ is $\lambda$-completely proper. Then $P * \dot{Q}$ satisfies the $\lambda$-iterability condition.*

*Proof* Let $M \in N$ be a pair of suitable models for $P * \dot{Q}$, $G \subseteq Q \cap M$ be

$(M, Q)$-generic, $\kappa < \lambda$, and $p_i$ $(i < \kappa)$ be totally $(N, Q)$-generic conditions which are lower bounds for $G$. Set $\bar{N}$ equal to the transitive collapse of $N$ and let $G^i$ denote the image of $\{p \in P \cap N : p_i \leq p\}$ under the collapsing map. Let $\hat{G}$ be a $V$-generic filter containing $G$. In $V[\hat{G}]$, we now have a $Q$-diagram $M[G] \to \bar{N}[G^i]$ $(i < \kappa)$ and therefore there is an $H \subseteq Q \cap M[G]$ which is $\overrightarrow{M[G]\bar{N}[G^i]}$-prebounded. It suffices to show that each $p_i$ forces that $H$ has a lower bound. To see this, let $\hat{G}^i$ be a $V$-generic filter containing $p_i$. Let $N^*$ be a suitable model containing $N$ as an element such that $N^* \cap \hat{G}^i$ is $(N^*, P)$-generic, $N$ and $G^i$ are in $N^*[\hat{G}^i]$. Since the inclusion map of $N[\hat{G}^i]$ into $N^*[\hat{G}^i]$ induces an embedding $\bar{N}[G^i] \to N^*[\hat{G}^i]$, it follows that $N^*[\hat{G}^i]$ must satisfy that $H$ has a lower bound. Since $\hat{G}^i$ was arbitrary subject to containing $p_i$, it must be that $p_i$ forces that $H$ has a lower bound. □

While we will not formulate the notion of being $\mathbb{D}$-*complete*, the connection to the present terminology is provided by the following proposition.

**Proposition 7.5** *[8] A $\lambda$-completely proper forcing is $\mathbb{D}$-complete with respect to a (specific) simple completeness system $\mathbb{D}$.*

In fact this proposition has a partial converse; see [8, 4.14].

Now we will return to our discussion of fusion schemes. It is not clear how to formulate a single condition to verify complete properness which works for all of our examples. Still there are two easy adaptations of (TP) which handle the main examples and even if these are not sufficient in a future application, it seems that a simple adaptation may work.

*Remark* 7.6 It should be noted that employing fusion schemes to verify complete properness is usually only warranted if one additionally wishes to verify that the forcing is $(< \omega_1)$-proper. In cases where this is not appropriate or necessary, a more direct approach is likely more efficient.

**Definition 7.7** If $k < \omega$, we say a fusion scheme $\leq_\sigma$ $(\sigma \in X^{<\omega})$ satisfies $(\mathrm{CP}_k)$ if whenever $M \to N_i$ $(i < k)$ is a $Q$-diagram, $p$ is in $Q \cap M$, $D \subseteq Q$ is dense and in $M$, and $\sigma_i$ is in $X^{N_i}$, there is a $q$ in $D \cap M$ such that for all $i < k$,

$$N_i \models q^{N_i} \leq_{\sigma_i} p^{N_i}.$$

(CP) will be used to abbreviate $(\mathrm{CP}_2)$. A fusion scheme satisfies $(\mathrm{CP}_{\aleph_1})$ if it satisfies $(\mathrm{CP}_k)$ for all $k$.

*Remark* 7.8 While it would perhaps seem more natural to define $(\mathrm{CP}_{\aleph_0})$ to mean that $(\mathrm{CP}_k)$ holds for all $k$, the above definition is chosen so that there is a correspondence between $(\mathrm{CP}_\lambda)$ and $\lambda$-complete properness. See Theorem 7.10 below.

**Definition 7.9** A fusion scheme satisfies $(\mathrm{CP}'_\lambda)$ if whenever $M \to N_i$ $(i < \kappa)$ is a $Q$-diagram for $\kappa < 1 + \lambda$, Player II has a strategy in $G(Q, \vec{\lesssim}, M)$ so that if $q_n$ $(n \in \omega)$ is the result of a legal play of the game by this strategy and $q_n$ $(n < \omega)$ generates an $(M, Q)$-generic filter $G$, then $G$ is $\overrightarrow{MN_i}$-prebounded for all $i < \kappa$.

Notice that if $M \in N$ are suitable models for $Q$ and $M \to N$ denotes the identity map, then any $\overrightarrow{MN}$-prebounded filter in $Q \cap M$ is actually bounded (since we may take $N^*$ a suitable model with $G, N \in N^*$ and $N \to N^*$ being the identity map). Hence $(\mathrm{CP}'_1)$ is a more restrictive property than $(TP)$.

**Theorem 7.10** *If $Q$ admits a fusion scheme satisfying either $(\mathrm{CP}_\lambda)$ or $(\mathrm{CP}'_\lambda)$, then $Q$ is $\lambda$-completely proper.*

*Proof* That $(\mathrm{CP}'_\lambda)$ implies $\lambda$-complete properness is a trivial modification of the proof of Theorem 6.3. The argument that $(\mathrm{CP}_k)$ implies $k$-complete properness of $Q$ for $k < \omega$ is similar except that Player I plays their 'book keeping' strategy (in $V$) against a team of $k$ many Player II's, each playing their winning strategy in a model $N_i$ for $i < k$. Player I's ability to follow this strategy is made possible by $(\mathrm{CP}_k)$. Let $q_n$ $(n \in \omega)$ be a resulting play of Player I and $G$ be the filter it generates. Notice that Player I has arranged that $G$ is $(M, Q)$-generic. We now must show that $G$ is $\overrightarrow{MN_i}$-prebounded for each $i < k$. To this end let $k$ be fixed and let $N_i \to N^*$ be such that $G$ is in $N^*$.

*Claim* 7.11 There is a sequence $q'_n, \sigma_n$ $(n \in \omega)$ in $N^*$ which is the result of a play by Player I against Player II playing their winning strategy in $N_i$ such that

$$G = \{p \in Q \cap M : \exists n(q'_n \leq p)\}$$

*Proof* Let $p_n$ $(n \in \omega)$ be an enumeration of $G$ which is in $N^*$. In $N^*$, let $T$ be the tree of all partial plays $\tau$ of $G(Q, \vec{\lesssim}, M)$ in which Player II's winning strategy is followed and Player I plays elements of $G$. We order $\tau \lhd \tau'$ if there is a play by Player I in $\tau'$ which extends $p_n$ where $n = |\tau|$. Clearly $T$ has an infinite branch if and only if the conclusion of the claim holds and from the outside, $T$ has an infinite branch as witnessed by $q_n$

($n < \omega$). Since $N^*$ is well founded and satisfies a sufficient fragment of ZFC, $N^*$ must also satisfy that $T$ has an infinite branch since otherwise $N^*$ would contain a strictly decreasing function from $T$ into its (well founded) ordinals. □

Since the team member for Player II who was in $N_i$ used a winning strategy (from the vantage point of $N_i$) and since $N_i$ is elementarily embedded into $N^*$, $N^*$ satisfies $G$ is bounded.

To handle the case $\lambda = \aleph_1$, suppose that $M \to N_i$ ($i < \omega$) is a $Q$-diagram. As before, Player I plays in $V$ against a team of Player II's playing from the models $N_i$ with their winning strategies. The difference is that the Player II playing from $N_i$ begins playing only in round $i$. The rest of the argument is as before. □

Now we will consider two more examples of forcings.

## 7.1 Destroying ♣-sequences

First, let us consider a forcing that destroys instances of ♣.

**Definition 7.12** Suppose $\vec{C}$ is a ladder system. Define $Q_{\vec{C}}$ to be the collection of all countable subsets $q$ of $\omega_1$ such that if $\delta \leq \sup(q)$ is a limit ordinal, then $C_\delta \not\subseteq q$. $Q$ is ordered by reverse end extension.

**Theorem 7.13** *$Q_{\vec{C}}$ admits a fusion scheme which satisfies* ($\mathrm{CP}_{\aleph_1}$) *and (A). In particular $Q_{\vec{C}}$ is both completely proper and ($< \omega_1$)-proper.*

*Remark* 7.14 The following proof actually shows that the forcing to add an uncountable subset of $\omega_1$ which is almost disjoint from every ladder is both completely proper and ($< \omega_1$)-proper.

*Proof* Let $X = \lim(\omega_1)$ and, for $\sigma \in X^{<\omega}$, define $q \leq_\sigma p$ if $q \setminus p \cap C_{\sigma(i)}$ is empty for all $i < |\sigma|$. To see that this defines a fusion scheme, let $M$ be a suitable model for $Q$. Player II plays $C_{M \cap \omega_1}$ in the first round of the game and arbitrarily after that. Suppose that $q_n$ ($n \in \omega$) is a play by Player I in which Player II followed this strategy and suppose that $q_n$ ($n \in \omega$) generates an $(M, Q)$-generic filter. Define $\bar{q} = \bigcup_n q_n$. It is trivial to verify that $\{q \in Q : \sup(q) > \alpha\}$ is dense for all $\alpha < \omega_1$ and hence $\sup(\bar{q}) = \delta$ where $\delta = M \cap \omega_1$. Since all proper initial parts of $\bar{q}$ are in $Q$, it is sufficient to check that $C_\delta$ is not contained in $\bar{q}$. But this follows from the fact that $C_\delta \cap (\bar{q} \setminus q_0) = \emptyset$, since $q_{n+1} \leq_{\langle\delta\rangle} q_n$ for all $n$.

Notice that this argument shows that Player II has a winning strategy in this game starting from any partial play of the game. Also, for each

$\delta < \omega_1$, $\{C_\alpha \cap \delta : \alpha \in \omega_1\}$ is countable. Finally, $\langle q_n : n < \omega\rangle \mapsto \bigcup_n q_n$ defines a conservative lower bound when $\langle q_n : n < \omega\rangle$ is descending and bounded. It follows that the fusion scheme satisfies (A).

We now claim that this fusion scheme satisfies ($\mathrm{CP}_{\aleph_1}$). To see this, suppose that $M \to N_i$ $(i < k)$ is a $Q$-diagram, $q$ is in $Q \cap M$, $D \subseteq Q$ is dense and in $M$, and $N_i \models \sigma_i \in \omega_1^{<\omega}$, Define

$$C = \bigcup_{i<k} \bigcup_{j<|\sigma_i|} C^{N_i}_{\sigma(j)}$$

and observe that the ordertype of $C$ (and hence of the closure of $C$) is less than $\omega^2$. Let $N \prec H(\aleph_2)$ be in $M$ such that everything relevant is in $N$ and $N \cap \omega_1$ is not in the closure of $C$. Let $\xi = \sup(C \cap N) + 1$ and define $q' = q \cup \{\xi\}$. $q'$ is in $N$ and therefore there is a $\bar{q} \leq q'$ in $D \cap N$. Notice that $q' \leq_{\sigma_i} q$ and, since $\bar{q} \setminus q'$ is contained in $N$ and bounded below by $\xi$. □

Now consider the variation $Q'_{\vec{C}}$ of $Q_{\vec{C}}$ in which the conditions are required to be closed sets. While the above proof shows that $Q'_{\vec{C}}$ is completely proper, we have already seen that $Q'_{\vec{C}}$ will almost never be $\omega$-proper. The reader should convince themselves that the analogous fusion scheme fails to have conservative lower bounds.

## 7.2 The forcing to measure a sequence of closed sets

In section we will consider the forcing associated with *measuring*.

**Definition 7.15** Let $\langle D_\alpha : \alpha < \omega_1\rangle$ be a sequence such that for each $\alpha < \omega_1$, $D_\alpha$ is a closed subset of $\alpha$. Define $Q_{\vec{D}}$ to be the set of all pairs $q = (x_q, E_q)$ such that:

1. $x_q$ is a countable closed set;
2. $E_q \subseteq \omega_1$ is a club with $\max(x_q) < \min(E_q)$;
3. if $\alpha \leq \max(x_q)$, then there is an $\alpha_0 < \alpha$ such that $x_q \cap (\alpha_0, \alpha)$ is contained in or disjoint from $D_\alpha$.

The order on $Q_{\vec{D}}$ is defined by $q \leq p$ if and only if $x_p$ is an initial part of $x_q$, $E_q \subseteq E_p$, and $x_q \setminus x_p \subseteq E_p$.

**Theorem 7.16** *The forcing $Q$ for measuring a sequence $\langle D_\alpha : \alpha < \omega_1\rangle$ admits a fusion scheme satisfying ($\mathrm{CP}'_{\aleph_1}$). In particular, it is completely proper.*

*Remark* 7.17 By remarks above, $Q_{\vec{D}}$ is not $\omega$-proper unless $\langle D \rangle$ is already measured. In this case, $Q_{\vec{D}}$ is countably closed.

*Proof* If $U \subseteq \omega_1$ is a countable open set, define $q \leq_U p$ if either $(x_q \setminus x_p) \cap \sup(U) \subseteq U$ or else $E_q \cap U = \emptyset$. If $\sigma$ is a finite sequence of countable open subsets of $\omega_1$, then $q \leq_\sigma p$ means that for each $i < |\sigma|$, $q \leq_U p$ where $U = \bigcap_{j \leq i} \sigma(j)$.

Let $M \to N_i$ $(i \in \omega)$ be a $Q$ diagram and let $\mathcal{U}$ be the collection of all open $M$-stationary subsets of $\delta = M \cap \omega_1$ which are in $N_i$ for some $i < \omega$. Construct a $\subseteq$-decreasing sequence $U_k$ $(k < \omega)$ of elements of $\mathcal{U}$ such that if $V$ is an open subset of $M \cap \omega_1$ in $N_i$ for some $i$, then there is a $k < \omega$ such that $U_k$ is either contained in $V$ or there is a club $E \subseteq \omega_1$ in $M$ such that $U_k \cap E$ is disjoint from $V$. It suffices to show that if Player II plays $U_k$ in round $i$ of $G(Q, \vec{\lesssim}, M)$, then this defines a strategy as required by $(\mathrm{CP}'_{\aleph_1})$. To see this, suppose that $G \subseteq M \cap Q$ is an $(M, Q)$-generic filter resulting from a play of the game by this strategy and $i < \omega$ is given.

Define $V = \delta \setminus D^{N_i}_\delta$ and let $k$ be such that $U_k \subseteq V$ or $V \cap U_k$ is disjoint from some club $E$ in $M$. If $U_k \subseteq V$, then by following the above strategy, Player II forces Player I to play so that $x_l \setminus x_k \subseteq U_k$ for all $l > k$. If $V \cap U_k$ is disjoint from some club $E$ in $M$, then by genericity of $G$, $E$ will contain $E_{p_l}$ for some $l_0 \geq k$. Now from this stage on, $x_l \setminus x_{l_0}$ is contained in $E \cap U_k$ and therefore disjoint from $V$. It follows that $G$ is $\overrightarrow{MN_i}$-prebounded. □

## 7.3 Adding subtrees to Aronszajn trees

Recall that an *Aronszajn tree (A-tree)* is an uncountable tree in which all levels and chains are countable. A Souslin tree is an A-tree in which, moreover, all antichains are countable. In this section, we will consider a forcing which adds a generic subtree to a given A-tree $T$. This subtree will have the property that the minimal elements of its complement will form an uncountable antichain and hence witness that $T$ is not Souslin in the generic extension.

We will first fix some notation. In order to simplify matters later on, assume without loss of generality that the $\alpha^{\text{th}}$-level of $T$ consists of functions from $\alpha$ into $\omega$ and that $T$ is ordered by extension. This equips $T$ with a canonical lexicographic order. Let $T^{[n]}$ denote the collection of all tuples $\langle t_i : i < n \rangle$ which all come from some level of $T$ and which are listed in non decreasing $\leq_{\text{lex}}$-order. For each $n$, $T^{[n]}$ is an A-tree when ordered by coordinate-wise extension and these trees will collectively be referred to as the *finite powers of* $T$. If $u$ is in $T^{[n]}$ and $A \subseteq T$, then we will write $u \subseteq A$ to mean that the range of $u$ is contained in $A$. If $t$ is in $T$ and $\alpha < \omega_1$, then $t \restriction \alpha$ is the element $s$ of $T$ of height $\alpha$ with $s \le t$ if $t$ has height at least $\alpha$ and $t \restriction \alpha = t$ otherwise. Similarly, if $u$ is in $T^{[n]}$ for some $n$, then $u \restriction \alpha$ is defined by coordinate-wise restriction (which agrees with the definition of restriction in $T^{[n]}$).

**Definition 7.18** Define $Q_T$ to be the set of all $q = (x_q, \mathcal{U}_q)$ for which:

- $x_q$ is a subtree of $T$ which has a last level $\alpha_q$.
- $\mathcal{U}_q$ is a countable collection of *pruned subtrees*[3] of some finite power of $T$.
- for all $U \in \mathcal{U}_q$ there is a $u$ in $U_{\alpha_q}$ with $u \subseteq x_q$.

$Q$ is ordered by declaring $q \le p$ to mean that $x_q \supseteq x_p$ and $\mathcal{U}_q \supseteq \mathcal{U}_p$.

**Theorem 7.19** *$Q_T$ admits a fusion scheme satisfying (A) and* (CP$_{\aleph_1}$)*. In particular, $Q_T$ is completely proper and $(< \omega_1)$-proper.*

By applying the Main Theorem and standard chain condition and book keeping arguments, we obtain the following corollary.

**Corollary 7.20** *Souslin's hypothesis is consistent with* CH.

*Remark* 7.21 Observe that each $U \in \mathcal{U}_p \cap \mathcal{P}(T)$ "promises" that the generic tree $\dot{S} = \bigcup_{q \in \dot{G}_Q} x_q$ added by $Q_T$ will intersect $U$ uncountably often. $Q_T$ is a simplification of a forcing of Shelah [10] that specializes $A$-trees without adding reals. We will show that $Q_T$ adds an uncountable antichain to $T$. It is not clear whether $Q_T$ necessarily specializes $T$.

*Proof* Define a fusion scheme on $Q = Q_T$ as follows. Set $X = T$ and if $\sigma$ is in $X^{<\omega}$, define $r \le_\sigma q$ to mean that $r \le q$ and for all $i < |\sigma|$, if $\sigma(i) \restriction \alpha_q \in x_q$ then $\sigma(i) \restriction \alpha_r \in x_r$. We leave it as an exercise to the reader to verify that this defines a fusion scheme and that in fact Player II has a winning strategy in $G(Q, \vec{\le}, M)$ from every initial position whenever $M$ is a suitable model for $Q$.

To see that this fusion scheme satisfies (A), observe that if $M$ is a suitable model and $\delta = M \cap \omega_1$, then

$$\le_\sigma \restriction (Q \cap M) = \le_{\sigma \restriction \delta} \restriction (Q \cap M).$$

Since $T$ is Aronszajn, it follows that $\{\le_\sigma \restriction (Q \cap M) : \sigma \in X^{<\omega}\}$ is

[3] A *subtree* of a tree is an initial part (*i.e.*, downward closed subset) of that tree. A tree is *pruned* if every element has uncountably many extensions.

countable. To see that $Q$ has conservative lower bounds, let $q_n$ $(n < \omega)$ be a descending sequence which has a lower bound. Set

$$\bar{\alpha} = \sup_n \alpha_{q_n} \qquad \mathcal{U}_{\bar{q}} = \bigcup_n \mathcal{U}_{q_n} \qquad x = \bigcup_n x_{q_n}$$

Define $x_{\bar{q}}$ to be the union of $x$ together with all $t$ in $T_{\bar{\alpha}}$ such that $t \restriction \xi$ is in $x$ for all $\xi < \bar{\alpha}$. It is straightforward to verify that $\bar{q} = (x_{\bar{q}}, \mathcal{U}_{\bar{q}})$ is a conservative lower bound for $q_n$ $(n < \omega)$.

**Lemma 7.22** *If $M$ is as usual, $q \in Q \cap M$, $D \subseteq Q$ is dense, $D \in M$, and $\sigma \in T^{<\omega}$, then there exists $r \leq_\sigma q$ such that $r \in D \cap M$. In particular, the fusion scheme define above satisfies* (CP$_{\aleph_1}$).

*Proof* We will begin by arguing that the second part of the lemma follows from the first. Observe that if $M \to N_i$ $(i < k)$ is a $Q$-diagram and for each $i < k$ $N_i \models \sigma_i \in T^{<\omega}$, then there is a model $M'$, suitable (enough) for $Q$ and containing $q$ and $D$. Let $\delta = M' \cap \omega_1$. Even though, for a given $i < k$, $\sigma_i$ is typically not in $N_j$ for $i \neq k$, $\sigma_i \restriction \delta$ is in each $M$ for all $i < k$. Therefore we can find a single $\sigma$ in $M$ which corresponds to the concatenation of these restrictions. Hence if we verify the first sentence in the lemma for $\sigma$ and $M' \to M$, we have verified the instance of (CP$_{\aleph_1}$) for $M \to N_i$ $(i < k)$, $\sigma_i$ $(i < k)$, $q$, and $D$. Hence we may now focus on the first sentence in the lemma.

Seeking a contradiction, suppose the first sentence of the lemma is false. Without loss of generality, $\sigma$ consists of entries $t$ such that $\mathrm{ht}(t) \leq \delta = M \cap \omega_1$ and $t \restriction \alpha_q \in x_q$. Since $T_\delta$ is countable, there is an $h \colon \omega_1 \to T^{<\omega}$ such that $h(\alpha) \in T_\alpha^{<\omega}$ and $h(\delta)$ is the entries of $\sigma$ of height $\delta$ (listed in the same order). We may assume $h \in M$ because $h(\delta)$ is in the Skolem hull $M \cup \delta$. Let $\Xi$ be the set of $\xi < \omega_1$ for which $h(\xi) \restriction \alpha_q = h(\delta) \restriction \alpha_q$ and, if $r \leq q$ with $r \in D$ and $\alpha_r < \xi$, then $r \not\leq_{h(\xi)} q$.

*Claim* 7.23 $\delta \in \Xi$, so $\Xi$ is uncountable.

*Proof* Seeking a contradiction, suppose $r$ witnesses that $\delta \notin \Xi$. Fix $\beta \in M \cap [\alpha_r, \omega_1)$. Observe that $r \leq_{h(\delta)} q$ if and only if $r \leq_{h(\delta) \restriction \beta} q$. By elementarity, there is an $r' \in M \cap D$ such that $\alpha_{r'} = \alpha_r$ and $r' \leq_{h(\delta) \restriction \beta} q$. Thus, $r' \leq_{h(\delta)} q$, in contradiction with our assumption that the lemma fails. □

Returning to the proof of the lemma, let $U \subseteq T^{[n]}$ be the set of $u$ for which uncountably $\xi \in \Xi$ satisfy $u \leq h(\xi)$. It follows that $U$ is pruned and $q' = \langle x_q, \mathcal{U}_q \cup \{U\} \rangle$ is a condition in $M$. Now $q$ has no extension in $D \cap M$, which is absurd. □

We leave the following as an exercise (see Lemma 5.7 of [8]).

*Exercise* 7.24 Prove that for every $\beta$, the set $\{q \in Q : \alpha_q \geq \beta\}$ is dense and that if $q \in Q$, then there is an $r \leq q$ such that for all $s$ in $x_q$ there is a $t$ in $T_{\alpha_r}$ such that $s \leq t$ and $t \notin x_r$. This proves that $\dot{S}$ does not contain any cone of $T$ (*i.e.*, a set of the form $\{t \in T : t \geq s\}$). Prove that this implies that $\dot{A}$, a name for the set of minimal elements of $\check{T} \setminus \dot{S}$, is forced to be an uncountable antichain of $\check{T}$.

This finishes the proof of the theorem. □

## 8 Proof of the Main Theorem

Now we will turn to the precise formulation of the concepts involved in the Main Theorem and provide its proof. First it will be helpful to define some notation.

**Definition 8.1** Suppose $G^* \in Gen(N, P)$ and $\dot{Q}$ is a $P$-name from $N$ for a totally proper notion of forcing. A sequence $\langle \dot{q}_n \rangle_{n<\omega}$ of names from $N^P$ is an $(N[G^*], \dot{Q})$-*generic sequence* if

- $\Vdash_P \dot{q}_n \in \dot{Q}$,
- $N[G^*] \models \dot{q}_{n+1} \leq \dot{q}_n$ (*i.e.*, there is a $p \in G^*$ such that $p \Vdash \dot{q}_{n+1} \leq \dot{q}_n$), and
- if $\dot{D} \in N^P$ is a name for a dense set in $\dot{Q}$, there exists $m$ such that $N[G^*] \models \dot{q}_m \in \dot{D}$.

Any lower bound for $G^*$ forces that the sequence of interpretations $\langle \dot{q}_n[\dot{G}_P] \rangle_{n<\omega}$ generates an element of $Gen(N[\dot{G}_P], \dot{Q})$. Notice that $G^*$ doesn't tell us whether $\langle \dot{q}_n[\dot{G}_P] \rangle_{n<\omega}$ has a lower bound. However, if $G^* \in Gen(M, P)$ where $N \in M$, then $G^*$ can determine if an $(N[G^* \cap N], \dot{Q})$-generic sequence in $M$ has a lower bound.

**Definition 8.2** Given an iteration $\langle P_\alpha, \dot{Q}_\alpha \rangle_{\alpha<\varepsilon}$, $\alpha < \varepsilon$, $p \in P_\alpha$, and $q \in P_\varepsilon$, we say that $q$ is a *completion* of $p$ if $q \restriction \alpha = p$.

**Definition 8.3** Given a totally proper forcing $P$, a $P$-name $\dot{Q}$ for a totally proper forcing, and $k \in \{2, 3, 4, \ldots, \omega, \omega_1\}$, we say that $\dot{Q}$ satisfies the *k-iterability condition over P* if the hypotheses below always imply the conclusion below.

Hypotheses:

- $N_0 \in N_1$ are as usual.

- $G^* \in Gen(N_0, P) \cap N_1$.
- $l^* < 1 + k$.
- for all $l < l^*$, $G^l \in Gen(N_1, P)$.
- for all $l < l^*$, $G^l \cap N_0 = G^*$.
- $\dot{q}$ is a $P$-name from $N_0$ for a condition in $\dot{Q}$.

Conclusion: There exists a $(N_0[G^*], \dot{Q})$-generic sequence $\langle \dot{q}_n \rangle_{n<\omega}$ such that $\Vdash_P \dot{q}_0 \leq \dot{q}$ and, whenever some $p \in P$ forces that $N_1 \cap \dot{G}_P = G^l$ for some $l < l^*$, $p$ also forces that $\langle \dot{q}_n \rangle_{n<\omega}$ has a lower bound. The *iterability condition* means the 2-*iterability condition.*

Now we recall the statement of the Main Theorem.

**Main Theorem** *If $\mathbb{P} = \langle P_\alpha, \dot{Q}_\alpha \rangle_{\alpha<\varepsilon}$ is a CS iteration of totally proper forcings, then $P_\varepsilon$ is totally proper provided that:*

1. $\Vdash_{P_\alpha} \dot{Q}_\alpha$ *is* $<\omega_1$*-proper.*
2. $\dot{Q}_\alpha$ *satisfies the iterability condition over* $P_\alpha$.

While this theorem is true as stated, we will only verify this for iterations such that $\dot{Q}_\alpha$ satisfies the $\aleph_1$-iterability condition over $P_\alpha$.

The following lemma is the key component of the proof of the Main Theorem.

**Lemma 8.4** *Suppose that $P$ is a totally proper forcing and $\dot{Q}$ is a name for a forcing such that $\dot{Q}$ satisfies the $\aleph_1$-iterability condition over $P$. Further suppose that $N_0$, $N_1$, $N_2$, $G^*$ and $p$ are such that the following conditions are satisfied:*

- $N_0 \in N_1 \in N_2$ *are as usual with* $P, \dot{Q} \in N_0$;
- $G^* \in Gen(N_0, P) \cap N_1$;
- *$p$ is $(N_2, P)$-generic, $p$ is $(N_1, P)$-generic, $p$ is totally $(N_0, P)$-generic, and $p$ is a lower bound of $G^*$;*
- *$\dot{q}$ is forced by $p$ to be an element of $\dot{Q} \cap N$.*

*Then there is a $\dot{s}$ such that $p * \dot{s}$ is a totally $(N_0, P * \dot{Q})$-generic condition extending $p * \dot{q}$.*

*Proof* It follows from our assumptions that $p \Vdash N_1 \cap \dot{G}_P \in N_2$ because $p \Vdash N_2[\dot{G}_P] \cap V = N_2$. Let $\langle G^l \rangle_{l<\omega}$ enumerate $N_2 \cap Gen(N_1, P)$. Every $p$ satisfying the above assumptions forces that $N_1 \cap \dot{G}_P \in \{G^l : l < \omega\}$. This is by elementarity of $N_2$ and because $P$ adds no new countable subsets to $V$. Hence, by the $\aleph_1$-iterability condition, there is a $(N_0[G^*], \dot{Q})$-generic sequence $\langle \dot{q}_n \rangle_{n<\omega}$ such that any $p$ satisfying the above assumptions forces that $\langle \dot{q}_n \rangle_{n<\omega}$ has a lower bound. Therefore, there is a $P$-name

$\dot{s}$ such that any such $p$ forces $\dot{s}$ to be a lower bound of $\langle \dot{q}_n \rangle_{n<\omega}$. Hence, for any such $p$, $p * \dot{s}$ is totally $(N_0, P * \dot{Q})$-generic. Note that $\{r * \dot{t} : r \in G^*, N_0[G^*] \models \exists n\ \dot{t} \geq \dot{q}_n\} \in Gen^+(N_0, P * \dot{Q})$. Thus, we can complete $G^*$ to a filter $G^\dagger \in Gen^+(N_0, P * \dot{Q})$ such that any $p$ satisfying the above assumptions can be completed to a lower bound $p * \dot{s}$ of $G^\dagger$. □

The Main Theorem is a corollary of the following claim.

*Claim* 8.5 Given $\mathbb{P} = \langle P_\xi, \dot{Q}_\xi \rangle_{\xi<\varepsilon}$ satisfying the hypotheses of the Main Theorem, the following hypotheses imply the following conclusion.

Hypotheses:

- $N$ is as usual with $\mathbb{P} \in N$.
- $\alpha \in \varepsilon \cap N$.
- $\mathcal{N} = \langle N_\xi : \xi \leq 2\varepsilon^* \rangle$ is a tower of models with $N_0 = N$ where $\beta^*$ denotes $\mathrm{otp}(N \cap \beta)$ whenever $\beta$ is an ordinal.
- $G^* \in Gen(N_0, P_\alpha) \cap N_{2\alpha^*+1}$.
- $p \in N \cap P_\varepsilon$.
- $p \restriction \alpha \in G^*$.

Conclusion: There is a $G^\dagger \in Gen(N_0, P_\varepsilon, p)$ such that any lower bound for $G^*$ that is $(N_\xi, P_\alpha)$-generic for all $\xi \in (2\alpha^*, 2\varepsilon^*]$ can be completed to a lower bound for $G^\dagger$.

*Proof* Proceed by induction on $\varepsilon$. First, consider the case $\varepsilon = \gamma + 1$. To avoid repetition, we will assume $\alpha < \gamma^*$ — the case $\alpha = \gamma^*$ is marginally simpler. Observe that $\varepsilon^* = \gamma^* + 1$ because $\mathbb{P} \in N$ implies $\varepsilon \in N$ which in turn implies $\gamma \in N$. Let $\alpha, \mathcal{N}, G^*, p$ be as given. By our induction hypothesis applied to $\alpha, \mathcal{N} \restriction (2\gamma^* + 1), G^*, p \restriction \gamma$, There is a $G' \in Gen(N_0, P_\gamma)$ such that any lower bound for $G^*$ that is $(N_\xi, P_\alpha)$-generic for all $\xi \in (2\alpha^*, 2\gamma^*]$ can be completed to a lower bound for $G'$. Moreover, by elementarity, such a $G'$ can be take to be in $N_{2\gamma*+1}$. Note that $N_0 \in N_{2\gamma^*+1} \in N_{2\varepsilon^*}$ and $G' \in Gen(N_0, P_\gamma) \cap N_{2\gamma^*+1}$. Therefore by Lemma 8.4, we can extend $G'$ to a filter $G^\dagger \in Gen(N_0, P_\varepsilon)$ such that any lower bound for $G'$ that is also $(N_{2\gamma^*+1}, P_\gamma)$-generic and $(N_{2\varepsilon^*}, P_\gamma)$-generic can be completed to a lower bound for $G^\dagger$.

*Subclaim* Suppose $q$ is a lower bound for $G^*$ and is $(N_\xi, P_\alpha)$-generic for all $\xi \in (2\alpha^*, 2\varepsilon^*]$. It follows that there is a $P_\alpha$-name $\dot{r}$ such that $q$ forces each of the following.

- $\dot{r} \in N_{2\gamma^*+1} \cap P_\gamma$.
- $\dot{r}$ is a lower bound for $G'$.
- $\dot{r} \restriction \alpha \in \dot{G}_{P_\alpha}$.

*Proof* Let $G \subseteq P_\alpha$ be generic with $q \in G$. We then have $N_0 \cap G = G^*$ and $G \cap N_\xi \in Gen^+(N_\xi, P_\alpha) \cap N_{\xi+1}$ for all $\xi \in (2\alpha^*, 2\varepsilon^*)$. Step inside $N_{2\gamma^*+1}$. By elementarity of $N_{2\gamma^*+1}$, $N_{2\gamma^*} \cap G$ has a lower bound $s$. Such an $s$ is totally $(N_\xi, P_\alpha)$-generic for all $\xi \in (2\alpha^*, 2\gamma^*]$. We can then complete $s$ to a lower bound $r \in P_\gamma$ for $G'$. Let $\dot{r}$ be a name for $r$, forced by $q$ to have our desired properties. □

We now have a $p \in P_\alpha$ such that $q \Vdash \dot{r} \in P_\gamma \cap N_{2\gamma^*+1}$ and $q \Vdash \dot{r} \restriction \alpha \in \dot{G}_{P_\alpha}$. By properness, we can complete $q$ to a condition $q' \in P_\gamma$ such that $q' \Vdash \dot{r} \in \dot{G}_{P_\gamma}$ and $q'$ is $(N_{2\gamma^*+1}, P_\gamma)$-generic and $(N_{2\varepsilon^*}, P_\gamma)$-generic. Therefore, to finish the successor case, it suffices to prove that $q'$ is a lower bound for $G'$. Seeking a contradiction, suppose it is not. Let $G \subseteq P_\gamma$ be generic with $q' \in G$. Since $q' \Vdash \dot{r} \in \dot{G}_{P_\gamma}$, we have $q' \leq \dot{r}[G]$. Since $q' \Vdash \dot{r} \leq G'$, we also have $\dot{r}[G] \leq G'$, so $q' \leq G'$, in contradiction with our assumption that $q' \nleq G'$.

Next, consider the case where $\varepsilon$ is a limit. Choose $\langle \alpha_n \rangle_{n<\omega}$ strictly increasing and cofinal in $N \cap \varepsilon$ with $\alpha_0 = \alpha$. Let $\langle D_n \rangle_{n<\omega}$ enumerate the dense subsets of $P_\varepsilon$ from $N$. Build $\langle p_n, G_n \rangle_{n<\omega}$ such that:

- $p_0 = p$ and $G_0 = G^*$.
- $p_n \geq p_{n+1} \in N_0 \cap D_n$.
- $G_n \in Gen(N_0, P_{\alpha_n}, p_n \restriction \alpha_n) \cap N_{2\alpha_n^*+1}$.
- $p_{n+1} \restriction \alpha_n \in G_n$.
- Any lower bound for $G_n$ that is $(N_\xi, P_{\alpha_n})$-generic for all $\xi \in (2\alpha_n^*, 2\alpha_{n+1}^*]$ can be completed to a lower bound for $G_{n+1}$.

For $n = 0$, there's nothing to do. Now assume we have $p_n$ and $G_n$. We can then find $p_{n+1} \in N_0 \cap D_n$ such that $p_{n+1} \leq p_n$ and $p_{n+1} \restriction \alpha_n \in G_n$. This is because the set of conditions in $P_{\alpha_n}$ that can be completed to an extension of $p_n$ in $D_n$ is dense below $p_n \restriction \alpha_n$, which is in $G_n$. Moreover, this dense set is in $N_0$. Since $G_n$ is in $Gen(N_0, P_{\alpha_n}, p_n \restriction \alpha_n)$, we can find $p_{n+1}$ as desired.

By our induction hypothesis applied to

$$\mathbb{P} \restriction \alpha_{n+1}, p_{n+1} \restriction \alpha_{n+1}, \alpha_n, G_n, \mathcal{N} \restriction (2\alpha_{n+1}^* + 1)$$

there is a filter

$$G_{n+1} \in Gen(N_0, P_{\alpha_{n+1}}, p_{n+1} \restriction \alpha_{n+1}) \cap N_{2\alpha_{n+1}^*+1}$$

such that any condition which is a lower bound for $G_n$ and is $(N_\xi, P_{\alpha_n})$-generic for all $\xi \in (2\alpha_n^*, 2\alpha_{n+1}^*]$ can be completed to a lower bound for

$G_{n+1}$, as desired. By elementarity, such a $G_{n+1}$ can be taken to be in $N_{2\alpha^*_{n+1}+1}$

Let $G^\dagger = \{r \in P_\varepsilon \cap N : \exists n \; r \geq p_n\}$. It follows that $G^\dagger \in Gen(N_0, P_\varepsilon, p)$. Let $q$ be a lower bound for $G^*$ that is $(N_\xi, P_\alpha)$-generic for all $\xi \in (2\alpha^*, 2\varepsilon^*]$. Define by induction a sequence $\langle q_n \rangle_{n<\omega}$ such that:

- $q_0 = q$.
- $q_n \in P_{\alpha_n}$.
- $q_{n+1} \restriction \alpha_n = q_n$.
- $q_n$ is a lower bound for $G_n$.
- $q_n$ is $(N_\xi, P_{\alpha_n})$-generic for all $\xi \in (2\alpha^*_n, 2\varepsilon^*]$.

Given $q_n$, let us find $q_{n+1}$ as follows. Arguing as in the successor case, there is a $P_{\alpha_n}$-name $\dot{r}$ such that $q_n$ forces each of the following:

- $\dot{r} \in N_{2\alpha^*_{n+1}+1} \cap P_{\alpha_{n+1}}$.
- $\dot{r}$ is a lower bound for $G_{n+1}$.
- $\dot{r} \restriction \alpha_n \in \dot{G}_{P_{\alpha_n}}$.

By $<\omega_1$-properness, we can complete $q_n$ to a condition $q_{n+1} \in P_{\alpha_{n+1}}$ such that $q_{n+1} \Vdash \dot{r} \in \dot{G}_{P_{\alpha_{n+1}}}$ and $q_{n+1}$ is $(N_\xi, P_{\alpha_{n+1}})$-generic for all $\xi \in (2\alpha^*_n, 2\varepsilon^*]$. Arguing as in the successor case, $q_{n+1}$ is a lower bound for $G_{n+1}$.

Set $q^\dagger = \bigcup_{n<\omega} q_n$ in $P_\varepsilon$. (Technically, this union has domain $\sup(N \cap \varepsilon)$, which may be less than $\varepsilon$, but we extend the domain to all of $\varepsilon$ without increasing the support.) It suffices to show that $q^\dagger \leq G^\dagger$, so it suffices to show that $q^\dagger \leq p_m$ for all $m$. Seeking a contradiction, suppose $q^\dagger \not\leq p_m$. It follows that for some $n \geq m$ we have $q_n = q^\dagger \restriction \alpha_n \not\leq p_m \restriction \alpha_n$. Hence, $q_n \not\leq p_{n+1} \restriction \alpha_n \in G_n$; hence, $q_n \not\leq G_n$, in contradiction with how $\langle q_n \rangle_{n<\omega}$ was constructed. □

## 9 The role of $(<\omega_1)$-properness

The point of this section is to present a proof of the following theorem of Shelah which shows that the hypothesis of $(<\omega_1)$-properness can not be dropped entirely from the Main Theorem.

**Theorem 9.1** *[10, XVIII.1.1] Assume $V = L$. There is a countable support iteration $\langle P_\xi; \dot{Q}_\xi : \xi < \omega^2 \rangle$ of forcings such that:*

*1. $\dot{Q}_\xi$ is forced to be an $\aleph_1$-completely proper forcing;*

2. *$\dot{Q}_\xi$ is forced to have cardinality $\aleph_1$;*
3. *$P_{\omega^2}$ introduces a new subset of $\omega$.*

*Remark* 9.2 In fact $P_{\omega^2}$ introduces a club which does not contain any infinite subset from $V$. It is not clear whether this is true in general.

Our presentation of Theorem 9.1 is somewhat different than in [10] in that we define the partial order explicitly and then work to prove its properties. It seems that the differences with the construction in [10] are, however, superficial.

Let $\mathcal{D}$ consist of all countable subsets of $\omega_1$ which are closed in their supremum. Fix a surjection $\phi : \omega \to H(\omega)$ which is in $L$ and satisfies that $\phi^{-1}(X)$ is infinite for all $X$ in $H(\omega)$. In all of the discussion which follows, we need to assume a minimum of CH. Fix a bijection $\mathrm{ind} : \omega_1 \to \mathcal{D}$. Additional assumptions on ind will be stated as they are needed.

Fix a ladder sequence $\vec{C} = \langle C_\alpha : \alpha \in \lim(\omega_1)\rangle$ for the moment.

**Definition 9.3** If $q$ is in $\mathcal{D}$, we say $q$ is *self coding* with respect to $\vec{C}$ if and only if whenever $\nu$ is a limit point of $q$, there is a well ordering $\lhd$ of $\omega$ and an $X \subseteq \omega$ such that

$$(\omega, \lhd, X) \simeq (\nu, \in, q \cap \nu)$$

and for all $m < \omega$ there is a $\nu_m < \nu$ such that whenever $\xi$ is in $q \cap (\nu_m, \nu)$ there is a $n > m$ such that

$$\phi(|C_\nu \cap \xi|) = (n, \lhd \restriction n, X \cap n).$$

The following fact states a key property of this definition.

*Fact* 9.4 If $p$ and $q$ are self coding with respect to ladder systems $\vec{C}^p$ and $\vec{C}^q$, respectively, $C^p_\delta = C^q_\delta$, and $p \cap \delta \neq q \cap \delta$, then $p \cap q$ is not cofinal in $\delta$.

*Proof* Suppose that $p \cap q \cap \delta$ is cofinal in $\delta$ for some limit ordinal $\delta$ and $C^p_\delta = C^q_\delta$. If $\xi_i$ $(i < \omega)$ is contained in $p \cap q$ and cofinal in $\delta$, then the structures

$$\phi(|C^p_\delta \cap \xi_i|) = (n_i, \lhd_i, X_i)$$

must converge in the sense that for any $m$, $n_i > m$ for all but finitely many $i$ the sequence $\{(m, \lhd_i \restriction m, X_i \cap m)\}_i$ is eventually constant. Let $\lhd$ be a relation on $\omega$ and $X$ be a subset of $\omega$ such that for any $m$, for all but finitely many $i$, $\lhd \restriction m = \lhd_i \restriction m$ and $X \cap m = X_i \cap m$. Now $(\omega, \lhd, X)$ is isomorphic to $(\delta, \in, p \cap \delta)$ and $(\delta, \in, q \cap \delta)$. Since the only automorphism of $(\delta, \in)$ is the identity, it must be that $p \cap \delta = q \cap \delta$. □

**Definition 9.5** If $E \subseteq \omega_1$ is a club and $q$ is in $\mathcal{D}$, then we say that $q$ is *E-fast* if whenever $\nu$ is a limit point of $q$,

$$\min(E \setminus (\nu + 1)) < \operatorname{ind}(q \cap \nu) < \min(q \setminus (\nu + 1))$$

(here we define the latter inequality to be vacuous if $\nu = \sup(q)$).

The following fact is our motivation for this definition.

*Fact* 9.6 Suppose that $E_i$ $(i < \omega)$ is a sequence of clubs such that for every $i < j$, all initial parts of $E_j$ are $E_i$-fast. Then whenever $\delta$ is in $\bigcap\{E_i : i < \omega\}$

$$\sup\{\operatorname{ind}(E_i \cap \delta) : i < \omega\} = \min(\bigcap\{E_i : i < \omega\} \setminus (\delta + 1))$$

(note that this ordinal is a limit point of $E_i$ for all $i < \omega$).

**Definition 9.7** Define $Q_{\vec{C},E}$ to be the collection of all elements of $\mathcal{D}$ which are $E$-fast and self coding with respect to $\vec{C}$. $Q_{\vec{C},E}$ is viewed as a forcing notion with the order of end extension.

In general we do not expect $Q_{\vec{C},E}$ to preserve $\omega_1$. Note, however, that it is trivial that

$$\{q \in Q_{\vec{C},E} : \exists\beta > \alpha(\beta \in q)\}$$

is dense for all $\alpha < \omega_1$ and hence every condition in $Q_{\vec{C},E}$ forces that the generic self coding set is cofinal in $\omega_1$. Recall that a ladder system $\langle C_\alpha : \alpha < \omega_1\rangle$ is a *strong club guessing sequence* if whenever $E \subseteq \omega_1$ is a club, $\{\delta < \omega_1 : C_\delta \subseteq_* E\}$ contains a club.

**Theorem 9.8** *Suppose that $\langle C_\alpha : \alpha < \omega_1\rangle$ is a strong club guessing sequence and E is a club. Then $Q_{\vec{C},E}$ is proper.*

*Proof* Let $Q = Q_{\vec{C},E}$ for brevity. We will actually prove something more precise. Let $\vec{C}$ be an arbitrary ladder sequence and let $S$ consist of all countable elementary submodels $M$ of $H(\omega_2)$ such that if $\delta = M \cap \omega_1$ and $E \subseteq \omega_1$ is a club in $M$, $C_\delta \setminus E$ is finite. We will prove that if $S' \subseteq S$ is stationary, then forcing with $Q$ preserves $S'$.

Let $M$ be as usual such that $M \cap H(\omega_2)$ is in $S$, let $q$ be in $Q \cap M$. Fix an enumeration $D_i$ $(i < \omega)$ of the dense subsets of $Q$ which are in $M$. Let $\zeta_i$ $(i < \omega)$ enumerate the ordinals which are at most $\zeta = \min(E \setminus (\delta + 1))$.

Fix a bijection $\pi : \omega \to \delta$ such that $|C_\delta \cap \pi(k)| \leq k$ and define $i \lhd j$ if $\pi(i) < \pi(j)$.

Construct a descending sequence $q_k$ $(k < \omega)$ in $Q \cap M$ by induction. Start by putting $q_0 = q$. Now suppose that $q_k$ has been constructed. Define $X_k = \pi^{-1}(q_k)$ and $n_k = |C_\delta \cap \sup(q_k)|$. By our choice of $\phi$, there are infinitely many $i$ such that $\phi(i) = (n_k, \lhd \restriction n_k, X_k \cap n_k)$. Our assumptions on $S$ imply that for all but finitely many $i$, there is a countable elementary submodel $N$ of $H(\omega_2)$ such that $q_k$ and $D_k$ are in $N$ and $|C_\delta \cap N| = i$. Therefore it is possible to find such an $N$ with

$$|C_\delta \cap \nu| > n_k$$

$$\phi(|C_\delta \cap \nu|) = (n_k, \lhd \restriction n_k, X_k \cap n_k)$$

where $\nu = N \cap \omega_1$. Let $q'_k = q_k \cup \{\sup(q_k), \xi\}$ where $\xi < \nu$ is such that $\sup(q_k) < \xi$ and $C_\delta \cap \nu = C_\delta \cap \xi$. Finally, let $q_{k+1}$ be an extension of $q'_k$ in $N$ such that $q_{k+1}$ is in $D_k$ and if $\bar{q}$ is in $Q$ with $\mathrm{ind}(\bar{q}) = \zeta_k$, then $q_{k+1}$ is not an initial part of $\bar{q}$. The key point here is that if $\eta$ is in $q_{k+1}$ with $\eta > \sup(q_k)$, then $C_\delta \cap \eta = C_\delta \cap \nu$. Furthermore, if $\bar{q}$ is any extension of $q_{k+1}$ in $M$, $\pi^{-1}(\bar{q}) \cap n_k = \pi^{-1}(q_k) \cap n_k$. Finish the construction by letting $\bar{q} = \bigcup\{q_k : k < \omega\}$.

Since $\{p \in Q : \sup(p) > \alpha\}$ is dense in $Q$ and in $M$ for all $\alpha < \delta$, we will necessarily have that $n_k \to \infty$. Also we have arranged that if $\xi$ is in $q_{k+1} \setminus q_k$, then

$$\phi(|C_\delta \cap \xi|) = (n_k, \lhd \restriction n_k, \pi^{-1}(\bar{q}) \cap n_k).$$

It follows that $\bar{q}$ is self coding with respect to $\vec{C}$. Furthermore, we have arranged that $\min(E \setminus (\delta+1)) < \mathrm{ind}(\bar{q})$ which, together with the fact that $\bar{q} \cap \xi$ is a condition in $Q$ for all $\xi < \delta$, implies that $\bar{q}$ is $E$-fast. Hence $\bar{q}$ is in $Q$ and we have clearly arranged that $\bar{q}$ is $(M, Q)$-generic. □

In order to prove that the forcing $Q_{\vec{C},E}$ is completely proper, we need to know that $\vec{C}$ satisfies the following strong condition for some $A \subseteq \omega_1$:

$(*)_A$ The following hold:

1. $L[A]$ contains $E$ and $\vec{C}$;
2. For every countable limit ordinal $\delta$, $L[A \cap \delta]$ satisfies $\delta$ is countable;
3. $L[A \cap \delta]$ satisfies $C_\delta$ is the $<_{L[A\cap\delta]}$-least ladder in $\delta$ such that whenever $L_\alpha[A \cap \delta]$ satisfies "$\delta$ is $\omega_1$ and every two closed unbounded subsets of $\delta$ intersect", $C_\delta$ is almost contained in every closed unbounded subset of $\delta$ in $L_\alpha[A \cap \delta]$.

The most stringent requirement we will need is on the function ind in proving the $\aleph_1$-complete properness of $Q_{\vec{C},E}$:

(∗∗) $\mathrm{ind}(q) = \xi$ if and only if $q$ is the $\xi^{\text{th}}$-least element of $\mathcal{D}$ in the $\lhd_L$-ordering.

**Proposition 9.9** *($V = L$) If $\vec{C}$ satisfies $(*)_A$ for some $A \subseteq \omega_1$ and* ind *satisfies (∗∗), then $Q_{\vec{C},E}$ is $\aleph_1$-completely proper.*

*Proof* Suppose that $M \to N_i$ $(i < \omega)$ is a $Q$-diagram and that $q \in Q \cap M$. Let $\zeta_i$ denote $\min(E^{N_i} \setminus (\delta + 1))$. While $\zeta_i$ depends on $i$, we can take the supremum $\zeta$ of this sequence. Working as in Proposition 9.8, we can build a sequence $q_k$ $(k < \omega)$ of extensions of $q$ such that, setting $\bar{q} = \bigcup\{q_k : k < \omega\}$, we have $\zeta < \mathrm{ind}(\bar{q})$ and $\bar{q}$ is self coding with respect to $\vec{C}$. Notice that if $N \to N'$ and the filter $G$ generated by $q_k$ $(k < \omega)$ is in $N'$, then so is $\bar{q}$. By absoluteness of $L_\alpha$, $\mathrm{ind}^{N'}$ is a restriction of ind. In particular, $N'$ satisfies $\bar{q}$ is $E^{N'}$-fast. Furthermore, while $\vec{C}^{N'}$ may not be an initial segment of $\vec{C}$, $\vec{C}^{N'} \restriction (\delta + 1) = \vec{C} \restriction (\delta + 1)$ by absoluteness of $L_\alpha[A \cap \delta]$ for $\alpha > \delta$ and this is the only portion of $\vec{C}^{N'}$ relevant in determining whether $\bar{q}$ is self coding with respect to $\vec{C}^{N'}$. Thus $\bar{q}$ is in $Q^{N'}$ and clearly $N'$ satisfies $\bar{q}$ is a lower bound for $G$. □

*Remark* 9.10 This is clearly against the spirit of complete properness. We do not expect in general that if, e.g., $\vec{C}$ is a ladder system in some suitable model $M$ and $M \to N_i$ $(i < 2)$, then $N_0$ and $N_1$ should agree about $C_\delta$. In fact this sort of behavior can be ruled out if, for instance, there is a measurable cardinal.

**Proposition 9.11** *($V = L$) Suppose that $\langle P_\xi; \dot{Q}_\xi : \xi < \omega^2\rangle$ is an iteration of forcings such that for all $\xi < \omega^2$:*

1. *$E_0 = \lim(\omega_1)$;*
2. *$\dot{Q}_\xi = Q_{\vec{C}^\xi, \dot{E}^\xi}$;*
3. *$\dot{E}_{\xi+1}$ is the $P_{\xi+1}$ name for the union of the generic filter for $\dot{Q}_\xi$;*
4. *if $\eta$ is a limit ordinal, then $\dot{E}_\eta$ is the $P_\eta$-name for $\bigcap\{\dot{E}_\xi : \xi < \eta\}$;*
5. *$\vec{C}^\xi$ is the $P_\xi$-name for the ladder system satisfying (∗) with respect to some $A$ coding $\langle(\vec{C}^\eta, E^\eta) : \eta < \xi\rangle$ in some canonical way;*
6. *$\dot{Q}_{\vec{C}^\xi, E_\xi}$ is computed using a fixed function* ind *in $L$.*

*Then $P_{\omega^2}$ introduces a new real.*

*Proof* In fact we will show that if

$$\alpha_0 = \min(\bigcap\{E_\xi : \xi < \omega^2\})$$

then $\langle E_\xi \cap \alpha_0 : \xi < \omega^2 \rangle$ is not in $L$. We will assume for contradiction that this is not the case and prove that $\langle E_\xi : \xi < \omega^2 \rangle$ is in $L$. Observe first that $\dot{Q}_\xi$ is a $P_\xi$-name for a subset of $\mathcal{D}$. In order to make statements in the forcing language easier to read, we will suppress "checks" on the names for ground model elements of $\mathcal{D}$. Define sequences $\alpha_\zeta$ $(\zeta < \omega_1)$ and $q_{\xi,\zeta}$ $(\xi < \omega^2; \zeta \in \lim(\omega_1) \cup \{0\})$ by recursion as follows:

$$q_{\xi,0} = E_\xi \cap \alpha_0$$

$$\alpha_{\zeta+k+1} = \sup\{\mathrm{ind}(q_{\omega \cdot k+i,\zeta}) : i < \omega\}$$

If $\zeta > 0$ then:

$$\alpha_\zeta = \sup_{\zeta' < \zeta} \alpha_{\zeta'}.$$

The next claim is used to handle the recursive definition of $q_{\xi,\zeta}$ for limit ordinals $\zeta$.

*Claim* 9.12 For each $\xi < \omega^2$ and $\zeta \in \lim(\omega_1)$, there is a unique element $q_{\xi,\zeta}$ of $\mathcal{D}$ such that:

1. $q_{\xi,\zeta}$ is a cofinal subset of $\alpha_\zeta$;
2. either:
   a. $\zeta$ is a limit of limit ordinals and $\{\alpha_{\zeta'} : \zeta' \in \zeta \cap \lim(\omega_1)\} \subseteq q_{\xi,\zeta}$ or
   b. $\zeta = \zeta_0 + \omega$ and there is a $k_0$ such that for all $k > k_0$, $\alpha_{\zeta_0+k}$ is in $q_{\xi,\zeta}$;
3. $\langle q_{\eta,\zeta} : \eta < \xi \rangle$ forces $q_{\xi,\zeta}$ is in $\dot{Q}_\xi$.

Moreover $q_{\xi,\zeta} = E_\xi \cap \alpha_\zeta$.

*Proof* This is proved by induction on the lexicographical order on $\lim(\omega_1) \times \omega^2$. Let $(\zeta, \xi)$ be in $\lim(\omega_1) \times \omega^2$ and suppose that the claim is true whenever $(\zeta', \xi') <_{\mathrm{lex}} (\zeta, \xi)$.

Case 1: $\zeta = \zeta_0 + \omega$ for some $\zeta_0$ in $\lim(\omega_1) \cup \{0\}$. By Fact 9.6, $\alpha_{\zeta_0+k}$ the least element of $\cap\{E_\xi : \xi < \omega \cdot k\}$ greater than $\alpha_\zeta$. By (*), $\langle q_{\xi',\zeta} : \xi' < \xi \rangle$ decides the element $C^\xi_{\alpha_\zeta}$ of $\vec{C}^\xi$. By Fact 9.4, $\langle q_{\xi',\zeta} : \xi' < \xi \rangle$ forces that $q = E_\xi \cap \alpha_\zeta$ is the unique element of $Q_\xi$ which contains all but finitely many elements of $\{\alpha_{\zeta_0+k} : k < \omega\}$. This finishes case 1.

Case 2: $\zeta \cap \lim(\omega_1)$ is cofinal in $\zeta$. Our induction hypothesis implies that for all $\zeta'$ in $\zeta \cap \lim(\omega_1)$, $\alpha_{\zeta'}$ is a limit point of $E_\xi$. Hence $\{\alpha_{\zeta'} : \zeta' \in \zeta \cap \lim(\omega_1)\}$ is contained in $E_\xi$. By (*), $\langle q_{\xi',\zeta} : \xi' < \xi \rangle$ decides the element $C^\xi_{\alpha_\zeta}$ of $\vec{C}^\xi$. By Fact 9.4, $\langle q_{\xi',\zeta} : \xi' < \xi \rangle$ forces that $q = E_\xi \cap \alpha_\zeta$ is the unique element of $Q_\xi$ which contains a tail of $\{\alpha_{\zeta'} : \zeta' \in \zeta \cap \lim(\omega_1)\}$. This finishes case 2 and the proof of the claim. □

With the claim in hand, we can apply the recursion theorem in $L$ to find objects $\langle \alpha_\zeta : \zeta < \omega_1 \rangle$ and $\langle q_{\xi,\zeta} : \xi < \omega^2; \zeta \in \lim(\omega_1) \rangle$ in $L$ which satisfy the equations of the recursion. By Claim 9.12, we moreover have that, for each $\xi < \omega^2$, $E_\xi = \bigcup\{q_{\xi,\zeta} : \zeta \in \lim(\omega_1)\}$ and hence $E_\xi$ is in $L$ for all $\xi$. Note, however, that if $q$ is in $Q_{\vec{C},E}$, then for all but countably many $\nu \in E$, $q \cup \{\nu\}$ is in $Q_{\vec{C},E}$ and, by genericity, $E_1$ does not contain any club from $L$, a contradiction. □

It is known that the length $\omega^2$ in this iteration is the shortest possible.

**Theorem 9.13** *A CS iteration of length less than $\omega^2$ of totally proper forcings satisfying the $\omega_1$-iterability condition does not add reals.*

## 10 Open problems

We will finish with some open problems. While the example discussed in Section 9 illustrates that we can not drop $(<\omega_1)$-properness from Theorem 4.5 entirely, it does not answer the following problem, which is ultimately what is of greatest interest.

**Problem 10.1** Assume it is consistent that there is a supercompact cardinal. Is the forcing axiom for completely proper forcings consistent with CH?

We have seen that a positive answer to this question implies that measuring is consistent with CH (modulo a large cardinal assumption).

**Problem 10.2** Assume there is a measurable cardinal. Is there a countable support iteration of completely proper forcings which adds a new real?

Of course if this question has a positive answer under the assumption of any large cardinal hypothesis, this would be of great interest. One might view that the "problem" with the example in Section 9 is that there are an insufficient number of embeddings $M \to N$ to give the definition of complete properness its intended strength. Since Woodin cardinals can be used to generate embeddings similar to $M \to N$ via the "countable tower" (see [7]), the existence of Woodin cardinals may be a natural hypothesis to consider in this context.

While club guessing on $\omega_1$ is easily seen to be preserved by countably closed forcings, strong club guessing is not. This can be used to show that $2^{<\omega_1}$ forces that $Q_{\vec{C},E}$ does not preserve stationary subsets of $\omega_1$ (where $Q_{\vec{C},E}$ is the forcing defined in Section 9.1). Shelah has proved the following iteration theorem in addition to Theorem 4.5.

**Theorem 10.3** *A countable support iteration of forcings which are:*

1. *completely proper and*
2. *remain proper in every totally proper forcing extension*

*does not add new reals.*

In the context of totally proper forcings, the requirement that the forcing remains proper in every totally proper forcing extension is usually met by forcings which have only a countable "working part" and no "side conditions." The example in Section 7.1 has no side conditions (and remains proper in every totally proper extension). The examples in Sections 7.2 and 7.3 do have side conditions. The side condition in the forcing for measure a sequence $\vec{D}$ of closed sets can be removed if either the map $\alpha \mapsto \text{otp}(D_\alpha)$ is regressive or if $D_\alpha$ is a clopen subset of $\alpha$ for each $\alpha < \omega_1$.

It should be remarked that Shelah's two iteration theorems can not (apparently) be combined to a single theorem where we require that each iterand satisfies one of the two hypotheses. While the forcing associated to measuring is cut from the same block as the example in Section 9.1, we do not expect that the forcing to destroy a Souslin tree should "cause problems."

**Problem 10.4** Is it true that countable support iteration of forcings which are:

1. completely proper and
2. remain proper after forcing with $2^{<\omega_1}$

does not add new reals?

The following is likely a closely related question.

**Problem 10.5** If $P$ is a countable support iteration of completely proper forcings which adds a new real, must $P \times 2^{<\omega_1}$ collapse $\omega_1$?

The answer to this question, however, is not clear even in the case of the example in Section 9.

**Problem 10.6** If $P$ is a countable support iteration of completely proper forcings which adds a new real, must $P$ add a club $E \subseteq \omega_1$ such that for some $\delta < \omega_1$, $E$ contains no ground model subsets of ordertype $\delta$?

If $P$ is example in Section 9, then $\bigcap_{\xi<\omega^2} E_\xi$ contains no ground model infinite set.

## References


[1] Uri Abraham and Stevo Todorčević. Partition properties of $\omega_1$ compatible with CH. *Fund. Math.*, 152(2):165–181, 1997.

[2] Keith Devlin and Saharon Shelah. A weak version of $\diamondsuit$ which follows from $2^{\aleph_0} < 2^{\aleph_1}$. *Israel Journal of Math*, 29(2–3):239–247, 1978.

[3] Keith J. Devlin and Håvard Johnsbråten. *The Souslin problem.* Lecture Notes in Mathematics, Vol. 405. Springer-Verlag, Berlin, 1974.

[4] T. Eisworth and J. Roitman. CH with no Ostaszewski spaces. *Trans. Amer. Math. Soc.*, 351(7):2675–2693, 1999.

[5] Todd Eisworth and Peter Nyikos. First countable, countably compact spaces and the continuum hypothesis. *Trans. Amer. Math. Soc.*, 357(11):4269–4299, 2005.

[6] Matthew Foreman, Menachem Magidor, and Saharon Shelah. Martin's Maximum, saturated ideals, and nonregular ultrafilters. I. *Ann. of Math. (2)*, 127(1):1–47, 1988.

[7] Paul B. Larson. *The stationary tower*, volume 32 of *University Lecture Series*. American Mathematical Society, Providence, RI, 2004. Notes on a course by W. Hugh Woodin.

[8] Justin Tatch Moore. $\omega_1$ and $\omega_1$ may be the only minimal uncountable order types. *Michigan Math. Journal*, 55(2):437–457, 2007.

[9] Adam Ostaszewski. On countably compact, perfectly normal spaces. *J. London Math. Soc.*, 14(3):505–516, 1976.

[10] Saharon Shelah. *Proper and improper forcing*. Springer-Verlag, Berlin, second edition, 1998.

[11] Robert Solovay and S. Tennenbaum. Iterated Cohen extensions and Souslin's problem. *Ann. of Math.*, 94:201–245, 1971.

[12] Stevo Todorčević. A dichotomy for P-ideals of countable sets. *Fund. Math.*, 166(3):251–267, 2000.

# 8
# Short extender forcing

Moti Gitik and Spencer Unger

---

The eleventh Appalachian Set Theory workshop was held at Carnegie Mellon University on April 3, 2010. The lecturer was Moti Gitik. As a graduate student Spencer Unger assisted in writing this chapter, which is based on the workshop lectures.

---

## 1 Introduction

The goal of these notes is to provide the reader with an introduction to the main ideas of a result due to Gitik [2].[1]

**Theorem 1.1** *Let $\langle \kappa_n \mid n < \omega \rangle$ be an increasing sequence with each $\kappa_n$ $\kappa_n^{+n+2}$-strong, and $\kappa =_{def} \sup_{n<\omega} \kappa_n$. There is a cardinal preserving forcing extension in which no bounded subsets of $\kappa$ are added and $\kappa^\omega = \kappa^{++}$.*

In order to present this result, we approach it by proving some preliminary theorems about different forcings which capture the main ideas in a simpler setting. For the entirety of the notes we work with an increasing sequence of large cardinals $\langle \kappa_n \mid n < \omega \rangle$ with $\kappa =_{def} \sup_{n<\omega} \kappa_n$. The large cardinal hypothesis that we use varies with the forcing. A recurring theme is the idea of a cell. A cell is a simple poset which is designed to be used together with other cells to form a large poset. Each

[1] During the workshop, Gitik also discussed the preparation forcing needed to add $\kappa^{+++}$ $\omega$-sequences to a singular cardinal $\kappa$, but we omit that discussion here. The original account of this forcing can be found in [3] and a revised version that is closer to the presentation given at the workshop can be found in [1].

of the forcings that we present has $\omega$-many cells which are put together in a canonical way to make the forcing.

First, we will present Diagonal Prikry Forcing which adds a single cofinal $\omega$-sequence to $\kappa$. The key property that we wish to present with this forcing is the *Prikry condition.* The Prikry condition is the property of the forcing that allows us to show that no bounded subsets of $\kappa$ are added. All subsequent forcings share this property. We also show that this forcing preserves cardinals and cofinalities above $\kappa$ using a chain condition argument.

Second, we present a forcing for adding $\lambda$-many $\omega$-sequences to $\kappa$ using long extenders. This forcing can be seen as both a more complicated version of Diagonal Prikry forcing and an approximation of the forcing used to prove Theorem 1.1. We want to repeat many of the arguments from the first poset, however each argument becomes more difficult. We sketch a proof of the Prikry condition and mention a strengthening needed to show that $\kappa^+$ is preserved. The chain condition argument is also more difficult and we sketch a proof of this too.

Third, we present the forcing from Theorem 1.1. To do this we present an attempt at a definition, which we ultimately refine to obtain the actual forcing. This is instructive because our attempt at a definition is similar to the forcing with long extenders and it is straightforward to see that they share many properties. In particular this attempted definition satisfies the Prikry condition. However there is a problem with the chain condition. We explain where the argument goes awry and present a revised definition which allows us to recover the chain condition.

Diagonal Prikry forcing first appeared in [7]. The second forcing originally appeared in [5]. A full presentation of the first and the second forcing as they appear in these notes can be found in [4]. Theorem 1.1 originally appeared in [2], but in a slightly different form. The presentation of the third forcing in these notes is closer to the presentation found in [6].

## 2 Adding a single $\omega$-sequence

For this section we assume that each $\kappa_n$ is measurable and let $U_n$ be a nonprincipal $\kappa_n$-complete ultrafilter on $\kappa_n$. As mentioned in the introduction our forcing is made up of $\omega$-many cells $Q_n$ for $n < \omega$. Each cell has two parts $Q_{n0}$ and $Q_{n1}$. Fixing $n < \omega$ we set $Q_{n1} =_{def} \kappa_n$ and $Q_{n0} =_{def} U_n$. While $Q_{n1}$ can be ordered by ordinal less than, we will not use this ordering. More importantly, we define an order on $Q_{n0}$, $\leq_{Q_{n0}}$ by $A \geq_{Q_{n0}} B$ if and only if $A \subseteq B$. Our forcing convention is $A \geq B$ means that $A$ is stronger than $B$.

Now we are ready to define the $n^{th}$ cell $Q_n$. We set $Q_n =_{def} Q_{n0} \cup Q_{n1}$ and define two orderings $\leq_n$ and $\leq_n^*$. $\leq_n$ is the ordering that we force with. $\leq_n^*$ will be a notion of *direct extension*, which is an important sub-ordering of $\leq_n$. We define direct extension first. For $p, q \in Q_n$, let $p \leq_n^* q$ if and only if either $p = q$ or $p, q \in Q_{n0}$ and $p \leq_{Q_{n0}} q$. Now we define $p \leq_n q$ if and only if either $p \leq_n^* q$ or $p \in Q_{n0}, q \in Q_{n1}$ and $q \in p$. There are two ways to extend in the $\leq_n$ relation. Starting with a measure one set we are allowed to pick another measure one set contained in the first, but then eventually we must pick an ordinal from the current measure one set. Once we pick an ordinal no further extensions are possible. Next we define the Prikry condition, which shows the purpose of the direct extension relation.

**Definition 2.1** A forcing $\langle P, \leq, \leq^* \rangle$ where $\leq^* \subseteq \leq$ satisfies the Prikry condition if and only if for all $p \in P$ and for all statements $\sigma$ in the forcing language, there is $q \ {}^*\!\geq p$ such that $q \parallel \sigma$.

**Proposition 2.2** $\langle Q_n, \leq_n, \leq_n^* \rangle$ *satisfies the Prikry condition.*

*Proof* Let $p \in Q_n$ and $\sigma$ be in the forcing language. If $p \in Q_{n1}$ then we are done since $p$ decides all statements in the forcing language. Suppose that $p \in Q_{n0}$. We have that $p \in U_n$. Let $A_0 = \{\nu \in p \mid \nu \Vdash \sigma\}$ and $A_1 = \{\nu \in p \mid \nu \Vdash \neg\sigma\}$. $A_0, A_1$ partition $p$. So one of them must be in $U_n$. Without loss of generality assume that $A_0 \in U_n$. Then $A_0 \ {}^*\!\geq p$ and $A_0 \Vdash \sigma$. □

Note that the forcing $Q_n$ is equivalent to the trivial forcing. We now put the $Q_n$'s together to define a forcing notion $P$.

**Definition 2.3** $p \in P$ if and only if $p = \langle p_n \mid n < \omega \rangle$ such that for all $n < \omega$, $p_n \in Q_n$ and there is $\ell(p) < \omega$ such that for all $n \geq \ell(p)$, $p_n \in Q_{n0}$ and for all $n < \ell(p)$, $p_n \in Q_{n1}$. Let $p = \langle p_n \mid n < \omega \rangle$ and $q = \langle q_n \mid n < \omega \rangle$ be members of $P$. Set $p \geq q$ if and only if for all $n < \omega$, $p_n \geq_n q_n$. And set $p \ {}^*\!\geq q$ if and only if for all $n < \omega$, $p_n \ {}^*\!\geq_n q_n$.

**Lemma 2.4** $\langle P, \leq, \leq^* \rangle$ *satisfies the Prikry condition.*

*Proof* Let $p \in P$ with $p = \langle p_n \mid n < \omega \rangle$. Let $\sigma$ be a statement in the forcing language. We need $p^* \ {}^*\!\geq p$ such that $p^* \parallel \sigma$. Suppose that there is no such $p^*$. We will construct a direct extension of $p$ so that no extension decides $\sigma$, a contradiction. Assume for simplicity that $\ell(p) = 0$, ie for all $n < \omega$, $p_n \in U_n$. By induction we construct an $\leq^*$-increasing sequence of conditions $\langle t(n) \mid n < \omega\rangle$ above $p$ with the property that for all $n$ and $s \geq t(n)$ with $\ell(s) = n$, $s$ does not decide $\sigma$.

For the base case we set $t(0) = p$ and note that our assumption for a contradiction shows that $t(0)$ has the desired property. Assume that we have constructed $t(n)$ as claimed. We work to construct $t(n+1)$. Let $\vec{\nu}$ be a sequence of length $n$ such that

$$t(n) \frown \vec{\nu} =_{def} \langle \nu_0, \nu_1, \ldots \nu_{n-1}, t(n)_n, t(n)_{n+1}, \ldots\rangle \geq t(n).$$

If $n = 0$, then the empty sequence is the only such $\vec{\nu}$. For each one point extension of $t(n) \frown \vec{\nu}$, we wish to capture a direct extension which decides $\sigma$, if one exists.

Let $\langle \nu_\eta \mid \eta < \kappa_n\rangle$ enumerate $t(n)_n$ in increasing order. We inductively construct an increasing sequence of direct extensions of $t(n)$, $\langle p_{\vec{\nu}}(\eta) \mid \eta < \kappa_n\rangle$. We set $p_{\vec{\nu}}(0) =_{def} t(n)$. Suppose that we have constructed $p_{\vec{\nu}}(\eta)$ for some $\eta$. If there is a direct extension of $p_{\vec{\nu}}(\eta) \frown \vec{\nu} \frown \nu_\eta$ that decides $\sigma$, then let $q$ be such an extension with $q = \langle q_k \mid k < \omega\rangle$. We set

$$p_{\vec{\nu}}(\eta + 1) =_{def} \langle t(n)_0, t(n)_1, \ldots t(n)_n, q_{n+1}, q_{n+2}, \ldots\rangle.$$

If $\gamma < \kappa_n$ is limit we let $p_{\vec{\nu}}(\gamma)_k =_{def} t(n)_k$ for $k \leq n$ and $p_{\vec{\nu}}(\gamma)_k =_{def} \bigcap_{\eta<\gamma} p_{\vec{\nu}}(\eta)_k$ for $k > n$. Note that for $k > n$, $U_k$ is $\kappa_k$-complete and $\kappa_k > \kappa_n > \gamma$. So $p(\gamma) \,{}^*\!\geq p(\eta)$ for all $\eta < \gamma$. This completes the inductive construction of the sequence $\langle p_{\vec{\nu}}(\eta) \mid \eta < \kappa_n\rangle$. Again using completeness of the relevant measures we can find $r(\vec{\nu}) \,{}^*\!\geq p_{\vec{\nu}}(\eta)$ for all $\eta < \kappa_n$.

We need to refine $t(n)_n$ and we do so as follows. For each $\vec{\nu}$ we can partition $t(n)_n$ into three sets

$$A_0^{\vec{\nu}} = \{\rho \in t(n)_n \mid r \frown \vec{\nu} \frown \rho \Vdash \sigma\},$$
$$A_1^{\vec{\nu}} = \{\rho \in t(n)_n \mid r \frown \vec{\nu} \frown \rho \Vdash \neg\sigma\}, \text{ and}$$
$$A_2^{\vec{\nu}} = \{\rho \in t(n)_n \mid r \frown \vec{\nu} \frown \rho \nVdash \sigma\}.$$

We claim that for all $\vec{\nu}$, $A_2^{\vec{\nu}} \in U_n$. Assume otherwise. Then without loss of generality there is $\vec{\nu}$ such that $A_0^{\vec{\nu}} \in U_n$. Then there is a direct extension of $t(n) \frown \vec{\nu}$ that decides $\sigma$, namely

$$\langle \nu_0, \nu_1, \ldots \nu_{n-1}, A_0^{\vec{\nu}}, r(\vec{\nu})_{n+1}, r(\vec{\nu})_{n+2}, \ldots\rangle.$$

This contradicts our inductive assumption in the construction of $\langle t(n) \mid n < \omega\rangle$. So for all $\vec{\nu}$, $A_2^{\vec{\nu}} \in U_n$. We are now ready to define $t(n+1)$. For $k < n$, we let $t(n+1)_k =_{def} t(n)_k$. For the other coordinates we wish to take the intersection of the relevant measure one sets. In particular, we let

$$t(n+1)_n =_{def} \bigcap_{\vec{\nu}} A_2^{\vec{\nu}}, \text{ and}$$

$$t(n+1)_k =_{def} \bigcap_{\vec{\nu}} r(\vec{\nu})_k \text{ for all } k > n.$$

There are at most $\kappa_{n-1}$ many possible $\vec{\nu}$. So by the completeness of each $U_k$ for $k \geq n$, each of the above sets is measure one for the appropriate measure. It follows that $t(n+1) \in P$ and $t(n+1) \,{}^*\!\geq t(n)$. Moreover, if $q =_{def} t(n+1) \frown \vec{\nu} \frown \nu$ is an $n+1$ step extension of $t(n+1)$, then $\nu \in A_2^{\vec{\nu}}$ and hence $q$ does not decide $\sigma$. This completes the construction of $\langle t(n) \mid n < \omega \rangle$.

We are now ready to conclude the proof that $P$ satisfies the Prikry condition. Let $p^* \,{}^*\!\geq t(n)$ for all $n < \omega$. It is easy to see that for all $n < \omega$, there is no $s \geq p^*$ with $\ell(s) = n$ such that $s$ decides $\sigma$. So in fact no extension of $p^*$ decides $\sigma$, a contradiction. □

Using the Prikry condition we can show that forcing with $P$ does not add bounded subsets of $\kappa$.

**Lemma 2.5** *$\langle P, \leq \rangle$ does not add bounded subsets of $\kappa$*

*Proof* Let $\underset{\sim}{a}$ be a $P$-name for a bounded subset of $\kappa$. We may assume that $\Vdash_P \underset{\sim}{a} \subseteq \underset{\sim}{\lambda}$ for some $\underset{\sim}{\lambda} < \kappa$. We choose $p \in P$ that decides the value of $\underset{\sim}{\lambda}$ to be $\lambda < \kappa$. There is $i < \omega$ such that $\kappa_i > \lambda$ and we choose $p' \geq p$ with $\ell(p') = i$. Now let $\sigma_\alpha \equiv$ "$\alpha \in a$." We inductively build an increasing sequence $\langle p(\alpha) \mid \alpha < \lambda \rangle$ of direct extensions of $p'$. Let $p(0) =_{def} p'$. At stage $\alpha + 1$ we use the Prikry condition to find $p(\alpha + 1) \,{}^*\!\geq p(\alpha)$ which decides $\sigma_\alpha$. Since the direct extension relation is $\kappa_i$ closed for conditions of length $i$, we can take upper bounds at both limit stages and for the whole construction. We let $p''$ be an upper bound for the sequence $\langle p(\alpha) \mid \alpha < \lambda \rangle$. Now $p''$ forces that $\underset{\sim}{a} = a$ for some $a \subseteq \lambda$ with $a \in V$. So there is a dense set of conditions which force $\underset{\sim}{a} \in V$. □

Let $G \subseteq P$ be $V$-generic over $P$. Define $t : \omega \to \kappa$ by $t(n) = \alpha$ if and only if there is $p \in G$ such that $\ell(p) > n$ and $p_n = \alpha$. Then we have that $t \in \prod_{n<\omega} \kappa_n$. Let $\underset{\sim}{t}$ be the canonical name for $t$. Using an easy density argument we can see that $t$ is bigger than any $s \in (\prod_{n<\omega} \kappa_n)^V$ mod finite, in particular $t \notin V$.

**Lemma 2.6** *For each $s \in \prod_{n<\omega} \kappa_n$ with $s \in V$, we have $s <^* t$.*

*Proof* Let $p \in P$ and $s \in \prod_{n<\omega} \kappa_n \cap V$. For each $n \geq \ell(p)$ replace $p_n$ by $p_n \setminus (s(n)+1)$. Let $q$ be the result. Then $q \Vdash \forall n \geq \ell(q), \underset{\sim}{t}(n) > s(n)$ □

Finally we show that cardinals and cofinalities above $\kappa$ are preserved.

**Lemma 2.7** *$\langle P, \leq \rangle$ has $\kappa^+$-cc*

*Proof* Assume that $\langle p_\alpha \mid \alpha < \kappa^+ \rangle$ is a sequence of pairwise incompatible members of $P$. Note that if two conditions $p, q \in P$ begin with the same sequence of ordinals, then they are compatible. To find an upper bound we just take the sequence of ordinals followed by $p_n \cap q_n$ for all $n \geq \ell(p) = \ell(q)$. There are only $\kappa$-many such finite sequences of ordinals. It follows that there are $\alpha, \beta < \kappa^+$ such that $p_\alpha$ is compatible with $p_\beta$, a contradiction. □

This completes the presentation of the forcing for adding a single $\omega$-sequence. The combination of the facts that $P$ adds no bounded subsets of $\kappa$ and that $P$ has the $\kappa^+$-cc shows that cardinals are preserved in the extension.

## 3 Adding many $\omega$-sequences with long extenders

In this section we present a forcing to add $\lambda$-many $\omega$-sequences to $\kappa$ and for this we use a stronger large cardinal assumption. This time we assume GCH in $V$ and let $\lambda > \kappa^+$ be regular. For this forcing we assume that for each $n < \omega$ there is a $(\kappa_n, \lambda+1)$-extender $E_n$ over $\kappa_n$. In particular we assume that for each $n < \omega$, there is $j_n : V \to M_n$ with $M_n$ transitive, ${}^{\kappa_n}M_n \subseteq M_n$, $M_n \supseteq V_{\lambda+1}$, $\mathrm{crit}(j_n) = \kappa_n$ and $j_n(\kappa_n) > \lambda$. Now for all $\alpha$ with $\kappa_n \leq \alpha < \lambda$, we define $E_{n\alpha} = \{X \subseteq \kappa_n \mid \alpha \in j_n(X)\}$, which is a nonprincipal, $\kappa_n$-complete ultrafilter on $\kappa_n$. Note that $E_{n\kappa_n}$ is normal.

Next we define an ordering that plays a key role in the definition of the ordering of the poset for this section. For $\alpha, \beta < \lambda$ define $\alpha \leq_{E_n} \beta$ if and only if there is $f : \kappa_n \to \kappa_n$ such that $j_n(f)(\beta) = \alpha$. It turns out that $\leq_{E_n}$ is $\kappa_n$-directed and this fact will play an important role in the proofs below.

**Lemma 3.1** *Fix $n < \omega$ and $\tau < \kappa_n$. Assume that $\langle \alpha_\nu \mid \nu < \tau \rangle$ is a sequence of ordinals less than $\lambda$. There are unboundedly many $\alpha < \lambda$ such that for all $\nu < \tau$, $\alpha_\nu \leq_{E_n} \alpha$.*

*Proof* To proceed with the proof we need a particular enumeration of small subsets of $\kappa_n$. Using GCH, we can construct $\langle a_\beta \mid \beta < \kappa_n \rangle$, an enumeration of $\kappa_n^{<\kappa_n}$ with the following property. For every regular $\delta < \kappa_n$, the sequence $\langle a_\beta \mid \beta < \delta \rangle$ enumerates $\delta^{<\delta}$ so that each $x \in \delta^{<\delta}$ appears unboundedly often in the sequence below $\delta$. We will be interested in the object $j_n(\langle a_\beta \mid \beta < \kappa_n \rangle)$, which when restricted to $\lambda$ enumerates $\lambda^{<\lambda}$ with the property given to $\delta$ above. We call this enumeration $\langle a_\beta \mid \beta < \lambda \rangle$. Fix $\alpha < \lambda$ such that $a_\alpha = \langle \alpha_\nu : \nu < \tau \rangle$. We claim that for all $\nu < \tau$, $\alpha_\nu \leq_{E_n} \alpha$. Recall that by the property of the enumeration there are $\lambda$ many such $\alpha$, so this will finish the proof of the lemma.

By the general theory of ultrapowers we have the diagram shown in Figure 8.1 for a fixed $\nu < \tau$.

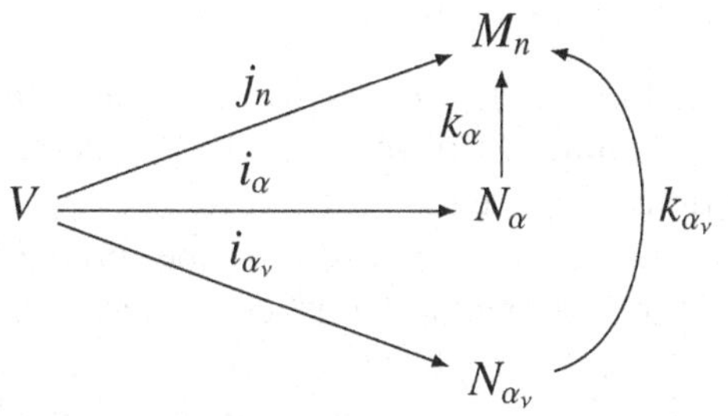


Figure 8.1

In the diagram $N_\alpha \simeq \mathrm{Ult}(V, E_{n\alpha})$ and $k_\alpha([f]_{E_{n\alpha}}) = j_n(f)(\alpha)$, and we have the same for $\alpha_\nu$ in place of $\alpha$. Each of the maps in the diagram is an elementary embedding. The diagram commutes, so in particular we have $j_n(\langle a_\beta \mid \beta < \kappa_n \rangle) = k_\alpha(i_\alpha(\langle a_\beta \mid \beta < \kappa_n \rangle))$. The $\alpha^{th}$ member of this sequence is $a_\alpha$. Moreover we can write $a_\alpha$ as the image of a $\tau$-sequence of ordinals in $N_\alpha$.

$$a_\alpha = j_n(\langle a_\beta \mid \beta < \kappa_n \rangle)(\alpha) = k_\alpha(i_\alpha(\langle a_\beta \mid \beta < \kappa_n \rangle)([id]_{E_{n\alpha}})).$$

Since $\operatorname{crit} k_\alpha \geq \kappa_n = \operatorname{crit}(j_n)$, we have that $i_\alpha(\langle a_\beta \mid \beta < \kappa_n \rangle)([id]_{E_{n\alpha}})$ is a $\tau$-sequence of ordinals in $N_\alpha$ by the elementarity of the map $k_\alpha$. We let $\alpha_\nu^*$ be the $\nu^{th}$ member of the $\tau$-sequence from $N_\alpha$. By elementarity $k_\alpha(\alpha_\nu^*) = \alpha_\nu$. This allows us to define an elementary embedding $k_{\alpha_\nu\alpha} : N_{\alpha_\nu} \to N_\alpha$ by the formula $k_{\alpha_\nu\alpha}([f]_{E_{n\alpha_\nu}}) = i_\alpha(f)(\alpha_\nu^*)$. The proofs that $k_{\alpha_\nu\alpha}$ is elementary and that the diagram shown in Figure 8.2 commutes are easy and will be omitted.

Using $k_{\alpha_\nu\alpha}$, we can define a map, $\pi_{\alpha\alpha_\nu}$ witnessing that $\alpha_\nu \leq_{E_n} \alpha$.

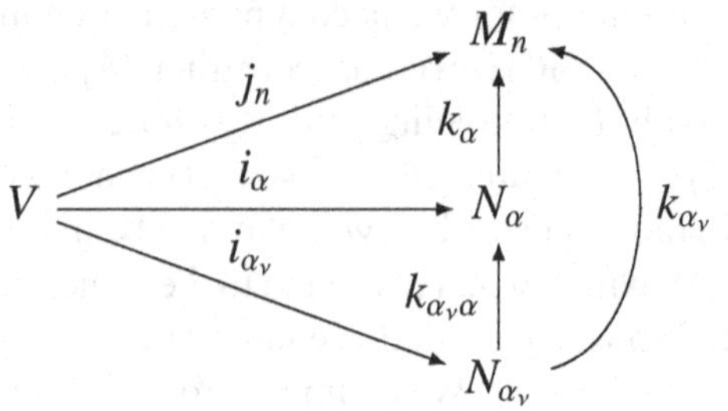


Figure 8.2

Let $\pi_{\alpha\alpha_\nu} : \kappa_n \to \kappa_n$ such that $[\pi_{\alpha\alpha_\nu}]_{E_{n\alpha}} = \alpha^*_\nu$. Then $j_n(\pi_{\alpha\alpha_\nu})(\alpha) = k_\alpha([\pi_{\alpha\alpha_\nu}]_{E_{n\alpha}}) = k_\alpha(\alpha^*_\nu) = \alpha_\nu$. So $\alpha_\nu \leq_{E_n} \alpha$. □

For all $\alpha, \beta < \lambda$ such that $\beta \leq_{E_n} \alpha$, we let $\pi_{\alpha\beta}$ be the projection as defined in the previous lemma. Also we let $\pi_{\alpha\alpha}$ be the identity map. We will use these projections to relate the Prikry sequences that we add.

We define a forcing that is similar in form to the forcing for adding a single $\omega$-sequence. First we define the $n^{th}$ cell $Q_n$ and its associated orderings $\leq_n$ and $\leq^*_n$. As before $Q_n$ has two parts $Q_{n1}$ and $Q_{n0}$. For this forcing $Q_{n1}$ is a kind of Cohen forcing that is restricted by $Q_{n0}$ and the definition of $Q_{n0}$ is significantly more complex than before.

**Definition 3.2** We define $Q_{n1} =_{def} \{f \mid f$ is a partial function from $\lambda$ to $\kappa_n$ with $|f| \leq \kappa\}$. We order $Q_{n1}$ by containment and call this ordering $\leq_{Q_{n1}}$. A triple $\langle a, A, f\rangle \in Q_{n0}$ if and only if all of the following conditions hold:

1. $f \in Q_{n1}$,
2. $a \subseteq \lambda$ such that
   a. $|a| < \kappa_n$,
   b. $a$ has a maximal element in the ordinal sense and $\max(a) \geq_{E_n} \beta$ for all $\beta \in a$ and
   c. $a \cap \operatorname{dom}(f) = \emptyset$,
3. $A \in E_{n\max(a)}$,
4. If $\alpha > \beta$ and $\alpha, \beta \in a$, then for all $\nu \in A$, $\pi_{\max(a)\alpha}(\nu) > \pi_{\max(a)\beta}(\nu)$ and
5. If $\alpha, \beta, \gamma \in a$ with $\alpha \geq_{E_n} \beta \geq_{E_n} \gamma$, then for all $\rho \in \pi_{\max(a)\alpha}$"$A$, $\pi_{\alpha\gamma}(\rho) = \pi_{\beta\gamma}(\pi_{\alpha\beta}(\rho))$.

Next we define the ordering $\leq_{Q_{n0}}$. $\langle a, A, f\rangle \geq_{Q_{n0}} \langle b, B, g\rangle$ if and only if

1. $f \supseteq g$,
2. $a \supseteq b$ and
3. $\pi_{\max(a)\max(b)}$"$A \subseteq B$.

We define $Q_n =_{def} Q_{n0} \cup Q_{n1}$ as before. We let $\leq_n^* =_{def} \leq_{Q_{n0}} \cup \leq_{Q_{n1}}$. For $p, q \in Q_n$, $p \leq_n q$ if and only if either $p \leq_n^* q$ or $p \in Q_{n0}$ with $p = \langle a, A, f \rangle$, $q \in Q_{n1}$ and

1. $q \supseteq f$,
2. $\mathrm{dom}(q) \supseteq a$,
3. $q(\max(a)) \in A$ and
4. $\forall \beta \in a\ q(\beta) = \pi_{\max(a)\beta}(q(\max(a)))$.

*Remark* 3.3 We collect some remarks on the definition on the forcing.

- It is not hard to show that $\leq_n$ is transitive.
- We call $f$ *the Cohen part* of $\langle a, A, f \rangle$.
- There are two technical lemmas here which show that (4) and (5) in the definition of definition of $Q_{n0}$ are possible. We omit them and refer the interested reader to [4] for their statements and proofs.
- The forcing $Q_n$ is equivalent to Cohen forcing.
- In contrast to the first forcing, a single condition can have many incompatible direct extensions.

Again we show that the $n^{th}$ cell satisfies the Prikry condition.

**Lemma 3.4** $\langle Q_n, \leq_n, \leq_n^* \rangle$ *satisfies the Prikry condition.*

*Proof* It is enough to consider conditions of the form $\langle a, A, f \rangle$, since conditions in $Q_{n1}$ can only be extended to stronger conditions in $Q_{n1}$ and this is a direct extension. Let $\sigma$ be a statement in the forcing language. For $\nu \in A$ let $\langle a, A, f \rangle \frown \nu =_{def} f \cup \{\langle \beta, \pi_{\max(a)\beta}(\nu) \mid \beta \in A\}$. We construct a $\leq_n^*$-increasing sequence of conditions in $Q_{n0}$ of length $\kappa_n$. At the induction step, we find a nondirect extension that decides $\sigma$ and then we extend the Cohen part of our condition to take this nondirect extension into account.

Let $\langle \nu_\eta \mid \eta < \kappa_n \rangle$ be an increasing enumeration of $A$ and let $q(0) =_{def} \langle a, A, f \rangle$. Assume that we have defined $q(\eta) =_{def} \langle a_\eta, A_\eta, f_\eta \rangle$ for some $\eta < \kappa_n$. We choose $p(\eta) \geq_{Q_{n1}} q(\eta) \frown \nu_\eta$ such that $p(\eta) \parallel \sigma$. We want a direct extension of $\langle a_\eta, A_\eta, f_\eta \rangle$ that takes $p(\eta)$ into account. We are going to fix the first two coordinates of $q(\eta)$ and extend the Cohen part. Let $f_{\eta+1} = p(\eta) \restriction (\mathrm{dom}(p(\eta)) \setminus a)$ and set $q(\eta+1) =_{def} \langle a, A, f_{\eta+1} \rangle$. Note that $q(\eta) \frown \nu_\eta = p(\eta)$. This finishes the successor step. Assume that $\gamma < \kappa_n$ is a limit ordinal and that for all $\eta < \gamma$ we have constructed $q(\eta)$. By induction we can assume that the first two coordinates of each $q(\eta)$ are fixed as $a, A$. On the third coordinate we just take the union of our increasing sequence of functions. Let $f_\gamma =_{def} \cup_{\eta<\gamma} f_{\eta+1}$ and define $q(\gamma) =_{def} \langle a, A, f_\gamma \rangle$. This process at limits gives a condition since the size of the union is less than or equal to $\kappa$ which is the allowed size of a condition in the Cohen part. Similar reasoning allows us to take an upper bound $\langle a, A, g \rangle$ for the whole sequence that we built. Also notice that for all $\nu \in A$, $\langle a, A, g \rangle \frown \nu \geq_{Q_n1} p(\nu)$. So for each $\nu$, $\langle a, A, g \rangle \frown \nu \parallel \sigma$.

For the next stage we want to shrink $A$ so that each $\nu$ gives the same decision. We define a partition of $A$ into

$$A_0 = \{\nu \in A \mid \langle a, A, g \rangle \frown \nu \Vdash \sigma\} \text{and}$$
$$A_1 = \{\nu \in A \mid \langle a, A, g \rangle \frown \nu \Vdash \neg\sigma\}.$$

Since $A \in E_{n\,\max(a)}$, we must have $A_0 \in E_{n\,\max(a)}$ or $A_1 \in E_{n\,\max(a)}$. Without loss of generality assume that $A_0 \in E_{n\,\max(a)}$. Then $\langle a, A_0, g \rangle \Vdash \sigma$, since every nondirect extension is below $p(\nu)$ for some $\nu \in A_0$ and each such $p(\nu)$ forces $\sigma$. □

We are now ready to define $\langle P, \leq, \leq^* \rangle$.

**Definition 3.5** $p \in P$ if and only if $p = \langle p_n \mid n < \omega \rangle$ with the following properties.

1. For all $n < \omega$, $p_n \in Q_n$.
2. There is $\ell(p) < \omega$ such that for all $n \geq \ell(p)$ $p_n \in Q_{n0}$ and for all $n < \ell(p)$, $p_n \in Q_{n1}$.
3. If $n \geq \ell(p)$ and $p_n = \langle a_n, A_n, f_n \rangle$, then $m \geq n$ implies $a_m \supseteq a_n$.

Suppose that $p = \langle p_n \mid n < \omega \rangle$ and $q = \langle q_n \mid n < \omega \rangle$ are in $P$. Then $p \geq q$ if and only if for all $n$, $p_n \geq_n q_n$ and $p \,{}^*\!\geq q$ if and only if for all $n$, $p_n \,{}^*\!\geq_n q_n$

This forcing is considerably more complex than the previous one. One way that we see this is that our new forcing only satisfies the $\kappa^{++}$-cc.

**Lemma 3.6** *$\langle P, \leq \rangle$ satisfies $\kappa^{++}$-cc.*

We will only sketch the proof here. Let $\langle p(\alpha) \mid \alpha < \kappa^{++} \rangle$ be a sequence of conditions from $P$. Without loss of generality we can assume that $\ell(p(\alpha))$ is constant for all $\alpha$ with value $\ell$. For $n < \ell$, the forcing is essentially Cohen forcing, so we can assume that the sets $\mathrm{dom}(p(\alpha)_n)$

form a delta system and that the conditions agree on the root. We can also form a delta system out of $\{\bigcup_{n\geq\ell}(a(\alpha)_n \cup f(\alpha)_n) \mid \alpha < \kappa^{++}\}$. With a little more work, we can refine further to find a $\kappa^{++}$-sequence of conditions with the property that any two have a common refinement. In this final argument, we use the fact that $\leq_{E_n}$ is $\kappa_n$-directed and the two lemmas that we omitted that show that certain conditions in the forcing are possible.

We can also show that $\langle P, \leq, \leq^* \rangle$ satisfies the Prikry condition. In fact it satisfies a stronger condition, which allows us to show that $\kappa^+$ is preserved. We omit the statement and proof of the stronger condition and give the basic idea of the proof that $P$ satisfies the Prikry condition.

**Lemma 3.7** *$\langle P, \leq, \leq^* \rangle$ satisfies the Prikry condition.*

The argument goes by combining the ideas from the proofs of Lemma 3.4 and Lemma 2.4. We take ideas from the proof that the first forcing satisfies the Prikry condition and the proof that the $n^{th}$ cell $Q_n$ of the current forcing satisfies the Prikry condition. The proof that the first forcing satisfies the Prikry condition shows us that we need to diagonalize over all possible non-direct extensions by using a direct extension that captures the information from the non-direct extension. The proof that $Q_n$ from the current forcing satisfies the Prikry condition shows us how to capture the information from a non-direct extension without increasing the length of the condition. The proof of the Prikry lemma for this forcing uses the closure of the Cohen conditions and the completeness of the measures heavily.

We now argue that we added $\lambda$-many new $\omega$-sequences. Let $G \subseteq P$ be $V$-generic. Let $n < \omega$ and define a function $F_n : \lambda \to \kappa_n$ by $F_n(\alpha) = \nu$ if and only if there is $p \in G$ such that $\ell(p) > n$, $\alpha \in \mathrm{dom}(p_n)$ and $p_n(\alpha) = \nu$. Note that $F_n$ is a function since $G$ is a filter and $F_n$ is defined on all of $\lambda$ by genericity. Let $t_\alpha = \langle F_n(\alpha) \mid n < \omega \rangle$ for all $\alpha < \lambda$. Then we have $t_\alpha \in \prod \kappa_n$ for all $\alpha < \lambda$. It is possible that for some $\alpha$, $t_\alpha \in V$, since it is possible that $t_\alpha$ is completely determined by a single condition. However the following lemma shows that the set $\{t_\alpha \mid \alpha < \lambda\}$ has size $\lambda$, which is enough to see that we added $\lambda$-many $\omega$-sequences to $\kappa$.

**Lemma 3.8** *Let $\beta < \lambda$. Then there is $\alpha$ with $\beta < \alpha < \lambda$ such that for all $\gamma < \alpha$, $t_\gamma <^* t_\alpha$*

*Proof* Work in $V$ and let $p = \langle p_n \mid n < \omega \rangle$ be a condition with $p_n = \langle a_n, A_n, f_n \rangle$ for $n \geq \ell(p)$. We work to define an extension of $p$ which forces the conclusion for a particular choice of $\alpha$. By the definition of conditions in our poset, we have $|\bigcup_{n<\omega} \mathrm{dom}(f_n)| \leq \kappa$. We choose $\alpha \in \lambda \setminus (\sup(\bigcup_{n<\omega} \mathrm{dom}(f_n)) \cup (\bigcup_{n<\omega} a_n))$. Now by Lemma 3.1, for each $n$ there is $\alpha_n^*$ such that $a_n \cup \{\alpha\} \leq_{E_n} \alpha_n^*$. Let $q \geq p$ be the condition given by taking $p$ and replacing each $a_n$ with $a_n \cup \{\alpha\} \cup \{\alpha_n^*\}$[2]. Let $q_n =_{def} \langle b_n, B_n, g_n \rangle$ for $n \geq \ell(q) = \ell(p)$. We claim that $q$ forces that $t_\gamma <^* t_\alpha$ for all $\gamma < \alpha$. We break the proof into two cases. In the first case we assume that $t_\gamma \in V$. We show that there is a dense set above $q$ which forces $t_\gamma <^* t_\alpha$. Let $r \geq q$ with $r_n = \langle c_n, C_n, h_n \rangle$ for $n \geq \ell(r)$. Define an extension of $r$ by finding a measure one set $D_n \subseteq C_n$ for each $n \geq \ell(r)$ so that for all $\zeta \in \pi_{\max(c_n)\alpha}$"$D_n$, $\zeta > t_\gamma(n) + 1$. Using the definition of the ordering, this extension forces $t_\gamma <^* t_\alpha$. Otherwise, we assume that $t_\gamma \notin V$. Then there is a dense set of $r \geq q$ such that $\gamma \in c_n$ for all $n \geq \ell(r)$ where $r_n = \langle c_n, C_n, h_n \rangle$. Then by condition 4 in the definition of the triple $\langle a, A, f \rangle$ and condition 4 in the definition of $\leq_n$ such a condition $r$ forces that $t_\gamma <^* t_\alpha$. □

## 4 Adding many $\omega$-sequences with short extenders

It is natural to ask whether the long extenders used in the last section are required. In the long extender forcing each $\omega$-sequence is controlled by $\omega$-many measure one sets, one from each extender. It seems natural to use extenders whose length is the number of $\omega$-sequences that we wish to add. In this way we only need to use each measure once. In this section we introduce a forcing that uses shorter extenders. The upshot is that we are required to use one measure from a given extender to control the values of more than one $\omega$-sequence.

We begin by describing a naive attempt to define a forcing that adds $\lambda$-many $\omega$-sequences using short extenders. The forcing resembles the long extender version with a few changes prompted by the discussion in the previous paragraph. It satisfies the strengthening of the Prikry property needed to see that $\kappa^+$ is preserved, however it collapses $\lambda$ to have size $\kappa^+$. More specifically the chain condition argument from before no longer works. We show that one can modify the definition and argue that a subforcing satisfies the $\kappa^{++}$-cc in the case when $\lambda = \kappa^{++}$. The subforcing that we identify is the forcing needed to prove Theorem 1.1.

[2] In order to obtain this condition we use the Lemmas mentioned in Remark 3.3.

## 4.1 A naive attempt

We continue to work with $\kappa$ singular of cofinality $\omega$ with $\langle\kappa_n \mid n < \omega\rangle$ increasing and cofinal in $\kappa$. This time we assume that for each $\kappa_n$ we have an extender of length $\kappa_n^{+n+2}$. In particular we assume that for each $n$ there is an elementary embedding $j_n : V \to M_n$ with $\mathrm{crit}(j_n) = \kappa_n$, ${}^{\kappa_n}M_n \subseteq M_n$, $j_n(\kappa_n) > \kappa_n^{+n+2}$ and $V_{\kappa_n^{+n+2}} \subseteq M_n$. Then we derive our extender as usual $E_{n\alpha} =_{def} \{X \subseteq \kappa_n \mid \alpha \in j_n(X)\}$. We let $\lambda > \kappa^+$ be regular and attempt to define a forcing to add $\lambda$-many cofinal $\omega$-sequences to $\kappa$.

We proceed as before by defining a cell for each $n$ and then putting them together. The definition of $Q_{n1}$ will be the same and the definition of $Q_{n0}$ will be slightly different.

**Definition 4.1** Fixing $n < \omega$, we define $Q_{n1} =_{def} \{f \mid f$ is a partial function from $\lambda$ to $\kappa_n$ with $|f| \leq \kappa\}$ and order it by extension, which we call $\leq_{Q_{n1}}$. Let $\langle a, A, f\rangle \in Q_{n0}$ if and only if

1. $f \in Q_{n1}$,
2. $a$ is a partial order preserving function from $\lambda$ to $\kappa_n^{+n+2}$ with $|a| < \kappa_n$ such that
   a. $\mathrm{dom}(a)$ has a maximal element in the ordinal sense,
   b. $a(\max(\mathrm{dom}(a))) = \max(\mathrm{rng}(a))$ and
   c. for all $\beta \in \mathrm{rng}(a)$, $\beta \leq_{E_n} a(\max(\mathrm{dom}(a)))$,
3. $A \in E_{na(\max(\mathrm{dom}(a)))}$,
4. If $\alpha > \beta$ and $\alpha, \beta \in \mathrm{dom}(a)$, then for all $\nu \in A$, $\pi_{a(\max(\mathrm{dom}(a)))a(\alpha)}(\nu) > \pi_{a(\max(\mathrm{dom}(a)))a(\beta)}(\nu)$and
5. if $\alpha, \beta, \gamma \in \mathrm{dom}(a)$ with $a(\alpha) \geq_{E_n} a(\beta) \geq_{E_n} a(\gamma)$, then for all $\rho \in \pi_{a(\max(\mathrm{dom}(a)))a(\alpha)}``A$, $\pi_{a(\alpha)a(\gamma)}(\rho) = \pi_{a(\beta)a(\gamma)}(\pi_{a(\alpha)a(\beta)}(\rho))$.

We define $\langle b, B, g\rangle \leq_{Q_{n0}} \langle a, A, f\rangle$ if and only if

1. $g \leq_{Q_{n1}} f$,
2. $b \subseteq a$ and
3. $\pi_{a(\max(\mathrm{dom}(a)))b(\max(\mathrm{dom}(b)))}``A \subseteq B$.

Let $\leq^*_{Q_n} =_{def} \leq_{Q_{n0}} \cup \leq_{Q_{n1}}$. Define $p \leq_{Q_n} q$ if and only if

1. $p \leq^*_{Q_n} q$ or
2. $p = \langle a, A, f\rangle \in Q_{n0}, q \in Q_{n1}$ and
   a. $f \leq_{Q_{n1}} q$,
   b. $\mathrm{dom}(a) \subseteq \mathrm{dom}(q)$,
   c. $q(\max(\mathrm{dom}(a))) \in A$ and
   d. for all $\beta \in \mathrm{dom}(a)$, $q(\beta) = \pi_{a(\max(\mathrm{dom}(a)))a(\beta)}(q(\max(\mathrm{dom}(a))))$.

*Remark* 4.2 As in the long extender forcing, we omit Lemmas which show that (4) and (5) in the definition of $Q_{n0}$ are possible.

The definition of the forcing $P$ from the cells $Q_n$ is exactly the same as in the long extender forcing. As we mentioned in the introduction to this section, our new forcing $P$ satisfies the Prikry condition and therefore adds no new bounded subsets to $\kappa$. The proof that $P$ satisfies the Prikry condition is similar to the proof for the long extender forcing. The key point is to deal with the fact that in the $n^{th}$ cell we have replaced the less than $\kappa_n$ sized subset of $\lambda$ with a partial order preserving function from $\lambda$ to $\kappa_n^{+n+2}$. In fact $P$ also satisfies the strengthening of the Prikry condition need to see that $\kappa^+$ is preserved.

**Theorem 4.3** *Forcing with $(P, \leq)$ preserves cardinals less than or equal to $\kappa^+$.*

The same chain condition argument from last time no longer works. It is instructive to see exactly what goes wrong. Let's say that we proceed as before and attempt to prove that the forcing is $\kappa^{++}$-cc. Given a $\kappa^{++}$-sequence of conditions, we can fix the length of the stem of all of the conditions. Again below the length of the stem we are essentially doing Cohen forcing, so this poses no problem. Working above the length of the stem, we form a $\Delta$-system out of the domains of our partial functions, the objects corresponding to $\mathrm{dom}(a) \cup \mathrm{dom}(f)$ in each condition. We would like to be able to amalgamate two functions $a$ and $b$ with disjoint domains. However, $a \cup b$ need not be an order preserving function. In fact we may assume that $\mathrm{otp}(\mathrm{dom}(a)) = \mathrm{otp}(\mathrm{dom}(b))$ and for all $i < \mathrm{otp}(\mathrm{dom}(a))$, $a(\alpha) = b(\beta)$ where $\alpha, \beta$ are the $i^{th}$ members of $\mathrm{dom}(a), \mathrm{dom}(b)$ respectively. Pictorially, we are in the situation shown in Figure 8.3.

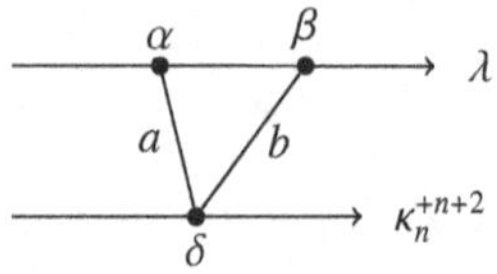


Figure 8.3

So our argument cannot proceed. In order to recover the chain condition, we want to find a $\delta' > \delta$ that is similar in some sense to $\delta$, so that we can instead map $\beta$ to $\delta'$. Note that our approach changes $b$ to a similar function, say $b'$. Again pictorially we have the situation shown in Figure 8.4.

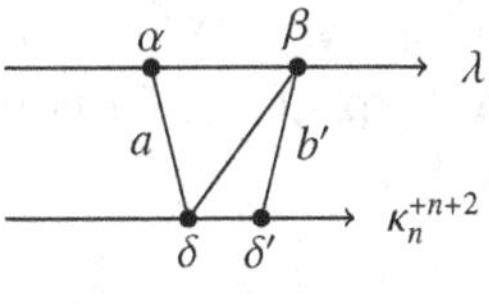


Figure 8.4

The similarity between $b$ and $b'$ induces an equivalence relation on conditions in a modified version of $P$. We'll use this equivalence relation to define a projection of the modified version of $P$. The projected forcing will be the forcing required for Theorem 1.1.

## 4.2 Gap 2: $\lambda = \kappa^{++}$

In this section we work towards the final definition of the forcing to add $\kappa^{++}$-many cofinal $\omega$-sequences to $\kappa$. We begin by clarifying the notion of similarity mentioned above, because it motivates the modified definition of the forcing and it will give us the projection map.

Let $n, k < \omega$ with $1 < k \leq n$ and define

$$\mathcal{A}_{n,k} =_{def} \langle H(\chi^{+k}), \in, <, \chi, E_n, \langle 0, 1, \ldots \tau, \ldots \mid \tau \leq \kappa_n^{+k} \rangle \rangle$$

where $\chi$ is a large regular cardinal, $<$ is a well-ordering of $H(\chi^{+k})$ and all of the other parameters mentioned are constants interpreted in the natural way. Let $\mathcal{L}_{n,k}$ be the language of $\mathcal{A}_{n,k}$. Let $\xi < \kappa_n^{+n+2}$ and define $\mathrm{tp}_{n,k}(\xi)$ be the type realized by $\xi$ in the model $\mathcal{A}_{n,k}$.

We are actually interested in a slight expansion of the above language. We choose this presentation to highlight a specific constant. Let $\mathcal{L}^*_{n,k} =_{def} \mathcal{L}_{n,k} \cup \{c\}$ where $c$ is a new constant. For a given ordinal $\delta$, let $\mathcal{A}^{\delta}_{n,k}$ be the expansion of $\mathcal{A}_{n,k}$ to $\mathcal{L}^*_{n,k}$ where $c$ is interpreted as $\delta$. Let $\mathrm{tp}_{n,k}(\delta, \xi)$ be the type realized by $\xi$ in $\mathcal{A}^{\delta}_{n,k}$.

**Lemma 4.4** *For a given $n < \omega$, the set $C =_{def} \{\beta < \kappa_n^{+n+2} \mid \forall \gamma < \beta\ \mathrm{tp}_{n,n}(\gamma, \beta)$ is realized stationarily often below $\kappa_n^{+n+2}\}$ contains a club.*

*Proof* Suppose for a contradiction that $S =_{def} \kappa_n^{+n+2} \setminus C = \{\beta < \kappa_n^{+n+2} \mid \exists \gamma < \beta\ \mathrm{tp}_{n,n}(\gamma, \beta)$ is not realized stationarily often below $\kappa_n^{+n+2}\}$ is stationary. By Fodor's Lemma there are a stationary set $S^* \subseteq S$ and an ordinal $\gamma^*$ such that for all $\beta \in S^*$, $\text{tp}_{n,n}(\gamma^*,\beta)$ is not realized stationarily often. A routine counting argument shows that there are only $\kappa_n^{+n+1}$-many possible types that each $\text{tp}_{n,n}(\gamma^*,\beta)$ could be. It follows that $\beta \mapsto \text{tp}_{n,n}(\gamma^*,\beta)$ is constant on a stationary set $S^{**} \subseteq S^*$. Let $\beta^* \in S^{**}$ be least. Then by the choice of $S^{**}$, $\text{tp}_{n,n}(\gamma^*,(\beta^*)$ is realized stationarily often. This contradicts that $\beta^* \in S^*$. □

**Definition 4.5** Let $n, k < \omega$ with $1 < k \leq n$ and $\beta < \kappa_n^{+n+2}$. $\beta$ is *k-good if and only if*

1. $\text{cf}(\beta) \geq \kappa_n^{++}$ *and*
2. *for all* $\gamma < \beta$, $\text{tp}_{n,k}(\gamma,\beta)$ *is realized stationarily often below* $\kappa_n^{+n+2}$.

We are now ready to modify our forcing $P$ from above. In addition to previous properties demanded of $a_n$ we require the following,

1. $a_n : \kappa^{++} \to \kappa_n^{+n+2}$ and for all $\alpha \in \text{dom}(a_n)$, $a_n(\alpha)$ is at least 2-good and
2. if for some $p = \langle p_n \mid n < \omega \rangle$ and $\alpha < \kappa^{++}$ there is $i \geq \ell(p)$ such that $\alpha \in \text{dom}(a_i)$ where $p_i = \langle a_i, A_i, f_i \rangle$, then there is a nondecreasing sequence $\langle k_m \mid i \leq m < \omega \rangle$ such that $k_m \to \infty$ as $m \to \infty$ and for every $m \geq i$, $a_m(\alpha)$ is $k_m$-good.

For ease of notation we will call this modified forcing $P$ as well. For this modified forcing we have the same theorems that we had before. In particular $P$ adds $\kappa^{++}$ many $\omega$-sequences to $\kappa$ and preserves cardinals less than or equal to $\kappa^+$.

We work towards the definition of our equivalence relation.

**Definition 4.6** Let $n, k < \omega$ with $1 < k \leq n$ and $\langle a, A, f \rangle, \langle b, B, g \rangle \in Q_{n0}$. Define $\langle a, A, f \rangle \leftrightarrow_{n,k} \langle b, B, g \rangle$ if and only if

1. $f = g$,
2. $A = B$,
3. $\text{dom}(a) = \text{dom}(b)$ and
4. $\text{rng}(a), \text{rng}(b)$ realize the same $\mathcal{A}_{n,k}$-type.

Using this definition we are ready to give the definition of the equivalence relation.

**Definition 4.7** Let $p = \langle p_n \mid n < \omega \rangle$ and $q = \langle q_n \mid n < \omega \rangle$ be members of $P$. Define $p \leftrightarrow q$ if and only if

1. $\ell(p) = \ell(q)$,
2. for all $n < \ell(p)$, $p_n = q_n$ and

3. there is a nondecreasing sequence $\langle k_m \mid \ell(p) \leq m < \omega \rangle$ such that $k_m \to \infty$ as $m \to \infty$ and for all $m \geq \ell(p)$, $p_m \leftrightarrow_{m,k_m} q_m$.

It is easy to see that this is an equivalence relation. What is more interesting is that it works well with the definition of the ordering on $P$.

**Lemma 4.8** *If $p, s, t \in P$ with $s \geq p \leftrightarrow t$, then there are $s' \geq s$, $t' \geq t$ such that $s' \leftrightarrow t'$.*

We can factor using this relation and get something nice, however we are going to explicitly describe the order that we will force with.

**Definition 4.9** Let $p, q \in P$. Define $p \to q$ if and only if there is an $m < \omega$ and a sequence of elements $\langle r_i \mid i < m \rangle$ such that

1. $r_0 = p$,
2. $r_{m-1} = q$ and
3. for each $i < m - 1$, either $r_i \leq r_{i+1}$ or $r_i \leftrightarrow r_{i+1}$.

Since both $\leq$ and $\leftrightarrow$ are transitive we can assume that use of $\leq$ and $\leftrightarrow$ alternates along the sequence $\langle r_i \mid i < m \rangle$. Using Lemma 4.8 we can prove a nice fact about the interaction between $\leq$ and $\to$.

**Lemma 4.10** *If $p \to q$, then there is $s \geq p$ such that $q \to s$*

Lemma 4.10 is precisely what we need to show that the identity map from $\langle P, \leq \rangle$ to $\langle P, \to \rangle$ is a projection. We will now sketch the proof that $\langle P, \to \rangle$ has good chain condition.

**Lemma 4.11** *$\langle P, \to \rangle$ satisfies $\kappa^{++}$-cc*

*Sketch* Let $\langle p(\alpha) \mid \alpha < \kappa^{++} \rangle$ be a sequence of conditions in $P$. We sketch the construction of a pair of conditions $q(\alpha) \geq p(\alpha)$ and $q(\beta) \geq p(\beta)$ such that $q(\alpha) \leftrightarrow q(\beta)$. This gives a contradiction since it follows that $p(\alpha), p(\beta)$ are compatible under $\to$. We may assume that for all $\alpha, \beta < \kappa^{++}$, $\ell(p(\alpha)) = \ell(p(\beta)) =_{def} l$ and for all $\alpha, \beta < \kappa^{++}$ and $n < l$ that $p(\alpha)_n \cup p(\beta)_n$ is a function.

What about above $l$? For $n \geq l$ let $p(\alpha)_n = \langle a(\alpha)_n, A(\alpha)_n, f(\alpha)_n \rangle$. We can assume that for all $n \geq l$, the sets $\mathrm{dom}(a(\alpha)_n) \cup \mathrm{dom}(f(\alpha)_n)$ form a $\Delta$-system where each function takes the same values on the kernel. We assume that for every $\alpha, \beta < \kappa^{++}$, $\mathrm{rng}(a(\alpha)_n) = \mathrm{rng}(a(\beta)_n)$. Now we have the situation shown in Figure 8.5.

For simplicity we assume that $\gamma = \min(\bigcup_{m \geq l} \mathrm{dom}(a(\beta)_m) \setminus (\mathit{kernel}))$ and $\gamma \in \mathrm{dom}(a(\beta)_n)$. From the picture we have $(a(\alpha)_n)(\delta) = (a(\beta)_n)(\gamma) = \rho$. We assume that $\rho$ is 5-good. Now we pick some $\rho' > \bigcup_{n \geq l} \mathrm{rng}(a(\alpha)_n)$

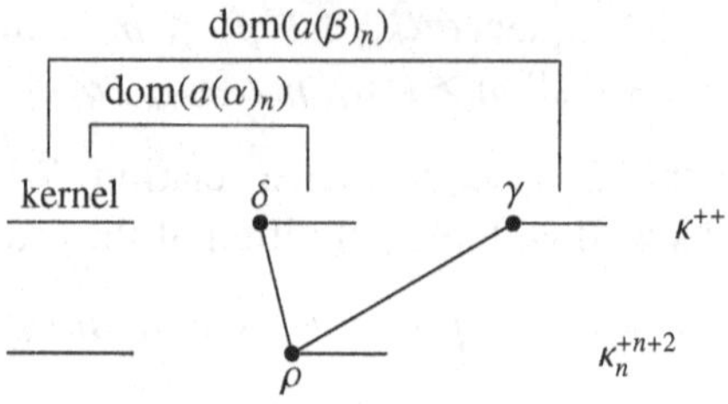


Figure 8.5

such that $\rho'$ realizes the same type over the image of the kernel as $\rho$ does. Here we are viewing the image of the kernel as coded by a single ordinal and taking types over it as discussed above.

We can now construct $q(\alpha)$. The key point is to construct the order preserving function part of each $q(\alpha)_n$. For this we take $a(\alpha)_n$ and add the off kernel part of $\mathrm{dom}(a(\beta)_n)$ to the domain and map this to a block that sits above $\rho'$ and realizes the same 3-type as $a(\alpha)_n$"$(\mathrm{dom}(a(\alpha)_n) \setminus (\mathit{kernel}))$. Figure 8.6 illustrates the order preserving function that results.

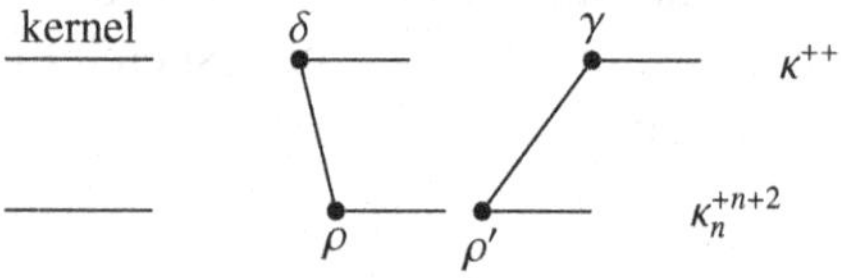


Figure 8.6

The construction of $q(\beta)$ is similar, but uses a fact that we have omitted. Namely if $\xi$ is $k$-good, then there are unboundedly many $k-1$-good ordinals less than $\xi$. We use this to choose an $\eta < \rho$ (and above the image of the kernel) so that there is a block above $\eta$ that realizes the same 3-type as the block beginning at $\rho$. We then add the off kernel part of $\mathrm{dom}(a(\alpha)_n)$ to the order preserving function $a(\beta)_n$ and map it to the block above $\eta$. The picture in Figure 8.7 is helpful.

It follows that $q(\alpha)_n \leftrightarrow_{n,3} q(\beta)_n$. Working in a similar fashion with all $m \geq n$ we can obtain $q(\alpha) \leftrightarrow q(\beta)$. This finishes our sketch of the proof. □

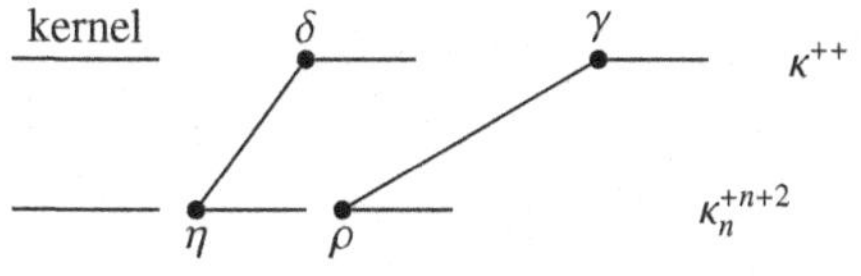


Figure 8.7

# References

[1] Moti Gitik. Short extender forcings 1. `http://www.math.tau.ac.il/~gitik/somepapers.html`.

[2] Moti Gitik. Blowing up the power of a singular cardinal. *Annals of Pure and Applied Logic*, 80(1):17 – 33, 1996.

[3] Moti Gitik. Blowing up power of a singular cardinal–wider gaps. *Annals of Pure and Applied Logic*, 116(1-3):1 – 38, 2002.

[4] Moti Gitik. Prikry-type forcings. In Matthew Foreman and Akihiro Kanamori, editors, *Handbook of Set Theory*, pages 1351–1447. Springer Netherlands, 2010.

[5] Moti Gitik and Menachem Magidor. Extender based forcings. *J. Symb. Logic*, 59:445–460, June 1994.

[6] Carmi Merimovich. The short extenders gap two forcing is of prikry type. *Archive for Mathematical Logic*, 48:737–747, 2009. 10.1007/s00153-009-0147-1.

[7] K. L. Prikry. Changing measurable into accessible cardinals. *Dissertationes Math. Rozprawy Mat.*, 68:55, 1970.

# 9

# The complexity of classification problems in ergodic theory

Alexander S. Kechris and Robin D. Tucker-Drob

---

The twelfth Appalachian Set Theory workshop was held at Vanderbilt University in Nashville on October 30, 2010. The lecturer was Alexander S. Kechris. As a graduate student Robin D. Tucker-Drob assisted in writing this chapter, which is based on the workshop lectures.

---

*Dedicated to the memory of Greg Hjorth (1963-2011)*

The last two decades have seen the emergence of a theory of set theoretic complexity of classification problems in mathematics. In these lectures we will discuss recent developments concerning the application of this theory to classification problems in ergodic theory.

The first lecture will be devoted to a general introduction to this area. The next two lectures will give the basics of Hjorth's theory of turbulence, a mixture of topological dynamics and descriptive set theory, which is a basic tool for proving strong non-classification theorems in various areas of mathematics.

In the last three lectures, we will show how these ideas can be applied in proving a strong non-classification theorem for orbit equivalence. Given a countable group $\Gamma$, two free, measure-preserving, ergodic actions of $\Gamma$ on standard probability spaces are called *orbit equivalent* if, roughly speaking, they have the same orbit spaces. More precisely this means that there is an isomorphism of the underlying measure spaces that takes the orbits of one action to the orbits of the other. A remarkable result of Dye and Ornstein-Weiss asserts that any two such actions of amenable groups are orbit equivalent. Our goal will be to outline a proof of a dichotomy theorem which states that for any non-amenable group, we have the opposite situation: The structure of its actions up to orbit equivalence is so complex that it is impossible, in a very strong sense, to classify them (Epstein-Ioana-Kechris-Tsankov).

Beyond the method of turbulence, an interesting aspect of this proof is the use of many diverse of tools from ergodic theory. These include: unitary representations and their associated Gaussian actions; rigidity properties of the action of $SL_2(\mathbb{Z})$ on the torus and separability arguments (Popa, Gaboriau-Popa, Ioana), Epstein's co-inducing construction for generating actions of a group from actions of another, quantitative aspects of inclusions of equivalence relations (Ioana-Kechris-Tsankov) and the use of percolation on Cayley graphs of groups and the theory of costs in proving a measure theoretic analog of the von Neumann Conjecture, concerning the "inclusion" of free groups in non-amenable ones (Gaboriau-Lyons). Most of these tools will be introduced as needed along the way and no prior knowledge of them is required.

*Acknowledgment.* Work in this paper was partially supported by NSF Grant DMS-0968710. We would like to thank Ernest Schimmerling and Greg Hjorth for many valuable comments on an earlier draft of this paper.

# Lecture I. A Survey

## Classification problems in ergodic theory

**Definition 1.1** A *standard measure space* is a measure space $(X, \mu)$, where $X$ is a standard Borel space and $\mu$ a non-atomic Borel probability measure on $X$.

All such spaces are isomorphic to the unit interval with Lebesgue measure.

**Definition 1.2** A *measure-preserving transformation* (mpt) on $(X, \mu)$ is a measurable bijection $T$ such that $\mu(T(A)) = \mu(A)$, for any Borel set $A$.

**Example 1.3**

- $X = \mathbb{T}$ with the usual measure; $T(z) = az$, where $a \in \mathbb{T}$, i.e., $T$ is a *rotation*.
- $X = 2^{\mathbb{Z}}, T(x)(n) = x(n-1)$, i.e., the *shift transformation*.

**Definition 1.4** A mpt $T$ is *ergodic* if every $T$-invariant measurable set has measure 0 or 1.

Any irrational, modulo $\pi$, rotation and the shift are ergodic. The *ergodic decomposition theorem* shows that every mpt can be canonically decomposed into a (generally continuous) direct sum of ergodic mpts.

In ergodic theory one is interested in classifying ergodic mpts up to various notions of equivalence. We will consider below two such standard notions.

- *Isomorphism or conjugacy*: A mpt $S$ on $(X, \mu)$ is isomorphic to a mpt $T$ on $(Y, \nu)$, in symbols $S \cong T$, if there is an isomorphism $\varphi$ of $(X, \mu)$ to $(Y, \nu)$ that sends $S$ to $T$, i.e., $S = \varphi^{-1} T \varphi$.
- Unitary isomorphism: To each mpt $T$ on $(X, \mu)$ we can assign the unitary (Koopman) operator $U_T : L^2(X, \mu) \to L^2(X, \mu)$ given by $U_T(f)(x) = f(T^{-1}(x))$. Then $S, T$ are unitarily isomorphic, in symbols $S \cong^u T$, if $U_S, U_T$ are isomorphic.

Clearly $\cong$ implies $\cong^u$ but the converse fails.

We state two classical classification theorems:

- (Halmos-von Neumann [HvN42]) An ergodic mpt has *discrete spectrum* if $U_T$ has discrete spectrum, i.e., there is a basis consisting of eigenvectors. In this case the eigenvalues are simple and form a (countable) subgroup of $\mathbb{T}$. It turns out that up to isomorphism these are exactly the ergodic rotations in compact metric groups $G$ : $T(g) = ag$, where $a \in G$ is such that $\{a^n : n \in \mathbb{Z}\}$ is dense in $G$. For such $T$, let $\Gamma_T \leq \mathbb{T}$ be its group of eigenvalues. Then we have:

  $$S \cong T \Leftrightarrow S \cong^u T \Leftrightarrow \Gamma_S = \Gamma_T.$$

- (Ornstein [Orn70]) Let $Y = \{1, \ldots, n\}$, $\bar{p} = (p_1, \cdots, p_n)$ a probability distribution on $Y$ and form the product space $X = Y^{\mathbb{Z}}$ with the product measure $\mu$. Consider the *Bernoulli shift* $T_{\bar{p}}$ on $X$. Its *entropy* is the real number $H(\bar{p}) = -\sum_i p_i \log p_i$. Then we have:

  $$T_{\bar{p}} \cong T_{\bar{q}} \Leftrightarrow H(\bar{p}) = H(\bar{q})$$

  (but all the shifts are unitarily isomorphic).

We will now consider the following question: Is it possible to classify, in any reasonable way, general ergodic mpts?

We will see how ideas from descriptive set theory can throw some light on this question.

## Complexity of classification

We will next give an introduction to recent work in set theory, developed primarily over the last two decades, concerning a theory of complexity of classification problems in mathematics, and then discuss its implications to the above problems.

A classification problem is given by:

- A collection of objects $X$.
- An equivalence relation $E$ on $X$.

A *complete classification* of $X$ up to $E$ consists of:

- A set of invariants $I$.
- A map $c : X \to I$ such that $xEy \Leftrightarrow c(x) = c(y)$.

For this to be of any interest both $I, c$ must be as explicit and concrete as possible.

**Example 1.5** Classification of Bernoulli shifts up to isomorphism (Ornstein).

INVARIANTS: Reals.

**Example 1.6** Classification of ergodic measure-preserving transformations with discrete spectrum up to isomorphism (Halmos-von Neumann).

INVARIANTS: Countable subsets of $\mathbb{T}$.

**Example 1.7** Classification of unitary operators on a separable Hilbert space up to isomorphism (Spectral Theorem).

INVARIANTS: Measure classes, i.e., probability Borel measures on a Polish space up to measure equivalence.

Most often the collection of objects we try to classify can be viewed as forming a "nice" space, namely a standard Borel space, and the equivalence relation $E$ turns out to be *Borel* or *analytic* (as a subset of $X^2$).

For example, in studying mpts the appropriate space is the Polish group $\mathrm{Aut}(X, \mu)$ of mpts of a fixed $(X, \mu)$, with the so-called weak topology. (As usual, we identify two mpts if they agree a.e.) Isomorphism then corresponds to conjugacy in that group, which is an analytic equivalence relation. Similarly unitary isomorphism is an analytic equivalence relation (in fact, it is Borel, using the Spectral Theorem). The ergodic mpts form a $G_\delta$ set in $\mathrm{Aut}(X, \mu)$.

The theory of equivalence relations studies the set-theoretic nature of possible (complete) invariants and develops a mathematical framework for measuring the complexity of classification problems.

The following simple concept is basic in organizing this study.

**Definition 1.8** Let $(X, E)$, $(Y, F)$ be equivalence relations. $E$ is *(Borel) reducible* to $F$, in symbols

$$E \leq_B F,$$

if there is Borel map $f : X \to Y$ such that

$$xEy \Leftrightarrow f(x)Ff(y).$$

Intuitively this means:

- The classification problem represented by $E$ is at most as complicated as that of $F$.
- $F$-classes are complete invariants for $E$.

**Definition 1.9** $E$ is *(Borel) bi-reducible* to $F$ if $E$ is reducible to $F$ and vice versa:

$$E \sim_B F \Leftrightarrow E \leq_B F \text{ and } F \leq_B E.$$

We also put:

**Definition 1.10**

$$E <_B F \Leftrightarrow E \leq_B F \text{ and } F \not\leq_B E.$$

**Example 1.11** (Isomorphism of Bernoulli shifts) $\sim_B$ $(=_{\mathbb{R}})$

**Example 1.12** (Isomorphism of ergodic discrete spectrum mpts)$\sim_B$ $E_c$, where $E_c$ is the equivalence relation on $\mathbb{T}^{\mathbb{N}}$ given by

$$(x_n) \; E_c \; (y_n) \Leftrightarrow \{x_n : n \in \mathbb{N}\} = \{y_n : n \in \mathbb{N}\}.$$

**Example 1.13** (Isomorphism of unitary operators)$\sim_B$ ME, where ME is the equivalence relation on the Polish space of probability Borel measures on $\mathbb{T}$ given by

$$\mu \text{ ME } \nu \Leftrightarrow \mu \ll \nu \text{ and } \mu \ll \nu.$$

The preceding concepts can be also interpreted as the basis of a "definable" or Borel cardinality theory for quotient spaces.

- $E \leq_B F$ means that there is a Borel injection of $X/E$ into $Y/F$, i.e., $X/E$ has Borel cardinality less than or equal to that of $Y/F$, in symbols

$$|X/E|_B \leq |Y/F|_B$$

(A map $f : X/E \to Y/F$ is called Borel if it has a Borel *lifting* $f^* : X \to Y$, i.e., $f([x]_E) = [f^*(x)]_F$.)

- $E \sim_B F$ means that $X/E$ and $Y/F$ have the same Borel cardinality, in symbols
$$|X/E|_B = |Y/F|_B$$
- $E <_B F$ means that $X/E$ has strictly smaller Borel cardinality than $Y/F$, in symbols
$$|X/E|_B < |Y/F|_B$$

## Non-classification results for isomorphism and unitary isomorphism

**Definition 1.14** An equivalence relation $E$ on $X$ is called *concretely classifiable* if $E \leq_B (=_Y)$, for some Polish space $Y$, i.e., there is a Borel map $f : X \to Y$ such that $xEy \Leftrightarrow f(x) = f(y)$.

Thus isomorphism of Bernoulli shifts is concretely classifiable. However in the 1970's Feldman showed that this fails for arbitrary mpts (in fact even for the so-called K-automorphisms, a more general class of mpts than Bernoulli shifts).

**Theorem 1.15** (Feldman [Fel74]) *Isomorphism of ergodic mpts is not concretely classifiable.*

One can also see that isomorphism of ergodic discrete spectrum mpts is not concretely classifiable.

An equivalence relation is called *classifiable by countable structures* if it can be Borel reduced to isomorphism of countable structures (of some given type, e.g., groups, graphs, linear orderings, etc.). More precisely, given a countable language $L$, denote by $X_L$ the space of $L$-structures with universe $\mathbb{N}$. This is a Polish space. Denote by $\cong$ the equivalence relation of isomorphism in $X_L$. We say that an equivalence relation is classifiable by countable structures if it is Borel reducible to isomorphism on $X_L$, for some $L$.

Such types of classification occur often, for example, in operator algebras, topological dynamics, etc.

It follows from the Halmos-von Neumann theorem that isomorphism (and unitary isomorphism) of ergodic discrete spectrum mpts is classifiable by countable structures. On the other hand we have:

**Theorem 1.16** (Kechris-Sofronidis [KS01]) ME *is not classifiable by countable structures and thus isomorphism of unitary operators is not classifiable by countable structures.*

**Theorem 1.17** (Hjorth [Hjo01]) *Isomorphism and unitary isomorphism of ergodic mpts cannot be classified by countable structures.*

This has more recently been strengthened as follows:

**Theorem 1.18** (Foreman-Weiss [FW04]) *Isomorphism and unitary isomorphism of ergodic mpts cannot be classified by countable structures on any generic class of ergodic mpts.*

One can now in fact calculate the exact complexity of unitary isomorphism.

**Theorem 1.19** (Kechris [Kec10])

*i) Unitary isomorphism of ergodic mpts is Borel bireducible to measure equivalence.*

*ii) Measure equivalence is Borel reducible to isomorphism of ergodic mpts.*

While isomorphism of ergodic mpts is clearly analytic, Foreman-Rudolph-Weiss also showed the following:

**Theorem 1.20** (Foreman-Rudolph-Weiss [FRW06]) *Isomorphism of ergodic mpts is not Borel.*

However recall that unitary isomorphism of mpts is Borel.

It follows from the last two theorems that

$$(\cong^u) <_B (\cong),$$

i.e., isomorphism of ergodic mpts is strictly more complicated than unitary isomorphism.

We have now seen that the complexity of unitary isomorphism of ergodic mpts can be calculated exactly and there are very strong lower bounds for isomorphism but its exact complexity is unknown. An obvious upper bound is the universal equivalence relation induced by a Borel action of the automorphism group of the measure space (see [BK96] for this concept).

**Problem 1.21** Is isomorphism of ergodic mpts Borel bireducible to the universal equivalence relation induced by a Borel action of the automorphism group of the measure space?

More generally one also considers in ergodic theory the problem of classifying measure-preserving actions of countable (discrete) groups $\Gamma$ on standard measure spaces. The case $\Gamma = \mathbb{Z}$ corresponds to the case of single transformations. We will now look at this problem from the point of view of the preceding theory.

We will consider again isomorphism (also called conjugacy) and unitary isomorphism of actions. Two actions of the group $\Gamma$ are isomorphic if there is a measure-preserving isomorphism of the underlying spaces that conjugates the actions. They are unitarily isomorphic if the corresponding unitary representations (the Koopman representations) are isomorphic.

We can form again in a canonical way a Polish space $A(\Gamma, X, \mu)$ of all measure-preserving actions of $\Gamma$ on $(X, \mu)$, in which the ergodic actions form again a $G_\delta$ subset of $A(\Gamma, X, \mu)$, and then isomorphism and unitary isomorphism become analytic equivalence relations on this space. We can therefore study their complexity using the concepts introduced earlier.

**Theorem 1.22** (Foreman-Weiss [FW04], Hjorth [Hjo97]) *For any infinite countable group $\Gamma$, isomorphism of free, ergodic, measure-preserving actions of $\Gamma$ is not classifiable by countable structures.*

**Theorem 1.23** (Kechris [Kec10]) *For any infinite countable group $\Gamma$, unitary isomorphism of free, ergodic, measure-preserving actions of $\Gamma$ is not classifiable by countable structures.*

Recall that an action $(\gamma, x) \mapsto \gamma \cdot x$ is *free* if for any $\gamma \in \Gamma \setminus \{1\}$, $\gamma \cdot x \neq x$, a.e.

Except for abelian $\Gamma$, where we have the same picture as for $\mathbb{Z}$, it is unknown however how isomorphism and unitary isomorphism relations compare with ME. However Hjorth and Tornquist have recently shown that unitary isomorphism is a Borel equivalence relation. Finally, it is again not known what is the precise complexity of these two equivalence relations. Is isomorphism Borel bireducible to the universal equivalence relation induced by a Borel action of the automorphism group of the measure space?

## Non-classification of orbit equivalence

There is an additional important concept of equivalence between actions, called orbit equivalence. The study of orbit equivalence is a very active area today that has its origins in the connections between ergodic theory and operator algebras and the pioneering work of Dye.

**Definition 1.24** Given an action of the group $\Gamma$ on $X$ we associate to it the orbit equivalence relation $E_\Gamma^X$, whose classes are the orbits of the action. Given measure-preserving actions of two groups $\Gamma$ and $\Delta$ on spaces $(X,\mu)$ and $(Y,\nu)$, resp., we say that they are *orbit equivalent* if there is an isomorphism of the underlying measure spaces that sends $E_\Gamma^X$ to $E_\Delta^Y$ (neglecting null sets as usual).

Thus isomorphism clearly implies orbit equivalence but not vice versa. Here we have the following classical result.

**Theorem 1.25** (Dye [Dye59, Dye63], Ornstein-Weiss [OW80]) *Every two free, ergodic, measure-preserving actions of amenable groups are orbit equivalent.*

Thus there is a single orbit equivalence class in the space of free, ergodic, measure-preserving actions of an amenable group $\Gamma$.

The situation for non-amenable groups has taken much longer to untangle. For simplicity, below "action" will mean "free, ergodic, measure-preserving action." Schmidt [Sch81], showed that every non-amenable group which does not have Kazhdan's property (T) admits at least two non-orbit equivalent actions and Hjorth [Hjo05] showed that every non-amenable group with property (T) has continuum many non-orbit equivalent actions. So every non-amenable group has at least two non-orbit equivalent actions.

For general non-amenable groups though very little was known about the question of how many non-orbit equivalent actions they might have. For example, until recently only finitely many distinct examples of non-orbit equivalent actions of the free, non-abelian groups were known. Gaboriau-Popa [GP05] finally showed that the free non-abelian groups have continuum many non-orbit equivalent actions (for an alternative treatment see Tornquist [Tor06] and the exposition in Hjorth [Hjo09]). In an important extension, Ioana [Ioa11] showed that every group that contains a free, non-abelian subgroup has continuum many such actions. However there are examples of non-amenable groups that contain no free, non-abelian subgroups (Ol'shanski [Ol'80]).

Finally, the question was completely resolved by Epstein.

**Theorem 1.26** (Epstein [Eps08]) *Every non-amenable group admits* *continuum many non-orbit equivalent free, ergodic, measure-preserving actions.*

This still leaves open however the possibility that there may be a concrete classification of actions of some non-amenable groups up to orbit equivalence. However the following has been now proved by combining very recent work of Ioana-Kechris-Tsankov and the work of Epstein.

**Theorem 1.27** (Epstein-Ioana-Kechris-Tsankov [IKT09]) *Orbit equivalence of free, ergodic, measure-preserving actions of any non-amenable group is not classifiable by countable structures.*

Thus we have a very strong dichotomy:

- If a group is amenable, it has exactly one action up to orbit equivalence.
- If it non-amenable, then orbit equivalence of its actions is unclassifiable in a strong sense.

In the rest of these lectures, we will give an outline of the proof of Theorem 1.27.

# Lecture II. Turbulence and classification by countable structures

## The space of countable structures

**Definition 2.1** A countable *signature* is a countable family

$$L = \{f_i\}_{i \in I} \cup \{R_j\}_{j \in J}$$

of *function symbols* $f_i$, with $f_i$ of arity $n_i \geq 0$, and *relation symbols* $R_j$, with $R_j$ of arity $m_j \geq 1$. A *structure* for $L$ or *L-structure* has the form

$$\mathcal{A} = \langle A, \{f_i^{\mathcal{A}}\}_{i \in I}, \{R_j^{\mathcal{A}}\}_{j \in J} \rangle,$$

where $A$ is a nonempty set, $f_i^{\mathcal{A}} : A^{n_i} \to A$, and $R_j^{\mathcal{A}} \subseteq A^{m_j}$.

**Example 2.2** If $L = \{\cdot, 1\}$, where $\cdot$ and 1 are binary and nullary function symbols respectively, then a group is any $L$-structure $\mathcal{G} = \langle G, \cdot^{\mathcal{G}}, 1^{\mathcal{G}} \rangle$ that satisfies the group axioms. Similarly, using various signatures, we can study structures that correspond to fields, graphs, etc.

We are interested in countably infinite structures here, so we can always take (up to isomorphism) $A = \mathbb{N}$.

**Definition 2.3** Denote by $X_L$ the *space of (countable) L-structures*, i.e.,

$$X_L = \prod_{i\in I} \mathbb{N}^{(\mathbb{N}^{n_i})} \times \prod_{j\in J} 2^{(\mathbb{N}^{m_j})}.$$

With the product topology ($\mathbb{N}$ and 2 being discrete) this is a Polish space.

**Definition 2.4** Let $S_\infty$ be the *infinite symmetric group* of all permutations of $\mathbb{N}$. It is a Polish group with the pointwise convergence topology. It acts continuously on $X_L$: If $\mathcal{A} \in X_L$, $g \in S_\infty$, then $g \cdot \mathcal{A}$ is the isomorphic copy of $\mathcal{A}$ obtained by applying $g$. For example, if $I = \emptyset$, $\{R_j\}_{j\in J}$ consists of a single binary relation symbol $R$, and $\mathcal{A} = \langle \mathbb{N}, R^{\mathcal{A}}\rangle$, then $g \cdot \mathcal{A} = \mathcal{B}$, where $(x, y) \in R^{\mathcal{B}} \Leftrightarrow (g^{-1}(x), g^{-1}(y)) \in R^{\mathcal{A}}$. This is called the *logic action* of $S_\infty$ on $X_L$.

Clearly, $\exists g(g \cdot \mathcal{A} = \mathcal{B}) \Leftrightarrow \mathcal{A} \cong \mathcal{B}$, i.e., the equivalence relation induced by this action is isomorphism.

Logic actions are universal among $S_\infty$-actions.

**Theorem 2.5** (Becker-Kechris [BK96, 2.7.3]) *There is a countable signature L such that for every Borel action of $S_\infty$ on a standard Borel space X there is a Borel equivariant injection $\pi : X \to X_L$ (i.e. $\pi(g \cdot x) = g \cdot \pi(x)$). Thus every Borel $S_\infty$-space is Borel isomorphic to the logic action on an isomorphism-invariant Borel class of L-structures.*

## Classification by countable structures

**Definition 2.6** Let $E$ be an equivalence relation on a standard Borel space $X$. We say the $E$ *admits classification by countable structures* if there is a countable signature $L$ and a Borel map $f : X \to X_L$ such that $xEy \Leftrightarrow f(x) \cong f(y)$, i.e., $E \leq_B \cong_L$ $( = \cong |X_L)$.

By Theorem 2.5 this is equivalent to the following: There is a Borel $S_\infty$-space $Y$ such that $E \leq_B E^Y_{S_\infty}$, where $xE^Y_{S_\infty}y \Leftrightarrow \exists g \in S_\infty(g \cdot x = y)$ is the equivalence relation induced by the $S_\infty$-action on $Y$.

**Example 2.7**

- If $E$ is concretely classifiable, then $E$ admits classification by countable structures.
- Let $X$ be an uncountable Polish space. Define $E_c$ on $X^{\mathbb{N}}$ by

$$(x_n)E_c(y_n) \Leftrightarrow \{x_n : n \in \mathbb{N}\} = \{y_n : n \in \mathbb{N}\}.$$

Then $E_c$ is classifiable by countable structures. ($E_c$ is, up to Borel isomorphism, independent of $X$.)

- (Giordano-Putnam-Skau [GPS95]) Topological orbit equivalence of minimal homeomorphisms of the Cantor set is classifiable by countable structures.
- (Kechris [Kec92]) If $G$ is Polish locally compact and $X$ is a Borel $G$-space, then $E_G^X$ admits classification by countable structures.

Hjorth developed in [Hjo00] a theory called *turbulence* that provides the basic method for showing that equivalence relations *do not* admit classification by countable structures. (Beyond [Hjo00] extensive expositions of this theory can be found in [Kec02], [Kan08] and [Gao09].)

**Theorem 2.8** (Hjorth [Hjo00, 3.19]) *Let G be a Polish group acting continuously on a Polish space X. If the action is* turbulent, *then $E_G^X$ cannot be classified by countable structures. In particular if E is an equivalence relation and $E_G^X \leq_B E$ for some turbulent action of a Polish group G on X, then E cannot be classified by countable structures.*

The rest of this lecture will be devoted to explaining the concept of turbulence and sketching some ideas in the proof the Theorem 2.8.

## Turbulence

Let $G$ be a Polish group acting continuously on a Polish space $X$. Below $U$ (with various embellishments) is a typical nonempty open set in $X$ and $V$ (with various embellishments) is a typical open symmetric nbhd of $1 \in G$.

**Definition 2.9** The $(U, V)$-*local graph* is given by

$$xR_{U,V}y \Leftrightarrow x, y \in U \,\&\, \exists g \in V\,(g \cdot x = y).$$

The $(U, V)$-*local orbit* of $x \in U$, denoted $\mathcal{O}(x, U, V)$, is the connected component of $x$ in this graph.

*Remark* 2.10 If $U = X$ and $V = G$, then $\mathcal{O}(x, U, V) = G \cdot x =$ the orbit of $x$.

**Definition 2.11** A point $x \in X$ is *turbulent* if $\forall U \ni x \forall V(\overline{\mathcal{O}(x, U, V)}$ has nonempty interior).

It is easy to check that this property depends only on the orbit of $x$, so we can talk about *turbulent orbits*.

The action is called *(generically) turbulent* if

(i) every orbit is meager,
(ii) there is a dense, turbulent orbit.

*Remark* 2.12 This implies that the set of dense, turbulent orbits is comeager (see [Kec02, 8.5]).

**Proposition 2.13** *Let $G$ be a Polish group acting continuously on a Polish space $X$ and let $x \in X$. Suppose there is a neighborhood basis $\mathcal{B}(x)$ for $x$ such that for all $U \in \mathcal{B}(x)$ and any open nonempty set $W \subseteq U$, there is a continuous path $(g_t)_{0 \le t \le 1}$ in $G$ with $g_0 = 1$, $g_1 \cdot x \in W$ and $g_t \cdot x \in U$ for each $t$. Then $x$ is turbulent.*

*Proof* It is enough to show that $O(x, U, V)$ is dense in $U$ for all $U \in \mathcal{B}(x)$ and open symmetric nbhds $V$ of $1 \in G$. Fix such a $U, V$ and take any nonempty open $W \subseteq U$ and let $(g_t)$ be as above. Using uniform continuity, we can find $t_0 = 0 < t_1 < \cdots < t_k = 1$ such that $g_{t_{i+1}} g_{t_i}^{-1} \in V$ for all $i < k$. Let $h_1 = g_{t_1} g_{t_0}^{-1} = g_{t_1}$, $h_2 = g_{t_2} g_{t_1}^{-1}, \ldots, h_k = g_{t_k} g_{t_{k-1}}^{-1}$. Then $h_i \in V$ and

$$h_i \cdot h_{i-1} \cdots h_1 \cdot x = g_{t_i} \cdot x \in U, \quad \forall 1 \le i \le k$$

and $g_1 \cdot x = h_k \cdot h_{k-1} \cdots h_1 \cdot x \in W$, so $g_1 \cdot x \in O(x, U, V) \cap W \neq \emptyset$. □

**Example 2.14** The action of the Polish group $(c_0, +)$ (with the sup-norm topology) on $(\mathbb{R}^{\mathbb{N}}, +)$ (with the product topology) by translation is turbulent. The orbits are the cosets of $c_0$ in $(\mathbb{R}^{\mathbb{N}}, +)$, so they are dense and meager. Also, any $x \in \mathbb{R}^{\mathbb{N}}$ is turbulent. Clearly the sets of the form $x + C$, with $C \subseteq \mathbb{R}^{\mathbb{N}}$ a convex open nbhd of 0, form a nbhd basis for $x \in \mathbb{R}^{\mathbb{N}}$. Now $c_0$ is dense in any such $C$, so $x + (c_0 \cap C)$ is dense in $x + C$. Let $g \in c_0 \cap C$. We will find a continuous path $(g_t)_{0 \le t \le 1}$ from 1 to $g$ in $c_0$ such that $x + g_t \in x + C$ for each $t$. Clearly $g_t = tg \in c_0 \cap C$ works.

*Remark* 2.15 There are groups that admit no turbulent action. For example Polish locally compact groups and $S_\infty$.

## Generic ergodicity

**Definition 2.16** Let $E$ be an equivalence relation on a Polish space $X$ and $F$ an equivalence relation on a Polish space $Y$. A *homomorphism* from $E$ to $F$ is a map $f : X \to Y$ such that $xEy \Rightarrow f(x)Ff(y)$.

We say that $E$ is *generically $F$-ergodic* if for every Baire measurable homomorphism $f$ there is a comeager set $A \subseteq X$ which $f$ maps into a single $F$-class.

**Example 2.17** Assume $E = E_G^X$ is induced by a continuous $G$-action of a Polish group $G$ on a Polish space $X$ with a dense orbit. Then $E$ is generically $=_Y$-ergodic, for any Polish space $Y$. (*Proof:* Let $f : X \to Y$ be a Baire measurable homomorphism from $E_G^X$ to $=_Y$ and let $(U_n)_{n\in\mathbb{N}}$ be a countable basis of nonempty open subsets of $Y$. For each $n$ the set $f^{-1}(U_n) \subseteq X$ is a $G$-invariant set and has the property of Baire. Thus, it is enough to show that if $A \subseteq X$ is $G$-invariant and has the property of Baire, then it is either meager or comeager. Otherwise there are open nonempty $U, U' \subseteq X$ with $A$ comeager in $U$ and $X \setminus A$ comeager in $U'$. Since there is a dense orbit, there is a $g \in G$ such that $W = g \cdot U \cap U' \neq \emptyset$ and since $g \cdot A = A$ we have that both $A$, $X \setminus A$ are comeager in $W$, a contradiction.)

In particular, if every $G$-orbit is also meager, then $E$ is not concretely classifiable.

Let as before $E_c$ be the equivalence relation on $(2^{\mathbb{N}})^{\mathbb{N}}$ given by

$$(x_n)E_c(y_n) \Leftrightarrow \{x_n : n \in \mathbb{N}\} = \{y_n : n \in \mathbb{N}\}.$$

(Note that this is Borel isomorphic to the one defined in 1.12.)

**Theorem 2.18** (Hjorth [Hjo00, 3.21]) *The following are equivalent:*

(i) *$E$ is generically $E_c$-ergodic.*
(ii) *$E$ is generically $E_{S_\infty}^Y$-ergodic for any Borel $S_\infty$-space $Y$.*

For another reference for the proof, see also [Kec02, 12.3].

It follows that if $E$ satisfies these properties and if every $E$-class is meager, then $E$ cannot be classified by countable structures.

**Theorem 2.19** (Hjorth [Hjo00]) *If a Polish group $G$ acts continuously on a Polish space $X$ and the action is turbulent, then $E_G^X$ is generically $E_c$-ergodic, so cannot be classified by countable structures.*

*Proof* (Following the presentation in [Kec02, 12.5]) Assume $f : X \to (2^{\mathbb{N}})^{\mathbb{N}}$ is a Baire measurable homomorphism from $E_G^X$ to $E_c$. Let $A(x) = \{f(x)_n : n \in \mathbb{N}\}$ so that $xE_G^X y \Rightarrow A(x) = A(y)$.

*Step 1.* Let $A = \{a \in 2^{\mathbb{N}} : \forall^* x\,(a \in A(x))\}$, where $\forall^* x$ means "on a comeager set of $x$." Then $A$ is countable.

(*Proof of Step 1:* The function $f$ is continuous on a dense $G_\delta$ set $C \subseteq X$. We have then $a \in A$ if and only if $\forall^* x \in C\,(a \in A(x))$. The set $B = \bigcup_n \{(x, a) \in C \times 2^{\mathbb{N}} : a = f(x)_n\}$ is Borel, and thus so is $A = \{a : \forall^* x \in C\,(x, a) \in B\}$. So if $A$ is uncountable, it contains a Cantor set $D$. Then $\forall^* a \in D\, \forall^* x(a \in A(x))$, so, by Kuratowski-Ulam, $\forall^* x \forall^* a \in D\,(a \in A(x))$, thus for some $x$, $A(x)$ is uncountable, which is obviously absurd. □[*Step 1*])

*Step 2.* $\forall^* x\,(A(x) = A)$, which completes the proof.

One proceeds by assuming that this fails, which implies that $\forall^* x(A \subsetneq A(x))$ and deriving a contradiction. Let $C$ be a dense $G_\delta$ set such that $f|C$ is continuous. Then one finds appropriate $a \notin A$, $l \in \mathbb{N}$ and (using genericity arguments) $z_i \in C$, $z \in C$ with $z_i \to z$ and $f(z_i)_l = a$ but $a \notin A(z)$. By continuity, $a = f(z_i)_l \to f(z)_l$, so $f(z)_l = a$, i.e., $a \in A(z)$, a contradiction. The point $z$ is found using the fact that $a \notin A$, so $\forall^* y(a \notin A(y))$ and the sequence $z_i$ is obtained from the turbulence condition. The detailed proof follows.

(*Proof of Step 2* Otherwise, since there is a dense orbit, the invariant set $\{x : A = A(x)\}$ (which has the Baire property) must be meager. It follows that $C_1 = \{x : A \subsetneq A(x)\}$ is comeager. Let

$$C_2 = \{x : \forall^* y \forall U \forall V\,(x \in U \Rightarrow G \cdot y \cap \overline{O(x, U, V)} \neq \emptyset)\},$$

so that $C_2$ is comeager as well, since it contains all turbulent points.

Next fix a comeager set $C_0 \subseteq X$ with $f|C_0$ continuous. For $B \subseteq X$ let

$$C_B = \{x : x \in B \Leftrightarrow \exists \text{ open nbhd } U \text{ of } x \text{ with } \forall^* y \in U(y \in B)\}.$$

Then, if $B$ has the Baire property, $C_B$ is comeager (see [Kec95, 8G]). Finally fix a countable dense subgroup $G_0 \subseteq G$ and find a countable collection $\mathcal{C}$ of comeager sets in $X$ with the following properties:

(i) $C_0, C_1, C_2 \in \mathcal{C}$;
(ii) $C \in \mathcal{C}, g \in G_0 \Rightarrow g \cdot C \in \mathcal{C}$;
(iii) $C \in \mathcal{C} \Rightarrow C^* = \{x : \forall^* g(g \cdot x \in C)\} \in \mathcal{C}$.
(iv) If $\{V_n\}$ enumerates a local basis of open symmetric nbhds of 1 in $G$, then, letting

$$A_{l,n} = \{x : \forall^* g \in V_n\,(f(x)_l = f(g \cdot x)_l)\},$$

we have that $C_{A_{l,n}} \in \mathcal{C}$.

(v) If $\{U_n\}$ enumerates a basis for $X$, and

$$C_{m,n,l} = \{x : x \notin U_m \text{ or } \forall^* g \in V_n\,(f(x)_l = f(g \cdot x)_l)\},$$

then $\mathcal{C}$ contains all $C_{m,n,l}$ which are comeager.

For simplicity, if $x \in \bigcap C$ (and there are comeager many such $x$), we call $x$ "generic".

So fix a generic $x$. Then there is $a \notin A$ so that $a \in A(x) = A(g \cdot x)$ for all $g$. So $\forall g \exists l(a = f(g \cdot x)_l)$, thus there is an $l \in \mathbb{N}$ and open nonempty $W \subseteq G$ so that $\forall^* g \in W\,(f(g \cdot x)_l = a)$. Fix $p_0 \in G_0 \cap W$ and $V$ a basic symmetric nbhd of 1 so that $Vp_0 \subseteq W$. Let $p_0 \cdot x = x_0$, so that $x_0$ is generic too, and $\forall^* g \in V\,(f(g \cdot x_0)_l = a)$. Now $\forall^* g \in V\,(g \cdot x_0 \in C_0)$, so we can find $g_i \in V$, $g_i \to 1$ with $g_i \cdot x_0 \in C_0$ and $f(g_i \cdot x_0)_l = a$, so as $g_i \cdot x_0 \to x_0 \in C_0$, by continuity we have $f(x_0)_l = a$. Also since $\forall^* g \in V\,(f(x_0)_l = f(g \cdot x_0)_l)$ and $x_0$ is generic, using (iv) we see that there is a basic open $U$ with $x_0 \in U$, such that

$$\forall^* z \in U \forall^* g \in V\,(f(z)_l = f(g \cdot z)_l),$$

i.e., if $U = U_m$, $V = V_n$, then $C_{m,n,l}$ is comeager, so by (v) it is in $C$. Since $a \notin A$, $\{y : a \notin A(y)\}$ is not meager, so choose $y$ generic with $a \notin A(y)$ and also

$$\forall \tilde{U}, \tilde{V}(x_0 \in \tilde{U} \Rightarrow G \cdot y \cap \overline{O(x_0, \tilde{U}, \tilde{V})} \neq \emptyset).$$

Thus we have $G \cdot y \cap \overline{O(x_0, U, V)} \neq \emptyset$. So choose $g_0, g_1, \cdots \in V$ so that if $g_i \cdot x_i = x_{i+1}$, then $x_i \in U$ and some subsequence of $(x_i)$ converges to some $y_1 \in G \cdot y$. Fix a compatible metric $d$ for $X$.

Since $\forall^* h(h \cdot x_0$ is generic) and $\forall^* g \in V\,(f(g \cdot x_0)_l = a)$, we can find $h_1$ so that $h_1 g_0 \in V$, $g_1 h_1^{-1} \in V$, $\overline{x}_1 = h_1 \cdot x_1 \in U$, $d(x_1, \overline{x}_1) < \frac{1}{2}$, $\overline{x}_1 = h_1 \cdot g_0 \cdot x_0$ is generic. Then $\forall^* g \in V(f(\overline{x}_1)_l = f(g \cdot \overline{x}_1)_l)$ (as $\overline{x}_1 \in C_{m,n.l}$), and $f(\overline{x}_1)_l = a$, so also $\forall^* g \in V\,(f(g \cdot \overline{x}_1)_l = a)$. Note that $g_1 h_1^{-1} \cdot \overline{x}_1 = x_2$ and $g_1 h_1^{-1} \in V$, so since $\forall^* h(h \cdot \overline{x}_1$ is generic) and $\forall^* g \in V\,(f(g \cdot \overline{x}_1)_l = a)$, we can find $h_2$ so that $h_2 g_1 h_1^{-1} \in V$, $g_2 h_2^{-1} \in V$, $\overline{x}_2 = h_2 \cdot x_2 \in U$, $d(x_2, \overline{x}_2) < \frac{1}{4}$, $\overline{x}_2 = h_2 g_1 h_1^{-1} \cdot \overline{x}_1$ is generic. Then $\forall^* g \in V\,(f(\overline{x}_2)_l = f(g \cdot \overline{x}_2)_l)$, and $f(\overline{x}_2)_l = a$, so $\forall^* g \in V(f(g \cdot \overline{x}_2) = a)$, etc.

Repeating this process, we get $x_0, \overline{x}_1, \overline{x}_2, \ldots$ generic and belonging to the $(U, V)$-local orbit of $x_0$, so that some subsequence of $\{\overline{x}_i\}$ converges to $y_1$ and $\forall^* g \in V\,(f(g \cdot \overline{x}_i)_l = a)$. Now

$$\forall^* g(g \cdot \overline{x}_i \in C_0), \quad \forall^* g(g \cdot y_1 \in C_0), \quad \forall^* g \in V(f(g \cdot \overline{x}_i)_l = a),$$

so fix $g$ satisfying all these conditions. Then for some subsequence $\{n_i\}$ we have $\overline{x}_{n_i} \to y_1$, so $g \cdot \overline{x}_{n_i} \to g \cdot y_1$ and $g \cdot \overline{x}_{n_i}, g.y_1 \in C_0$, so by continuity,

$$a = f(g \cdot \overline{x}_{n_i})_l \to f(g \cdot y_1)_l,$$

so $f(g \cdot y_1)_l = a$, i.e.,

$$a \in A(g \cdot y_1) = A(y_1) = A(y),$$

a contradiction. □[*Step 2*])

□

## Lecture III. Turbulence in the irreducible representations

Let $H$ be a separable complex Hilbert space. We denote by $U(H)$ the *unitary group* of $H$, i.e., the group of Hilbert space automorphisms of $H$. The strong topology on $U(H)$ is generated by the maps $T \in U(H) \mapsto T(x) \in H$ $(x \in H)$, and it is the same as the *weak topology* generated by the maps $T \in U(H) \mapsto \langle T(x), y \rangle \in \mathbb{C}$ $(x, y \in H)$. With this topology $U(H)$ is a Polish group.

If now $\Gamma$ is a countable (discrete) group, $\text{Rep}(\Gamma, H)$ is the space of unitary representations of $\Gamma$ on $H$, i.e., homomorphisms of $\Gamma$ into $U(H)$ or equivalently actions of $\Gamma$ on $H$ by unitary transformations. It is a closed subspace of $U(H)^\Gamma$, equipped with the product topology, so it is a Polish space. The group $U(H)$ acts continuously on $\text{Rep}(\Gamma, H)$ via conjugacy $T \cdot \pi = T\pi T^{-1}$ (where $T\pi T^{-1}(\gamma) = T \circ \pi(\gamma) \circ T^{-1}$) and the equivalence relation induced by this action is *isomorphism* of unitary representations: $\pi \cong \rho$.

A representation $\pi \in \text{Rep}(\Gamma, H)$ is *irreducible* if it has no non-trivial invariant closed subspaces. Let $\text{Irr}(\Gamma, H) \subseteq \text{Rep}(\Gamma, H)$ be the space of irreducible representations. Then it can be shown that $\text{Irr}(\Gamma, H)$ is a $G_\delta$ subset of $\text{Rep}(\Gamma, H)$, and thus also a Polish space in the relative topology (see [Kec10, H.5]).

From now on we assume that $H$ is $\infty$-dimensional.

Thoma [Tho64] has shown that if $\Gamma$ is abelian-by-finite then we have $\text{Irr}(\Gamma, H) = \emptyset$, but if $\Gamma$ is not abelian-by-finite then $\cong | \text{Irr}(\Gamma, H)$ is not concretely classifiable. Hjorth [Hjo97] extended this by showing that it is not even classifiable by countable structures. This is proved by showing that the conjugacy action is turbulent on an appropriate conjugacy invariant closed subspace of $\text{Irr}(\Gamma, H)$ (see [Kec10, H.9]). We will need below this result for the case $\Gamma = \mathbb{F}_2 =$ the free group with two generators, so we will state and sketch the proof of the following stronger result for this case.

**Theorem 3.1** (Hjorth [Hjo00]) *The conjugacy action of the group $U(H)$ on $\mathrm{Irr}(\mathbb{F}_2, H)$ is turbulent.*

*Proof* Note that we can identify $\mathrm{Rep}(\mathbb{F}_2, H)$ with $U(H)^2$ and the action of $U(H)$ on $U(H)^2$ becomes

$$T \cdot (U_1, U_2) = (TU_1T^{-1}, TU_2T^{-1}).$$

*Step 1:* $\mathrm{Irr}(\mathbb{F}_2, H)$ is dense $G_\delta$ in $\mathrm{Rep}(\mathbb{F}_2, H) = U(H)^2$.

(*Proof of Step 1* Note that if $U(n) = U(\mathbb{C}^n)$, then, with some canonical identifications, $U(1) \subseteq U(2) \subseteq \cdots \subseteq U(H)$ and $\overline{\bigcup_n U(n)} = U(H)$. Now $U(n)$ is compact, connected, so, by a result of Schreier-Ulam [SU35], the set of $(g, h) \in U(n)^2$ such that $\overline{\langle g, h\rangle} = U(n)$ is a dense $G_\delta$ in $U(n)^2$. By Baire Category this shows that the set of $(g, h) \in U(H)^2$ with $\overline{\langle g, h\rangle} = U(H)$ is dense $G_\delta$, since for each nonempty open set $N$ in $U(H)$, the set of $(g, h) \in U(H)^2$ that generate a subgroup intersecting $N$ is open dense. Thus the generic pair $(g, h) \in U(H)^2$ generates a dense subgroup of $U(H)$ so, viewing $(g, h)$ as a representation, any $(g, h)$-invariant closed subspace $A$ is invariant for all of $\overline{\langle g, h\rangle} = U(H)$, hence $A = \{0\}$ or $A = H$ and the generic pair is irreducible.

□[*Step 1*])

*Step 2:* Every orbit is meager.

(*Proof of Step 2* It is enough to show that every conjugacy class in $U(H)$ is meager. This is a classical result but here is a simple proof recently found by Rosendal. (This proof is general enough so it works in other Polish groups.)

For each infinite $I \subseteq \mathbb{N}$, let $A(I) = \{T \in U(H) : \exists i \in I\, (T^i = 1)\}$. It is easy to check that $A(I)$ is dense in $U(H)$. (Use the fact that $\bigcup_n U(n)$ is dense in $U(H)$. Then it is enough to approximate elements of each $U(n)$ by $A(I)$ and this can be easily done using the fact that elements of $U(n)$ are conjugate in $U(n)$ to diagonal unitaries.) Let now $V_0 \supseteq V_1 \supseteq \cdots$ be a basis of open nbhds of 1 in $U(H)$ and put $B(I, k) = \{T \in U(H) : \exists i \in I(i > k \text{ and } T^i \in V_k)\}$. This contains $A(I \setminus \{0, \ldots, k\})$, so is open dense. Thus

$$C(I) = \bigcap_k B(I, k) = \{T : \exists (i_n) \in I^{\mathbb{N}}\, (T^{i_n} \to 1)\}$$

is comeager and conjugacy invariant. If a conjugacy class $C$ is non-meager, it will thus be contained in all $C(I)$, $I \subseteq \mathbb{N}$ infinite. Thus $T \in C \Rightarrow T^n \to 1$, so letting $d$ be a left invariant metric for $U(H)$

we have for $T \in C$, $d(T, 1) = d(T^{n+1}, T^n) \to 0$, whence $T = 1$, a contradiction.

□[*Step 2*])

*Step 3:* There is a dense conjugacy class in $\mathrm{Irr}(\mathbb{F}_2, H)$ (so the set of all $\pi \in \mathrm{Irr}(\mathbb{F}_2, H)$ with dense conjugacy class is dense $G_\delta$ in $\mathrm{Irr}(\mathbb{F}_2, H)$).

(*Proof of Step 3* As $\mathrm{Irr}(\mathbb{F}_2, H)$ is dense $G_\delta$ in $\mathrm{Rep}(\mathbb{F}_2, H)$ it is enough to find $\pi \in \mathrm{Rep}(\mathbb{F}_2, H)$ with dense conjugacy class in $\mathrm{Rep}(\mathbb{F}_2, H)$ – then the set of all such $\pi$'s is dense $G_\delta$ so intersects $\mathrm{Irr}(\mathbb{F}_2, H)$. Let $(\pi_n)$ be dense in $\mathrm{Rep}(\mathbb{F}_2, H)$ and let $\pi \cong \bigoplus_n \pi_n$, $\pi \in \mathrm{Rep}(\mathbb{F}_2, H)$. This $\pi$ easily works. □[*Step 3*])

*Remark* 3.2 The also gives an easy proof of a result of Yoshizawa: There exists an irreducible representation of $\mathbb{F}_2$ which weakly contains any representation of $\mathbb{F}_2$.

*Step 4:* Let $\pi \in \mathrm{Irr}(\mathbb{F}_2, H)$ have dense conjugacy class. Then $\pi$ is turbulent.

Thus by Steps 2,3,4, the conjugacy action of $U(H)$ on $\mathrm{Irr}(\mathbb{F}_2, H)$ is turbulent.

(*Proof of Step 4*

**Lemma 3.3** *For $\rho, \sigma \in \mathrm{Irr}(\mathbb{F}_2, H)$, the following are equivalent:*

*(i)* $\rho \cong \sigma$,
*(ii)* $\exists (T_n) \in U(H)^{\mathbb{N}}$ *such that* $T_n \cdot \rho = T_n \rho T_n^{-1} \to \sigma$ *and no subsequence of* $(T_n)$ *converges in the weak topology of* $B_1(H) = \{T \in B(H) : \|T\| \leq 1\}$ *(a compact metrizable space) to* 0.

(Here $B(H)$ is the set of bounded linear operators on $H$.)

(*Proof of Lemma 3.3* Use the compactness of the unit ball $B_1(H)$ and Schur's Lemma (see, e.g., [Fol95, 3.5 (b)]): if $\pi_1, \pi_2 \in \mathrm{Irr}(\mathbb{F}_2, H)$, then $\pi_1 \cong \pi_2 \Leftrightarrow (\exists S \in B(H) \setminus \{0\}$ such that $\forall \gamma \in \mathbb{F}_2\, (S\pi_1(\gamma) = \pi_2(\gamma)S))$.
□[*Lemma 3.3*])

**Lemma 3.4** *Given any nonempty open $W \subseteq \mathrm{Irr}(\mathbb{F}_2, H)$ (open in the relative topology) and orthonormal $e_1, \ldots, e_p \in H$, there are orthonormal $e_1, \ldots, e_p, e_{p+1}, \ldots, e_q$ and $T \in U(H)$ such that*

*(i)* $T(e_i) \perp e_j$, $\forall i, j \leq q$;
*(ii)* $T^2(e_i) = -e_i$, $\forall i \leq q$;
*(iii)* $T = \mathrm{id}$ *on* $(H_0 \oplus T(H_0))^\perp$, *where* $H_0 = \langle e_1, \ldots, e_q \rangle$;

*(iv)* $T \cdot \pi \in W$.

(*Proof of Lemma 3.4* We can assume that

$$W = \{\rho \in \mathrm{Irr}(\mathbb{F}_2, H) : \forall \gamma \in F\ \forall i, j \leq q$$
$$|\langle \rho(\gamma)(e_i), e_j\rangle - \langle \sigma(\gamma)(e_i), (e_j)\rangle| < \epsilon\}$$

for some $e_1, \ldots, e_p, e_{p+1}, \ldots, e_q$, $F \subseteq \mathbb{F}_2$ finite, $\epsilon > 0$, and $\sigma \in \mathrm{Irr}(\mathbb{F}_2, H) \setminus U(H) \cdot \pi$ (since this set is comeager, hence dense). So, by Lemma 3.3, there is a sequence $(T_n) \in U(H)^{\mathbb{N}}$ with $T_n \cdot \pi \to \sigma$, $T_n \xrightarrow{w} 0$, so also $T_n^{-1} \xrightarrow{w} 0$. Thus for all large enough $n$, $e_1, \ldots, e_q, T_n^{-1}(-e_1), \ldots, T_n^{-1}(-e_q)$ are linearly independent. So apply Gram-Schmidt to get an orthonormal set $e_1, \ldots, e_q, f_1^{(n)}, \ldots, f_q^{(n)}$ with the same span. Then for all $i \leq q$, $\|T_n^{-1}(-e_i) - f_i^{(n)}\| \to 0$ (as $\langle T_n^{-1}(-e_i), e_j\rangle \to 0$, $\forall i, j \leq q$). Define $S_n \in U(H)$ by $S_n(e_i) = f_i^{(n)}$, $S_n(f_i^{(n)}) = -e_i$, $\forall i \leq q$, and $S_n =$ id on $\langle e_1, \ldots, e_q, f_1^{(n)}, \ldots, f_q^{(n)}\rangle^{\perp}$. Then if $n$ is large enough, $T = S_n$ works. □[*Lemma 3.4*])

We now show that $\pi$ is turbulent. We will apply Proposition 2.13 of Lecture II. Fix a basic nbhd of $\pi$ of the form

$$U = \{\rho \in \mathrm{Irr}(\mathbb{F}_2, H) : \forall \gamma \in F\ \forall i, j \leq k$$
$$|\langle \rho(\gamma)(e_i), e_j\rangle - \langle \pi(\gamma)(e_i), (e_j)\rangle| < \epsilon\},$$

$\epsilon > 0$, $F \subseteq \mathbb{F}_2$ finite, $e_1, \ldots, e_k$ orthonormal. Let $e_1, \ldots, e_k, e_{k+1}, \ldots, e_p$ be an orthonormal basis for the span of $\{e_1, \ldots, e_k\} \cup \{\pi(\gamma)(e_i) : \gamma \in F,\ 1 \leq i \leq k\}$, and let $W \subseteq U$ be an arbitrary nonempty open set. Then let $e_1, \ldots, e_p, e_{p+1}, \ldots, e_q$ and $T$ be as in Lemma 3.4 (so that $T \cdot \pi \in W$). It is enough to find a continuous path $(T_\theta)_{0 \leq \theta \leq \pi/2}$ in $U(H)$ with $T_0 = 1$, $T_{\pi/2} = T$, and $T_\theta \cdot \pi \in U$ for all $\theta$. Take

$$T_\theta(e_i) = (\cos\theta)e_i + (\sin\theta)T(e_i)$$
$$T_\theta(T(e_i)) = (-\sin\theta)e_i + (\cos\theta)T(e_i),$$

for $i = 1, \ldots, q$ and let $T_\theta =$ id on $(H_0 \oplus T(H_0))^{\perp}$, where $H_0 = \langle e_1, \ldots, e_q\rangle$. Then one can easily see that $T_\theta \cdot \pi \in U$, for all $\theta$.

□[*Step 4*])

□

# Lecture IV. Non-classification of orbit equivalence by countable structures, Part A: Outline of the proof and Gaussian actions.

Our goal in the remaining three lectures is to prove the following result.

**Theorem 4.1** (Epstein-Ioana-Kechris-Tsankov [IKT09, 3.12]) *Let* $\Gamma$ *be a countable non-amenable group. Then orbit equivalence for measure-preserving, free, ergodic (in fact mixing) actions of* $\Gamma$ *is not classifiable by countable structures.*

We will start by giving a *very rough* idea of the proof and then discussing the (rather extensive) set of results needed to implement it.

## Definitions

A *standard measure space* $(X, \mu)$ is a standard Borel space $X$ with a non-atomic probability Borel measure $\mu$. All such spaces are isomorphic to $[0, 1]$ with Lebesgue measure on the Borel sets.

The measure algebra $\mathrm{MALG}_\mu$ of $\mu$ is the algebra of Borel sets of $X$, modulo null sets, with the topology induced by the metric $d(A, B) = \mu(A \Delta B)$. Let $\mathrm{Aut}(X, \mu)$ be the group of measure-preserving automorphisms of $(X, \mu)$ (again modulo null sets) with the *weak topology*, i.e., the one generated by the maps $T \mapsto T(A)$ $(A \in \mathrm{MALG}_\mu)$. It is a Polish group.

If $\Gamma$ is a countable group, denote by $A(\Gamma, X, \mu)$ the space of measure-preserving actions of $\Gamma$ on $(X, \mu)$ or equivalently, homomorphisms of $\Gamma$ into $\mathrm{Aut}(X, \mu)$. It is a closed subspace of $\mathrm{Aut}(X, \mu)^\Gamma$ with the product topology, so also a Polish space. We say that $a \in A(\Gamma, X, \mu)$ is *free* if $\forall \gamma \neq 1\, (\gamma \cdot x \neq x, \text{a.e.})$, and it is *ergodic* if every invariant Borel set $A \subseteq X$ is either null or conull. We say that $a \in A(\Gamma, X, \mu)$, $b \in A(\Gamma, Y, \nu)$ are *orbit equivalent*, $a \,\text{Œ}\, b$, if, denoting by $E_a$, $E_b$ the equivalence relations induced by $a, b$ respectively, $E_a$ is isomorphic to $E_b$, in the sense that there is a measure-preserving isomorphism of $(X, \mu)$ to $(Y, \nu)$ that sends $E_a$ to $E_b$ (modulo null sets). Thus Œ is an equivalence relation on $A(\Gamma, X, \mu)$ and Theorem 4.1 asserts that it cannot be classified by countable structures if $\Gamma$ is not amenable.

## Idea of the proof

We start with the following fact.

**Theorem 4.2** *To each $\pi \in \mathrm{Rep}(\Gamma, H)$, we can assign in a Borel way an action $a_\pi \in A(\Gamma, X, \mu)$ (on some standard measure space $(X, \mu)$), called the* Gaussian action *associated to $\pi$, such that*

*(i)* $\pi \cong \rho \Rightarrow a_\pi \cong a_\rho$,
*(ii) If $\kappa_0^{a_\pi}$ is the Koopman representation on $L_0^2(X, \mu)$ associated to $a_\pi$, then $\pi \leq \kappa_0^{a_\pi}$.*

*Moreover, if $\pi$ is irreducible, then $a_\pi$ is weak mixing.*

(*Explanations:*

(i) $a, b \in A(\Gamma, X, \mu)$ are *isomorphic*, $a \cong b$ if there is a $T \in \mathrm{Aut}(X, \mu)$ taking $a$ to $b$,

$$T\gamma^a T^{-1} = \gamma^b, \quad \forall \gamma \in \Gamma.$$

(Here we let $\gamma^a = a(\gamma)$.)
(ii) Let $a \in A(\Gamma, X, \mu)$. Let $L_0^2(X, \mu) = \{f \in L^2(X, \mu) : \int f = 0\} = \mathbb{C}^\perp$. The *Koopman representation* $\kappa_0^a$ of $\Gamma$ on $L_0^2(X, \mu)$ is given by $(\gamma \cdot f)(x) = f(\gamma^{-1} \cdot x)$.
(iii) If $\pi \in \mathrm{Rep}(\Gamma, H)$, $\rho \in \mathrm{Rep}(\Gamma, H')$, then $\pi \leq \rho$ iff $\pi$ is isomorphic to a subrepresentation of $\rho$, i.e., the restriction of $\rho$ to an invariant, closed subspace of $H'$.)

So let $\pi \in \mathrm{Irr}(\mathbb{F}_2, H)$ and look at $a_\pi \in A(\mathbb{F}_2, X, \mu)$. We will then modify $a_\pi$, in a Borel and isomorphism preserving way, to another action $a(\pi) \in A(\mathbb{F}_2, Y, \nu)$ (for reasons to be explained later) and finally apply a construction of Epstein to "co-induce" appropriately $a(\pi)$, in a Borel and isomorphism preserving way, to an action $b(\pi) \in A(\Gamma, Z, \rho)$, which will turn out to be free and ergodic (in fact *mixing* i.e., $\mu(\gamma \cdot A \cap B) \to \mu(A)\mu(B)$ as $\gamma \to \infty$, for every Borel $A, B$). So we finally have a Borel function $\pi \in \mathrm{Irr}(\mathbb{F}_2, H) \mapsto b(\pi) \in A(\Gamma, Z, \rho)$.

Put

$$\pi R \rho \Leftrightarrow b(\pi) \,\text{Œ}\, b(\rho).$$

Then $R$ is an equivalence relation on $\mathrm{Irr}(\mathbb{F}_2, H)$, and $\pi \cong \rho \Rightarrow \pi R \rho$ (since $\pi \cong \rho \Rightarrow a_\pi \cong a_\rho \Rightarrow a(\pi) \cong a(\rho) \Rightarrow b(\pi) \cong b(\rho)$).

*Fact:* $R$ has countable index over $\cong$ (i.e., every $R$-class contains only countably many $\cong$-classes).

If now Œ on $A(\Gamma, Z, \rho)$ admitted classification by countable structures, so would $R$ on $\mathrm{Irr}(\mathbb{F}_2, H)$. So let $F : \mathrm{Irr}(\mathbb{F}_2, H) \to X_L$ be Borel, where

$X_L$ is the standard Borel space of countable structures for a signature $L$, with

$$\pi R \rho \Leftrightarrow F(\pi) \cong F(\rho).$$

Therefore $\pi \cong \rho \Rightarrow F(\pi) \cong F(\rho)$. By Theorem 3.1 and Theorems 2.18, 2.19, there is a comeager set $A \subseteq \mathrm{Irr}(\mathbb{F}_2, H)$ and $\mathcal{A}_0 \in X_L$ such that $F(\pi) \cong \mathcal{A}_0$, $\forall \pi \in A$. But every $\cong$-class in $\mathrm{Irr}(\mathbb{F}_2, H)$ is meager, so by the previous fact every $R$-class in $\mathrm{Irr}(\mathbb{F}_2, H)$ is meager, so there are $R$-inequivalent $\pi, \rho \in A$, and thus $F(\pi) \not\cong F(\rho)$, a contradiction.

(*Sketch of proof of Theorem 4.2* (See [Kec10, Appendix E]) For simplicity we will discuss the case of real Hilbert spaces, the complex case being handled by appropriate complexifications.

Let $H$ be an infinite-dimensional, separable, real Hilbert space. Consider the product space $(\mathbb{R}^{\mathbb{N}}, \mu^{\mathbb{N}})$, where $\mu$ is the normalized, centered Gaussian measure on $\mathbb{R}$ with density $\frac{1}{\sqrt{2\pi}} e^{-x^2/2}$. Let $p_i : \mathbb{R}^{\mathbb{N}} \to \mathbb{R}$, $i \in \mathbb{N}$, be the projection functions. The closed linear space $\langle p_i \rangle \subseteq L_0^2(\mathbb{R}^{\mathbb{N}}, \mu^{\mathbb{N}})$ (real valued) has countable infinite dimension, so we can assume that $H = \langle p_i \rangle \subseteq L_0^2(\mathbb{R}^{\mathbb{N}}, \mu^{\mathbb{N}})$.

**Lemma 4.3** *If $S \in O(H)$ = the orthogonal group of $H$ (i.e., the group of Hilbert space automorphisms of $H$), then we can extend uniquely $S$ to $\overline{S} \in \mathrm{Aut}(\mathbb{R}^{\mathbb{N}}, \mu^{\mathbb{N}})$ in the sense that the Koopman operator $O_{\overline{S}} : L_0^2(\mathbb{R}^{\mathbb{N}}, \mu^{\mathbb{N}}) \to L_0^2(\mathbb{R}^{\mathbb{N}}, \mu^{\mathbb{N}})$ defined by $O_{\overline{S}}(f) = f \circ (\overline{S})^{-1}$ extends $S$, i.e., $O_{\overline{S}}|H = S$.*

Thus if $\pi \in \mathrm{Rep}(\Gamma, H)$, we can extend each $\pi(\gamma) \in O(H)$ to $\overline{\pi(\gamma)} \in \mathrm{Aut}(\mathbb{R}^{\mathbb{N}}, \mu^{\mathbb{N}})$. Let $a_\pi \in A(\Gamma, \mathbb{R}^{\mathbb{N}}, \mu^{\mathbb{N}})$ be defined by $a_\pi(\gamma) = \overline{\pi(\gamma)}$. This clearly works.

(*Proof of Lemma 4.3:* The $p_i$'s form an orthonormal basis for $H$. Let $S(p_i) = q_i \in H$. Then let $\theta : \mathbb{R}^{\mathbb{N}} \to \mathbb{R}^{\mathbb{N}}$ be defined by

$$\theta(x) = (q_0(x), q_1(x), \dots).$$

Then $\theta$ is 1-1, since the $\sigma$-algebra generated by $(q_i)$ is the Borel $\sigma$-algebra, modulo null sets, so $(q_i)$ separates points modulo null sets. Moreover $\theta$ preserves $\mu^{\mathbb{N}}$. This follows from the fact that every $f \in \langle p_i \rangle$ (including $q_i$) has centered Gaussian distribution and the $q_i$ are independent, since $\mathbb{E}(q_i q_j) = \langle q_i, q_j \rangle = \langle p_i, p_j \rangle = \delta_{ij}$. Thus $\theta \in \mathrm{Aut}(\mathbb{R}^{\mathbb{N}}, \mu^{\mathbb{N}})$. Now put $\overline{S} = \theta^{-1}$. □[*Lemma 4.3*])

□[*Theorem 4.2*])

# Lecture V. Non-classification of orbit equivalence by countable structures, Part B: An action of $\mathbb{F}_2$ on $\mathbb{T}^2$ and a separability argument.

Recall that our plan consists of the three steps

$$\pi \xrightarrow{(1)} a_\pi \xrightarrow{(2)} a(\pi) \xrightarrow{(3)} b(\pi),$$

where $\pi$ is an irreducible unitary representation of $\mathbb{F}_2$, $a_\pi$ is the corresponding Gaussian action of $\mathbb{F}_2$, (2) is the "perturbation" to a new action of $\mathbb{F}_2$ and (3) is the co-inducing construction, which from the $\mathbb{F}_2$-action $a(\pi)$ produces a $\Gamma$-action $b(\pi)$.

We already discussed step (1).

## Properties of the co-induced action

Let's summarize next the key properties of the co-inducing construction (3) that we will need and discuss this construction in Lecture VI. Below we write $\gamma^a \cdot x$ for $\gamma^a(x)$.

**Theorem 5.1** *Let $\Gamma$ be non-amenable. Given $a \in A(\mathbb{F}_2, Y, \nu)$ we can construct $b \in A(\Gamma, Z, \rho)$ and $a' \in A(\mathbb{F}_2, Z, \rho)$ ($Z, \rho$ independent of $a$) with the following properties:*

*(i)* $E_{a'} \subseteq E_b$,

*(ii)* *$b$ is free and ergodic (in fact mixing),*

*(iii)* *$a'$ is free,*

*(iv)* *$a$ is a factor of $a'$ via a map $f : Z \to Y$ ($f$ independent of $a$), i.e., $f(\delta^{a'} \cdot z) = \delta^a \cdot f(z)$ ($\delta \in \mathbb{F}_2$), and $f_*\rho = \nu$, and in fact if $a$ is ergodic, then for every $a'$-invariant Borel set $A \subseteq Z$ of positive measure, if $\rho_A = \frac{\rho|A}{\rho(A)}$, then $a$ is a factor of $(a'|A, \rho_A)$ via $f$.*

*(v)* *If a free action $\overline{a} \in A(\mathbb{F}_2, \overline{Y}, \overline{\nu})$ is a factor of the action $a$ via $g : Y \to \overline{Y}$*

$$\begin{array}{ccc} Y & \xrightarrow{g} & \overline{Y} \\ \uparrow{\scriptstyle f} & & \\ Z & & \end{array}$$

*then for $\gamma \in \Gamma \setminus \{1\}$, $gf(\gamma^b \cdot z) \neq gf(z)$, $\rho$-a.e.*

*Moreover the map $a \mapsto b$ is Borel and preserves isomorphism.*

## A separability argument

We now deal with construction (2). The key here is a particular action of $\mathbb{F}_2$ on $\mathbb{T}^2$ utilized first to a great effect by Popa, Gaboriau-Popa [GP05] and then Ioana [Ioa11]. The group $\mathrm{SL}_2(\mathbb{Z})$ acts on $(\mathbb{T}^2, \lambda)$ ($\lambda$ is Lebesgue measure) in the usual way by matrix multiplication

$$A \cdot (z_1, z_2) = (A^{-1})^t \begin{pmatrix} z_1 \\ z_2 \end{pmatrix}.$$

This is free, measure-preserving, and ergodic, in fact weak mixing.

Fix also a copy of $\mathbb{F}_2$ with finite index in $\mathrm{SL}_2(\mathbb{Z})$ (see, e.g., [New72, VIII.2]) and denote the restriction of this action to $\mathbb{F}_2$ by $\alpha_0$. It is also free, measure-preserving and weak mixing. For any $c \in A(\mathbb{F}_2, X, \mu)$, we let $a(c) \in A(\mathbb{F}_2, \mathbb{T}^2 \times X, \lambda \times \mu)$ be the product action

$$a(c) = \alpha_0 \times c$$

(i.e. $\gamma^{a(c)} \cdot (z, x) = (\gamma^{\alpha_0} \cdot z, \gamma^c \cdot x)$). Then in our case we take

$$a(\pi) = a(a_\pi) = \alpha_0 \times a_\pi.$$

Then $a(\pi)$ is also weak mixing.

The key property of the passage from $c$ to $a(c)$ is the following separability result established by Ioana (in a somewhat different context – but his proof works as well here).

Below, for each $c \in A(\mathbb{F}_2, X, \mu)$, with $a(c) = \alpha_0 \times c \in A(\mathbb{F}_2, Y, \nu)$, where $Y = \mathbb{T}^2 \times X$, $\nu = \lambda \times \mu$, we let $b(c) \in A(\Gamma, Z, \rho)$, $a'(c) \in A(\mathbb{F}_2, Z, \rho)$ come from $a(c)$ via Theorem 5.1.

**Theorem 5.2** (Ioana [Ioa11]) *If $(c_i)_{i \in I}$ is an uncountable family of actions in $A(\mathbb{F}_2, X, \mu)$ and $(b(c_i))_{i \in I}$ are mutually orbit equivalent, then there is uncountable $J \subseteq I$ such that if $i, j \in J$, we can find Borel sets $A_i, A_j \subseteq Z$ of positive measure which are respectively $a'(c_i), a'(c_j)$-invariant and $a'(c_i)|A_i \cong a'(c_j)|A_j$ with respect to the normalized measures $\rho_{A_i}, \rho_{A_j}$.*

*Proof* We have the following situation, letting $b_i = b(c_i)$: $(b_i)_{i \in I}$ is an uncountable family of pairwise orbit equivalent free, ergodic actions in $A(\Gamma, Z, \rho)$, $a'_i = a'(c_i)$ are free in $A(\mathbb{F}_2, Z, \rho)$, with $E_{a'_i} \subseteq E_{b_i}$, and $\alpha_0$ is a factor of $a'_i$ via a map $p : Z \to \mathbb{T}^2$ such that

$$(*) \qquad \text{for } \gamma \in \Gamma \setminus \{1\},\ i \in I,\ \ p(\gamma^{b_i} \cdot z) \neq p(z),\ \ \rho\text{-a.e.}$$

Here $p = \mathrm{proj} \circ f$, where $f$ is as in (iv) of Theorem 5.1 and $\mathrm{proj} : Y =$

$\mathbb{T}^2 \times X \to \mathbb{T}^2$ is the projection. This follows from (v) of Theorem 5.1 with $\overline{a} = \alpha_0 \in A(\mathbb{F}_2, \mathbb{T}^2, \lambda)$ and $g = \text{proj}$.

$$\begin{array}{ccc} Z & & a'_i \\ \downarrow{\scriptstyle p} & & \big\downarrow \\ \mathbb{T}^2 & & \alpha_0 \end{array}$$

By applying to each $b_i, a'_i$ a measure preserving transformation $T_i \in \text{Aut}(Z, \rho)$, i.e., replacing $b_i$ by $Tb_iT^{-1}$ and $a'_i$ by $Ta'_iT^{-1}$, which we just call again $b_i$ and $a'_i$ by abuse of notation, we can clearly assume that there is $E$ such that $E_{b_i} = E$ for each $i \in I$. Then $\alpha_0$ is a factor of $a'_i$ via $p_i = p \circ T_i^{-1}$ and $(*)$ holds as well for $p_i$ instead of $p$.

Consider now the $\sigma$-finite measure space $(E, P)$ where for each Borel set $A \subseteq E$, $P(A) = \int |A_z| \, d\rho(z)$.

The action of $\text{SL}_2(\mathbb{Z})$ on $\mathbb{Z}^2$ by matrix multiplication gives a semidirect product $\text{SL}_2(\mathbb{Z}) \ltimes \mathbb{Z}^2$, and similarly $\mathbb{F}_2 \ltimes \mathbb{Z}^2$ (as we view $\mathbb{F}_2$ as a subgroup of $\text{SL}_2(\mathbb{Z})$). The key point is that $(\mathbb{F}_2 \ltimes \mathbb{Z}^2, \mathbb{Z}^2)$ has the so-called *relative property* (T):

$\exists$ finite $Q \subseteq \mathbb{F}_2 \ltimes \mathbb{Z}^2$, $\epsilon > 0$, such that for any unitary representation $\pi \in \text{Rep}(\mathbb{F}_2 \ltimes \mathbb{Z}^2, H)$, if $v$ is a $(Q, \epsilon)$-invariant unit vector (i.e., $\|\pi(q)(v) - v\| < \epsilon$, $\forall q \in Q$), then there is a $\mathbb{Z}^2$-invariant vector $w$ with $\|v - w\| < 1$

(see [BHV08, 4.2] and also [Hjo09, 2.3]).

Given now $i, j \in I$, we will define a representation $\pi_{i,j} \in \text{Rep}(\mathbb{F}_2 \ltimes \mathbb{Z}^2, L^2(E, P))$. Identify $\mathbb{Z}^2$ with $\widehat{\mathbb{T}^2}$ = the group of characters of $\mathbb{T}^2$ so that $\bar{m} = (m_1, m_2) \in \mathbb{Z}^2$ is identified with $\chi_{\bar{m}}(z_1, z_2) = z_1^{m_1} z_2^{m_2}$. Via this identification, the action of $\mathbb{F}_2$ on $\mathbb{Z}^2$ by matrix multiplication is identified with the shift action of $\mathbb{F}_2$ on $\widehat{\mathbb{T}^2}$: $\delta \cdot \chi(t) = \chi(\delta^{-1} \cdot t)$, $\delta \in \mathbb{F}_2, \chi \in \widehat{\mathbb{T}^2}$, $t \in \mathbb{T}^2$. Then the semidirect product $\mathbb{F}_2 \ltimes \mathbb{Z}^2$ is identified with $\mathbb{F}_2 \ltimes \widehat{\mathbb{T}^2}$ and multiplication is given by

$$(\delta_1, \chi_1)(\delta_2, \chi_2) = (\delta_1\delta_2, \chi_1(\delta_1 \cdot \chi_2)).$$

If $\chi \in \widehat{\mathbb{T}^2}$, let $\eta^i_\chi = \chi \circ p_i : Z \to \mathbb{T}$. Then define $\pi_{i,j}$ as follows

$$\pi_{i,j}(\delta, \chi)(f)(x, y) = \eta^i_\chi(x)\overline{\eta^j_\chi(y)} f((\delta^{-1})^{a'_i} \cdot x, (\delta^{-1})^{a'_j} \cdot y)$$

for $\delta \in \mathbb{F}_2, \chi \in \widehat{\mathbb{T}^2}$. To check that this is a representation note that

$$(**) \qquad \begin{aligned} \eta^i_{\delta\cdot\chi}(x) &= (\delta\cdot\chi)(p_i(x)) \\ &= \chi((\delta^{-1})^{\alpha_0}\cdot p_i(x)) = \chi(p_i((\delta^{-1})^{a'_i}\cdot x)) \\ &= \eta^i_\chi((\delta^{-1})^{a'_i}\cdot x) \end{aligned}$$

and similarly for $j$.

Then if $v = 1_\Delta$, where $\Delta = \{(z,z) : z \in Z\}$, using the separability of $L^2(Z,\rho)$ and $L^2(E,P)$ one can find $J \subseteq I$ uncountable such that if $i, j \in J$, then

$$\|\pi_{i,j}(q)(v) - v\|_{L^2(E,P)} < \epsilon,$$

$\forall q \in Q$. Indeed, note that

$$\|\pi_{i,j}(\delta,\chi)(v) - v\|^2_{L^2(E,P)} = 2\int \Re(1 - fg)\mathrm{d}\rho,$$

where $f = \eta^i_\chi\overline{\eta^j_\chi}$ and $g = 1_{\{z:(\delta^{-1})^{a'_i}\cdot z=(\delta^{-1})^{a'_j}\cdot z\}}$. Since $1 - fg = (1-f) + (1-g) - (1-f)(1-g)$, this is bounded by

$$2(\|1-f\|_{L^2(Z,\rho)} + \|1-g\|_{L^2(Z,\rho)} + \|1-f\|_{L^2(Z,\rho)}\|1-g\|_{L^2(Z,\rho)}).$$

Now

$$\|1-f\|_{L^2(Z,\rho)} = \|\eta^i_\chi - \eta^j_\chi\|_{L^2(Z,\rho)}$$

and $|1-g| = 1_{\{z:(\delta^{-1})^{a'_i}\cdot z\neq(\delta^{-1})^{a'_j}\cdot z\}}$, so denoting by $f_i^\delta$ the characteristic function of the graph of $(\delta^{-1})^{a'_i}$ and similarly for $f_j^\delta$,

$$\|1-g\|_{L^2(Z,\rho)} = \frac{1}{2}\|f_i^\delta - f_j^\delta\|_{L^2(E,P)}$$

and so by the separability of $L^2(Z,\rho)$ and $L^2(E,P)$, there is $J \subseteq I$ uncountable, so that if $i, j \in J$, then

$$\|\pi_{i,j}(q)(v) - v\|_{L^2(E,P)} < \epsilon, \quad \forall q \in Q.$$

So by relative property (T), there is $f \in L^2(E,P)$ with $\|f - 1_\Delta\| < 1$ and $f$ is $\mathbb{Z}^2$-invariant (for $\pi_{i,j}$), i.e.,

$$f(x,y) = \eta^i_\chi(x)\overline{\eta^j_\chi(y)}f(x,y), \quad \forall\chi \in \widehat{\mathbb{T}^2}.$$

Since $f \neq 0$ (as $\|1_\Delta\| = 1$ and $\|f - 1_\Delta\| < 1$) the set

$$S = \{(x,y) \in E : \eta^i_\chi(x) = \eta^j_\chi(y), \forall\chi \in \widehat{\mathbb{T}^2}\}$$

has positive $P$-measure.

Using the fact that $E$ is generated by $b_j$, and the fact that characters separate points, it follows that for almost all $x \in X$, there is at most one $y$ such that $(x, y) \in S$. Indeed, fix $x$ such that $(x, y_1) \in S$, $(x, y_2) \in S$ with $y_1 \neq y_2$. Then $y_1 = \gamma_1^{b_j} \cdot x$, $y_2 = \gamma_2^{b_j} \cdot x$, $\gamma_1 \neq \gamma_2$. But also

$$\forall \chi(\eta_\chi^i(x) = \eta_\chi^j(\gamma_1^{b_j} \cdot x) = \eta_\chi^j(\gamma_2^{b_j} \cdot x)).$$

But $\widehat{\mathbb{T}^2}$ separates points, so

$$p_j(\gamma_1^{b_j} \cdot x) = p_j(\gamma_2^{b_j} \cdot x).$$

Let $\gamma_1 \gamma_2^{-1} = \gamma \neq 1$. Then

$$p_j(\gamma^{b_j} \cdot \gamma_2^{b_j} \cdot x) = p_j(\gamma_2^{b_j} \cdot x)$$

which happens only on a null set by $(*)$ for $p_j$.

Let

$$A_i = \{x : \exists \text{ unique } y\ (x, y) \in S\}, \quad A_j = \{y : \exists x \in A_i\ (x, y) \in S\}.$$

Then $\rho(A_i) > 0$ (as $P(S) > 0$). Now if $(x, y) \in S$ then by $(**)$

$$(\delta^{a'_i} \cdot x, \delta^{a'_j} \cdot y) \in S, \quad \forall \delta \in \mathbb{F}_2,$$

so $A_i$ and $A_j$ are respectively $a'_i$-invariant and $a'_j$-invariant sets of positive measure. Let $\varphi : A_i \to A_j$ be defined by $\varphi(x) = y \Leftrightarrow (x, y) \in S$. Then

$$\begin{aligned} \varphi(x) = y &\Leftrightarrow (x, y) \in S \\ &\Leftrightarrow (\delta^{a'_i} \cdot x, \delta^{a'_j} \cdot y) \in S \\ &\Leftrightarrow \varphi(\delta^{a'_i} \cdot x) = \delta^{a'_j} \cdot y, \end{aligned}$$

i.e., $\varphi$ shows that $a'_i | A_i \cong a'_j | A_j$. □

### Completion of the proof

We now complete the proof of Theorem 4.1.

We have $a'(\pi), b(\pi)$ as in Theorem 5.1 coming from $a(\pi)$ in step (3). Recall that we defined

$$\pi R \rho \Leftrightarrow b(\pi)\, \text{Œ}\, b(\rho).$$

To complete the proof of Theorem 4.1, we only had to show that $R$ has countable index over $\cong$. We now prove this:

Assume, toward a contradiction, that $(\pi_i)_{i\in I}$ is an uncountable family of pairwise non-isomorphic representations in $\mathrm{Irr}(\mathbb{F}_2, H)$ such that if $b_i = b(\pi_i)$, then $(b_i)$ are pairwise orbit equivalent. Recall the chain:

$$\pi_i \to a_{\pi_i} = c_i \to a(\pi_i) = a(c_i) = \alpha_0 \times c_i \to b(c_i) = b_i,\ a'(c_i) = a'_i.$$

By Theorem 5.2, we can find uncountable $J \subseteq I$ such that if $i, j \in J$, there are $a'_i, a'_j$-resp. invariant Borel sets $A_i, A_j$ of positive measure, so that $a'_i|A_i \cong a'_j|A_j$. Moreover by property (iv) of Theorem 5.1, $a(c_i)$ is a factor of $a'_i|A_i$ and similarly for $a(c_j)$, $a'_j|A_j$.

Fix $i_0 \in J$. Then for any $j \in J$, fix $A_{i_0}, A_j$ as in Theorem 5.2. We have

$$\pi_j \le \kappa_0^{a_{\pi_j}} \left(= \kappa_0^{c_j}\right) \le \kappa_0^{\alpha_0\times c_j} \left(= \kappa_0^{a(c_j)}\right) \le \kappa_0^{a'_j|A_j} \cong \kappa_0^{a'_{i_0}|A_{i_0}} \le \kappa_0^{a'_{i_0}}.$$

Thus $(\pi_j)$ is (up to isomorphism) an uncountable family of pairwise non-isomorphic *irreducible* subrepresentations of $\kappa_0^{a'_{i_0}}$, a contradiction.

# Lecture VI. Non-classification of orbit equivalence by countable structures, Part C: Co-induced actions

It only remains to prove Theorem 5.1 from Lecture V. This is based on a co-inducing construction due to Epstein [Eps08].

## Co-induced actions

We have two countable groups $\Delta$ and $\Gamma$ (in our case $\Delta$ will be $\mathbb{F}_2$) and are given a *free*, *ergodic*, measure-preserving action $a_0$ of $\Delta$ on $(\Omega, \omega)$ and a measure-preserving action $b_0$ of $\Gamma$ on $(\Omega, \omega)$ with $E_{a_0} \subseteq E_{b_0}$ (note that $b_0$ is also ergodic). Let $N = [E_{b_0} : E_{a_0}] =$ (the number of $E_{a_0}$-classes in each $E_{b_0}$-class) $\in \{1, 2, 3, \ldots, \aleph_0\}$. Work below with $N = \aleph_0$ the other cases being similar.

Given these data we will describe Epstein's *co-inducing* construction that, given any $a \in A(\Delta, Y, \nu)$, will produce $b \in A(\Gamma, Z, \rho)$, where $Z = \Omega \times Y^{\mathbb{N}}$, $\rho = \omega \times \nu^{\mathbb{N}}$, called the *co-induced action of a, modulo* $(a_0, b_0)$,

$$b = \mathrm{CInd}(a_0, b_0)_\Delta^\Gamma(a),$$

which will satisfy Theorem 5.1 of Lecture V. See [IKT09, §3] for more details.

Put $E = E_{a_0}$, $F = E_{b_0}$. We can then find a sequence $(C_n) \in \mathrm{Aut}(\Omega, \omega)^{\mathbb{N}}$

of *choice functions*, i.e., $C_0 = \text{id}$ and $\{C_n(w)\}$ is a transversal for the $E$-classes contained in $[w]_F$.

To prove this, define first a sequence $(D_n)$ of Borel choice functions as follows: define the equivalence relation on $\Gamma$

$$\gamma \sim_w \delta \Leftrightarrow (\gamma^{b_0} \cdot w) E (\delta^{b_0} \cdot w).$$

Let $\{\gamma_{n,w}\}$ be a transversal for $\sim_w$ with $\gamma_{0,w} = 1$ and put $D_n(w) = \gamma_{n,w}^{b_0} \cdot w$.

We can then use the ergodicity of $E$ to modify $(D_n)$ to a sequence $(C_n)$ of 1-1 choice functions (which are then in $\text{Aut}(\Omega, \omega)$) as follows; see [IKT09, 2.1]. Fix $n \in \mathbb{N}$ and consider $D_n$. As it is countable-to-1, let $\Omega = \bigsqcup_{k=1}^{\infty} Y_k$ be a Borel partition such that $D_n|Y_k$ is 1-1. Let then $Z_k = D_n(Y_k)$, so that $\mu(Z_k) = \mu(Y_k)$. Since $E$ is ergodic, there is $T_k \in \text{Aut}(\Omega, \omega)$ with $T_k(w)Ew$, a.e., such that $T_k(Z_k) = Y_k$ (see, e.g., [KM04, 7.10]). Let then $C_n(w) = T_k(D_n(w))$, if $w \in Y_k$. We have $C_n(w) E D_n(w)$ and $C_n$ is 1-1. So $\{C_n\}$ are choice functions and each $C_n$ is 1-1.

Using the $\{C_n\}$ we can define the *index cocycle*

$$\varphi_{E,F} = \varphi : F \to S_\infty = \text{the symmetric group of } \mathbb{N}$$

given by

$$\varphi(w_1, w_2)(k) = n \Leftrightarrow [C_k(w_1)]_E = [C_n(w_2)]_E$$

(cocycle means: $\varphi(w_2, w_3)\varphi(w_1, w_2) = \varphi(w_1, w_3)$ whenever $w_1 F w_2 F w_3$). Define also for each $(w_1, w_2) \in F$, $\vec{\delta}(w_1, w_2) \in \Delta^{\mathbb{N}}$ by

$$(\vec{\delta}(w_1, w_2)_n)^{a_0} \cdot C_{\varphi(w_1,w_2)^{-1}(n)}(w_1) = C_n(w_2).$$

Now $S_\infty$ acts on the product group $\Delta^{\mathbb{N}}$ by shift: $(\sigma \cdot \vec{\delta})_n = \vec{\delta}_{\sigma^{-1}(n)}$, where $\vec{\delta} = (\vec{\delta}_n) \in \Delta^{\mathbb{N}}$. So we can form the semi-direct product $S_\infty \ltimes \Delta^{\mathbb{N}}$, with multiplication

$$(\sigma_1, \vec{\delta}_1)(\sigma_2, \vec{\delta}_2) = (\sigma_1\sigma_2, \vec{\delta}_1(\sigma_1 \cdot \vec{\delta}_2)).$$

Given $a \in A(\Delta, Y, \nu)$, we then have a measure-preserving action of $S_\infty \ltimes \Delta^{\mathbb{N}}$ on $(Y^{\mathbb{N}}, \nu^{\mathbb{N}})$ by

$$((\sigma, \vec{\delta}) \cdot \vec{y})_n = (\vec{\delta}_n)^a \cdot \vec{y}_{\sigma^{-1}(n)}.$$

Finally we have a cocycle for the action $b_0$, $\psi : \Gamma \times \Omega \to S_\infty \ltimes \Delta^{\mathbb{N}}$ given by

$$(*) \qquad \psi(\gamma, w) = (\varphi(w, \gamma^{b_0} \cdot w), \vec{\delta}(w, \gamma^{b_0} \cdot w))$$

(cocycle means: $\psi(\gamma_1\gamma_2, w) = \psi(\gamma_1, \gamma_2 \cdot w)\psi(\gamma_2, w)$). Finally let

$$b = \text{CInd}(a_0, b_0)_\Delta^\Gamma(a)$$

be the skew product $b = b_0 \ltimes_{\psi} (Y^{\mathbb{N}}, \mu^{\mathbb{N}})$, i.e., for $\gamma \in \Gamma$

$$\begin{aligned}\gamma^b \cdot (w, \vec{y}) &= (\gamma^{b_0} \cdot w, \psi(\gamma, w) \cdot \vec{y})\\ &= (\gamma^{b_0} \cdot w, (n \mapsto (\vec{\delta}(w, \gamma^{b_0} \cdot w)_n)^a \cdot \vec{y}_{\varphi(w,\gamma^{b_0}\cdot w)^{-1}(n)})).\end{aligned}$$

We also let $a' = a_0 \ltimes_{\psi'} (Y^{\mathbb{N}}, \mu^{\mathbb{N}})$, where $\psi'$ is the cocycle for the action $a_0$ given by replacing $b_0$ by $a_0$ in $(*)$. Thus for $\delta \in \Delta$

$$\delta^{a'} \cdot (w, \vec{y}) = (\delta^{a_0} \cdot w, (n \mapsto (\vec{\delta}(w, \delta^{a_0} \cdot w)_n)^a \cdot \vec{y}_{\varphi(w,\delta^{a_0}\cdot w)^{-1}(n)}).$$

We verify some properties of $a', b$ needed in Theorem 5.1:

(i) $E_{a'} \subseteq E_b$: trivial as $E_{a_0} \subseteq E_{b_0}$.
(ii) $b$ is free: trivial as $b_0$ is free.
(iii) $a'$ is free: trivial as $a_0$ is free.
(iv) Let $f : Z = \Omega \times Y^{\mathbb{N}} \to Y$ be given by $f(w, \vec{y}) = \vec{y}_0$. Then $a$ is a factor of $a'$ via $f$ (this follows from $C_0(w) = w$).

  Next we show that if $a$ is ergodic and $A \subseteq \Omega \times Y^{\mathbb{N}}$ has positive measure and is $a'$-invariant, then $f_* \rho_A = \nu$. Let $B \subseteq Y$, $\nu(B) = 1$ be $a$-invariant such that $\nu | B$ is the unique $a$-invariant probability measure on $B$. Then $\rho(f^{-1}(B)) = 1$, so $f_* \rho_A$ lives on $B$ and then $f_* \rho_A = \nu$.
(v) If $\overline{a} \in A(\Delta, \overline{Y}, \overline{\nu})$ is a free action which is a factor of $a$ via $g : Y \to \overline{Y}$, then for $\gamma \in \Gamma \setminus \{1\}$, $gf(\gamma^b \cdot z) \neq gf(z)$, $\rho$-a.e: Fix $\gamma \in \Gamma \setminus \{1\}$. We need to show that

$$(**) \qquad g((\vec{\delta}(w, \gamma^{b_0} \cdot w)_0)^a \cdot \vec{y}_{\varphi(w,\gamma^{b_0}\cdot w)^{-1}(0)}) \neq g(\vec{y}_0)$$

  for almost all $w, \vec{y}$. Assume not, i.e. for positively many $w, \vec{y}$ $(**)$ fails. We can also assume that $\varphi(w, \gamma^{b_0} \cdot w)^{-1}(0) = k$ is fixed for the $w, \vec{y}$ and $\vec{\delta}(w, \gamma^b \cdot w)_0 = \delta$ is also fixed. Thus $g(\delta^a \cdot \vec{y}_k) = \delta^{\overline{a}} \cdot g(\vec{y}_k) = g(\vec{y}_0)$ on a set of positive measure of $w, \vec{y}$. If $k \neq 0$ this is false using Fubini. If $k = 0$, then $\delta = 1$ by the freeness of $\overline{a}$, so $w = \gamma^{b_0} \cdot w$ for a positive set of $w$, contradicting the freeness of $b_0$.

## Small subequivalence relations

In general it is not clear that the second part of (ii) in 5.1, i.e., "$b$ is ergodic" is true. However if $E = E_{a_0}$ is "small" in $F = E_{b_0}$ in the sense to be described below, then this will be the case and in fact $b$ will be mixing.

For $\gamma \in \Gamma$, let

$$|\gamma|_E = \omega(\{w : (w, \gamma \cdot w) \in E\}) \in [0, 1].$$

We say that $E = E_{a_0}$ is *small* in $F = E_{b_0}$, if $|\gamma|_E \to 0$ as $\gamma \to \infty$.

**Theorem 6.1** (Ioana-Kechris-Tsankov [IKT09, 3.3]) *In the above notation and assuming also that $b_0$ is mixing, if $E$ is small in $F$, then* for any $a \in A(\Delta, Y, \nu)$, $b = \mathrm{CInd}(a_0, b_0)_\Delta^\Gamma(a)$ *is mixing.*

We will omit the somewhat technical proof.

## A measure theoretic version of the von Neumann Conjecture

Thus to complete the proof of Theorem 5.1 we will need to show that for $\Delta = \mathbb{F}_2$, $\Gamma$ non-amenable, there are free, ergodic $a_0 \in A(\Delta, \Omega, \omega)$, and $b_0 \in A(\Gamma, \Omega, \omega)$ mixing with $E_{a_0} \subseteq E_{b_0}$ and $E_{a_0}$ small in $E_{b_0}$.

This is based on a construction of Gaboriau-Lyons [GL09], using ideas from probability theory as well as the theory of costs, who proved that there are such $a_0, b_0$ without considering the smallness condition, which was later established by Ioana-Kechris-Tsankov. The Gaboriau-Lyons result provided an affirmative answer to a measure theoretic version of von Neumann's Conjecture.

Since a full account of the background theory needed in this construction will take us too far afield, we will only give a very rough sketch of the ideas involved. For more details see [GL09] or Houdayer [Hou11].

By some simple manipulations this result can be reduced to the case where $\Gamma$ is non-amenable and finitely generated (note that every non-amenable countable group contains a non-amenable finitely generated one).

For a fixed finite set of generators $S \subseteq \Gamma$, we denote by $\mathrm{Cay}(\Gamma, S)$ its Cayley graph with (oriented) edge set $\mathcal{E}$ (and of course vertex set $\Gamma$): $(\gamma, \delta) \in \mathcal{E}$ iff $\exists s \in S\ (\delta = \gamma s)$. $\Gamma$ acts freely on this graph by left multiplication and thus acts on $\Omega = \{0,1\}^{\mathcal{E}}$ by shift. We can view $w \in \{0,1\}^{\mathcal{E}}$ as the subgraph with vertex set $\Gamma$ and edges $e$ being those $e \in \mathcal{E}$ with $w(e) = 1$. The connected components of this graph are called the *clusters* of $w$.

On $\Omega = \{0,1\}^{\mathcal{E}}$ we put the product measure $\mu_p = \nu_p^{\mathcal{E}}$, where $0 < p < 1$, and $\nu_p(\{1\}) = p$. It is invariant under the action of $\Gamma$ and is called the *Bernoulli bond percolation*. This action is also mixing and free. For an appropriate choice of $p$, we will take $\omega = \mu_p$ and $b_0 =$ this Bernoulli action.

We now define a subequivalence relation $E^{cl} \subseteq E_{b_0} = F$ (called the *cluster equivalence relation*) by

$$(w_1, w_2) \in E^{cl} \Leftrightarrow \exists \gamma (\gamma^{-1} \cdot w_1 = w_2\ \&\ \gamma \text{ is in the cluster of 1 in } w_1).$$

Each $E^{cl}$-class $[w]_{E^{cl}}$ carries in a natural way a graph structure isomorphic to the cluster of 1 in $w$.

Now Pak and Smirnova-Nagnibeda [PSN00] show that one can choose $S$ and $p$ so that $\mu_p$-a.e. the subgraph given by $w$ has infinitely many infinite clusters each with infinitely many ends. (A connected, locally finite graph has *infinitely many ends* if for every $k$ there is a finite set of vertices which upon removal leave at least $k$ infinite connected components in the remaining graph.) It follows that the set $U^\infty \subseteq \Omega$ given by

$$w \in U^\infty \Leftrightarrow [w]_{E^{cl}} \text{ is infinite}$$

has positive $\omega (= \mu_p)$-measure and by a result of Gaboriau [Gab00, IV.24(2)] $E^{cl}|U^\infty$ has normalized cost that is finite but greater than 1. (For the theory of cost see Gaboriau [Gab00], [Gab10] and also Kechris-Miller [KM04], [Hjo09].) Also it turns out that $E^{cl}|U^\infty$ is ergodic. By a standard extension process this gives a subequivalence relation $E' \subseteq F$ such that $E'$ is ergodic and has finite cost > 1. Using the theory of cost and results of Kechris-Miller and independently Pichot (see [KM04, 28.11]) and Hjorth [Hjo06] (see also [KM04, 28.2]), this gives a free, ergodic action $a_0 \in A(\mathbb{F}_2, \Omega, \omega)$ with $E_{a_0} \subseteq E' \subseteq F = E_{b_0}$.

To make sure now that $E_{a_0}$ is small in $E_{b_0}$ one can either choose above $p$ with more care or else one starts with any $a_0$, $b_0$ as above and co-induces by $(a_0, b_0)$ an appropriate Bernoulli percolation $\overline{a}$ of $\mathbb{F}_2$ to get $\overline{b}_0$ and then shows that one can find a small subequivalence relation $E_{\overline{a}_0} \subseteq E_{\overline{b}_0}$ generated by a free, ergodic action $\overline{a}_0$ of $\mathbb{F}_2$.

## References

[BHV08] B. Bekka, P. de la Harpe, and A. Valette, *Kazhdan's property* (T), Cambridge Univ. Press, 2008.

[BK96] H. Becker and A.S. Kechris, *The descriptive set theory of Polish group actions*, Cambridge Univ. Press, 1996.

[Dye59] H.A. Dye, *On groups of measure preserving transformations, I*, Amer. J. Math. **81** (1959), 119–159.

[Dye63] ———, *On groups of measure preserving transformations, II*, Amer. J. Math. **85** (1963), no. 4, 551–576.

[Eps08] I. Epstein, *Orbit inequivalent actions of non-amenable groups*, arXiv e-print (2008), math.GR/0707.4215v2.

[Fel74] J. Feldman, *Borel structures and invariants for measurable transformations*, Proc. Amer. Math. Soc. **46** (1974), 383–394.

[Fol95] G.B. Folland, *A course in abstract harmonic analysis*, CRC Press, 1995.

[FRW06] M. Foreman, D.J. Rudolph, and B. Weiss, *On the conjugacy relation in ergodic theory*, C.R. Math. Acad. Sci. Paris **343** (2006), no. 10, 653–656.

[FW04] M. Foreman and B. Weiss, *An anti-classification theorem for ergodic measure preserving transformations*, J. Eur. Math. Soc. **6** (2004), 277–292.

[Gab00] D. Gaboriau, *Coût des relations d'equivalence et des groupes*, Inv. Math. **139** (2000), 41–98.

[Gab10] ______, *What is cost?*, Notices Amer. Math. Soc. **57(10)** (2010), 1295–1296.

[Gao09] S. Gao, *Invariant descriptive set theory*, CRC Press, 2009.

[GL09] D. Gaboriau and R. Lyons, *A measurable-group-theoretic solution to von Neumann's problem*, Inv. Math. **177** (2009), no. 3, 533–540.

[GP05] D. Gaboriau and S. Popa, *An uncountable family of nonorbit equivalent actions of $F_n$*, J. Amer. Math. Soc. **18** (2005), 547–559.

[GPS95] T. Giordano, I.F. Putnam, and C. Skau, *Topological orbit equivalence and $C^*$-crossed products*, J. Reine Angew. Math. **469** (1995), 51–111.

[Hjo97] G. Hjorth, *Non-smooth infinite dimensional group representations*, Notes at: http://www.math.ucla.edu/greg/, 1997.

[Hjo00] ______, *Classification and orbit equivalence relations*, Amer. Math. Society, 2000.

[Hjo01] ______, *On invariants for measure preserving transformations*, Fund. Math. **169** (2001), 1058–1073.

[Hjo05] ______, *A converse to Dye's theorem*, Trans. Amer. Math. Soc. **357** (2005), 3083–3103.

[Hjo06] ______, *A lemma for cost attained*, Ann. Pure Appl. Logic **143** (2006), no. 1-3, 87–102.

[Hjo09] ______, *Countable Borel equivalence relations, Borel reducibility, and orbit equivalence*, Notes at: http://www.math.ucla.edu/greg/, 2009.

[Hou11] C. Houdayer, *Invariant percolation and measured theory of nonamenable groups*, arXiv e-print (2011), math.GR/1106.5337.

[HvN42] P.R. Halmos and J. von Neumann, *Operator methods in classical mechanics, II*, Ann. of Math. **43** (1942), no. 2, 332–350.

[IKT09] A. Ioana, A.S. Kechris, and T. Tsankov, *Subequivalence relations and positive-definite functions*, Groups, Geom. and Dynam. **3** (2009), no. 4, 579–625.

[Ioa11] A. Ioana, *Orbit inequivalent actions for groups containing a copy of $F_2$*, Inv. Math. **185** (2011), no. 1, 55–73.

[Kan08] V. Kanovei, *Borel equivalence relations:structure and classification*, Amer. Math. Soc., 2008.

[Kec92] A.S. Kechris, *Countable sections for locally compact actions*, Erg. Theory and Dynam. Syst. **12** (1992), 283–295.

[Kec95] ______, *Classical descriptive set theory*, Springer, 1995.

[Kec02] ______, *Actions of Polish groups and classification problems*, Analysis and Logic, Cambridge Univ. Press, 2002.

[Kec10] ______, *Global aspects of ergodic group actions*, Amer. Math. Soc., 2010.

[KM04] A.S. Kechris and B.D. Miller, *Topics in orbit equivalence*, Springer, 2004.

[KS01] A.S. Kechris and N.E. Sofronidis, *A strong generic ergodicity property of unitary and self-adjoint operators*, Erg. Theory and Dynam. Syst. **21** (2001), 1459–1479.

[New72] M. Newman, *Integral matrices*, Academic Press, 1972.

[Ol'80] A. Yu. Ol'shanskii, *On the question of the existence of an invariant mean on a group*, Uspekhi Matem. Nauk **35** (1980), no. 4, 199–200.

[Orn70] D. Ornstein, *Bernoulli shifts with the same entropy are isomorphic*, Adv. in Math. **4** (1970), 337–352.

[OW80] D. Ornstein and B. Weiss, *Ergodic theory and amenable group actions I: The Rohlin lemma*, Bull. Amer. Math. Soc. **2** (1980), 161–164.

[PSN00] I. Pak and T. Smirnova-Nagnibeda, *On non-uniqueness of percolation in nonamenable Cayley graphs*, Acad. Sci. Paris. Sér. I. Math. **330** (2000), no. 6, 495–500.

[Sch81] K. Schmidt, *Amenability, Kazhdan's property T, strong ergodicity and invariant means for ergodic group actions*, Erg. Theory and Dynam. Syst. **1** (1981), 223–236.

[SU35] J. Schreier and S. Ulam, *Sur le nombre des générateurs d'un groupe topologique compact et connexe*, Fund. Math. **24** (1935), 302–304.

[Tho64] E. Thoma, *Über unitäre Darstellungen abzählbarer discreter Gruppen*, Math. Ann. **153** (1964), 111–138.

[Tor06] A. Tornquist, *Orbit equivalence and actions of* $\mathbb{F}_n$, J. Symb. Logic **71** (2006), 265–282.

# 10
# On the strengths and weaknesses of weak squares

Menachem Magidor and Chris Lambie-Hanson

The thirteenth Appalachian Set Theory workshop was held at Carnegie Mellon University in Pittsburgh on March 19, 2011. The lecturer was Menachem Magidor. As a graduate student Chris Lambie-Hanson assisted in writing this chapter, which is based on the workshop lectures.

## 1 Introduction

The term "square" refers not just to one but to an entire family of combinatorial principles. The strongest is denoted by "$\square$" or by "Global $\square$," and there are many interesting weakenings of this notion. Before introducing any particular square principle, we provide some motivating applications. In this section, the term "square" will serve as a generic term for "some particular square principle."

- Jensen introduced square principles based on work regarding the fine structure of $L$. In his first application, he showed that, in $L$, there exist $\kappa$-Suslin trees for every uncountable cardinal $\kappa$ that is not weakly compact.
- Let $T$ be a countable theory with a distinguished predicate $R$. A model of $T$ is said to be of type $(\lambda, \mu)$ if the cardinality of the model is $\lambda$ and the cardinality of the model's interpretation of $R$ is $\mu$. For cardinals $\alpha, \beta, \gamma$, and $\delta$, $(\alpha,\beta) \to (\gamma,\delta)$ is the assertion that for every countable theory $T$, if $T$ has a model of type $(\alpha,\beta)$, then it has a model of type $(\gamma,\delta)$. Chang showed that under GCH, $(\aleph_1,\aleph_0) \to (\kappa^+,\kappa)$ holds for

every regular cardinal $\kappa$. Jensen later showed that under GCH+square, $(\aleph_1, \aleph_0) \to (\kappa^+, \kappa)$ holds for every singular cardinal $\kappa$ as well.

- Square can be used to produce examples of incompactness, i.e. structures such that every substructure of a smaller cardinality has a certain property but the entire structure does not:
  - Square allows for the construction of a family of countable sets such that every subfamily of smaller cardinality has a transversal (i.e. a $1-1$ choice function) but the entire family does not.
  - Assuming square, one can construct a first countable topological space such that every subspace of smaller cardinality is metrizable but the entire space is not.
  - We say that an abelian group $G$ is free if, for some index set $I$,

$$G \approx \sum_{i \in I} \mathbb{Z}$$

    where $\sum$ denotes the direct sum. Square can be used to construct a group $G$ such that $G$ is not free but every subgroup of smaller cardinality is.
  - We say that an abelian group $G$ is free$^+$ if, for some index set $I$,

$$G \subseteq \prod_{i \in I} \mathbb{Z}$$

    where $\prod$ denotes the direct product. Square can be used to construct a group $G$ such that $G$ is not free$^+$ but every subgroup of smaller cardinality is.

This chapter will further explore these and other applications, as well as the consistency strengths of the failures of certain square principles. In sections 2 and 3, we introduce basic square principles and derive some immediate consequences thereof. In section 4, we present forcing arguments to separate the strengths of different square principles. Section 5 deals with scales and their interactions with squares. In section 6, we provide two examples of incompactness that can be derived from square principles. In section 7, we present a stronger version of Jensen's original construction of Suslin trees from squares. In section 8, we consider the consistency strengths of the failures of square principles. Section 9 contains results regarding weak squares at singular cardinals.

## 2 Jensen's original square principle

**Definition 2.1** Let $\kappa$ be a cardinal. $\square_\kappa$ is the assertion that there exists a sequence $\langle C_\alpha \mid \alpha \text{ limit}, \kappa < \alpha < \kappa^+\rangle$ such that for all $\alpha, \beta$ limit with $\kappa < \alpha < \beta < \kappa^+$, we have the following:

1. $C_\alpha$ is a closed, unbounded subset of $\alpha$
2. $\text{otp}(C_\alpha) < \alpha$
3. (Coherence) If $\alpha$ is a limit point of $C_\beta$, then $C_\beta \cap \alpha = C_\alpha$.

Such a sequence is called a $\square_\kappa$-sequence and can be thought of as a canonical way of witnessing that the ordinals between $\kappa$ and $\kappa^+$ are singular.

We start with a few easy observations about $\square_\kappa$-sequences.

**Proposition 2.2** *If $\square_\kappa$ holds, then there is a $\square_\kappa$-sequence*

$$\langle D_\alpha \mid \alpha \textit{ limit}, \kappa < \alpha < \kappa^+\rangle$$

*such that for all $\alpha$, $\text{otp}(D_\alpha) \leq \kappa$. In addition, if $\kappa$ is singular, then we can require that for all $\alpha$, $otp(D_\alpha) < \kappa$.*

*Proof* Suppose that $\langle C_\alpha \mid \alpha \text{ limit}, \kappa < \alpha < \kappa^+\rangle$ is a $\square_\kappa$-sequence. We will define $\langle D_\alpha \mid \alpha \text{ limit}, \alpha < \kappa^+\rangle$ so that $\langle D_\alpha \mid \alpha \text{ limit}, \kappa < \alpha < \kappa^+\rangle$ works. For $\kappa < \alpha < \kappa^+$, let $C^*_\alpha = C_\alpha \setminus \kappa$. We first define $D_\kappa$ to be any club subset of $\kappa$ of order-type $\text{cf}(\kappa)$ (if $\kappa$ is regular, we can let $D_\kappa = \kappa$). If $\delta$ is a limit point of $D_\kappa$, let $D_\delta = D_\kappa \cap \delta$. For all other limit ordinals $\delta < \kappa$, let $D_\delta = \delta \setminus \sup(D_\kappa \cap \delta)$. Recursively define $D_\alpha \subseteq C^*_\alpha$ for $\kappa < \alpha < \kappa^+$ by letting $D_\alpha = \{\gamma \mid \gamma \in C^*_\alpha, \text{otp}(C_\alpha \cap \gamma) \in D_{\text{otp}(C^*_\alpha)}\}$. It is easy to check by induction on $\alpha$ that $\langle D_\alpha \mid \alpha \text{ limit}, \kappa < \alpha < \kappa^+\rangle$ is as desired. □

Notice that, if $\langle D_\alpha \mid \alpha \text{ limit}, \kappa < \alpha < \kappa^+\rangle$ is a $\square_\kappa$-sequence as given in Proposition 2.1, if we let $D^*_\alpha = \alpha$ for limit $\alpha \leq \kappa$ and $D^*_\alpha = D_\alpha \setminus \kappa$ for $\kappa < \alpha < \kappa^+$, $\alpha$ limit, then $\langle D^*_\alpha \mid \alpha \text{ limit}, \alpha < \kappa^+\rangle$ satisfies, for all limit $\alpha < \beta < \kappa^+$:

1. $D^*_\alpha$ is a club in $\alpha$
2. $\text{otp}(D^*_\alpha) \leq \kappa$
3. If $\alpha$ is a limit point of $D^*_\beta$, then $D^*_\beta \cap \alpha = D^*_\alpha$.

Therefore, $\square_\kappa$ is equivalent to the existence of such a sequence $\langle D^*_\alpha \mid \alpha \text{ limit}, \alpha < \kappa^+\rangle$, and we will sometimes refer to such a sequence as a $\square_\kappa$-sequence.

Soon after introducing this square principle, Jensen showed that, in $L$, $\Box_\kappa$ holds for every infinite cardinal $\kappa$. In fact, it is the case that in certain other canonical inner models (all Mitchell-Steel core models, for example), $\Box_\kappa$ holds for every infinite cardinal $\kappa$. The proof that $\Box_\kappa$ holds in $L$ can be found in [4] and [7]. For more recent work concerning other inner models, see [10].

## 3 Weak squares

A natural question to ask is whether one can weaken the square principle and still get interesting combinatorial results. One such weakening of square is given by the following notion, introduced by Schimmerling.

**Definition 3.1** $\Box_{\kappa,\lambda}$ is the assertion that there exists a sequence $\langle \mathcal{C}_\alpha \mid \alpha \text{ limit}, \kappa < \alpha < \kappa^+\rangle$ such that for all $\alpha$, $|\mathcal{C}_\alpha| \leq \lambda$ and for every $C \in \mathcal{C}_\alpha$,

1. $C$ is a club in $\alpha$
2. $\text{otp}(C) \leq \kappa$
3. If $\beta$ is a limit point of $C$, then $C \cap \beta \in \mathcal{C}_\beta$

$\Box_{\kappa,<\lambda}$ is defined similarly, except, for each $\alpha$, we require $|\mathcal{C}_\alpha| < \lambda$.

Note that $\Box_{\kappa,\lambda}$ weakens as $\lambda$ grows. $\Box_{\kappa,1}$ is simply $\Box_\kappa$. $\Box_{\kappa,\kappa}$ is often called *weak square* and written as $\Box^*_\kappa$. $\Box_{\kappa,\kappa^+}$ is often called *silly square*. It is a theorem of ZFC that $\Box_{\kappa,\kappa^+}$ holds for every infinite cardinal $\kappa$: for every limit $\alpha$ such that $\kappa < \alpha < \kappa^+$, let $C_\alpha$ be a club in $\alpha$. For limit $\beta$ such that $\kappa < \beta < \kappa^+$, let $\mathcal{C}_\beta = \{C_\alpha \cap \beta \mid \alpha \text{ limit}, \beta < \alpha < \kappa^+\}$. It is easy to verify that $\langle \mathcal{C}_\beta \mid \beta \text{ limit}, \kappa < \beta < \kappa^+\rangle$ is a $\Box_{\kappa,\kappa^+}$-sequence.

**Definition 3.2** Let $\kappa$ be an infinite cardinal. A $\kappa^+$-Aronszajn tree $T$ is *special* if there is a function $f : T \to \kappa$ such that, for all $x, y \in T$, if $x <_T y$, then $f(x) \neq f(y)$.

**Theorem 3.3** *There is a special $\kappa^+$-Aronszajn tree if and only if $\Box^*_\kappa$ holds.*

*Proof* We will prove only the forward direction. The proof of the reverse direction can be found in [1].

Let $T$ be a special $\kappa^+$-Aronszajn tree, as witnessed by $f : T \to \kappa$. Let $U_\alpha$ denote the nodes of $T$ in level $\alpha$. By thinning out the tree if necessary, we can assume without loss of generality that the nodes in a branch below a limit level $\beta$ uniquely determine the node of the branch at level $\beta$. For $\alpha$ limit, $\kappa < \alpha < \kappa^+$, we define $C_\alpha$ as follows.

Let $x \in U_\alpha$. We will construct, for some $\gamma \leq \kappa$, $\langle x_\beta \mid \beta < \gamma \rangle$, an increasing sequence in the tree such that, for every $\beta < \gamma$, $x_\beta <_T x$. Let $x_0$ be the root of the tree. If $x_\alpha$ has been chosen and $x_\alpha <_T x$, let $\delta_\alpha$=min($\{f(y) \mid x_\alpha <_T y <_T x\}$. Let $x_{\alpha+1}$ be the unique $y$ such that $x_\alpha < y < x$ and $f(y) = \delta_\alpha$. If $\alpha$ is a limit ordinal and $x_\beta$ has been chosen for every $\beta < \alpha$, let $x_\alpha$ be the least upper bound of $\{x_\beta \mid \beta < \alpha\}$. Continue this construction as long as $x_\alpha < x$ and sup($\{\delta_\beta \mid \beta < \alpha\}$) $< \kappa$. In fact, we claim that if $x_\alpha <_T x$, then sup($\{\delta_\beta \mid \beta < \alpha\} < \kappa$. Suppose for sake of contradiction that there is $\alpha$ such that $x_\alpha <_T x$ but sup($\{\delta_\beta \mid \beta < \alpha\} = \kappa$. Then $f(x_\alpha) < \delta_\beta$ for some $\beta < \alpha$, contradicting the choice of $\delta_\beta$. Therefore, we can continue the construction until we reach $\gamma \leq \kappa$ such that $x_\gamma = x$.

Now let $C_x = \{\text{level}(x_\alpha) \mid \alpha < \gamma\}$. It is easy to verify that $C_x$ is a club in $\alpha$ and that otp($C_x$) $= \gamma \leq \kappa$. Let $\mathcal{C}_\alpha = \{C_x \mid x \in U_\alpha\}$. Since $T$ is Aronszajn, $|\mathcal{C}_\alpha| \leq \kappa$. It remains to check the coherence condition. Let $\beta$ be a limit point of $C_x$. Then $\beta$ is the level of some $x_\beta$, where $\langle x_\alpha \mid \alpha < \gamma \rangle$ is the sequence leading up to $x$ used to define $C_x$. Let $\langle y_\alpha \mid \alpha < \gamma' \rangle$ be the sequence leading up to $x_\beta$ used to define $C_{x_\beta}$. Notice that when defining $\langle y_\alpha \mid \alpha < \gamma' \rangle$, we went through the same steps as we went through when defining $\langle x_\alpha \mid \alpha < \gamma \rangle$, so it is easy to check by induction that, for all $\alpha < \beta$, $x_\alpha = y_\alpha$, so $C_{x_\beta} = C_x \cap \beta$, so $C_x \cap \beta \in \mathcal{C}_\beta$. □

**Definition 3.4** A $\kappa^+$-tree $T$ is *normal* if it satisfies the following properties:

1. $T$ has a unique least element.
2. For every $x \in T$, $x$ has $\kappa$-many immediate successors in $T$.
3. For every $\alpha < \beta < \kappa^+$ and every $x$ in level $\alpha$ of $T$, there is a $y$ in level $\beta$ of $T$ such that $x <_T y$.
4. For every limit ordinal $\beta < \kappa^+$, if $x$ and $y$ are in level $\beta$ of $T$ and $\{z \mid z <_T x\} = \{z \mid z <_T y\}$, then $x = y$.

We now show that if $\kappa$ is a regular cardinal, then $\Box^*_\kappa$ automatically holds under sufficient cardinal arithmetic assumptions.

**Theorem 3.5** *Suppose that $\kappa^{<\kappa} = \kappa$. Then there is a normal special $\kappa^+$-Aronszajn tree.*

*Proof* Let $Q$ be the set ${}^{<\omega}\kappa$ equipped with the lexicographic ordering. That is, if $s, t \in Q$, then $s <_l t$ iff

1. There is $n \in \mathrm{dom}(s) \cap \mathrm{dom}(t)$ such that $s(n) < t(n)$ and $s \restriction n = t \restriction n$ or
2. $\mathrm{dom}(s) < \mathrm{dom}(t)$ and $t \restriction \mathrm{dom}(s) = s$.

We will construct a special $\kappa^+$-Aronszajn tree $T$. For $\alpha < \kappa^+$, the $\alpha$-th level of the tree will be denoted $U_\alpha$. For all $\alpha < \kappa^+$, $U_\alpha$ will consist of increasing sequences from $Q$ of length $\alpha + 1$. The tree will be ordered so that for all $x, y \in T$, $x \leq_T y$ iff $x \subseteq y$. $T$ cannot have a branch of length $\kappa^+$, as such a branch would correspond to an increasing sequence from $Q$ of length $\kappa^+$. This is a contradiction, since $|Q| = \kappa$. Thus, $T$ will be an Aronszajn tree provided that $U_\alpha \neq \emptyset$ and $|U_\alpha| \leq \kappa$ for all $\alpha < \kappa^+$. It will also follow that $T$ is special: Fix a bijection $F$ between $Q$ and $\kappa$. If $x \in U_\alpha$ for some $\alpha < \kappa^+$, let $f(x) = F(x(\alpha))$. Then $f$ witnesses that $T$ is special.

We will construct $U_\alpha$ by recursion on $\alpha < \kappa^+$ so that each $U_\alpha$ satisfies the following conditions:

1. $|U_\alpha| \leq \kappa$.
2. For every $\beta < \alpha$ and $x \in U_\beta$, if $|x(\beta)| = n + 1$, there is $y \in U_\alpha$ such that $x \subset y$ and $y(\alpha) \leq_l (x(\beta) \restriction n) ^\frown \langle x(\beta)(n) + 1 \rangle$.

Let $U_0 = \{\langle\langle 0 \rangle\rangle\}$. If $\alpha = \beta + 1$, let $U_\alpha = \{x ^\frown s \mid x \in U_\beta,\ x(\beta) <_l s\}$. It is clear that $U_\alpha$ satisfies conditions 1 and 2.

Suppose $\alpha$ is a limit ordinal of cofinality $< \kappa$. Let $T_\alpha$ denote the tree below level $\alpha$. We say $b$ is a branch through $T_\alpha$ if $b$ is an increasing $\alpha$-sequence from $Q$ such that, for all $\beta < \alpha$, $b \restriction (\beta + 1) \in T_\alpha$ and such that there exists $\sup(\mathrm{ran}(b)) \in Q$. Let $U_\alpha = \{b ^\frown \langle s \rangle \mid b$ is a branch through $T_\alpha$ and $\sup(\mathrm{ran}(b)) = s\}$. $U_\alpha$ satisfies condition 1 because $\kappa^{<\kappa} = \kappa$, so there are at most $\kappa$ many branches through $T_\alpha$. We claim that $U_\alpha$ also satisfies condition 2.

To show this, fix $\beta < \alpha$ and $x \in U_\beta$ with $|x(\beta)| = n + 1$. Fix an increasing, continuous sequence of ordinals $\langle \alpha_\gamma \mid \gamma < \mathrm{cf}(\alpha) \rangle$ cofinal in $\alpha$ such that $\alpha_0 = \beta$. For $\gamma < \mathrm{cf}(\alpha)$, let $s_\gamma = x(\beta) ^\frown \langle \gamma \rangle$. Note that $\langle s_\gamma \mid \gamma < \mathrm{cf}(\alpha) \rangle$ is strictly increasing and $s = \sup(\{s_\gamma \mid \gamma < \mathrm{cf}(\alpha)\}) = x(\beta) ^\frown \langle \mathrm{cf}(\alpha) \rangle$. Now we will define a sequence $\langle x_\gamma \mid \gamma < \mathrm{cf}(\alpha) \rangle$ such that:

1. For all $\gamma < \mathrm{cf}(\alpha)$, $x_\gamma \in U_{\alpha_\gamma}$ or $x_\gamma \in U_{\alpha_\gamma + 1}$.
2. For all $\gamma < \mathrm{cf}(\alpha)$, $x_\gamma(\alpha_\gamma) = s_\gamma$ or $x_\gamma(\alpha_\gamma + 1) = s_\gamma$.
3. For all $\delta < \gamma < \mathrm{cf}(\alpha)$, $x_\delta \subset x_\gamma$.

We go by recursion on $\gamma < \mathrm{cf}(\alpha)$. Let $x_0 = x ^\frown \langle s_0 \rangle$. If $\gamma = \gamma' + 1$, then let $\bar{x}_\gamma \in U_{\alpha_\gamma}$ be such that $x_{\gamma'} \subset \bar{x}_\gamma$ and $\bar{x}_\gamma(\alpha_\gamma) \leq_l s_\gamma$. Such an $\bar{x}_\gamma$ exists because $U_{\alpha_\gamma}$ satisfies condition 2. If $\bar{x}_\gamma(\alpha_\gamma) = s_\gamma$, let $x_\gamma = \bar{x}_\gamma$. Otherwise, let $x_\gamma = \bar{x}_\gamma {}^\frown \langle s_\gamma \rangle$. If $\gamma$ is a limit ordinal, then $\bigcup_{\delta<\gamma} x_\delta$ is a branch through $T_\gamma$, and $\sup(\mathrm{ran}(\bigcup_{\delta<\gamma} x_\delta)) = s_\gamma$. By the way we constructed $U_\gamma$, $(\bigcup_{\delta<\gamma} x_\delta)^\frown \langle s_\gamma \rangle \in U_\gamma$. Let $x_\gamma = (\bigcup_{\delta<\gamma} x_\delta)^\frown \langle s_\gamma \rangle$.

Now $b = \bigcup_{\gamma<\mathrm{cf}(\alpha)} x_\gamma$ is a branch through $T_\alpha$ and $\sup(\mathrm{ran}(b)) = s$. Let $y = b^\frown \langle s \rangle$. It is easy to see that $y$ is as desired, so $U_\alpha$ satisfies condition 2.

Finally, suppose $\alpha$ is a limit ordinal of cofinality $\kappa$. Note that we can not extend all branches through $T_\alpha$, as there are possibly more than $\kappa$ many of them. We claim that for each $\beta < \alpha$ and $x \in U_\beta$, if $|x(\beta)| = n+1$, there is a branch $b$ through $T_\alpha$ such that $x \subset b$ and $\sup(\mathrm{ran}(b)) = (x(\beta) \restriction n)^\frown \langle x(\beta)(n) + 1 \rangle$. To show this, fix an increasing, continuous sequence of ordinals $\langle \alpha_\gamma \mid \gamma < \kappa \rangle$ cofinal in $\alpha$ such that $\alpha_0 = \beta$. For $\gamma < \kappa$, let $s_\gamma = x(\beta)^\frown \langle \gamma \rangle$. $\langle s_\gamma \mid \gamma < \kappa \rangle$ is increasing and $s = \sup(\{s_\gamma \mid \gamma < \kappa\}) = (x(\beta) \restriction n)^\frown \langle x(\beta)(n) + 1 \rangle$. Exactly as above, define a sequence $\langle x_\gamma \mid \gamma < \kappa \rangle$ such that:

1. For all $\gamma < \kappa$, $x_\gamma \in U_{\alpha_\gamma}$ or $x_\gamma \in U_{\alpha_\gamma+1}$.
2. For all $\gamma < \kappa$, $x_\gamma(\alpha_\gamma) = s_\gamma$ or $x_\gamma(\alpha_\gamma + 1) = s_\gamma$.
3. For all $\delta < \gamma < \kappa$, $x_\delta \subset x_\gamma$.

Then $b = \bigcup_{\gamma<\kappa} x_\gamma$ is a branch through $T_\alpha$ such that $x \subset b$ and $\sup(\mathrm{ran}(b)) = s$. Now, for each $x \in T_\alpha$, choose such a branch, $b_x$. Let $U_\alpha = \{b_x{}^\frown \langle s \rangle \mid x \in T_\alpha, \sup(\mathrm{ran}(b_x)) = s\}$. By construction, $U_\alpha$ is easily seen to satisfy conditions 1 and 2. This completes the construction of $T$. It is easy to see that $T$ is in fact a normal tree, thus concluding the proof of the theorem. □

We would like to understand the extent to which these weak squares are sufficient to obtain some of the implications of the original square principle. We are interested in particular in some combinatorial principles that serve as intermediaries between the square principles and their applications in algebra, topology, and other fields. A basic example of such a combinatorial principle is given by stationary reflection.

**Definition 3.6** Let $\mu$ be an uncountable, regular cardinal, and let $S \subseteq \mu$ be stationary. We say that $S$ *reflects* at $\alpha$ if $\alpha < \mu$, $\mathrm{cf}(\alpha) > \omega$, and $S \cap \alpha$ is stationary in $\alpha$. $S$ *does not reflect* if there is no $\alpha < \mu$ such that $S$ reflects at $\alpha$.

**Proposition 3.7** *Suppose that $\square_\kappa$ holds. Then for every stationary $S \subseteq \kappa^+$, there is a stationary $S^* \subseteq S$ such that $S^*$ does not reflect.*

*Proof* Let $\langle C_\alpha \mid \alpha \text{ limit}, \kappa < \alpha < \kappa^+ \rangle$ be a $\square_\kappa$-sequence. Let $S \subseteq \kappa^+$ be stationary. By thinning out $S$ if necessary, we may assume that $S$ consists entirely of limit ordinals and that $S \subseteq \kappa^+ \setminus \kappa$. Define a function $f : S \to \kappa$ by letting $f(\alpha) = \mathrm{otp}(C_\alpha)$ for all $\alpha \in S$. Then $f$ is a regressive function, so, by Fodor's Lemma, there is a stationary $S^* \subseteq S$ and a $\mu \leq \kappa$ such that for all $\alpha \in S^*$, $\mathrm{otp}(C_\alpha) = \mu$. Now suppose for sake of contradiction that there is $\beta < \kappa^+$ such that $\mathrm{cf}(\beta) > \omega$ and $S^* \cap \beta$ is stationary. Let $C'_\beta$ be the set of limit points of $C_\beta$. Then, since $C'_\beta$ is a club in $\beta$, $C'_\beta \cap S^*$ is unbounded in $\beta$. Let $\gamma_1 < \gamma_2 \in C'_\beta \cap S^*$. $C_\beta \cap \gamma_1 = C_{\gamma_1}$ and $C_\beta \cap \gamma_2 = C_{\gamma_2}$, so $C_{\gamma_1} \subsetneq C_{\gamma_2}$. But this is a contradiction, since $\mathrm{otp}(C_{\gamma_1}) = \mathrm{otp}(C_{\gamma_2})$. $\square$

Notice that we have actually shown something more: for every limit $\alpha$ such that $\kappa < \alpha < \kappa^+$, $C'_\alpha \cap S^*$ consists of at most one point. Note also that if, for every limit $\alpha$ such that $\kappa < \alpha < \kappa^+$, we define $D_\alpha = C'_\alpha \setminus \gamma$ if $\gamma \in C'_\alpha \cap S^*$ and $D_\alpha = C'_\alpha$ otherwise, then $\langle D_\alpha \mid \alpha \text{ limit}, \kappa < \alpha < \kappa^+ \rangle$ is a $\square_\kappa$-sequence. We thus obtain the following corollary, which plays an important role in Jensen's proof that, in $L$, $\kappa^+$ Suslin trees exist for every infinite cardinal $\kappa$:

**Corollary 3.8** *Suppose that $\square_\kappa$ holds. Then for every stationary $S \subseteq \kappa^+$, there is a non-reflecting stationary $S^* \subseteq S$ and a $\square_\kappa$-sequence $\langle D_\alpha \mid \alpha \text{ limit}, \kappa < \alpha < \kappa^+ \rangle$ such that, for every $\alpha$, $D_\alpha \cap S^* = \emptyset$.*

# 4 Separating squares

In this section, we show that, for an uncountable cardinal $\kappa$ and cardinals $\mu, \nu$ such that $1 \leq \mu < \nu \leq \kappa$, $\square_{\kappa,\mu}$ and $\square_{\kappa,\nu}$ are in fact distinct principles. We first introduce two forcing posets.

The first, denoted $\mathbb{S}(\kappa, \lambda)$, adds a $\square_{\kappa,\lambda}$-sequence while preserving all cardinals up to and including $\kappa^+$, where $\kappa$ is an uncountable cardinal and $1 \leq \lambda \leq \kappa$. Conditions of $\mathbb{S}(\kappa, \lambda)$ are functions $s$ such that:

1. $\mathrm{dom}(s) = \{\beta \leq \alpha \mid \beta \text{ is a limit ordinal}\}$ for some limit ordinal $\alpha < \kappa^+$.
2. For all $\beta \in \mathrm{dom}(s)$, $1 \leq |s(\beta)| \leq \lambda$.
3. For all $\beta \in \mathrm{dom}(s)$, $s(\beta)$ is a set of clubs in $\beta$ of order type $\leq \kappa$. If $\mathrm{cf}(\beta) < \kappa$, then $s(\beta)$ is a set of clubs $\beta$ of order type $< \kappa$.

4. For all $\beta \in \mathrm{dom}(s)$, if $C \in s(\beta)$ and $\gamma$ is a limit point of $C$, then $C \cap \gamma \in s(\gamma)$.

For all $s, t \in \mathbb{S}(\kappa, \lambda)$, $t \leq s$ iff $t$ end-extends $s$ (i.e. $s \subseteq t$).

*Fact* 4.1 $\mathbb{S}(\kappa, \lambda)$ is $\kappa^+$-distributive.

We next introduce a forcing poset that kills a square sequence.

**Definition 4.2** Let $\vec{C} = \langle C_\alpha \mid \alpha < \kappa^+ \rangle$ be a $\square_{\kappa,\lambda}$-sequence in $V$. Let $W$ be an outer model of $V$. Then $C \in W$ *threads* $\vec{C}$ iff $C$ is a club in $\kappa^+$ and for every limit point $\alpha$ of $C$, $C \cap \alpha \in C_\alpha$. It is clear from order-type considerations that if there is $C \in W$ such that $C$ threads $\vec{C}$, then $\vec{C}$ is not a $\square_{\kappa,\lambda}$-sequence in $W$.

Given a $\square_{\kappa,\lambda}$-sequence $\vec{C} = \langle C_\alpha \mid \alpha < \kappa^+ \rangle$, let $\gamma$ be a regular cardinal such that $\gamma \leq \kappa$. We will define a threading poset $\mathbb{T}_\gamma(\vec{C})$. Conditions of the poset are sets $c$ such that:

1. $c$ is a closed, bounded subset of $\kappa^+$.
2. $c$ has order type $< \gamma$.
3. For all limit points $\beta$ of $c$, $c \cap \beta \in C_\beta$.

For all $c, d \in \mathbb{T}_\gamma(\vec{C})$, $d \leq c$ iff $d$ end-extends $c$ (i.e. $d \cap (\max(c)+1) = d$).

If $\vec{C}$ is introduced by forcing with $\mathbb{S}(\kappa, \lambda)$, then $\mathbb{T}_\gamma(\vec{C})$ behaves quite nicely.

**Lemma 4.3** *Suppose $\kappa$ is an uncountable cardinal, $\lambda$ is a cardinal such that $1 \leq \lambda \leq \kappa$, and $\gamma$ is a regular cardinal $\leq \kappa$. Let $\mathbb{S} = \mathbb{S}(\kappa, \lambda)$, and let $\mathbb{T} = \mathbb{T}_\gamma(\vec{C})_{V^{\mathbb{S}}}$, where $\vec{C}$ is the $\square_{\kappa,\lambda}$-sequence added by forcing with $\mathbb{S}$. Then*

1. *$\mathbb{S} * \mathbb{T}$ has a dense $\gamma$-closed subset.*
2. *$\mathbb{T}$ adds a set of order type $\gamma$ which threads $\vec{C}$, and $(\kappa^+)^V$ has cofinality $\gamma$ in $V^{\mathbb{S}*\mathbb{T}}$.*

Namely, the dense $\gamma$-closed subset of $\mathbb{S} * \mathbb{T}$ is the set of conditions $(s, \dot{c})$ such that, for some $c \in V$, $s \Vdash \dot{c} = \check{c}$ and $\max(\mathrm{dom}(s)) = \max(c)$. The proof of the above Lemma can be found in [2]. We will also need the following Lemma:

**Lemma 4.4** *Let $\rho$, $\kappa$, and $\lambda$ be cardinals such that $\rho$ is regular and $\rho < \kappa < \lambda$. Suppose that, in $V^{\mathrm{Coll}(\rho,<\kappa)}$, $\mathbb{P}$ is a $\rho$-closed poset and $|\mathbb{P}| < \lambda$. Let $i$ be the canonical complete embedding of* $\mathrm{Coll}(\rho, < \kappa)$ *into* $\mathrm{Coll}(\rho, < \lambda)$

*(namely, $i$ is the identity map). Then $i$ can be extended to a complete embedding $j$ of* $\mathrm{Coll}(\rho, <\kappa) * \mathbb{P}$ *into* $\mathrm{Coll}(\rho, <\lambda)$ *so that the quotient forcing,* $\mathrm{Coll}(\rho, <\lambda)/j[\mathrm{Coll}(\rho, <\kappa) * \mathbb{P}]$ *is $\rho$-closed in* $V^{j[\mathrm{Coll}(\rho,<\kappa)*\mathbb{P}]}$..

**Theorem 4.5** *Let $\rho$ be a regular, uncountable cardinal and let $\mu > \rho$ be Mahlo. Then $\square_{\rho,<\rho}$ fails in* $V^{\mathrm{Coll}(\rho,<\mu)}$.

*Proof* Let $G$ be $\mathrm{Coll}(\rho, <\mu)$-generic over $V$ and suppose for sake of contradiction that $\vec{C} = \langle C_\alpha \mid \alpha < \mu \rangle$ is a $\square_{\rho,<\rho}$-sequence in $V[G]$. For $\alpha < \mu$, let $G \restriction \alpha$ denote the pointwise image of $G$ under the canonical projection from $\mathrm{Coll}(\rho, <\mu)$ onto $\mathrm{Coll}(\rho, <\alpha)$. By a standard nice names argument, the set $\{\alpha < \mu \mid \text{for all } \beta < \alpha, C_\beta \in V[G \restriction \alpha]\}$ is club in $\mu$. Thus, since $\mu$ is Mahlo, there is an inaccessible $\kappa < \mu$ such that for every $\beta < \kappa$, $C_\beta \in V[G \restriction \kappa]$. Since $G \restriction \kappa$ is $\mathrm{Coll}(\rho, <\kappa)$-generic over $V$, $\kappa = \rho^+$ in $V[G \restriction \kappa]$. It can easily be verified that $\langle C_\beta \mid \beta < \kappa \rangle$ is a $\square_{\rho,<\rho}$-sequence in $V[G \restriction \kappa]$. Note that the quotient forcing $\mathrm{Coll}(\rho, <\mu)/\mathrm{Coll}(\rho, <\kappa)$ is $\rho$-closed. Note also that the sequence $\langle C_\beta \mid \beta < \kappa \rangle$ is threaded in $V[G]$, namely by any element of $C_\kappa$. The following Lemma therefore suffices to prove the theorem:

**Lemma 4.6** *Suppose $\lambda$ is a regular, uncountable cardinal, $\vec{\mathcal{D}} = \langle \mathcal{D}_\alpha \mid \alpha < \lambda^+ \rangle$ is a $\square_{\lambda,<\lambda}$-sequence, and $\mathbb{P}$ is a $\lambda$-closed forcing poset. Then $\mathbb{P}$ does not add a thread through $\vec{\mathcal{D}}$.*

*Proof* Assume for sake of contradiction that $\dot{D}$ is a $\mathbb{P}$-name such that $\Vdash_{\mathbb{P}}$ "$\dot{D}$ is club in $\lambda^+$ and for all limit points $\alpha \in \dot{D}$, $\dot{D} \cap \alpha \in \mathcal{D}_\alpha$". First suppose that $\lambda$ is not strongly inaccessible. Let $\gamma$ be the least cardinal such that $2^\gamma \geq \lambda$. We will construct $\langle p_s \mid s \in {}^{\leq\gamma}2 \rangle$ and $\langle \alpha_\beta \mid \beta \leq \gamma \rangle$ such that:

1. For all $s, t \in {}^{\leq\gamma}2$ such that $s \subseteq t$, we have $p_s, p_t \in \mathbb{P}$ and $p_t \leq p_s$.
2. $\langle \alpha_\beta \mid \beta \leq \gamma \rangle$ is a strictly increasing, continuous sequence of ordinals less than $\lambda^+$.
3. For all $s \in {}^{<\gamma}2$, there is $\alpha < \alpha_{|s|+1}$ such that $p_{s^\frown\langle 0 \rangle}$ and $p_{s^\frown\langle 1 \rangle}$ decide the statement "$\alpha \in \dot{D}$" in opposite ways.
4. For all limit ordinals $\beta \leq \gamma$ and all $s \in {}^\beta 2$,

$$p_s \Vdash \text{“}\alpha_\beta \text{ is a limit point of } \dot{D}\text{”},$$

   and there is $D_s \in \mathcal{D}_{\alpha_\beta}$ such that $p_s \Vdash$ "$\dot{D} \cap \alpha_\beta = D_s$".

Assume for a moment that we have successfully constructed these sequences. For all $s \in {}^{\gamma}2$, there is $D_s \in \mathcal{D}_{\alpha_\gamma}$ such that

$$p_s \Vdash \text{“}\alpha_\gamma \text{ is a limit point of } \dot{D} \text{ and } \dot{D} \cap \alpha_\beta = D_s\text{”}.$$

But if $s, t \in {}^{\gamma}2$, $s \neq t$, then there is $\alpha < \alpha_\gamma$ such that $p_s$ and $p_t$ decide the statement “$\alpha \in \dot{D}$” in opposite ways, so $D_s \neq D_t$. But, since $2^\gamma \geq \lambda$, this contradicts the fact that $|\mathcal{D}_{\alpha_\gamma}| < \lambda$.

We now turn to the construction of $\langle p_s \mid s \in {}^{\leq\gamma}2\rangle$ and $\langle \alpha_\beta \mid \beta \leq \gamma\rangle$. Let $p_{\langle\rangle} = 1_{\mathbb{P}}$ and $\alpha_0 = 0$. Fix $\beta < \gamma$ and suppose that $\langle p_s \mid s \in {}^{\beta}2\rangle$ and $\alpha_\beta$ are given. Fix $s \in {}^{\beta}2$. Since $\Vdash_{\mathbb{P}}$ “$\dot{D}$ is club in $\lambda^+$”, we can find $p'_s \leq p_s$ and $\alpha > \alpha_\beta$ such that $p'_s \Vdash$ “$\alpha \in \dot{D}$”. Since $\Vdash_{\mathbb{P}}$ “$\dot{D} \notin V$”, we can find $\alpha_s > \alpha$ and $p_0, p_1 \leq p'$ such that $p_0$ and $p_1$ decide the statement “$\alpha_s \in \dot{D}$” in opposite ways. Let $p_{s^\frown\langle 0\rangle} = p_0$ and $p_{s^\frown\langle 1\rangle} = p_1$. Do this for all $s \in {}^{\beta}2$, and let $\alpha_\beta = \sup\{\alpha_s \mid s \in {}^{\beta}2\}$. $2^\beta < \lambda$, so $\alpha_\beta < \lambda^+$.

Suppose $\beta \leq \gamma$ is a limit ordinal and that $\langle p_s \mid s \in {}^{<\beta}2\rangle$ and $\langle \alpha_\delta \mid \delta < \beta\rangle$ have been constructed. Let $\alpha_\beta = \sup\{\alpha_\delta \mid \delta < \beta\rangle\}$. Fix $s \in {}^{\beta}2$. As $\mathbb{P}$ is $\lambda$-closed, we can find a $p \in \mathbb{P}$ such that, for every $\delta < \beta$, $p \leq p_{s\restriction\delta}$. Note that for every $\delta < \beta$, there is $\alpha > \alpha_\delta$ such that $p_{s\restriction\delta+1} \Vdash$ “$\alpha \in \dot{D}$”. Thus, $p \Vdash$ “$\alpha_\beta$ is a limit point of $\dot{D}$”, so $p \Vdash$ “$\dot{D} \cap \alpha_\beta \in \mathcal{D}_{\alpha_\beta}$”. Find $p' \leq p$ and $D_s \in \mathcal{D}_{\alpha_\beta}$ such that $p' \Vdash$ “$\dot{D} \cap \alpha_\beta = D_s$”. Let $p_s = p'$. It is easy to see that this is as desired.

Now suppose that $\lambda$ is strongly inaccessible. We modify the previous argument slightly. First, use Fodor's Lemma to fix a $\gamma < \lambda$ and a stationary $S \subseteq \lambda^+$ such that, for every $\alpha \in S$, $|\mathcal{D}_\alpha| \leq \gamma$. Construct sequences $\langle p_s \mid s \in {}^{\leq\gamma}2\rangle$ and $\langle \alpha_\beta \mid \beta \leq \gamma\rangle$ exactly as in the previous case. For each $s \in {}^{\gamma}2$, let $E_s = \{\alpha > \alpha_\gamma \mid$ there is $q \leq p_s$ such that $q \Vdash$ “$\alpha$ is a limit point of $\dot{D}$”$\}$. Since $\Vdash_{\mathbb{P}}$ “$\dot{D}$ is club in $\lambda^+$”, each $E_s$ contains a club, so $E = \bigcap_{s \in {}^{\gamma}2} E_s$ contains a club in $\lambda^+$. Fix $\alpha \in E \cap S$. For each $s \in {}^{\gamma}2$, find $D'_s \in \mathcal{D}_\alpha$ and $q_s \leq p_s$ such that $q_s \Vdash$ “$\dot{D} \cap \alpha = D'_s$. If $s, t \in {}^{\gamma}2$, $s \neq t$, then, as in the previous case, $D'_s \neq D'_t$, but this contradicts the fact that, since $\alpha \in S$ $|\mathcal{D}_\alpha| \leq \gamma$. This finishes the proof of the lemma and hence of the theorem. □

□

Note that if GCH holds in $V$, then $(\rho \text{ is regular and } \rho^{<\rho} = \rho)^{V^{\mathrm{Coll}(\rho,<\mu)}}$. Thus, by theorems 3.1 and 3.2, $\square^*_\rho$ holds in $V^{\mathrm{Coll}(\rho,<\mu)}$, so we have the following consistency result:

**Corollary 4.7** *Suppose $\mu$ is a Mahlo cardinal, $\rho < \mu$ is a regular, uncountable cardinal, and GCH holds in $V$. Then there is a generic extension in which*

1. *All cardinals less than or equal to* $\rho$ *are preserved and* $\mu = \rho^+$.
2. $\square^*_\rho$ *holds.*
3. $\square_{\rho,<\rho}$ *fails.*

*Remark* 4.8 Mitchell [8] showed that if $\rho > \omega_1$ is regular and there is a Mahlo cardinal $\mu > \rho$, then there is a forcing extension in which all cardinals $\leq \rho$ are preserved and there are no special $\rho^+$-Aronszajn trees (and hence $\square^*_\rho$ fails).

We will now prove another specific instance of the consistency of the separation of different square principles. This theorem is due to Jensen, who proved the result using a Mahlo cardinal rather than a measurable [6].

**Theorem 4.9** *Suppose* $\kappa$ *is a measurable cardinal and* $\rho < \kappa$ *is a regular, uncountable cardinal. Then there is a generic extension in which*

1. *All cardinals less than or equal to* $\rho$ *are preserved and* $\kappa = \rho^+$.
2. $\square_{\rho,2}$ *holds.*
3. $\square_\rho$ *fails.*

*Proof* Let $\mathbb{P} = \text{Coll}(\rho, <\kappa)$. Let $\mathbb{S} = \mathbb{S}(\rho, 2)_{V^{\mathbb{P}}}$ and let $\mathbb{T} = \mathbb{T}_\rho(\vec{C})_{V^{\mathbb{P}*\mathbb{S}}}$, where $\vec{C}$ is the $\square_{\rho,2}$-sequence added by $\mathbb{S}$. $V^{\mathbb{P}*\mathbb{S}}$ will be the model in which the desired conclusion will hold.

Fix an elementary embedding $j : V \to M$ witnessing that $\kappa$ is measurable. $j \restriction \mathbb{P}$ is the identity map and thus gives the natural complete embedding of $\mathbb{P}$ into $j(\mathbb{P}) = \text{Coll}(\rho, < j(\kappa))$. In $V^{\mathbb{P}}$, $|\mathbb{S} * \mathbb{T}| < j(\kappa)$ and, by Lemma 4.2, $\mathbb{S} * \mathbb{T}$ has a dense $\rho$-closed subset. Thus, by Lemma 4.3, we can extend $j \restriction \mathbb{P}$ to a complete embedding of $\mathbb{P} * \mathbb{S} * \mathbb{T}$ into $j(\mathbb{P})$ so that the quotient forcing $j(\mathbb{P})/\mathbb{P} * \mathbb{S} * \mathbb{T}$ is $\rho$-closed in $V^{\mathbb{P}*\mathbb{S}*\mathbb{T}}$.

Now let $G$ be $\mathbb{P}$-generic over $V$, let $H$ be $\mathbb{S}$-generic over $V[G]$, let $I$ be $\mathbb{T}$-generic over $V[G * H]$, and let $J$ be $j(\mathbb{P})/G * H * I$-generic over $V[G * H * I]$. Then, by letting $j(\tau_G) = j(\tau)_{G*H*I}$ for all $\mathbb{P}$-names $\tau$, we can extend $j$ to $j : V[G] \to M[G * H * I * J]$. We now show how to further extend $j$ so that its domain is $V[G * H]$.

Let $\vec{C} = \langle C_\alpha \mid \alpha \text{ limit}, \alpha < \kappa\rangle = \bigcup_{s \in H} s$ (so $\vec{C}$ is the $\square_{\rho,2}$-sequence added by $H$). Let $C$ be the club in $\kappa$ added by $I$. Note that for all $s \in H$, $j(s) = s$, and $j''C = C$. $\vec{C}$ is not a condition in $j(\mathbb{S})$, since it has no top element. However, it is easy to see that $S = \vec{C} \cup \{(\kappa, \{C\})\}$ is a condition and that $S \leq s = j(s)$ for every $s \in H$.

Now let $K$ be $j(\mathbb{S})$-generic over $V[G * H * I * J]$ such that $S \in K$. $j"H \subseteq K$, so we can further extend $j$ to $j : V[G*H] \to M[G*H*I*J*K]$.

Suppose for sake of contradiction that $\vec{D} = \langle D_\alpha \mid \alpha \text{ limit}, \alpha < \kappa \rangle$ is a $\square_\rho$-sequence in $V[G * H]$.

*Claim* 4.10 In $V[G*H*I*J]$, there is a club $F \subseteq \kappa$ such that for every limit point $\alpha$ of $F$, $F \cap \alpha = D_\alpha$.

Let $j(\vec{D}) = \vec{E} = \langle E_\alpha \mid \alpha \text{ limit}, \alpha < j(\kappa) \rangle$. Let $F = E_\kappa$. $F \in M[G * H * I * J * K]$, but since $j(\mathbb{S})$ is $j(\kappa)$-distributive, we have $F \in M[G*H*I*J]$. For all $\alpha < \kappa$, $D_\alpha = E_\alpha$, so $F \cap \alpha = D_\alpha$ for every limit point $\alpha$ of $F$. Thus, $F$ is as desired.

Note that, since $j(\mathbb{P})/G * H * I$ is $\rho$-closed, by Lemma 4.5 we may assume that $F \in V[G * H * I]$.

*Claim* 4.11 $F \in V[G * H]$.

Suppose not. Then there is an $\mathbb{S} * \mathbb{T}$-name $\dot{F} \in V[G]$ such that $\dot{F}^{H*I} = F$ and $\Vdash^{V[G]}_{\mathbb{S}*\mathbb{T}}$ "$\dot{F} \notin V[G][G_{\mathbb{S}}]$". We claim that for all $(s, t) \in \mathbb{S} * \mathbb{T}$, there are $s' \le s$, $t_0, t_1$, and $\alpha$ such that $(s', t_0), (s', t_1) \le (s, t)$ and the conditions $(s', t_0)$ and $(s', t_1)$ decide the statement "$\alpha \in \dot{F}$" in opposite ways. For, if not, we can define in $V[G]$ an $\mathbb{S}$-name $\dot{F}'$ such that for all $s' \le s$ and all $\alpha < \rho^+$, $s' \Vdash^{V[G]}$ "$\alpha \in \dot{F}'$" if and only if there is $t'$ such that $(s', t') \le (s', t)$ and $(s', t') \Vdash^{V[G]}$ "$\alpha \in \dot{F}$". Then $(s, t) \Vdash^{V[G]}$ "$\dot{F}' = \dot{F}$", contradicting the assumption that $F \notin V[G * H]$.

Fix a condition $(s, t)$ such that

$$(s, t) \Vdash^{V[G]} \text{"For every limit point } \alpha \text{ of } \dot{F}, \dot{F} \cap \alpha = D_\alpha\text{"}.$$

Fix $s' \le s$, $t_0, t_1 \le t$, and $\alpha < \rho^+$ such that $(s', t_0), (s', t_1) \le (s, t)$ and $(s', t_0)$ and $(s', t_1)$ decide the statement "$\alpha \in \dot{F}$" in opposite ways. Now recursively construct $s^i_j$, $t^i_j$, and $\alpha^i_j$ for $i \in \omega$ and $j \in \{0, 1\}$ such that:

1. $s^0_0 \le s'$ and, for all $i \in \omega$, $s^{i+1}_0 \le s^i_1 \le s^i_0$.
2. $\alpha < \alpha^0_0$ and, for all $i \in \omega$, $\alpha^i_0 < \alpha^i_1 < \alpha^{i+1}_0$.
3. For each $j \in \{0, 1\}$, $(s^0_j) \le (s', t_j)$ and, for all $i \in \omega$, $(s^{i+1}_j, t^{i+1}_j) \le (s^i_j, t^i_j)$.
4. For each $i \in \omega$ and $j \in \{0, 1\}$, $(s^i_j, t^i_j) \Vdash^{V[G]}$ "$\alpha^i_j \in \dot{F}$".

The construction is straightforward. Now let $\alpha^* = \sup\{\alpha^i_j \mid i \in \omega, j \in \{0, 1\}\}$. For $j \in \{0, 1\}$, let $t^*_j = \bigcup_i t^i_j \cup \{\alpha^*\}$, and let $s^* = \bigcup_{i,j} s^i_j \cup \{(\alpha^*, \{t^*_j \cap \alpha^* \mid j \in \{0, 1\}\})\}$ (note that each $t^*_j \cap \alpha^* \in V[G]$, since $\mathbb{S}$ is $\rho^+$-distributive in $V[G]$). Now $s^* \in \mathbb{S}$ and $(s^*, t^*_j) \in \mathbb{S} * \mathbb{T}$ for $j \in \{0, 1\}$. Find $\bar{s} \le s^*$

such that $\bar{s}$ decides the value of $D_{\alpha^*}$. For each $j \in \{0,1\}$, $(\bar{s}, t_j^*) \Vdash^{V[G]}$ "$\alpha^*$ is a limit point of $\dot{F}$, so $\dot{F} \cap \alpha^* = D_{\alpha^*}$". But $\alpha < \alpha^*$, and $(\bar{s}, t_0^*)$ and $(\bar{s}, t_1^*)$ decide the statement "$\alpha \in \dot{F}$" in opposite ways. Contradiction.

Thus, $F \in V[G * H]$. But $F$ threads $\vec{D}$, which was supposed to be a $\square_\rho$-sequence in $V[G*H]$. This is a contradiction, so $\square_\rho$ fails in $V[G*H]$, thus proving the theorem. □

Slight modifications of this proof will yield separation results for any $\square_{\rho,\mu}$ and $\square_{\rho,<\mu}$ where $\rho$ is regular and $1 < \mu < \rho$. Cummings, Foreman, and Magidor, in [2], provided a further modification to obtain a similar result at singular cardinals. Their result is specifically about $\aleph_\omega$, but similar methods work at other singular cardinals:

**Theorem 4.12** *Suppose $\kappa$ is a supercompact cardinal and $2^{\kappa^{+\omega}} = \kappa^{+\omega+1}$. Let $\mu$ and $\nu$ be cardinals such that $1 \leq \mu < \nu < \aleph_\omega$. Then there is a generic extension in which:*

1. *All cardinals less than or equal to $\nu$ are preserved.*
2. $\aleph_\omega = \kappa_V^{+\omega}$.
3. $\square_{\aleph_\omega,\nu}$ *holds.*
4. $\square_{\aleph_\omega,\mu}$ *fails.*

## 5 Scales

We now introduce another intermediary combinatorial principle which has useful applications and follows from weakenings of square.

Let $\lambda$ be a singular cardinal. Let $\vec{\mu} = \langle \mu_i \mid i < \mathrm{cf}(\lambda)\rangle$ be an increasing sequence of regular cardinals cofinal in $\lambda$. For $f$ and $g$ in $\prod_{i<\mathrm{cf}(\kappa)} \mu_i$, we say that $f <^* g$ if $\{j < \mathrm{cf}(\lambda) \mid f(j) \geq g(j)\}$ is bounded in $\mathrm{cf}(\lambda)$. Similarly, $f \leq^* g$ if $\{j < \mathrm{cf}(\lambda) \mid f(j) > g(j)\}$ is bounded in $\mathrm{cf}(\lambda)$.

**Definition 5.1** If $\lambda$ and $\vec{\mu}$ are as above, a *$(\lambda^+, \vec{\mu})$-scale* is a sequence $\langle f_\alpha \mid \alpha < \lambda^+\rangle$ such that:

1. For every $\alpha < \lambda^+$, $f_\alpha \in \prod_{i<cf(\kappa)} \mu_i$
2. For every $\alpha < \beta < \lambda^+$, $f_\alpha <^* f_\beta$
3. For every $g \in \prod_{i<cf(\kappa)} \mu_i$, there is $\alpha < \lambda^+$ such that $g <^* f_\alpha$

Shelah, as part of PCF theory, proved the following [12]:

**Theorem 5.2** *If $\lambda$ is a singular cardinal, then there is a sequence $\vec{\mu}$ such that there is a $(\lambda^+, \vec{\mu})$-scale.*

**Definition 5.3** Let $D$ be a set of ordinals and let $\langle f_\delta \mid \delta \in D\rangle$ be a sequence of functions in ${}^{\mathrm{cf}(\lambda)}\mathrm{OR}$ such that, for all $\delta, \delta' \in D$, if $\delta < \delta'$, then $f_\delta <^* f_{\delta'}$. The sequence is said to be *strongly increasing* if, for each $\delta \in D$, there is an $i_\delta \in \mathrm{cf}(\lambda)$ such that, for all $\delta, \delta' \in D$, if $\delta < \delta'$ and $j \geq i_\delta, i_{\delta'}$, then $f_\delta(j) < f_{\delta'}(j)$.

The following are useful strengthenings of the notion of a scale:

**Definition 5.4**
1. A $\lambda^+$-scale $\langle f_\alpha \mid \alpha < \lambda^+\rangle$ is *good* if, for every limit ordinal $\alpha < \lambda^+$, there is $D_\alpha \subsetneq \alpha$ such that $D_\alpha$ is cofinal in $\alpha$ and $\langle f_\beta \mid \beta \in D_\alpha\rangle$ is strongly increasing.
2. A $\lambda^+$-scale $\langle f_\alpha \mid \alpha < \lambda^+\rangle$ is *better* if in the definition of a good scale one can assume in addition that each $D_\alpha$ is club in $\alpha$.
3. A $\lambda^+$-scale $\langle f_\alpha \mid \alpha < \lambda^+\rangle$ is *very good* if in the definition of a good scale one can assume in addition that each $D_\alpha$ is club in $\alpha$ and that there is a $j \in \mathrm{cf}(\lambda)$ such that, if $i \geq j$, $\beta, \gamma \in D_\alpha$, and $\beta < \gamma$, then $f_\beta(i) < f_\gamma(i)$.

There is a relationship between square principles and the existence of good scales. For example, the following theorem, a proof of which can be found in [2], provides a sufficient condition for the existence of very good scales.

**Theorem 5.5** *If $\lambda$ is singular, $\kappa < \lambda$, and $\square_{\lambda,\kappa}$ holds, then there is a very good $\lambda^+$-scale.*

We give the proof here of an analogous theorem, also from [2], relating weak square and the existence of better scales.

**Theorem 5.6** *If $\lambda$ is singular and $\square^*_\lambda$ holds, then there is a better $\lambda^+$-scale.*

*Proof* Let $\langle f_\alpha \mid \alpha < \lambda^+\rangle$ be a $(\lambda^+, \vec{\mu})$-scale for some $\vec{\mu} = \langle \mu_i \mid i < \mathrm{cf}(\lambda)\rangle$. We will improve this scale to a better $(\lambda^+, \vec{\mu})$-scale, $\langle g_\alpha \mid \alpha < \lambda^+\rangle$. Fix a $\square^*_\lambda$-sequence, $\langle \mathcal{C}_\alpha \mid \alpha \text{ limit}, \alpha < \lambda^+\rangle$ such that for all $\alpha$ and all $C \in \mathcal{C}_\alpha$, $\mathrm{otp}(C) < \lambda$. We will define $\langle g_\alpha \mid \alpha < \lambda^+\rangle$ by induction.

If $g_\alpha$ has been defined, choose $g_{\alpha+1}$ such that $f_{\alpha+1} \leq^* g_{\alpha+1}$ and $g_\alpha <^* g_{\alpha+1}$.

Suppose $\alpha$ is a limit ordinal and $g_\beta$ has been defined for all $\beta < \alpha$. For each $C \in \mathcal{C}_\alpha$, define $h_C \in \prod \mu_i$ so that

$$h_C(i) = \begin{cases} 0 & : \mu_i \leq \mathrm{otp}(C) \\ \sup_{\beta \in C}(g_\beta(i)) & : \mathrm{otp}(C) < \mu_i \end{cases}$$

Since $|C_\alpha| \leq \lambda$, we can choose $g_\alpha$ such that $f_\alpha \leq^* g_\alpha$ and $h_C <^* g_\alpha$ for every $C \in \mathcal{C}_\alpha$.

It is immediate from the construction that $\langle g_\alpha \mid \alpha < \lambda^+ \rangle$ is a $(\lambda^+, \vec{\mu})$-scale. We claim that it is in fact a better scale. To show this, let $\alpha < \lambda^+$ be a limit ordinal. If $\mathrm{cf}(\alpha) = \omega$, then any $D$ which has order type $\omega$, is cofinal in $\alpha$, and consists of successor ordinals witnesses that $\langle g_\alpha \mid \alpha < \lambda^+ \rangle$ is a better scale. So, suppose that $\mathrm{cf}(\alpha) > \omega$. Pick $C \in \mathcal{C}_\alpha$. Let $D$ be the club subset of $\alpha$ consisting of the limit points of $C$. For $\beta \in D$, $C \cap \beta \in \mathcal{C}_\beta$. Thus, in defining $g_\beta$, we considered the function $h_{C\cap\beta}$, so $h_{C\cap\beta} \leq^* g_\beta$. Pick $i_\beta < \mathrm{cf}(\lambda)$ such that for all $i_\beta < j < \mathrm{cf}(\lambda)$, $\mathrm{otp}(C) < \mu_j$ and $h_{C\cap\beta}(j) \leq g_\beta(j)$. Now let $\beta, \beta' \in D$ with $\beta < \beta'$. If $j \geq i_\beta, i_{\beta'}$, then $g_\beta(j) < h_{C\cap\beta'}(j) \leq g_{\beta'}(j)$. Thus, $D$ witnesses that $\langle g_\alpha \mid \alpha < \lambda^+ \rangle$ is a better scale. □

Scales can be useful as tools for constructing interesting objects. An example is given by the following [2]:

**Theorem 5.7** *If $\lambda$ is a singular cardinal and there exists a better $\lambda^+$-scale, then there is a sequence $\langle A_\alpha \mid \alpha < \lambda^+ \rangle$ such that:*

1. *For each $\alpha < \lambda^+$, $|A_\alpha| = \mathrm{cf}(\lambda)$.*
2. *For each $\alpha < \lambda^+$, $A_\alpha$ is a cofinal subset of $\lambda$.*
3. *For each $\beta < \lambda^+$, there is a function $g_\beta : \beta \to \lambda$ such that $\{A_\alpha \setminus g_\beta(\alpha) \mid \alpha < \beta\}$ consists of mutually disjoint sets.*

*Remark* 5.8 Note that there can be no function $g : \lambda^+ \to \lambda$ such that $\{A_\alpha \setminus g(\alpha) \mid \alpha < \lambda^+\}$ consists of disjoint sets. This theorem therefore gives an example of incompactness.

*Proof* Let $\langle f_\alpha \mid \alpha < \lambda^+ \rangle$ be a better $(\lambda^+, \vec{\mu})$-scale. For each $\alpha < \lambda^+$, let $A_\alpha$ be a subset of $\lambda$ which codes $f_\alpha$ in a canonical way. By induction on $\beta$, we will show that for every $\beta < \lambda^+$, there is a function $g_\beta : \beta \to \lambda$ such that $\{A_\alpha \setminus g_\beta(\alpha) \mid \alpha < \beta\}$ consists of pairwise disjoint sets.

First, suppose that $\beta = \beta' + 1$. Let $g_\beta(\beta') = 0$. If $\alpha < \beta'$, let $k_\alpha \in \mathrm{cf}(\lambda)$ be large enough so that $\mu_{k_\alpha} > g_{\beta'}(\alpha)$ and, if $j \geq k_\alpha$, then $f_\alpha(j) < f_{\beta'}(j)$. Then, let $g_\beta(\alpha) = \mu_{k_\alpha}$. It is clear that this $g_\beta$ is as required.

Now suppose that $\beta$ is a limit ordinal. Since $\langle f_\alpha \mid \alpha < \lambda^+ \rangle$ is a better scale, there is $D$, a club in $\beta$, such that, for each $\gamma \in D$, there is an $i_\gamma < \mathrm{cf}(\lambda)$ such that, for every $\gamma < \gamma'$ in $D$, if $j \geq i_\gamma, i_{\gamma'}$, then $f_\gamma(j) < f_{\gamma'}(j)$. Let $\alpha < \beta$. Then there is a unique $\gamma \in D$ such that $\gamma \leq \alpha < \bar{\gamma}$, where $\bar{\gamma}$ denotes the smallest ordinal of $D$ larger than $\gamma$. Define $k_\alpha \in \mathrm{cf}(\lambda)$ such that

- $k_\alpha > i_\gamma, i_{\bar{\gamma}}$
- If $j \geq k_\alpha$, then $f_\gamma(j) < f_\alpha(j) < f_{\bar{\gamma}}(j)$, and
- $\mu_{k_\alpha} > g_{\bar{\gamma}}(\alpha)$

Then, let $g_\beta(\alpha) = \mu_{k_\alpha}$. We claim that this $g_\beta$ works. To show this, take $\alpha < \alpha' < \beta$. If $\alpha$ and $\alpha'$ belong to the same interval of $D$ (i.e., if there is $\gamma \in D$ such that $\gamma < \alpha < \alpha' < \bar{\gamma}$), then $g_\beta(\alpha) > g_{\bar{\gamma}}(\alpha)$ and $g_\beta(\alpha') > g_{\bar{\gamma}}(\alpha')$, so $((A_\alpha \backslash g_\beta(\alpha)) \cap (A_{\alpha'} \backslash g_\beta(\alpha'))) \subseteq ((A_\alpha \backslash g_{\bar{\gamma}}(\alpha)) \cap (A_{\alpha'} \backslash g_{\bar{\gamma}}(\alpha'))) = \emptyset$.

Suppose that $\alpha$ and $\alpha'$ do not belong to the same interval. Let $\gamma, \gamma' \in D$ be such that $\gamma < \alpha < \bar{\gamma}$ and $\gamma' < \alpha' < \bar{\gamma}'$. Note that $\bar{\gamma} \leq \gamma'$. Now, if $\mu_j > g_\beta(\alpha), g_\beta(\alpha')$, then $f_\alpha(j) < f_{\bar{\gamma}}(j) \leq f_{\gamma'}(j) < f_{\alpha'}(j)$. Thus, $g_\beta$ is as required. □

## 6 Examples of incompactness

We will now use the result of Theorem 5.4 to construct two concrete examples of incompactness, one of a topological nature and the other algebraic.

**Theorem 6.1** *Let $\lambda$ be a singular cardinal with* cf$(\lambda) = \omega$. *If $\square^*_\lambda$ holds, then there is a first countable topological space $X$ such that $X$ is not metrizable, but every subspace $Y \subset X$ with $|Y| < \lambda^+$ is metrizable.*

*Proof* Since $\square^*_\lambda$ holds and cf$(\lambda) = \omega$, there is a sequence $\langle A_\beta \mid \lambda < \beta < \lambda^+ \rangle$ such that, for every $\beta$,

1. $A_\beta$ is a cofinal subset of $\lambda$
2. $A_\beta$ is countable
3. There is a function $g_\beta : \beta \to \lambda$ such that $\{A_\alpha \setminus g_\beta(\alpha) \mid \lambda < \alpha < \beta\}$ consists of pairwise disjoint sets.

We define a topological space $X = \lambda \cup (\lambda, \lambda^+)$. $\lambda$ is endowed with the discrete topology. In general, a subset $U$ of $X$ is open if for all $\alpha$ such that $\lambda < \alpha < \lambda^+$, if $\alpha \in U$, then $A_\alpha \setminus U$ is finite. Note that $X$ is first countable: if $\alpha < \lambda$, then $\{\alpha\}$ is a neighborhood base for $\alpha$. If $\alpha \in (\lambda, \lambda^+)$, then the cofinite subsets of $A_\alpha$ form a neighborhood base.

We show that every subspace $Y \subset X$ such that $|Y| < \lambda^+$ is metrizable. First note that every such subspace $Y$ is contained in $\lambda \cup (\lambda, \beta)$ for some $\beta < \lambda^+$. It thus suffices to prove that $\lambda \cup (\lambda, \beta)$ is metrizable for every $\beta < \lambda^+$. Fix such a $\beta$. Pick $g_\beta : \beta \to \lambda$ such that $\{A_\alpha \setminus g_\beta(\alpha) \mid \lambda < \alpha < \beta\}$ consists of mutually disjoint sets. For each $\lambda < \alpha < \beta$, enumerate $A_\alpha$ as

$\{\eta_n^\alpha \mid n < \omega\}$. Set $d(\alpha, \eta_n^\alpha) = 1/n$ if $\eta_n^\alpha \in A_\alpha \setminus g_\beta(\alpha)$ and $d(\alpha, \gamma) = 1$ in all other cases. It is routine to check that $d$ is a metric and induces the subspace topology on $\lambda \cup (\lambda, \beta)$.

Finally, we show that $X$ is not metrizable. Suppose for sake of contradiction that $d$ is a metric compatible with $X$. Note that, if $\lambda < \alpha < \lambda^+$, then $\{\alpha\} \cup A_\alpha$ is an open set. Thus, there is an $n_\alpha < \omega$ such that if $d(\alpha, x) < 1/n_\alpha$, then $x \in A_\alpha$. Also, as $\alpha = \lim_{k\to\infty} \eta_k^\alpha$, there is an $\eta_\alpha \in A_\alpha$ such that $d(\alpha, \eta_\alpha) < 1/(2n_\alpha)$. Find $\alpha < \alpha'$ such that $n_\alpha = n_{\alpha'} = n$ and $\eta_\alpha = \eta_{\alpha'} = \eta$. Then $d(\alpha, \eta) < 1/(2n)$ and $d(\alpha', \eta) < 1/(2n)$, so $d(\alpha, \alpha') < 1/n$. But this means that $\alpha' \in A_\alpha$, which is a contradiction, since $\alpha' \notin \lambda$. □

**Theorem 6.2** *Let $\kappa$ be a singular cardinal with* $\mathrm{cf}(\kappa) = \omega$. *If $\Box_\kappa^*$ holds, then there is an abelian group $G$ of cardinality $\kappa^+$ such that every subgroup of $G$ of cardinality $< \kappa^+$ is free but $G$ is not free itself.*

*Proof* As before, fix a sequence $\langle A_\beta \mid \kappa < \beta < \kappa^+\rangle$ such that, for all $\beta$,

1. $A_\beta$ is a cofinal subset of $\kappa$
2. $A_\beta$ is countable
3. There is a function $g_\beta : \beta \to \kappa$ such that $\{A_\alpha \setminus g_\beta(\alpha) \mid \kappa < \alpha < \beta\}$ consists of pairwise disjoint sets.

Enumerate each $A_\beta$ as $\langle \eta_\beta^n \mid n < \omega\rangle$. Let $G$ be the abelian group generated by elements $\{X_\eta \mid \eta < \kappa\} \cup \{Z_\beta^n \mid n < \omega, \kappa < \beta < \kappa^+\}$ subject to the relations $2Z_\beta^{n+1} - Z_\beta^n = X_{\eta_\beta^n}$ for every $n < \omega$ and $\kappa < \beta < \kappa^+$. $G$ can be thought of us the quotient of the free abelian group, $F$, generated by $\{X_\eta \mid \eta < \kappa\} \cup \{Z_\beta^n \mid n < \omega, \kappa < \beta < \kappa^+\}$ with respect to these relations (so $G$ consists of cosets of $F$). To simplify notation, we will use $X_\eta$ and $Z_\beta^n$ to refer to the cosets of $F$ in $G$ containing $X_\eta$ and $Z_\beta^n$, respectively.

*Claim* 6.3 If $H$ is a subgroup of $G$ and $|H| < \kappa^+$, then $H$ is free.

Because a subgroup of a free group is necessarily free, it suffices to prove that if $H$ is generated by $\{X_\eta \mid \eta < \kappa\} \cup \{Z_\alpha^n \mid n < \omega, \kappa < \alpha < \beta\}$ for some $\beta < \kappa^+$, then $H$ is free. For each $\alpha < \beta$, let $k_\alpha = g_\beta(\alpha)$, and let $A_\alpha^* = \{\eta_\alpha^i \mid i \geq k_\alpha\}$ (so $\langle A_\alpha^* \mid \kappa < \alpha < \beta\rangle$ is a sequence of pairwise disjoint sets). We claim that $H$ is generated freely by $S = \{X_\eta \mid \eta \notin \bigcup_{\alpha<\beta} A_\alpha^*\} \cup \{Z_\alpha^i \mid \kappa < \alpha < \beta,\ i \geq k_\alpha\}$.

Let $H'$ be the group generated by $S$. We will show that $H'$=$H$. First, fix $\eta < \kappa$. If $\eta \notin \bigcup_{\alpha<\beta} A_\beta^*$, then $X_\eta$ is a generator of $H'$. So, suppose that $\eta \in A_\alpha^*$ for some $\alpha < \beta$. Then $\eta = \eta_\alpha^i$ for some $i \geq k_\alpha$. But then $Z_\alpha^{i+1}$ and

$Z^i_\alpha$ are in $S$, so, since $2Z^{i+1}_\alpha - Z^i_\alpha = X_{\eta^i_\alpha}$, we have that $X_\eta \in H'$. Thus, $X_\eta \in H'$ for every $\eta < \kappa$. Now fix $\alpha$ such that $\kappa < \alpha < \beta$. $Z^{k_\alpha - 1}_\alpha \in H'$, since $2Z^{k_\alpha}_\alpha - Z^{k_\alpha - 1}_\alpha = X_{\eta^{k_\alpha - 1}_\alpha}$ and both $Z^{k_\alpha}_\alpha$ and $X_{\eta^{k_\alpha-1}_\alpha}$ are in $H'$. Continuing inductively in this way, one shows that $Z^i_\alpha \in H'$ for every $\kappa < \alpha < \beta$ and $i < \omega$. Thus, $H \subseteq H'$, so in fact $H = H'$.

We now check that $S$ generates $H$ freely. To do this, suppose we have a relation $\sum r_i Z^{\ell_i}_{\beta_i} + \sum s_j X_{\eta_j} = 0$ which holds in $H$ (and hence in $G$), where all $Z^{\ell_i}_{\beta_i}$ and $X_{\eta_j}$ are from S. Then, by the construction of $G$, it must be the case that this relation is a linear combination of our basic relations of the form $2Z^{n+1}_\alpha - Z^n_\alpha - X_{\eta^n_\alpha} = 0$ for $n < \omega$ and $\kappa < \alpha < \kappa^+$. Say that $\sum r_i Z^{\ell_i}_{\beta_i} + \sum s_j X_{\eta_j} = \sum t_k R_k$, where the $R_k$ are of the form $2Z^{n+1}_\alpha - Z^n_\alpha - X_{\eta^n_\alpha}$. Let LHS denote $\sum r_i Z^{\ell_i}_{\beta_i} + \sum s_j X_{\eta_j}$ and RHS denote $\sum t_k R_k$.

*Subclaim* If $\kappa < \alpha < \kappa^+$ and $i < \omega$ are such that $Z^i_\alpha$ is not in $S$, then $2Z^{i+1}_\alpha - Z^i_\alpha - X_{\eta^i_\alpha}$ cannot appear in the RHS.

First note that if $Z^i_\alpha \notin S$, then $Z^j_\alpha \notin S$ for all $j < i$. Now suppose for sake of contradiction that $Z^i_\alpha \notin S$ but $2Z^{i+1}_\alpha - Z^i_\alpha - X_{\eta^i_\alpha}$ does appear in the RHS. Then, since $Z^i_\alpha$ does not appear in the LHS, it must be canceled by another term in the RHS. But the only term that can do this is $2Z^i_\alpha - Z^{i-1}_\alpha - X_{\eta^{i-1}_\alpha}$, so this term must appear in the RHS. But then, continuing inductively, we find that $2Z^1_\alpha - Z^0_\alpha - X_{\eta^0_\alpha}$ must appear in the RHS. $Z^0_\alpha \notin S$, so it doesn't appear in the LHS. However, there is nothing that can cancel it in the RHS. This is a contradiction and proves the subclaim.

We now claim that the LHS is not of the form $\sum s_j X_{\eta_j}$ (where at least one $s_j$ is nonzero). To show this, suppose for sake of contradiction that it is of this form. Suppose $\eta$ is such that $X_\eta$ appears in the LHS. Then $X_\eta$ must appear in the RHS. Then there is $\kappa < \alpha < \kappa^+$ and $i < \omega$ such that $\eta = \eta^i_\alpha$ and $2Z^{i+1}_\alpha - Z^i_\alpha - X_{\eta^i_\alpha}$ appears in the RHS. But $Z^i_\alpha$ does not appear in the LHS, so something must cancel it in the RHS. By the same argument as in the subclaim, we arrive at a contradiction.

Now suppose that some $r_i$ in the LHS is non-zero. Fix $\alpha$ such that $Z^{\ell_i}_\alpha$ appears in the LHS for some $\ell_i$. Let $\ell$ be smallest such that $Z^\ell_\alpha$ appears in the LHS. Note that, by the subclaim, $2Z^\ell_\alpha - Z^{\ell-1}_\alpha - X_{\eta^{\ell-1}_\alpha}$ cannot appear in the RHS. Thus, $2Z^{\ell+1}_\alpha - Z^\ell_\alpha - X_{\eta^\ell_\alpha}$ appears in the RHS. $\eta^\ell_\alpha \in A^*_\alpha$, so $X_{\eta^\ell_\alpha} \notin S$, so it does not appear in the LHS. It must therefore be canceled in the LHS. This implies that there is $\gamma \neq \alpha$ and $j < \omega$ such that $\eta^\ell_\alpha = \eta^j_\gamma$

and $2Z_\gamma^{j+1} - Z_\gamma^j - X_{\eta_\gamma^j}$ appears on the RHS. But, since $\gamma \neq \alpha$, either $\gamma \geq \beta$ or $\gamma < \beta$ and $\eta_\gamma^j \notin A_\gamma^*$ (so $j < k_\gamma$). In either case, $Z_\gamma^j \notin S$, contradicting the subclaim. Thus, the relation is trivial, so $S$ generates $H$ freely.

*Claim* 6.4 $G$ is not free.

Suppose for sake of contradiction that $G$ is free. Fix a set of $T$ of elements of $G$ such that $T$ generates $G$ freely. By the regularity of $\kappa^+$, we can find a $\beta < \kappa^+$ such that, if $H$ is the subgroup generated by $\{X_\eta \mid \eta < \kappa\} \cup \{Z_\alpha^n \mid n < \omega,\ \alpha < \beta\}$, then $H$ is generated freely by $T \cap H$. It follows that the quotient group $G/H$ is free.

Now, in $G/H$, we have that $2Z_\beta^{n+1} - Z_\beta^n = 0$ for all $n < \omega$. Thus, for all $n < \omega$, $Z_\beta^0 = 2^n Z_\beta^n$. In particular, $Z_\beta^0$ is infinitely divisible. Since $G/H$ is free, this means that, in $G/H$, $Z_\beta^0 = 0$. This implies that $Z_\beta^0 = \sum k_i Z_{\alpha_i}^{n_i} + \sum \ell_j X_{\eta_j}$, where each $\alpha_i < \beta$. Thus, the relation $Z_\beta^0 - \sum k_i Z_{\alpha_i}^{n_i} - \sum \ell_j X_{\eta_j}$ must hold in $G$, so this relation must be a linear combination of basic relations of the form $2Z_\alpha^{n+1} - Z_\alpha^n - X_{\eta_\alpha^n} = 0$. But this is impossible, since, to account for the $Z_\beta^0$ term, any such linear combination must contain some $Z_\beta^n$, where $n > 0$. Thus, $G$ is not free, and, in light of the fact that every subgroup of $G$ of cardinality $< \kappa^+$ is free, we get also that $|G| = \kappa^+$. □

## 7 Suslin trees

**Definition 7.1** If $\kappa$ is an infinite cardinal, then a *Suslin tree* on $\kappa$ is a tree $T$ such that the nodes of $T$ are ordinals less than $\kappa$ and every branch and every antichain of $T$ has cardinality $< \kappa$.

One of the first applications of the square principle was the following theorem of Jensen [7]:

**Theorem 7.2** *If V=L, then, for all infinite cardinals $\kappa$, there is a Suslin tree on $\kappa^+$.*

The proof of this theorem actually shows that, if there are $\vec{C}$ and $S$ such that $\vec{C} = \langle C_\alpha \mid \alpha \text{ limit}, \alpha < \kappa^+\rangle$ is a $\square_\kappa$-sequence, $S \subseteq \kappa^+$ is stationary such that, for all $\alpha$ limit, $\alpha < \kappa$, $C'_\alpha \cap S = \emptyset$ (where $C'_\alpha$ denotes the limit points of $C_\alpha$), and $\diamondsuit(S)$ holds, then there is a Suslin tree on $\kappa^+$.

We are interested in determining the minimal assumptions required to guarantee the existence of a Suslin tree. The situation is rather complex for successors of singular cardinals. For example, if $\kappa$ is a singular cardinal, it is unknown whether one can obtain a model in which there are no Suslin trees on $\kappa^+$ without killing all $\kappa^+$-Aronszajn trees.

The following result of Shelah [11] provides a slightly better result than Jensen's original theorem:

**Theorem 7.3** *If $\kappa$ is an infinite cardinal, $2^\kappa = \kappa^+$, and $S \subseteq \kappa^+$ is stationary such that, for all $\alpha \in S$, $\mathrm{cf}(\alpha) \neq \mathrm{cf}(\kappa)$, then $\diamondsuit(S)$ holds.*

**Corollary 7.4** *If $\kappa$ is an infinite cardinal, $\square_\kappa$ holds, and $2^\kappa = \kappa^+$, then there is a Suslin tree on $\kappa^+$.*

We prove here a strengthening of this result, showing that one can obtain a Suslin tree on $\kappa^+$ from weaker assumptions.

**Theorem 7.5** *If $\kappa$ is an infinite cardinal, $\square_{\kappa,<\mathrm{cf}(\kappa)}$ holds, and $2^\kappa = \kappa^+$, then there is a Suslin tree on $\kappa^+$.*

*Proof* We begin with the following claim:

*Claim* 7.6 Suppose $\langle \mathcal{C}_\alpha \mid \alpha \text{ limit}, \alpha < \kappa^+\rangle$ is a $\square_{\kappa,<\mathrm{cf}(\kappa)}$-sequence. Then, for every stationary $S \subseteq \kappa^+$, there is a stationary $S^* \subseteq S$ and a $\square_{\kappa,<\mathrm{cf}(\kappa)}$-sequence $\langle \mathcal{C}^*_\alpha \mid \alpha \text{ limit}, \alpha < \kappa^+\rangle$ such that for all $\alpha$, if $C \in \mathcal{C}^*_\alpha$, then $C' \cap S^* = \emptyset$, where $C'$ denotes the limit points of $C$.

We will prove this claim in parallel for singular and regular $\kappa$. If $\kappa$ is singular, let $\langle \kappa_i \mid i < \mathrm{cf}(\kappa)\rangle$ be a sequence of regular cardinals cofinal in $\kappa$ such that, for all $i$, $\mathrm{cf}(\kappa) < \kappa_i$. If $\kappa$ is regular, let $\kappa_i = \kappa$ for all $i < \kappa$. We will now define, by induction on $\alpha < \kappa^+$, a sequence $\langle f_\alpha \mid \alpha < \kappa^+\rangle$ (not necessarily a scale) such that, for all $\alpha, \alpha' < \kappa^+$, we have $f_\alpha \in \prod \kappa_i$ and, if $\alpha < \alpha'$, then $f_\alpha <^* f'_{\alpha'}$.

If $f_\alpha$ has been defined, we simply let $f_{\alpha+1}$ be such that $f_\alpha <^* f_{\alpha+1}$. Suppose that $\alpha$ is a limit ordinal and $\langle f_\beta \mid \beta < \alpha\rangle$ has been defined. If $\mathrm{cf}(\alpha) < \kappa$, let $s(\alpha) = \sup\{\mathrm{otp}(C) \mid C \in \mathcal{C}_\alpha, \mathrm{otp}(C) < \kappa\}$. Note that, since $|\mathcal{C}_\alpha| < \mathrm{cf}(\kappa)$, we have $s(\alpha) < \kappa$. Now, for each $C \in \mathcal{C}_\alpha$, define $h_C \in \prod \kappa_i$ by

$$h_C(i) = \begin{cases} 0 & : \kappa_i \leq \mathrm{otp}(C) \\ \sup_{\beta \in C}\{f_\beta(i) + 1\} & : \mathrm{otp}(C) < \kappa_i \end{cases}$$

If $\kappa$ is singular, then, for all $i < \mathrm{cf}(\kappa)$, let $f_\alpha(i) = \sup_{C \in \mathcal{C}_\alpha}\{h_C(i)\}$. If $\kappa$ is regular and $\mathrm{cf}(\alpha) < \kappa$, then let $f_\alpha(i) = \sup_{C \in \mathcal{C}_\alpha}\{h_C(i)\}$. If $\kappa$ is regular and $\mathrm{cf}(\alpha) = \kappa$, then simply let $f_\alpha$ be any $<^*$ bound for $\langle f_\beta \mid \beta < \alpha\rangle$.

Let $S \subseteq \kappa^+$ be stationary. Assume that, for all $\alpha \in S$, $\mathrm{cf}(\alpha) < \kappa$. By

Fodor's Lemma, we can find a stationary $\bar{S} \subseteq S$ and a $\mu < \kappa$ such that $s(\alpha) = \mu$ for all $\alpha \in \bar{S}$. Fix $i$ such that $\mu < \kappa_i$. Apply Fodor's Lemma again to obtain a stationary $S^* \subseteq \bar{S}$ and an $\eta < \kappa_i$ such that $f_\alpha(i) = \eta$ for every $\alpha \in S^*$.

Let $\beta < \kappa^+$ be such that $\mathrm{cf}(\beta) > \omega$. We claim that for every $C \in \mathcal{C}_\beta$, $C' \cap S^*$ contains at most one point. Suppose for sake of contradiction that $\alpha < \alpha'$ are such that $\alpha, \alpha' \in C' \cap S^*$. Then $C \cap \alpha' \in \mathcal{C}_{\alpha'}$, so we considered $C \cap \alpha'$ when we defined $f_{\alpha'}$. Since $\mathrm{otp}(C \cap \alpha') \leq \mu < \kappa_i$, $h_{C \cap \alpha'} = \sup_{\gamma \in C \cap \alpha'} \{f_\gamma(i) + 1\}$. Then $f_{\alpha'}(i) = \sup_{D \in \mathcal{C}_{\alpha'}} \{h_D(i)\} \geq h_{C \cap \alpha'}(i) > f_\alpha(i)$. But this contradicts the fact that $\alpha, \alpha' \in S^*$. Thus, $C' \cap S^*$ contains at most one point, so, as before, we can adjust the $\square_{\kappa, <\mathrm{cf}(\kappa)}$-sequence so that it avoids $S^*$. This finishes the claim.

We will now sketch the construction of a $\kappa^+$-Suslin tree. The construction is very much like Jensen's original construction, which can be found in [7]. The reader is directed there for more details.

By the claim, we can assume that $\langle \mathcal{C}_\alpha \mid \alpha \text{ limit}, \alpha < \kappa^+ \rangle$ is a $\square_{\kappa, <\mathrm{cf}(\kappa)}$ -sequence, $S \subseteq \kappa^+$ is stationary such that $\mathrm{cf}(\alpha) \neq \mathrm{cf}(\kappa)$ for all $\alpha \in S$ and, for all limit $\alpha < \kappa^+$ and $C \in \mathcal{C}_\alpha$, we have $C' \cap S = \emptyset$. By the above theorem of Shelah, $\diamondsuit(S)$ holds, i.e., there is a sequence $\langle B_\alpha \mid \alpha \in S \rangle$ such that, for all $X \subseteq \kappa^+$, $\{\alpha \mid \alpha \in S,\ X \cap \alpha = B_\alpha\}$ is stationary in $\kappa^+$.

We will define a Suslin tree on $\kappa^+$ by recursion on the levels of the tree. At the successor stage, we will simply split above each node, so that every node on level $\alpha$ of the tree has two immediate successors in level $\alpha + 1$. If $\alpha$ is a limit ordinal, we define level $\alpha$ of the tree as follows. Let $T_\alpha$ be the tree up to level $\alpha$. For every $x \in T_\alpha$ and every $C \in \mathcal{C}_\alpha$, we will define a branch in $T_\alpha$, $b_{x,C}$, that will be continued. Let $\mathrm{lev}(x)$ denote the level of $x$ in $T_\alpha$.

Suppose first that $\alpha \notin S$. Let $x_0 = x$. Let $x_1$ be the least ordinal in $T_\alpha$ above $x_0$ in level $\beta_0$, where $\beta_0$ is the least $\beta \in C$ such that $\beta > \mathrm{lev}(x_0)$. If $x_\gamma$ has been defined, let $x_{\gamma+1}$ be the least ordinal above $x_\gamma$ in level $\beta_\gamma$ of the tree, where $\beta_\gamma$ is the least $\beta \in C$ such that $\beta > \mathrm{lev}(x_\gamma)$. If $\gamma$ is a limit ordinal, let $x_\gamma$ be the least ordinal in level $\sup_{\eta < \gamma} \{\beta_\eta\}$ of the tree such that $x_\gamma$ is above $x_\eta$ for every $\eta < \gamma$. Continue in this manner until reaching a stage $\delta$ such that $\{\mathrm{lev}(x_\gamma) \mid \gamma < \delta\}$ is cofinal in $\alpha$. By the same argument used in Jensen's original proof, the coherence of the square sequence ensures that the construction will not break down before this point. Let $b_{x,C}$ be the downward closure of $\{x_\gamma \mid \gamma < \delta\}$, and place one node above $b_{x,C}$ in level $\alpha$ of the tree.

If $\alpha \in S$, then, if possible, let $x_0$ be the least ordinal above $x$ in $T_\alpha$ such that $x_0 \in B_\alpha$ and then continue defining $b_{x,C}$ as above.

It is routine to check by induction on $\alpha < \kappa^+$ that $|T_\alpha| \leq \kappa$. The rest of the argument that $T$ is a Suslin tree is exactly as in Jensen's original proof. □

## 8 The failure of square

In this section, we investigate the consistency strength of the failure of various square principles. We start with the following proposition of Burke and Kanamori (see [9]).

**Proposition 8.1** *Suppose $\kappa$ is a strongly compact cardinal, $\mu$ is a regular cardinal, and $\kappa \leq \mu$. Then, for all stationary $S \subseteq \mu$ such that $\mathrm{cf}(\alpha) < \kappa$ for all $\alpha \in S$, $S$ reflects to some $\beta < \mu$.*

**Corollary 8.2** *If $\kappa$ is a strongly compact cardinal, then $\square_{\lambda,<\mathrm{cf}(\lambda)}$ fails for every $\lambda \geq \kappa$.*

The following result of Shelah provides a stronger result for singular cardinals above a strongly compact.

**Theorem 8.3** *Suppose $\kappa$ is a strongly compact cardinal, $\lambda$ is a singular cardinal, and $\mathrm{cf}(\lambda) < \kappa$. Then there is no good $\lambda^+$-scale.*

**Corollary 8.4** *If $\kappa$ is strongly compact, $\lambda$ is a singular cardinal, and $\mathrm{cf}(\lambda) < \kappa$, then $\square^*_\lambda$ fails.*

However, a result of Cummings, Foreman, and Magidor [2] limits the extent to which the preceding results can be strengthened:

**Theorem 8.5** *Suppose the existence of a supercompact cardinal is consistent. Then it is consistent that there is a supercompact cardinal $\kappa$ such that $\square_{\lambda,\mathrm{cf}(\lambda)}$ holds for all singular cardinals $\lambda$ such that $\mathrm{cf}(\lambda) \geq \kappa$.*

We showed in Section 4 how to force to obtain the failure of square at a regular cardinal. Forcing to obtain the failure of square at a singular cardinal is more difficult. The following large cardinal notion will be of help in achieving this goal.

**Definition 8.6** A cardinal $\kappa$ is *subcompact* if, for all $A \subseteq H_{\kappa^+}$, there is a $\mu < \kappa$, a $B \subseteq H_{\mu^+}$, and a $\pi : \langle H_{\mu^+}, \in, B\rangle \to \langle H_{\kappa^+}, \in, A\rangle$ such that $\pi$ is an elementary embedding with critical point $\mu$.

**Proposition 8.7** *If $\kappa$ is a subcompact cardinal, then $\square_{\kappa,<\kappa}$ fails.*

*Proof* Suppose for sake of contradiction that $\langle C_\alpha \mid \alpha \text{ limit}, \alpha < \kappa^+\rangle$ is a $\square_{\kappa,<\kappa}$-sequence. We can code this sequence in a canonical way as a subset $A$ of $H_{\kappa^+}$. By subcompactness, there is a $\mu < \kappa$, a $B \subseteq H_{\mu^+}$, and a $\pi : \langle H_{\mu^+}, \in, B\rangle \to \langle H_{\kappa^+}, \in, A\rangle$ such that $\pi$ is elementary with critical point $\mu$. By absoluteness of our coding, $B$ codes a $\square_{\mu,<\mu}$-sequence, $\langle C^*_\beta \mid \beta \text{ limit}, \beta < \mu^+\rangle$. Let $D = \{\pi(\rho) \mid \rho < \mu^+\}$, and let $\eta = \sup(D)$. Let $C \in \mathcal{C}_\eta$, and let $E = C \cap D$. Note that $E$ is a $< \mu$-closed, unbounded subset of $\eta$. Now, for every limit $\alpha < \mu^+$, since $|\mathcal{C}^*_\alpha| < \mu$, $\pi[\mathcal{C}^*_\alpha] = \mathcal{C}_{\pi(\alpha)}$. Thus, if $\pi(\alpha) \in E$, then $C \cap \pi(\alpha)$ is in the range of $\pi$. Therefore, if $F = \bigcup\{\pi^{-1}(C \cap \pi(\alpha)) \mid \pi(\alpha) \in E\}$, then $F$ is an unbounded subset of $\mu^+$ such that $F \cap \alpha \in \mathcal{C}^*_\alpha$ for every $\alpha$ that is a limit point of $E$. But this contradicts the fact that $\langle \mathcal{C}^*_\alpha \mid \alpha \text{ limit}, \alpha < \mu^+\rangle$ is a $\square_{\mu,<\mu}$-sequence. Thus $\square_{\kappa,<\kappa}$ fails. □

Another notion that will be of use to us is that of Prikry forcing. Let $\kappa$ be a measurable cardinal, and fix a normal measure $U$ on $\kappa$. The Prikry forcing poset $\mathbb{P}_\kappa$ consists of conditions of the form $\langle \vec{\beta}, A\rangle$, where $\vec{\beta}$ is a finite, increasing sequence from $\kappa$ and $A \in U$. We say that $\langle \vec{\beta}^*, A^*\rangle \leq \langle \vec{\beta}, A\rangle$ if and only if $A^* \subseteq A$, $\vec{\beta}^*$ is an end extension of $\vec{\beta}$, and $\vec{\beta}^* \setminus \vec{\beta} \subseteq A$. In $V^{\mathbb{P}_\kappa}$, $\kappa$ is a singular cardinal of countable cofinality. An important feature of this forcing is that it has the Prikry property: Given a statement $\Phi$ in the forcing language and a condition $\langle \vec{\beta}, A\rangle$, there is an $A^* \subseteq A$ such that $\langle \vec{\beta}, A^*\rangle$ decides the truth value of $\Phi$.

We now present a result, due to Zeman, on the consistency of the failure of square at singular cardinals of countable cofinality.

**Theorem 8.8** *Suppose $\kappa$ is a subcompact measurable cardinal, and let $\mathbb{P}_\kappa$ be Prikry forcing for $\kappa$ with respect to a normal measure $U$. Then $\square_\kappa$ fails in $V^{\mathbb{P}_\kappa}$.*

*Proof* Suppose for sake of contradiction that $\square_\kappa$ holds in $V^{\mathbb{P}_\kappa}$. Let $\langle \dot{C}_\alpha \mid \alpha \text{ limit}, \alpha < \kappa^+\rangle$ be a sequence of $\mathbb{P}_\kappa$-names forced to be a $\square_\kappa$ sequence. $\mathbb{P}_\kappa$ and $\langle \dot{C}_\alpha \mid \alpha \text{ limit}, \alpha < \kappa^+\rangle$ can be coded by a single set $A \subseteq H_{\kappa^+}$. As $\kappa$ is subcompact, there are $\mu < \kappa$, $\bar{A} \subseteq H_{\mu^+}$, and $\pi : \langle H_{\mu^+}, \in, \bar{A}\rangle \to \langle H_{\kappa^+}, \in, A\rangle$ such that $\pi$ is elementary and $\mu = \text{crit}(\pi)$. By decoding $\bar{A}$, we obtain a forcing poset $\mathbb{P}_\mu$ and a sequence of $\mathbb{P}_\mu$-names, $\langle \dot{\bar{C}}_\alpha \mid \alpha \text{ limit}, \alpha < \mu^+\rangle$. By the elementarity of $\pi$, we may assume that every member of $\mathbb{P}_\mu$ is of the form $\langle \vec{\beta}, B\rangle$, where $\vec{\beta} \in {}^{<\omega}\mu$ and $B \subseteq \mu$ is such that $\pi(B) \in U$.

For $\alpha < \mu^+$ of countable cofinality, fix a condition $\langle \vec{\beta}_\alpha, B_\alpha\rangle \in \mathbb{P}_\mu$

and an $\eta_\alpha < \mu$ such that $\langle \vec{\beta}_\alpha, B_\alpha \rangle \Vdash \mathrm{otp}(\dot{\check{C}}_\alpha) = \check{\eta}_\alpha$. By Fodor's Lemma, we get a fixed $\vec{\beta}$ and $\eta$ such that $S = \{\alpha \mid \mathrm{cf}(\alpha) = \omega, \exists B_\alpha(\langle \vec{\beta}, B_\alpha \rangle \Vdash \mathrm{otp}(\dot{\check{C}}_\alpha) = \check{\eta})\}$ is stationary in $\mu^+$. Note that, for any $\alpha, \alpha' \in S$, $\langle \vec{\beta}, B_\alpha \rangle$ and $\langle \vec{\beta}, B_{\alpha'} \rangle$ are compatible. Let $\rho = \sup \pi''\mu^+$. $\pi''\mu^+$ is $\omega$-closed and cofinal in $\rho$, so, as $S$ is stationary in $\mu^+$, $\pi''S$ is a stationary subset of $\rho$.

Let $D$ be the set

$$\{\gamma \mid \gamma < \rho,\ \mathrm{cf}(\gamma) = \omega,\ \exists B \in U(\langle \vec{\beta}, B \rangle \Vdash \text{“}\gamma \text{ is a limit point of } \dot{C}_\rho\text{”})\}.$$

We claim first that $D$ is $\omega$-closed. To show this, let $\langle \gamma_i \mid i < \omega \rangle$ be an increasing sequence from $D$. For each $i < \omega$, there is $B_i \in U$ such that $\langle \vec{\beta}, B_i \rangle \Vdash \check{\gamma}_i \in \dot{C}_\rho$. Then $\langle \vec{\beta}, \bigcap_{i<\omega} B_i \rangle \Vdash \sup(\check{\gamma}_i) \in \dot{C}_\rho$.

We next claim that $D$ is unbounded in $\rho$. Suppose for sake of contradiction that $D$ is bounded. Let $F$ be a club in $\rho$ such that $\mathrm{otp}(F) = \mu^+ < \kappa$ and, for every $\delta \in F$, $\sup(D) < \delta$. Then for every $\delta \in F$, there is a $B_\delta \in U$ such that $\langle \vec{\beta}, B_\delta \rangle \Vdash$ “$\check{\delta}$ is not a limit point of $\dot{C}_\rho$”. Then $\langle \vec{\beta}, \bigcap_{\delta \in F} B_\delta \rangle \Vdash \check{F} \cap \dot{C}_\delta = \emptyset$. But $\dot{C}_\rho$ is forced to be a club in $\rho$, and $\mathrm{cf}(\rho)^{V^{\mathbb{P}_\kappa}} = \mu^+$, so $F$ is a club subset of $\rho$ in $V^{\mathbb{P}_\kappa}$. This is a contradiction.

Thus, $D$ is an unbounded, $\omega$-closed subset of $\rho$. Since $\mathrm{cf}(\alpha) = \omega$ for all $\alpha \in \pi''S$, we know that $\pi''S \cap D$ is unbounded in $\rho$. Let $\gamma_1, \gamma_2 < \mu^+$ be such that $\pi(\gamma_1), \pi(\gamma_2) \in \pi''S \cap D$. We know that there are $B_1^*$ and $B_2^*$ such that $\langle \vec{\beta}, B_1^* \rangle \Vdash_{\mathbb{P}_\mu} \mathrm{otp}(\dot{\check{C}}_{\gamma_1}) = \check{\eta}$ and $\langle \vec{\beta}, B_2^* \rangle \Vdash_{\mathbb{P}_\mu} \mathrm{otp}(\dot{\check{C}}_{\gamma_2}) = \check{\eta}$. Thus, appealing to the elementarity of $\pi$, there are $B_1, B_2 \in U$ such that $\langle \vec{\beta}, B_1 \rangle \Vdash \mathrm{otp}(\dot{C}_{\pi(\gamma_1)}) = \check{\eta}$ and $\langle \vec{\beta}, B_2 \rangle \Vdash \mathrm{otp}(\dot{C}_{\pi(\gamma_2)}) = \check{\eta}$. Also, there are $B_3, B_4 \in U$ such that $\langle \vec{\beta}, B_3 \rangle \Vdash \pi(\check{\gamma}_1)$ is a limit point of $\dot{C}_\rho$ and $\langle \vec{\beta}, B_4 \rangle \Vdash \pi(\check{\gamma}_2)$ is a limit point of $\dot{C}_\rho$. But then $\langle \vec{\beta}, B_1 \cap B_2 \cap B_3 \cap B_4 \rangle \Vdash \dot{C}_{\pi(\gamma_1)} = \dot{C}_\rho \cap \pi(\gamma_1)$, $\dot{C}_{\pi(\gamma_2)} = \dot{C}_\rho \cap \pi(\gamma_2)$, and $\mathrm{otp}(\dot{C}_{\pi(\gamma_1)}) = \mathrm{otp}(\dot{C}_{\pi(\gamma_2)})$. This is a contradiction. Thus, $\square_\kappa$ fails in $V^{\mathbb{P}_\kappa}$. □

There is a limit to how far we can extend this result, though, as evidenced by the following theorem of Cummings and Schimmerling [3].

**Theorem 8.9** *Suppose that $\kappa$ is a measurable cardinal and $\mathbb{P}_\kappa$ is Prikry forcing for $\kappa$. Then $\square_{\kappa,\omega}$ holds in $V^{\mathbb{P}_\kappa}$.*

## 9 Weak squares at singular cardinals

We end with a result showing that it is difficult to avoid weak squares at singular cardinals. The theorems in this section are due both to Gitik and to Dzamonja and Shelah. We start with a definition.

**Definition 9.1** Let $S \subseteq \kappa^+$ be a set of ordinals. We say that $\langle C_\alpha \mid \alpha \text{ limit}, \alpha \in S\rangle$ is a *partial square sequence* if, for all limit $\alpha \in S$:

1. $C_\alpha$ is a club in $\alpha$.
2. If $\beta$ is a limit point of $C_\alpha$, then $\beta \in S$ and $C_\beta = C_\alpha \cap \beta$.
3. $\text{otp}(C_\alpha) \leq \kappa$.

If such a sequence exists, then we say that S carries a partial square sequence.

The following fact is due to Shelah and can be found in [12].

**Proposition 9.2** *Suppose $\kappa$ is a regular cardinal and $\kappa > \aleph_1$. Then there is a sequence of sets $\langle S_i \mid i < \kappa\rangle$ such that*

1. $\bigcup_{i<\kappa} S_i = \{\alpha < \kappa^+ \mid \text{cf}(\alpha) < \kappa\}$.
2. *For each $i < \kappa$, $S_i$ carries a partial square sequence, $\langle C^i_\alpha \mid \alpha$ limit, $\alpha \in S_i\rangle$.*

*Moreover, if $\kappa$ is weakly inaccessible, then for every $i < \kappa$, there is $\mu_i < \kappa$ such that for all limit $\alpha \in S_i$, $\text{otp}(C^i_\alpha) < \mu_i$.*

**Theorem 9.3** *Suppose W is an outer model of V, $\kappa$ is an inaccessible cardinal in V and a singular cardinal in W, and $(\kappa^+)^V = (\kappa^+)^W$. In V, let $\langle D_\alpha \mid \alpha < \kappa^+\rangle$ be a sequence of clubs in $\kappa$. Then, in W, there is a sequence $\langle \delta_i \mid i < \text{cf}(\kappa)\rangle$ cofinal in $\kappa$ such that, for each $\alpha < \kappa^+$, $\{\delta_i \mid i < \text{cf}(\kappa)\rangle \setminus D_\alpha$ is bounded in $\kappa$. Moreover, if $\mu < \kappa$, then we may assume that for every $i < \text{cf}(\kappa)$, $\text{cf}(\delta_i) \geq \mu$.*

*Remark* 9.4 We will omit the proof of the "Moreover" clause at the end of the theorem and refer the interested reader to [5]. We give the proof of the remainder of the theorem here.

*Proof* Let $\text{cf}(\kappa)^W = \mu$. In $V$, let $\langle S_i \mid i < \kappa\rangle$ be as given by Proposition 9.1. For each $i < \kappa$, let $\langle C^i_\alpha \mid \alpha \text{ limit}, \alpha \in S_i\rangle$ be a partial square sequence and let $\mu_i < \kappa$ be such that for all limit $\alpha \in S_i$, $\text{otp}(C^i_\alpha) < \mu_i$. Since $(\kappa^+)^V = (\kappa^+)^W$ and the relevant notions are absolute, the following holds in $W$:

1. $\bigcup_{i<\kappa} S_i \supseteq \{\alpha < \kappa^+ \mid \text{cf}(\alpha) \neq \mu\}$.

2. For each $i < \kappa$, $\langle C^i_\alpha \mid \alpha$ limit, $\alpha \in S_i\rangle$ is a partial square sequence.
3. For each $i < \kappa$, for all limit $\alpha \in S_i$, $\text{otp}(C^i_\alpha) < \mu_i$.

We now work in $W$.

*Claim* 9.5 There is an $i^* < \kappa$ such that, if $C \in W$ is a club in $\kappa^+$, then for stationarily many $\alpha \in S_{i^*}$, $C^{i^*}_\alpha \cap C$ is a club in $\alpha$ and $\text{cf}(\alpha) = \mu^+$.

$S^{\kappa^+}_{\mu^+} = \{\alpha < \kappa^+ \mid \text{cf}(\alpha) = \mu^+\}$ is stationary in $\kappa^+$ and $S^{\kappa^+}_{\mu^+} \subseteq \bigcup_{i<\kappa} S_i$, so we can find an $i^* < \kappa$ such that $S_{i^*} \cap S^{\kappa^+}_{\mu^+}$ is stationary. But then, if $C$ is a club in $\kappa^+$ and $\alpha \in S_{i^*} \cap S^{\kappa^+}_{\mu^+} \cap C'$, then $C \cap C^{i^*}_\alpha$ is a club in $\alpha$ and $\text{cf}(\alpha) = \mu^+$.

*Claim* 9.6 In $V$, there is a sequence $\langle D^*_\alpha \mid \alpha < \kappa^+\rangle$ such that:

1. For each $\alpha < \kappa^+$, $D^*_\alpha$ is a club in $\kappa$ and $D^*_\alpha \subseteq D_\alpha$.
2. If $\alpha < \beta < \kappa^+$, then $|D^*_\beta \setminus D^*_\alpha| < \kappa$.
3. If $\beta \in C^{i^*}_\alpha$, then $D^*_\alpha \subseteq D^*_\beta$.

Work in $V$. We will prove this claim by recursion on $\alpha < \kappa^+$. Let $D^*_0 = D_0$. Suppose $\langle D^*_\beta \mid \beta < \alpha\rangle$ has been defined. Let $D^\alpha = \Delta_{\beta<\alpha} D^*_\beta$. If $\alpha \in S_{i^*}$, let $D^*_\alpha = D_\alpha \cap D^\alpha \cap \bigcap_{\beta\in C^{i^*}_\alpha} D^*_\beta$. If $\alpha \notin S_{i^*}$, let $D^*_\alpha = D_\alpha \cap D^\alpha$. Note that $D^*_\alpha$ is a club in $\kappa$ and that the sequence $\langle D^*_\alpha \mid \alpha < \kappa^+\rangle$ is as required.

Move back to $W$. Let $\langle \rho_\gamma \mid \gamma < \mu\rangle$ be a strictly increasing sequence cofinal in $\kappa$. Assume moreover that the sequence is anti-continuous, i.e., for every limit $\gamma < \mu$, $\sup_{\gamma'<\gamma} \rho_{\gamma'} < \rho_\gamma$. Let $F_\gamma$ be the interval $(\sup_{\gamma'<\gamma} \rho_{\gamma'}, \rho_\gamma)$. For every $\alpha < \kappa^+$, if $D^*_\alpha \cap F_\gamma \neq \emptyset$, let $\rho^\alpha_\gamma = \sup(D^*_\alpha \cap F_\gamma)$. If $D^*_\alpha \cap F_\gamma = \emptyset$, then $\rho^\alpha_\gamma$ is not defined. Let $d_\alpha = \{\gamma \mid D^*_\alpha \cap F_\gamma \neq \emptyset\}$. Let $E_\alpha = \{\rho^\alpha_\gamma \mid \gamma \in d_\alpha\}$. Note that if $\alpha < \alpha'$, then $D^*_{\alpha'} \setminus D^*_\alpha$ is bounded in $\kappa$, so $d_{\alpha'} \setminus d_\alpha$ is bounded in $\mu$.

Suppose there is an $\alpha < \kappa^+$ such that for every $\alpha < \alpha' < \kappa^+$, $|E_{\alpha'} \Delta E_\alpha| < \mu$. Then it is easy to verify that, if $\langle \delta_i \mid i < \mu\rangle$ is an enumeration of $E_\alpha$, then $\langle \delta_i \mid i < \mu\rangle$ is as required, and we are done.

If $\alpha < \alpha'$, we say that there is a *major change* between $\alpha$ and $\alpha'$ if $|E_\alpha \Delta E_{\alpha'}| = \mu$. Note that if there is a major change between $\alpha$ and $\alpha'$ and $\alpha' < \alpha''$, then there is a major change between $\alpha$ and $\alpha''$.

Suppose now that for every $\alpha < \kappa^+$, there is an $f(\alpha) > \alpha$ such that there is a major change between $\alpha$ and $f(\alpha)$. Let $C = \{\alpha < \kappa^+ \mid \alpha$ is closed under $f\}$. $C$ is a club in $\kappa^+$, so there is an $\alpha^* \in S_{i^*}$ such that $C^{i^*}_{\alpha^*} \cap C$ is a club in $\alpha^*$ and $\text{cf}(\alpha^*) = \mu^+$. Let $\langle \alpha_\xi \mid \xi < \mu^+\rangle$ be an increasing enumeration of a cofinal subsequence of the limit points of

$C^{i^*}_{\alpha^*}$. Note that for $\xi < \xi' < \mu^+$, $D^*_{\alpha_{\xi'}} \subseteq D^*_{\alpha_\xi}$, so $d_{\alpha_{\xi'}} \subseteq d_{\alpha_\xi}$. Thus, since $\langle d_{\alpha_\xi} \mid \xi < \mu^+ \rangle$ is a decreasing sequence of length $\mu^+$ of subsets of $\mu$, there is a fixed $d \subset \mu$ such that $d_{\alpha_\xi} = d$ for sufficiently large $\xi$.

If $\gamma \in d$ and $\xi < \xi' < \mu^+$, then, since $D^*_{\alpha_{\xi'}} \subseteq D^*_{\alpha_\xi}$, $\rho_\gamma^{\alpha_{\xi'}} \leq \rho_\gamma^{\alpha_\xi}$. Thus, for every $\gamma \in d$, there is a $\xi_\gamma < \mu^+$ and a $\bar{\rho}_\gamma$ such that for all $\xi > \xi_\gamma$, $\rho^{\alpha_\xi} = \bar{\rho}_\gamma$. Let $\xi^* = \sup\{\xi_\gamma \mid \gamma \in d\}$. Then, for every $\xi > \xi^*$ and every $\gamma \in d$, we have $\rho_\gamma^{\alpha_\xi} = \bar{\rho}_\gamma$, so there is a fixed $E$ such that, for all $\xi > \xi^*$, we have $E_{\alpha_\xi} = E$.

Now let $\xi^* < \xi < \xi'$ be such that $\xi, \xi' \in C$. Then $f(\alpha_\xi) < \alpha_{\xi'}$, so there is a major change between $\alpha_\xi$ and $\alpha_{\xi'}$. But $E_{\alpha_\xi} = E = E_{\alpha_{\xi'}}$. This is a contradiction, and we are finished. □

**Theorem 9.7** *Suppose $W$ is an outer model of $V$, $\kappa$ is an inaccessible cardinal in $V$ and a singular cardinal of countable cofinality in $W$, and $(\kappa^+)^V = (\kappa^+)^W$. Then $\square_{\kappa,\omega}$ holds in $W$.*

*Proof* We will define a sequence $\langle C_\alpha \mid \alpha \text{ limit}, \kappa < \alpha < \kappa^+ \rangle$ in $W$ witnessing $\square_{\kappa,\omega}$. We define $C_\alpha$ if and only if $\mathrm{cf}(\alpha)^V \neq \kappa$. If $\mathrm{cf}(\alpha)^V = \kappa$, then $\mathrm{cf}(\alpha)^W = \omega$, and the definition of a suitable $C_\alpha$ is trivial.

Let $\chi$ be a sufficiently large regular cardinal, and let $<_\chi$ be a well-ordering of $H_\chi$. Work in $V$. For $\alpha$ limit, $\kappa < \alpha < \kappa^+$, let $\langle M^\alpha_\gamma \mid \gamma < \kappa \rangle$ be a continuous $\subseteq$-increasing sequence of elementary submodels of $H_\chi$ such that:

1. $\alpha, \kappa \in M^\alpha_0$.
2. For every $\gamma < \kappa$, $|M^\alpha_\gamma| < \kappa$.
3. For every $\gamma < \kappa$, $M^\alpha_\gamma \cap \kappa$ is an ordinal.

Note that $\alpha \subseteq \bigcup_{\gamma<\kappa} M^\alpha_\gamma$, since $\kappa \subseteq \bigcup_{\gamma<\kappa} M^\alpha_\gamma$ and there is a function in $H_\chi$ mapping $\kappa$ onto $\alpha$.

For each limit ordinal $\alpha$ with $\kappa < \alpha < \kappa^+$, let $D_\alpha = \{M^\alpha_\gamma \cap \kappa \mid \gamma \leq \kappa\}$. $D_\alpha$ is a club in $\kappa$, so, by Theorem 9.2, there is in $W$ a sequence $\langle \delta_n \mid n < \omega \rangle$ cofinal in $\kappa$ such that, for every limit ordinal $\alpha$ with $\kappa < \alpha < \kappa^+$, $\{\delta_n \mid n < \omega\} \setminus D_\alpha$ is finite and, for every $n < \omega$, $\mathrm{cf}(\delta_n) > \omega$.

*Claim* 9.8 If $\mathrm{cf}(\gamma) > \omega$ and $M^\alpha_\gamma \cap \alpha$ is cofinal in $\alpha$, then $M^\alpha_\gamma \cap \alpha$ is $\omega$-closed.

Suppose $\langle \beta_n \mid n < \omega \rangle$ is an increasing sequence from $M^\alpha_\gamma \cap \alpha$ with $\beta_\omega = \sup_{n<\omega}(\beta_n) < \alpha$. Suppose for sake of contradiction that $\beta_\omega \notin M^\alpha_\gamma$. Let $\bar{\beta}_\omega$ be the minimal element of $M^\alpha_\gamma \cap \alpha$ above $\beta_\omega$. It is easy to see that $\mathrm{cf}(\bar{\beta}_\omega) = \kappa$: If $\mathrm{cf}(\bar{\beta}_\omega) < \kappa$, then $\mathrm{cf}(\bar{\beta}_\omega) + 1 \subseteq M^\alpha_\gamma$, so $M^\alpha_\gamma$ is cofinal in $\bar{\beta}_\omega$.

But this is a contradiction, since there are no points in $M^\alpha_\gamma$ between $\beta_\omega$ and $\bar{\beta}_\omega$. Thus, there is $E \in M^\alpha_\gamma$ such that $E$ has order type $\kappa$ and is cofinal in $\bar{\beta}_\omega$. Then, for each $n < \omega$, there is $\bar{\beta}_n \in M^\alpha_\gamma \cap E$ such that $\bar{\beta}_n \geq \beta_n$. But $\langle \bar{\beta}_n \mid n < \omega \rangle$ can not be cofinal in $M^\alpha_\gamma \cap E$, because $M^\alpha_\gamma \cap E$ has order type $M^\alpha_\gamma \cap \kappa$, and $\mathrm{cf}(M^\alpha_\gamma \cap \kappa) = \mathrm{cf}(\gamma) > \omega$. Thus, there is $\beta \in M^\alpha_\gamma \cap E$ such that $\beta > \beta_\omega$. But this contradicts our choice of $\bar{\beta}_\omega$, thus proving the claim.

If $\alpha$ is a limit ordinal, $\kappa < \alpha < \kappa^+$, and $\mathrm{cf}(\alpha) \neq \kappa$, then there is $\gamma < \kappa$ such that for all $\gamma' \geq \gamma$, $M^\alpha_{\gamma'}$ is cofinal in $\alpha$. Now let $C_\alpha = \{\overline{M^\beta_\gamma \cap \alpha} \mid \beta \geq \alpha,\ \gamma < \kappa,\ M^\beta_\gamma \cap \alpha$ is cofinal in $\alpha$, $M^\beta_\gamma \cap \beta$ is cofinal in $\beta$, and $M^\beta_\gamma \cap \kappa = \delta_n$ for some $n < \omega\}$, where $\overline{M^\beta_\gamma \cap \alpha}$ denotes the closure of $M^\beta_\gamma \cap \alpha$. By construction, $C_\alpha$ consists of clubs in $\alpha$ of order type $< \kappa$, and, by the choice of $\langle \delta_n \mid n < \omega \rangle$, each $C_\alpha$ is nonempty. Also, if $\delta < \alpha$ is a limit point of $\overline{M^\beta_\gamma \cap \alpha}$, then it is immediate from our construction that $\overline{M^\beta_\gamma \cap \delta} \in C_\delta$, so the coherence property holds.

It remains to show that $|C_\alpha| \leq \omega$. Suppose we are given $\beta, \beta', \gamma$, and $\gamma'$ such that, for some $n < \omega$, $M^\beta_\gamma \cap \kappa = \delta_n = M^{\beta'}_{\gamma'} \cap \kappa$ and both $M^\beta_\gamma$ and $M^{\beta'}_{\gamma'}$ are cofinal in $\alpha$. Notice that, since $\mathrm{cf}(\delta_n) > \omega$, it must be that $\mathrm{cf}(\gamma) > \omega$, so, by our claim, $M^\beta_\gamma \cap \beta$ is $\omega$-closed. We claim that $M^\beta_\gamma \cap \alpha = M^{\beta'}_{\gamma'} \cap \alpha$. Note that this claim implies $|C_\alpha| \leq \omega$, thus finishing the proof of the theorem.

First, suppose $\mathrm{cf}(\alpha) = \omega$. Then $\alpha \in M^\beta_\gamma, M^{\beta'}_{\gamma'}$. But then, since $M^\beta_\gamma$ and $M^{\beta'}_{\gamma'}$ are elementary submodels of $H_\chi$ having the same intersection with $\kappa$, they also have the same functions from $\kappa$ to $\alpha$, so $M^\beta_\gamma \cap \alpha = M^{\beta'}_{\gamma'} \cap \alpha$.

Finally, suppose $\mathrm{cf}(\alpha) > \omega$. If $\delta \in M^\beta_\gamma \cap M^{\beta'}_{\gamma'} \cap \alpha$ then, by the argument of the previous paragraph, $M^\beta_\gamma \cap \delta = M^{\beta'}_{\gamma'} \cap \delta$. However, since $M^\beta_\gamma \cap \alpha$ and $M^{\beta'}_{\gamma'} \cap \alpha$ are $\omega$-closed and cofinal in $\alpha$, $M^\beta_\gamma \cap M^{\beta'}_{\gamma'} \cap \alpha$ is also $\omega$-closed and cofinal in $\alpha$, so $M^\beta_\gamma \cap \alpha = M^{\beta'}_{\gamma'} \cap \alpha$. □

## References

[1] James Cummings. Notes on singular cardinal combinatorics. *Notre Dame J. Formal Logic*, 46(3):251-282, 2005.

[2] James Cummings, Matthew Foreman, and Menachem Magidor. Squares, scales, and stationary reflection. *J. Math. Log.*, 1(1): 35-98, 2001.

[3] James Cummings and Ernest Schimmerling. Indexed squares. *Israel J. Math.*, 131:61-99, 2002.

[4] Keith J. Devlin. *Constructibility*. Perspectives in Mathematical Logic. Springer-Verlag, Berlin, 1984.

[5] Mirna Džamonja and Saharon Shelah. On squares, outside guessing of clubs and $I_{<f}[\lambda]$. *Fund. Math.*, 148(2):165-198, 1995.

[6] Ronald B. Jensen. Some remarks on $\square$ below $0^{\P}$. Circulated notes.

[7] Ronald B. Jensen. The fine structure of the constructible hierarchy. *Ann. Math. Logic*, 4(3):229-308, 1972.

[8] William Mitchell. Aronszajn trees and the independence of the transfer property. *Ann. Math. Logic*, 5(1):21-46, 1972.

[9] Ernest Schimmerling. Combinatorial principles in the core model for one Woodin cardinal. *Ann. Pure Appl. Logic*, 74(2):153-201, 1995.

[10] Ernest Schimmerling and Martin Zeman. Square in core models. *Bull. Symbolic Logic*, 7(3):305-314, 2001.

[11] Saharon Shelah. On the successors of singular cardinals. In *Logic Colloquium 78*, volume 97 of *Stud. Logic Foundations Math*, pages 357-380. North-Holland, Amsterdam-New York, 1979.

[12] Saharon Shelah. *Cardinal Arithmetic*, volume 29 of *Oxford Logic Guides*. Oxford University Press, New York, 1994.

# 11

# Proper forcing remastered

Boban Veličković and Giorgio Venturi

---

The fourteenth Appalachian Set Theory workshop was held at the University of Illinois in Chicago on October 15, 2011. The lecturer was Boban Veličković. As a graduate student Giorgio Venturi assisted in writing this chapter, which is based on the workshop lectures.

---

## Abstract

We present the method introduced by Neeman of generalized side conditions with two types of models. We then discuss some applications: a variation of the Friedman-Mitchell poset for adding a club with finite conditions, the consistency of the existence of an $\omega_2$ increasing chain in $(\omega_1^{\omega_1}, <_{\text{fin}})$, originally proved by Koszmider, and the existence of a thin very tall superatomic Boolean algebra, originally proved by Baumgartner-Shelah. We expect that the present method will have many more applications.

## Introduction

We present a generalization of the method of model as side conditions. Generally speaking a poset that uses models as side conditions is a notion of forcing whose elements are pairs, consisting of a working part which is some partial information about the object we wish to add and a finite $\in$-chain of countable elementary substructures of $H(\theta)$, for some cardinal $\theta$ i.e. the structure consisting of sets whose transitive closure has cardinality less than $\theta$. The models in the side condition are used to control the extension of the working part. This is crucial in showing some general property of the forcing such as properness.

The generalization we now present amounts to allowing also certain uncountable models in the side conditions. This is used to show that the forcing preserves both $\aleph_1$ and $\aleph_2$. This approach was introduced by Neeman [9] who used it to give an alternative proof of the consistency of PFA and also to obtain generalizations of PFA to higher cardinals. In §1 we present the two-type poset of pure side conditions from [9], in the case of countable models and approachable models of size $\omega_1$, and work out the details of some of its main properties that were mentioned in [9]. The remainder of the paper is devoted to applications. We will be primarily interested in adding certain combinatorial objects of size $\aleph_2$. These results were known by other methods but we believe that the present method is more efficient and will have other applications. In §2 we present a version of the forcing for adding a club in $\omega_2$ with finite conditions, preserving $\omega_1$ and $\omega_2$. This fact has been shown to be consistent with ZFC independently by Friedman ([3]) and Mitchell ([7]) using more complicated notions of forcing. In §3 we show how to add a chain of length $\omega_2$ in the structure $(\omega_1^{\omega_1}, <_{\text{fin}})$. This result is originally due to Koszmider [5]. Finally, in §4 we give another proof of a result of Baumgartner and Shelah [2] by using side condition forcing to add a thin very tall superatomic Boolean algebra.

## 1 The forcing $\mathbb{M}$

In this section we present the forcing consisting of pure side conditions. Our presentation follows [9], but we only consider side conditions consisting of models which are either countable or of size $\aleph_1$. We consider the structure $(H(\aleph_2), \in, \trianglelefteq)$ equipped with a fixed well-ordering $\trianglelefteq$. In this way we have definable Skolem functions, so if $M$ and $N$ are elementary submodels of $H(\aleph_2)$ then so is $M \cap N$.

**Definition 1.1** Let $P$ an elementary submodel of $H(\aleph_2)$ of size $\aleph_1$. We say that $P$ is *internally approachable* if it can be written as the union of an increasing continuous $\in$-chain $\langle P_\xi : \xi < \omega_1 \rangle$ of countable elementary submodels of $H(\aleph_2)$ such that $\langle P_\xi : \xi < \eta \rangle \in P_{\eta+1}$, for every ordinal $\eta < \omega_1$.

If $P$ is internally approachable of size $\aleph_1$ we let $\vec{P}$ denote the least $\trianglelefteq$-chain witnessing this fact and we write $P_\xi$ for the $\xi$-th element of this chain. Note also that in this case $\omega_1 \subseteq P$.

**Definition 1.2** We let $\mathcal{E}_0^2$ denote the collection of all countable elementary submodels of $H(\aleph_2)$ and $\mathcal{E}_1^2$ the collection of all internally approachable elementary submodels of $H(\aleph_2)$ of size $\aleph_1$. We let $\mathcal{E}^2 = \mathcal{E}_0^2 \cup \mathcal{E}_1^2$.

The following fact is well known.

*Fact* 1.3 The set $\mathcal{E}_1^2$ is stationary in $[H(\aleph_2)]^{\aleph_1}$.

We are now ready to define the forcing notion $\mathbb{M}$ consisting of pure side conditions.

**Definition 1.4** The forcing notion $\mathbb{M}$ consists of finite $\in$-chains $p = \mathcal{M}_p$ of models in $\mathcal{E}^2$ closed under intersection. The order on $\mathbb{M}$ is reverse inclusion, i.e. $q \leq p$ if $\mathcal{M}_p \subseteq \mathcal{M}_q$.

Suppose $M$ and $N$ are elements of $\mathcal{E}^2$ with $M \in N$. If $|M| \leq |N|$ then $M \subseteq N$. However, if $M$ is of size $\aleph_1$ and $N$ is countable then the $\trianglelefteq$-least chain $\vec{M}$ witnessing that $M$ is internally approachable belongs to $N$ and so $M \cap N = M_{\delta_N}$, where $\delta_N = N \cap \omega_1$ and $M_{\delta_N}$ is the $\delta_N$-th member of $\vec{M}$.

We can split every condition in $\mathbb{M}$ in two parts: the models of size $\aleph_0$ and the models of size $\aleph_1$.

**Definition 1.5** For $p \in \mathbb{M}$ let $\pi_0(p) = p \cap \mathcal{E}_0^2$ and $\pi_1(p) = p \cap \mathcal{E}_1^2$.

Let us see some structural property of the elements of $\mathbb{M}$. First, let $\in^*$ be the transitive closure of the $\in$ relation, i.e. $x \in^* y$ if $x \in \mathrm{tcl}(y)$. Clearly, if $p \in \mathbb{M}$ then $\in^*$ is a total ordering on $\mathcal{M}_p$. Given $M, N \in \mathcal{M}_p \cup \{\emptyset, H(\aleph_2)\}$ with $M \in^* N$ let

$$(M, N)_p = \{P \in \mathcal{M}_p : M \in^* P \in^* N\}.$$

We let $(M, N]_p = (M, N)_p \cup \{N\}$, $[M, N)_p = (M, N)_p \cup \{M\}$ and $[M, N]_p = (M, N)_p \cup \{M, N\}$. Given a condition $p \in \mathbb{M}$ and $M \in p$ we let $p \restriction M$ denote the restriction of $p$ to $M$, i.e. $\mathcal{M}_p \cap M$.

*Fact* 1.6 Suppose $p \in \mathbb{M}$ and $N \in \pi_1(p)$. Then $\mathcal{M}_p \cap N = (\emptyset, N)_p$. Therefore, $p \cap N \in \mathbb{M}$.

*Fact* 1.7 Suppose $p \in \mathbb{M}$ and $M \in \pi_0(p)$. Then

$$\mathcal{M}_p \cap M = \mathcal{M}_p \setminus \bigcup \{[M \cap N, N)_p : N \in (\pi_1(p) \cap M) \cup \{H(\aleph_2)\}\}.$$

Therefore, $p \cap M \in \mathbb{M}$.

The next lemma will be used in the proof of properness of $\mathbb{M}$.

**Lemma 1.8** *Suppose $M \in \mathcal{E}^2$ and $p \in \mathbb{M} \cap M$. Then there is a new condition $p^M$, which is the smallest element of $\mathbb{M}$ extending $p$ and containing $M$ as an element.*

*Proof* If $M \in \mathcal{E}^2_1$ we can simply let

$$p^M = \mathcal{M}_{p^M} = \mathcal{M}_p \cup \{M\}.$$

If $M \in \mathcal{E}^2_0$ we close $\mathcal{M}_p \cup \{M\}$ under intersections and show that it is still an $\in$-chain. First of all notice that, since $p$ is finite and belongs to $M$, we have $\mathcal{M}_p \subseteq M$. For this reason if $P \in \pi_0(p)$, then $P \cap M = P$. On the other hand, if $P \in \pi_1(p)$, by the internal approachability of $P$ and the fact that $P \in M$ we have that $P \cap M \in P$. Now, if $N \in P$ is the $\in^*$-greatest element of $\mathcal{M}_p$ below $P$, then $N \in P \cap M$, since $\mathcal{M}_p \subseteq M$. Finally the $\in^*$-greatest element of $\mathcal{M}_p$ belongs to $M$, since $\mathcal{M}_p$ does. □

Let $\mathcal{P}$ be a forcing notion. We say that a set $M$ is *adequate* for $\mathcal{P}$ if for every $p, q \in M \cap \mathcal{P}$ if $p$ and $q$ are compatible then there is $r \in \mathcal{P} \cap M$ such that $r \leq p, q$. Note that we do not require that $\mathcal{P}$ belongs to $M$. In the forcing notions we consider if two conditions $p$ and $q$ are compatible then this will be witnessed by a condition $r$ which is $\Sigma_0$-definable from $p$ and $q$. Thus, all elements of $\mathcal{E}^2$ will be adequate for the appropriate forcing notions.

**Definition 1.9** Suppose $\mathcal{P}$ is a forcing notion and $M$ is adequate for $\mathcal{P}$. We say that a condition $p$ is $(M, \mathcal{P})$*-strongly generic* if $p$ forces that $\dot{G} \cap M$ is a $V$-generic subset of $\mathcal{P} \cap M$, where $\dot{G}$ is the canonical name for the $V$-generic filter over $\mathcal{P}$.

In order to check that a condition is strongly generic over a set $M$ we can use the following characterization, see [8] for a proof.

*Fact* 1.10 Suppose $\mathcal{P}$ a notion of forcing and $M$ is adequate for $\mathcal{P}$. A condition $p$ is $(M, \mathcal{P})$-strongly generic if and only if for every $r \leq p$ in $\mathcal{P}$ there is a condition $r \mid M \in \mathcal{P} \cap M$ such that any condition $q \leq r \mid M$ in $M$ is compatible with $r$.

**Definition 1.11** Suppose $\mathcal{P}$ is a forcing notion and $\mathcal{S}$ is a collection of sets adequate for $\mathcal{P}$. We say that $\mathcal{P}$ is $\mathcal{S}$-*strongly proper*, if for every $M \in \mathcal{S}$, every condition $p \in \mathcal{P} \cap M$ can be extended to an $(M, \mathcal{P})$-strongly generic condition $q$.

Our goal is to show that $\mathbb{M}$ is $\mathcal{E}^2$-strongly proper. We will need the following.

**Lemma 1.12** *Suppose $r \in \mathbb{M}$ and $M \in \mathcal{M}_r$. Let $q \in M$ be such that $q \leq r \cap M$. Then $q$ and $r$ are compatible.*

*Proof* If $M$ is uncountable then one can easily check that $\mathcal{M}_s = \mathcal{M}_q \cup \mathcal{M}_r$ is an $\in$-chain which is closed under intersection. Therefore $s = \mathcal{M}_s$ is a common extension of $q$ and $r$. Suppose now $M$ is countable. We first check that $\mathcal{M}_q \cup \mathcal{M}_r$ is an $\in$-chain, then we close this chain under intersections and show that the resulting set is still an $\in$-chain.

*Claim* 1.13 The set $\mathcal{M}_q \cup \mathcal{M}_r$ is an $\in$-chain.

*Proof* Note that any model of $\mathcal{M}_r \setminus M$ is either in $[M, H(\aleph_2))_r$ or belongs to an interval of the form $[N \cap M, N)_r$, for some $N \in \pi_1(r \restriction M)$. Consider one such interval $[N \cap M, N)_r$. Since $N \in r \restriction M$ and $q \leq r \restriction M$ we have that $N \in \mathcal{M}_q$. The models in $\mathcal{M}_r \cap [N \cap M, N)_r$ are an $\in$-chain. The least model on this chain is $N \cap M$ and the last one belongs to $N$. Consider the $\in^*$-largest model $P$ of $\mathcal{M}_q$ below $N$. Since $q \in M$ we have that $P \in M$. Moreover, since $\mathcal{M}_q$ is an $\in$-chain we have that also $P \in N$, therefore $P \in N \cap M$. Similarly, the least model of $\mathcal{M}_r$ in $[M, H(\aleph_2))_r$ is $M$ and it contains the top model of $\mathcal{M}_q$. Therefore, $\mathcal{M}_q \cup \mathcal{M}_r$ is an $\in$-chain. □

We now close $\mathcal{M}_q \cup \mathcal{M}_r$ under intersections and check that it is still an $\in$-chain. We let $Q \in \mathcal{M}_q \setminus \mathcal{M}_r$ and consider models of the form $Q \cap R$, for $R \in \mathcal{M}_r$.

*Case 1*: $Q \in \pi_0(q)$. We show by $\in^*$-induction on $R$ that $Q \cap R$ is already on the chain $\mathcal{M}_q$. Since $Q \in M$ and $Q$ is countable we have that $Q \subseteq M$. Therefore, $Q \cap R = Q \cap (R \cap M)$. We know that $R, M \in \mathcal{M}_r$ and $\mathcal{M}_r$ is closed under intersections, so $R \cap M \in \mathcal{M}_r$. By replacing $R$ by $R \cap M$ we may assume that $R$ is countable and below $M$ in $\mathcal{M}_r$. If $R \in M$ then $R \in \mathcal{M}_q$ and $\mathcal{M}_q$ is closed under intersection, so $Q \cap R \in \mathcal{M}_q$. If $R \in \mathcal{M}_r \setminus M$ then it belongs to an interval of the form $[N \cap M, N)_r$, for some $N \in \pi_1(r \restriction M)$. Since $N$ is uncountable and $R \in^* N$ it follows that $R \subseteq N$. If there is no uncountable model in the interval $[N \cap M, R)_r$ then we have that $N \cap M \subseteq R \subseteq N$. It follows that

$$Q \cap (N \cap M) \subseteq Q \cap R \subseteq Q \cap N.$$

However, $Q$ is a subset of $M$ and so $Q \cap (N \cap M) = Q \cap N$. Therefore, $Q \cap R = Q \cap N$ and since $Q, N \in \mathcal{M}_q$ we have again that $Q \cap N \in \mathcal{M}_q$. Now, suppose there is an uncountable model in $[N \cap M, R)_r$ and let $S$ be the $\in^*$-largest such model. Since all the models in the interval $(S, R)_r$ are countable we have that $S \in R$. On the other hand, $S$ is uncountable and above $N \cap M$ in $\mathcal{M}_r$. It follows that $N \cap M \subseteq S$. Now, consider the model $R^* = R \cap S$. It is below $S$ in $\mathcal{M}_r$. We claim that $Q \cap R = Q \cap R^*$. To see this note that, since $Q \subseteq M$ and $R \subseteq N$, we have

$$Q \cap R \subseteq Q \cap (N \cap M) \subseteq Q \cap S.$$

Therefore, $Q \cap R^* = Q \cap (R \cap S) = Q \cap R$. Since $R^*$ is below $R$ in $\mathcal{M}_r$, by the inductive assumption, we have that $Q \cap R^* \in \mathcal{M}_q$.

*Case 2*: $Q \in \pi_1(q)$. We first show that the largest element of $\mathcal{M}_q \cup \mathcal{M}_r$ below $Q$ is in $\mathcal{M}_q$. To see this note that by Fact 1.7 any model, say $S$, in $\mathcal{M}_r \setminus M$ which is below $M$ under $\in^*$ belongs to an interval of the form $[N \cap M, N)_r$, for some $N \in \pi_1(r \restriction M)$. By our assumption, $Q \in \mathcal{M}_q \setminus \mathcal{M}_r$ so $N$ is distinct from $Q$. Since $N, Q \in \mathcal{M}_q$ and they are both uncountable it follows that either $Q \in N$ or $N \in Q$. In the first case, $Q \in N \cap M$, i.e. $Q$ is $\in^*$-below $S$. In the second case, $S \in^* N \in^* Q$ and $N \in M$.

We now consider models of the form $Q \cap R$, for $R \in \mathcal{M}_r$. If $R$ is uncountable then either $Q \subseteq R$ or $R \subseteq Q$ so $Q \cap R$ is in $\mathcal{M}_q \cup \mathcal{M}_r$. If $R$ is countable and below $Q$ on the chain $\mathcal{M}_q \cup \mathcal{M}_r$ then $R \subseteq Q$, so $Q \cap R = R$. If $R \in \mathcal{M}_r \cap M$ then $R \in \mathcal{M}_q$ and since $\mathcal{M}_q$ is closed under intersections we have that $Q \cap R \in \mathcal{M}_q$. So, suppose $R \in \pi_0(r) \setminus M$. By Fact 1.7 we know that $R$ is either in $[M, H(\aleph_2))_r$ or in $[N \cap M, N)_r$, for some $N \in \pi_1(r \restriction M)$. We show by $\in^*$-induction that $Q \cap R$ is either in $\mathcal{M}_q \cup \mathcal{M}_r$ or is equal to $Q_{\delta_R}$ and moreover $\delta_R \geq \delta_M$. Consider the case $R \in [M, H(\aleph_2))_r$. If there is no uncountable $S$ in the interval $(M, R)_r$ then $M \subseteq R$. Therefore, $Q \in R$ and $\delta_R \geq \delta_M$. Since $Q \in R$ then $Q \cap R = Q_{\delta_R}$. If there is an uncountable model in the interval $(M, R)_r$ let $S$ be the largest such model. Since $Q$ is below $S$ in the $\mathcal{M}_q \cup \mathcal{M}_r$ chain we have $Q \subseteq S$, so if we let $R^* = R \cap S$, then $Q \cap R^* = Q \cap R$, and moreover $\delta_{R^*} = \delta_R$. Therefore, we can use the inductive hypothesis for $R^*$. The case when $R$ belongs to an interval of the form $[N \cap M, N)_r$, for some $N \in \pi_1(r \restriction M) \cup \{H(\aleph_2)\}$ is treated in the same way.

The upshot of all of this is that when we close $\mathcal{M}_q \cup \mathcal{M}_r$ under intersections the only new models we add are of the form $Q_\xi$, for $Q \in$

$\pi_1(\mathcal{M}_q \setminus \mathcal{M}_r)$, and finitely many countable ordinals $\xi \geq \delta_M$. These models form an $\in$-chain, say $C_Q$. In particular, the case $R = M$ falls under the last case of the previous paragraph, therefore $Q_{\delta_M} = Q \cap M$ is the $\in^*$-least member of $C_Q$. Moreover, if $Q'$ is the predecessor of $Q$ in $\mathcal{M}_q \cup \mathcal{M}_r$, then $Q'$ belongs to both $Q$ and $M$ and hence it belongs to $Q_{\delta_M}$. The largest member of $C_Q$ is a member of $Q$ since it is of the form $Q_\xi$, for some countable $\xi$. Thus, adding all these chains to $\mathcal{M}_q \cup \mathcal{M}_r$ we preserve the fact that we have an $\in$-chain. □

As an immediate consequence of Lemma 1.12 we have the following.

**Theorem 1.14** *$\mathbb{M}$ is $\mathcal{E}^2$-strongly proper.*

*Proof* Suppose $M \in \mathcal{E}^2$ and $p \in M \cap \mathbb{M}$. We shall show that $p^M$ is $(M, \mathbb{M})$-strongly generic. To see this we for every condition $r \leq p^M$ we have to define a condition $r \mid M \in \mathbb{M} \cap M$ such that for every $q \in \mathbb{M} \cap M$ if $q \leq r \mid M$ then $q$ and $r$ are compatible. If we let $r \mid M$ simply be $r \cap M$ this is precisely the statement of Lemma 1.12. □

**Corollary 1.15** *The forcing $\mathbb{M}$ is proper and preserves $\omega_2$.*

## 2 Adding a club in $\omega_2$ with finite conditions

We now present a version of the Friedman-Mitchell (see [3] and [7]) forcing for adding a club to $\omega_2$ with finite conditions. This will be achieved by adding a working part to the side conditions.

**Definition 2.1** Let $\mathbb{M}_2$ be the forcing notion whose elements are triples $p = (F_p, A_p, \mathcal{M}_p)$, where $F_p \in [\omega_2]^{<\omega}$, $A_p$ is a finite collection of intervals of the form $(\alpha, \beta]$, for some $\alpha, \beta < \omega_2$, $\mathcal{M}_p \in \mathbb{M}$, and

1. $F_p \cap \bigcup A_p = \emptyset$,
2. if $M \in \mathcal{M}_p$ and $I \in A_p$, then either $I \in M$ or $I \cap M = \emptyset$.

The order on $\mathbb{M}_2$ is coordinatewise reverse inclusion, i.e. $q \leq p$ if $F_p \subseteq F_q$, $A_p \subseteq A_q$ and $\mathcal{M}_p \subseteq \mathcal{M}_q$.

The information carried by a condition $p$ is the following. The points of $F_p$ are going to be in the generic club, and the intervals in $A_p$ are a partial description of the complement of that club. The side conditions are there to ensure that the forcing is $\mathcal{E}^2$-strongly proper. It should be pointed our that a condition $r$ may force some ordinals to be in the generic club even though they are not explicitly in $F_r$. The reason is that we may not be able to exclude them by intervals which satisfy conditions (1) and (2) of Definition 2.1.

*Fact* 2.2 If $p \in \mathbb{M}_2$ and $M \in \mathcal{M}_p$ then $\sup(M \cap \omega_2) \notin \bigcup A_p$.

*Proof* Any interval $I$ which contains $\sup(M \cap \omega_2)$ would have to intersect $M$ without being an element of $M$. This contradicts condition (2) of Definition 2.1. □

*Fact* 2.3 Suppose $p \in \mathbb{M}_2$, $M \in \mathcal{M}_p$ and $\gamma \in F_p$. Then

$$\min(M \setminus \gamma), \sup(M \cap \gamma) \notin \bigcup A_p.$$

*Proof* Suppose $\gamma \in F_p$ and let $I \in A_p$. Then $I$ is of the form $(\alpha, \beta]$, for some ordinals $\alpha, \beta < \omega_2$. Since $p$ is a condition we know that $\gamma \notin I$. By condition (2) of Definition 2.1 we know that either $I \cap M = \emptyset$ or $I \in M$. If $I \cap M = \emptyset$ then $\sup(M \cap \gamma), \min(M \setminus \gamma) \notin I$. Assume now that $I \in M$. Since $\gamma \notin I$ we have that either $\gamma \leq \alpha$ or $\gamma > \beta$. Suppose first that $\gamma \leq \alpha$. Since $\alpha \in M$ it follows $\min(M \setminus \gamma) \leq \alpha$ and so $\min(M \setminus \gamma) \notin I$. Clearly, also $\sup(M \cap \gamma) \notin I$. Suppose now $\gamma > \beta$. In that case, clearly, $\min(M \setminus \gamma) \notin I$. Also, since $\beta \in M$ it follows that $\beta < \sup(M \cap \gamma)$ and so $\sup(M \cap \gamma) \notin I$. □

**Definition 2.4** Suppose $p \in \mathbb{M}_2$ and $M \in \mathcal{M}_p$. We say that $p$ is *$M$-complete* if

1. $\sup(N \cap \omega_2) \in F_p$, for all $N \in \mathcal{M}_p$,
2. $\min(M \setminus \gamma), \sup(M \cap \gamma) \in F_p$, for all $\gamma \in F_p$.

We say that $p$ is *complete* if it is $M$-complete, for all $M \in \mathcal{M}_p$.

The following is straightforward.

*Fact* 2.5 Suppose $p \in \mathbb{M}_2$ and $M \in \mathcal{M}_p$. Then there is an $M$-complete condition $q$ which is equivalent to $p$. We call the least, under inclusion, such condition the *$M$-completion* of $p$.

*Proof* First let $F^* = F_p \cup \{\sup(N \cap \omega_2) : N \in \mathcal{M}_p\}$. Then let

$$F_q = F^* \cup \{\sup(M \cap \gamma) : \gamma \in F^*\} \cup \{\min(M \setminus \gamma) : \gamma \in F^*\}.$$

Let $A_q = A_p$ and $\mathcal{M}_q = \mathcal{M}_p$. It is straightforward to check that $q = (F_q, A_q, \mathcal{M}_q)$ is a condition equivalent to $p$ and $M$-complete. □

*Remark* 2.6 Note that in the above fact $q$ is $M$-complete for a single

$M \in \mathcal{M}_p$. We may not be able find $q$ which is complete, i.e. $M$-complete, for all $M \in \mathcal{M}_q$. To see this, suppose there are $M, N \in \mathcal{M}_p$ such that

$$\lim(M \cap N \cap \omega_2) \neq \lim(M \cap \omega_2) \cap \lim(N \cap \omega_2).$$

Note that if $\gamma \in M \cap N$ then either $M \cap \gamma \subseteq N$ or $N \cap \gamma \subseteq M$. Therefore, the least common limit of $M$ and $N$ which is not a limit of $M \cap N$ is above $\sup(M \cap N)$. If $q$ is an extension of $p$ which is complete then $\sup(M \cap N) \in F_q$, because $M \cap N \in \mathcal{M}_q$. Now, $\sup(M \cap N) \notin M \cap N$. Let us assume, for concreteness, that $\sup(M \cap N) \notin M$. We can define inductively a strictly increasing sequence $(\gamma_n)_n$ by setting $\gamma_0 = \sup(M \cap N)$ and

$$\gamma_{n+1} = \begin{cases} \min(M \setminus \gamma_n) & \text{if } n \text{ is even} \\ \min(N \setminus \gamma_n) & \text{if } n \text{ is odd.} \end{cases}$$

Since, $q$ was assumed to be both $M$-complete and $N$-complete we would have that $\gamma_n \in F_q$, for all $n$. This means that $F_q$ would have to be infinite, which is a contradiction. We do not know if such a pair of models can exist in a condition in $\mathbb{M}$. Nevertheless, we will later present a variation of $\mathbb{M}_2$ in which this situation does not occur and in which the set of fully complete conditions is dense.

We now come back to Lemma 1.8 and observe that it is valid also for $\mathbb{M}_2$.

**Lemma 2.7** *Let $M \in \mathcal{E}^2$ and let $p \in \mathbb{M}_2 \cap M$. Then there is a new condition, which we will call $p^M$, that is the smallest element of $\mathbb{M}_2$ extending $p$ such that $M \in \mathcal{M}_{p^M}$.*

*Proof* If $M \in \mathcal{E}^2_1$ then simply let $p^M = (F_p, A_p, \mathcal{M}_p \cup \{M\})$. If $M \in \mathcal{E}^2_0$, then, as in Lemma 1.8, we let $\mathcal{M}_{p^M}$ be the closure of $\mathcal{M}_p \cup \{M\}$ under intersection. We also let $F_{p^M} = F_p$ and $A_{p^M} = A_p$. We need to check that conditions (1) and (2) of Definition 2.1 are satisfied for $p^M$, but this is straightforward. □

Our next goal is to show that $\mathbb{M}_2$ is $\mathcal{E}^2$-strongly proper. We first establish the following.

**Lemma 2.8** *Suppose $p \in \mathbb{M}_2$ and $M \in \mathcal{M}_p$. Then $p$ is $(M, \mathbb{M}_2)$-strongly generic.*

*Proof* We need to define, for each $r \leq p$ a restriction $r \mid M \in M$ such that for every $q \in M$ if $q \leq r \mid M$ then $q$ and $r$ are compatible. So, suppose $r \leq p$. By replacing $r$ with its $M$-completion we may assume that $r$ is $M$-complete. We define

$$r \mid M = (F_r \cap M, A_r \cap M, \mathcal{M}_r \cap M).$$

By Facts 1.7 or 1.6 according to whether $M$ is countable or not we have that $\mathcal{M}_r \cap M \in \mathbb{M}$ and therefore $r \mid M \in \mathbb{M}_2 \cap M$. We need to show that for every $q \in M$ if $q \leq r \mid M$ then $q$ and $r$ are compatible.

If $M \in \mathcal{E}_1^2$ we already know that $\mathcal{M}_s = \mathcal{M}_q \cup \mathcal{M}_r$ is an $\in$-chain closed under intersection. Let $F_s = F_q \cup F_r$ and $A_s = A_q \cup A_r$. Finally, let $s = (F_s, A_s, \mathcal{M}_s)$. It is straightforward to check that $s$ is a condition and $s \leq r, q$.

We now concentrate on the case $M \in \mathcal{E}_0^2$. We define a condition $s$ as follows. We let $F_s = F_q \cup F_r$, $A_s = A_q \cup A_r$ and

$$\mathcal{M}_s = \mathcal{M}_q \cup \mathcal{M}_r \cup \{Q \cap R : Q \in \mathcal{M}_q, R \in \mathcal{M}_r\}.$$

We need to check that $s \in \mathbb{M}_2$. By Lemma 1.13 we know that $\mathcal{M}_s$ is an $\in$-chain closed under intersection. Therefore we only need to check that (1) and (2) of Definition 2.1 are satisfied for $s$. First we check (1).

*Claim* 2.9 $F_s \cap \bigcup A_s = \emptyset$.

*Proof* It suffices to check that $F_q \cap \bigcup A_r = \emptyset$ and $F_r \cap \bigcup A_q = \emptyset$. Suppose first $\gamma \in F_q$ and $I \in A_r$. Since $M \in \mathcal{M}_r$ we have, by (2) of Definition 2.1, that either $I \cap M = \emptyset$ or $I \in M$. If $I \cap M = \emptyset$ then, since $\gamma \in M$, we have that $\gamma \notin I$. If $I \in M$ then $I \in A_r \cap M$ and, since $q \leq r \mid M$, it follows that $I \in A_q$. Now, $q$ is a condition, so $\gamma \notin I$.

Suppose now $\gamma \in F_r$ and $I \in A_q$. If $\gamma \in F_r \cap M$ then $\gamma \in F_q$. Therefore $\gamma \notin I$. Suppose now $\gamma \in F \setminus M$. Since $r$ is $M$-complete $\gamma^* = \min(M \setminus \gamma) \in F_r$. Then $\gamma^* \in F_r \cap M$ and so $\gamma^* \in F_q$. Now, $I \in M$ and so if $\gamma \in I$ then $\gamma^* \in I$, which would be a contradiction. Therefore $\gamma \notin I$. □

We now turn to condition (2) of Definition 2.1.

*Claim* 2.10 If $Q \in \mathcal{M}_q$ and $I \in A_r$ then either $I \in Q$ or $I \cap Q = \emptyset$.

*Proof* Since $M \in \mathcal{M}_r$ we have that either $I \in M$ or $I \cap M = \emptyset$. If $I \in M$ then $I \in A_r \cap M$ and so $I \in A_q$. Since $q$ is a condition we have that either $I \in Q$ or $I \cap Q = \emptyset$. So, suppose $I \cap M = \emptyset$. If $Q \in \mathcal{E}_0^2$ then $Q \subseteq M$ and so $Q \cap I = \emptyset$, as well. If $Q \in \mathcal{E}_1^2$ then $Q \cap \omega_2$ is an initial segment of $\omega_2$, say $\gamma$. Now, if $I \cap Q \neq \emptyset$ and $I \notin Q$ we would have that $\gamma \in I$. Since $\gamma \in M$ this contradicts the fact that $I \cap M = \emptyset$. □

*Claim* 2.11 If $R \in \mathcal{M}_r$ and $I \in A_q$ then either $I \in R$ or $I \cap R = \emptyset$.

*Proof* Assume first that $R \in \mathcal{E}_1^2$. Then $R \cap \omega_2$ is an initial segment of $\omega_2$, say $\gamma$. If $I \cap R \neq \emptyset$ and $I \notin R$ then $\gamma \in I$. Now, since $r$ is $M$-complete we have that $\gamma \in F_r$. If $\gamma \in M$ then $\gamma \in F_q$ and this would contradict the fact that $q$ is a condition. If $\gamma \notin M$ let $\gamma^* = \min(M \setminus \gamma)$. Then, again by $M$-completeness of $r$, we have that $\gamma^* \in F_r$. However, $\gamma^* \in M$ and therefore $\gamma^* \in F_q$. Since $I \in A_q$ and $q \in M$ we have that $I \in M$. If $\gamma \in I$ we would also have that $\gamma^* \in I$, which contradicts the fact that $q$ is a condition.

We now consider the case $R \in \mathcal{E}_0^2$. We will show by $\in^*$-induction on the chain $\mathcal{M}_r$ that either $I \cap R = \emptyset$ or $I \in R$. If $R \in M$ then $R \in \mathcal{M}_q$ so this is clear. If $R \notin M$ then $R$ either belongs to $[M, H(\aleph_2))_r$ or else belongs to $[N \cap M, N)_r$, for some uncountable $N \in \mathcal{M}_r \cap M$.

Suppose $R \in [N \cap M, N)_r$, for some $N \in \pi_1(\mathcal{M}_r \cap M)$. Since $I \in A_q$ and $N \in \mathcal{M}_q$ we have that $I \in N$ or $I \cap N = \emptyset$. On the other hand, $R \subseteq N$ so if $I \cap N = \emptyset$ then also $I \cap R = \emptyset$. If $I \in N$ then, since $q \in M$ and $I \in A_q$, we have that $I \in M$ and so $I \in N \cap M$. If there are no uncountable models in the interval $[N \cap M, R)_r$ then $N \cap M \subseteq R$ and so $I \in R$. If there is an uncountable model in this interval let $S$ be the largest such model. Now, $N \cap M \subseteq S$ and so $I \in S$ and $I \subseteq S$. It follows that if $I \cap R \neq \emptyset$ then also $I \cap R \cap S \neq \emptyset$. Let $R^* = R \cap S$. Then $R^* \in \mathcal{M}_r$ and $R^*$ is below $R$ in the $\in^*$-ordering. By the inductive assumption we would have that $I \in R^*$ and so $I \in R$. The case when $R \in [M, H(\aleph_2))_r$ is treated in the same way. □

Finally, suppose $Q \in \mathcal{M}_q$, $R \in \mathcal{M}_r$ and $I \in A_q \cup A_r$. Consider the relation between the model $Q \cap R$ and $I$. If $I$ belongs to both $Q$ and $R$ then it belongs to $Q \cap R$. If $I$ is disjoint from $Q$ or $R$ it is also disjoint from $Q \cap R$. This completes the proof that $s$ is a condition. Since $s \leq q, r$ it follows that $q$ and $r$ are compatible. □

Now, by Lemmas 2.7 and 2.8 we have the following.

**Theorem 2.12** *The forcing* $\mathbb{M}_2$ *is* $\mathcal{E}^2$*-strongly proper. Hence it is proper and preserves* $\omega_2$.

Suppose now $G$ is $V$-generic filter for the forcing notion $\mathbb{M}_2$. We can define

$$C_G = \bigcup\{F_p : p \in G\} \text{ and } U_G = \bigcup\bigcup\{A_p : p \in G\}.$$

Then $C_G \cap U_G = \emptyset$. Moreover, by genericity, $C_G \cup U_G = \omega_2$. Since $U_G$ is a union of open intervals it is open in the order topology. Therefore, $C_G$ closed and, again by genericity, it is unbounded in $\omega_2$. Unfortunately, we cannot say much about the generic club $C_G$. For reasons explained in Remark 2.6, we cannot even say that it does not contain infinite subsets which are in the ground model. In order to circumvent this problem, we now define a variation of the forcing notion $\mathbb{M}_2$. We start by some definitions.

**Definition 2.13** Suppose $M, N \in \mathcal{E}^2$. We say that $M$ and $N$ are lim-compatible if

$$\lim(M \cap N \cap \omega_2) = \lim(M \cap \omega_2) \cap \lim(N \cap \omega_2).$$

*Remark* 2.14 Clearly, this conditions is non trivial only if both $M$ and $N$ are countable. We will abuse notation and write $\lim(M)$ for $\lim(M \cap \omega_2)$.

We now define a version of the forcing notion $\mathbb{M}$.

**Definition 2.15** Let $\mathbb{M}^*$ be the suborder of $\mathbb{M}$ consisting of conditions $p = \mathcal{M}_p$ such that any two models in $\mathcal{M}_p$ are lim-compatible.

We have the following version of Lemma 1.8.

**Lemma 2.16** *Let $M \in \mathcal{E}^2$ and let $p \in \mathbb{M}^* \cap M$. Then there is a new condition, which we will call $p^M$, that is the smallest element of $\mathbb{M}^*$ extending $p$ such that $M \in \mathcal{M}_{p^M}$.*

*Proof* If $M \in \mathcal{E}^2_1$ then simply let $p^M = \mathcal{M}_p \cup \{M\}$. If $M \in \mathcal{E}^2_0$, then we let $\mathcal{M}_{p^M}$ be the closure of $\mathcal{M}_p \cup \{M\}$ under intersection. Then, thanks to Lemma 1.8, we just need to check that the models in $\mathcal{M}_{p^M}$ are lim-compatible. Suppose $P \in \pi_0(p)$. Then $P \in M$ and hence $P \subseteq M$. Therefore, $P$ and $M$ are lim-compatible. Suppose now $P \in \pi_1(p)$. Then $P \cap \omega_2$ is an initial segment of $\omega_2$, say $\gamma$. Therefore

$$\lim(M \cap P) = \lim(M \cap \gamma) = \lim(M) \cap (\gamma + 1) = \lim(M) \cap \lim(P),$$

and so $P$ and $M$ are lim-compatible. We also need to check that, for any $P, Q \in \mathcal{M}_p$, the models $P \cap M$ and $Q \cap M$, as well as $P \cap M$ and $Q$ are lim-compatible, but this is straightforward. □

We now have a version of Lemma 1.12.

**Lemma 2.17** *Suppose $r \in \mathbb{M}^*$ and $M \in \mathcal{M}_r$. Let $q \in \mathbb{M}^* \cap M$ be such that $q \leq r \cap M$. Then $q$ and $r$ are compatible in $\mathbb{M}^*$.*

*Proof* If $M$ is uncountable then one can easily check that $\mathcal{M}_s = \mathcal{M}_q \cup \mathcal{M}_r$ is $\in$-chain closed under intersection and that any two models in $\mathcal{M}_s$ are lim-compatible.

Suppose now $M$ is countable and let

$$\mathcal{M}_s = \mathcal{M}_q \cup \mathcal{M}_r \cup \{Q \cap R : Q \in \mathcal{M}_q, R \in \mathcal{M}_r\}.$$

Thanks to Lemma 1.12 we know that $\mathcal{M}_s$ is an $\in$-chain closed under intersection. It remains to check that any two models in $\mathcal{M}_s$ are lim-compatible.

*Claim* 2.18 If $Q \in \pi_0(\mathcal{M}_q)$ and $R \in \pi_0(\mathcal{M}_r)$, then $Q$ and $R$ are lim-compatible.

*Proof* We show this by $\in^*$-induction on $R$. Since $Q \in \mathcal{M}_q$ then $Q \in M$ and, since $Q$ is countable, we have that $\lim(Q) \subseteq M$. Moreover, since $R$ and $M$ are both in $\mathcal{M}_r$, we have that $\lim(R \cap M) = \lim(R) \cap \lim(M)$, and so

$$\lim(Q) \cap \lim(R) = \lim(Q) \cap \lim(R) \cap \lim(M) = \lim(Q) \cap \lim(R \cap M).$$

Hence, without loss of generality we can assume $R$ to be $\in^*$-below $M$. If $R \in M$ then $R \in \mathcal{M}_q$ and so $Q$ and $R$ are lim-compatible. Assume now, $R \notin M$. Then by Fact 1.7 there is $N \in \pi_1(\mathcal{M}_r \cap M)$ such that $R \in [N \cap M, N)_r$. We may also assume $Q$ is $\in^*$-below $N$, otherwise we could replace $Q$ by $Q \cap N$. Hence $Q \subseteq N \cap M$. If there are no uncountable model in the interval $[N \cap M, R)_r$, then $N \cap M \subseteq R$ and since $Q \in N \cap M$ we have $Q \in R$. Therefore, $Q$ and $R$ are lim-compatible. Otherwise, let $S$ be the $\in^*$-largest uncountable model in $[N \cap M, R)_r$. Then $Q \in S$ and $S \cap \omega_2$ is an initial segment of $\omega_2$. Let $R^* = R \cap S$. It follows that $\lim(R) \cap \lim(Q) = \lim(R^*) \cap \lim(Q)$. By the inductive assumption we have that $\lim(R^*) \cap \lim(Q) = \lim(R^* \cap Q)$ and hence $\lim(R) \cap \lim(Q) = \lim(R \cap Q)$. □

Now, we need to check that any two models in $\mathcal{M}_s$ are lim-compatible. So, suppose $S, S^* \in \mathcal{M}_s$. We may assume $S$ and $S^*$ are both countable and of the form $S = Q \cap R$, $S^* = Q^* \cap R^*$, for $Q, Q^* \in \mathcal{M}_q$ and $R, R^* \in \mathcal{M}_r$. Then

$$\lim((Q \cap R) \cap (Q^* \cap R^*)) = \lim((Q \cap Q^*) \cap (R \cap R^*))$$

and by Claim 2.18

$$\lim((Q \cap Q^*) \cap (R \cap R^*)) = \lim(Q \cap Q^*) \cap \lim(R \cap R^*),$$

because $Q \cap Q^* \in \mathcal{M}_q$ and $R \cap R^* \in \mathcal{M}_r$. Moreover, we have $= \lim(Q \cap Q^*) = \lim(Q) \cap \lim(Q^*)$ and $\lim(R \cap R^*) = \lim(R) \cap \lim(R^*)$, since the elements of $\mathcal{M}_q$, respectively $\mathcal{M}_r$, are lim-compatible. Finally, again by Claim 2.18, we have

$$\lim(Q) \cap \lim(R) \cap \lim(Q^*) \cap \lim(R^*) = \lim(Q \cap R) \cap \lim(Q^* \cap R^*).$$

□

We now define a variation of the forcing $\mathbb{M}_2$ which will have some additional properties.

**Definition 2.19** Let $\mathbb{M}_2^*$ be the forcing notion whose elements are triples $p = (F_p, A_p, \mathcal{M}_p)$, where $F_p \in [\omega_2]^{<\omega}$, $A_p$ is a finite collection of intervals of the form $(\alpha, \beta]$, for some $\alpha, \beta < \omega_2$, $\mathcal{M}_p \in \mathbb{M}^*$, and

1. $F_p \cap \bigcup A_p = \emptyset$,
2. if $M \in \mathcal{M}_p$ and $I \in A_p$, then either $I \in M$ or $I \cap M = \emptyset$,

The order on $\mathbb{M}_2^*$ is coordinatewise reverse inclusion, i.e. $q \leq p$ if $F_p \subseteq F_q$, $A_p \subseteq A_q$ and $\mathcal{M}_p \subseteq \mathcal{M}_q$.

*Remark* 2.20 Note that the only difference between $\mathbb{M}_2^*$ and $\mathbb{M}_2$ is that for $p$ to be in $\mathbb{M}_2^*$ we require that $\mathcal{M}_p \in \mathbb{M}^*$, i.e. the models in $\mathcal{M}_p$ are pairwise lim-compatible.

We can now use Lemmas 2.18 and 2.17 to prove the analogs of Lemmas 2.7 and 2.8 for $\mathbb{M}_2^*$. We then obtain the following.

**Theorem 2.21** *The forcing notion $\mathbb{M}_2^*$ is $\mathcal{E}^2$-strongly proper. Hence, it is proper and preserves $\omega_2$.*

Let $G^*$ be a $V$-generic filter for $\mathbb{M}_2^*$. As in the case of the forcing $\mathbb{M}_2$, we define

$$C_G^* = \bigcup\{F_p : p \in G^*\} \text{ and } U_G^* = \bigcup\bigcup\{A_p : p \in G^*\}.$$

As before $C_G^*$ is forced to be a club in $\omega_2$. Our goal now is to show that it does not contain any infinite subset from the ground model. For this we will need the following lemma which explains the reason for the requirement of lim-compatibility for models $\mathcal{M}_p$, for conditions $p$ in $\mathbb{M}_2^*$.

**Lemma 2.22** *The set of complete conditions is dense in $\mathbb{M}_2^*$.*

*Proof* Consider a condition $p \in \mathbb{M}_2^*$. For each $M \in \mathcal{M}_p$ we consider functions $\mu_M, \sigma_M : \omega_2 \to \omega_2$ defined as follows:

$$\mu_M(\alpha) = \min(M \setminus \alpha) \text{ and } \sigma_M(\alpha) = \sup(M \cap \alpha).$$

To obtain a complete condition extending $p$ we first define:

$$F_p^* = F_p \cup \{\sup(M \cap \omega_2) : M \in \mathcal{M}_p\}.$$

We then let $\bar{F}_p$ be the closure of $F_p^*$ under the functions $\mu_M$ and $\sigma_M$, for $M \in \mathcal{M}_p$. Then $q = (\bar{F}_p, A_p, \mathcal{M}_p)$ will be the required complete condition extending $p$. The main point is to show the following.

*Claim* 2.23 $\bar{F}_p$ is finite.

*Proof* Let $L = \bigcup\{\lim(M) : M \in \mathcal{M}_p\}$. For each $\gamma \in L$ let

$$Y(p,\gamma) = \{M \in \mathcal{M}_p : \gamma \in \lim(M)\}.$$

and let $M(p,\gamma) = \bigcap Y(p,\gamma)$. Then, since $\mathcal{M}_p$ is closed under intersection $M(p,\gamma) \in \mathcal{M}_p$. Since the models in $\mathcal{M}_p$ are lim-compatible it follows that $\gamma \in \lim(M(p,\gamma))$. Thus, $M(p,\gamma)$ is the least (under inclusion) model in $\mathcal{M}_p$ which has $\gamma$ as its limit point. For each $\gamma \in L$ pick an ordinal $f(\gamma) \in M(p,\gamma) \cap \gamma$ above $\sup(F_p^* \cap \gamma)$ and $\sup(M \cap \gamma)$, for all $M \in \mathcal{M}_p \setminus Y(p,\gamma)$. For a limit $\gamma \in \omega_2 \setminus L$ let

$$f(\gamma) = \sup\{\sup(M \cap \gamma) : M \in \mathcal{M}_p\}.$$

Notice now that for any limit $\gamma$ and any $M \in \mathcal{M}_p$, if $\xi \notin (f(\gamma),\gamma)$ then $\mu_M(\xi), \sigma_M(\xi) \notin (f(\gamma),\gamma)$. Since $\bar{F}_p$ is the closure of $F_p^*$ under the functions $\mu_M$ and $\sigma_M$, for $M \in \mathcal{M}_p$, and $F_p^* \cap (f(\gamma),\gamma) = \emptyset$, for all limit $\gamma$, it follows that $\bar{F}_p \cap (f(\gamma),\gamma) = \emptyset$, for all limit $\gamma$. This means that $\bar{F}_p$ has no limit points and therefore is finite. □

□

**Lemma 2.24** *Let $p \in \mathbb{M}_2^*$ be a complete condition, and let $\gamma \in \omega_2 \setminus F_p$. Then there is a condition $q \leq p$ such that $\gamma \in I$, for some $I \in A_q$.*

*Proof* Without loss of generality we can assume that there is an $M \in \mathcal{M}_p$ such that $\sup(M \cap \omega_2) > \gamma$, otherwise we could let

$$q = (F_p, A_p \cup \{(\eta,\gamma]\}, \mathcal{M}_p),$$

for some $\eta < \gamma$ sufficiently large so that $(\eta,\gamma]$ does not intersect any model in $\mathcal{M}_p$.

Now, since $\sup(M \cap \omega_2) \in F_p$, for every $M \in \mathcal{M}_p$, the set $F_p \setminus \gamma$ is nonempty. Let $\tau$ be $\min(F_p \setminus \gamma)$. Notice that for every model $M \in \mathcal{M}_p$ either $\sup(M \cap \tau) < \gamma$, or $\tau \in \lim(M)$, because

$$\gamma < \sup(M \cap \tau) < \tau,$$

would contradict the minimality of $\tau$. Moreover, if $\sup(M \cap \tau) = \gamma$, then $\gamma$ would be in $F_p$, contrary to the hypothesis of the lemma.

Let

$$Y = \{M \in \mathcal{M}_p : \tau \in \lim(M)\}.$$

Without loss of generality we can assume $Y \neq \emptyset$, because otherwise we can let

$$q = (F_p, A_p \cup \{(\eta, \gamma]\}, \mathcal{M}_p)$$

for some $\eta$ sufficiently large so that $(\eta, \gamma]$ avoids $\sup(M \cap \tau)$, for every $M \in \mathcal{M}_p$. Let $M_0 = \bigcap Y$. Since $\mathcal{M}_p$ is closed under intersection $M_0 \in \mathcal{M}_p$. Moreover, since any two models in $\mathcal{M}_p$ are lim-compatible we have that $\tau \in \lim M_0$. Thus, $M_0$ is itself in $Y$ and is contained in any member of $Y$. Therefore, if an interval $I$ belongs to $M_0$, then it belongs to every model in $Y$. Let $\eta = \min(M_0 \setminus \gamma)$. Since $\tau \in \lim(M_0)$ we have $\gamma \leq \eta < \tau$. Since $\tau$ is the least element of $F_p$ above $\gamma$ it follows that $\eta \notin F_p$.

*Claim* 2.25 $\sup(M_0 \cap \gamma) > \sup(F_p \cap \gamma)$.

*Proof* Suppose $\xi$ is an element of $F_p \cap \gamma$. Since $p$ is $M_0$-complete, we also have $\min(M_0 \setminus \xi) \in F_p$. Notice that $\min(M_0 \setminus \xi) \neq \eta$, since $\eta \notin F_p$. Then

$$\xi \leq \min(M_0 \setminus \xi) < \gamma,$$

and so $\sup(M_0 \cap \gamma) > \xi$. □

Consider now some $M \in \mathcal{M}_p \setminus Y$. Then $\tau \notin \lim(M)$ and, since $p$ is $M$-complete, we have that $\sup(M \cap \tau) \in F_p$. Since $\tau$ is the least element of $F_p$ above $\gamma$ it follows that $\sup(M \cap \tau) \in F_p \cap \gamma$. Now, pick an element $\eta' \in M_0$ above $\sup(F_p \cap \gamma)$ and let $I = (\eta', \eta]$. It follows that $I \in M$, for all $M \in Y$ and $I \cap M = \emptyset$, for all $M \in \mathcal{M}_p \setminus Y$. Therefore,

$$q = (F_p, A_p \cup \{I\}, \mathcal{M}_p)$$

is a condition stronger than $p$ and $\gamma \in I$. Thus, $q$ is as required. □

**Corollary 2.26** *If $G^*$ is a V-generic filter over $\mathbb{M}_2^*$, then the generic club $C_G^*$ does not contain any infinite subset which is in V.*

## 3 Strong chains of uncountable functions

We now consider the partial order $(\omega_1^{\omega_1}, <_{\text{fin}})$ of all functions from $\omega_1$ to $\omega_1$ ordered by $f <_{\text{fin}} g$ iff $\{\xi : f(\xi) \geq g(\xi)\}$ is finite. In [5] Koszmider constructed a forcing notion which preserves cardinals and adds an $\omega_2$ chain in $(\omega_1^{\omega_1}, <_{\text{fin}})$. The construction uses an $(\omega_1, 1)$-morass which is a stationary coding set and is quite involved. In this section we present a streamlined version of this forcing which uses generalizes side conditions and is based on the presentation of Mitchell [6]. Before that we show that Chang's conjecture implies that there is no such chain. The argument is inspired by a proof of Shelah from [10]. A similar argument appears in [4].

**Proposition 3.1** *Assume Chang's conjecture. Then there is no chain in $(\omega_1^{\omega_1}, <_{\text{fin}})$ of length $\omega_2$.*

*Proof* Assume towards contradiction that Chang's conjecture holds and $\{f_\alpha : \alpha < \omega_2\}$ is a chain in $(\omega_1^{\omega_1}, <_{\text{fin}})$. Given a function $g : I \to \omega_1$ and $\eta < \omega_1$ we let $\min(g, \eta)$ be the function defined by:

$$\min(g, \eta)(\zeta) = \min(g(\zeta), \eta).$$

For each $\alpha < \omega_2$, and $\xi, \eta < \omega_1$ we define a function $f_\alpha^{\xi,\eta}$ by:

$$f_\alpha^{\xi,\eta} = \min(f_\alpha \restriction [\xi, \xi + \omega), \eta).$$

Given $\xi, \eta < \omega_1$, the sequence $\{f_\alpha^{\xi,\eta} : \alpha < \omega_2\}$ is $\leq_{\text{fin}}$-increasing. We define a club $C^{\xi,\eta} \subseteq \omega_2$ as follows.

*Case 1:* If the sequence $\{f_\alpha^{\xi,\eta} : \alpha < \omega_2\}$ eventually stabilizes under $=_{\text{fin}}$ we let $C^{\xi,\eta} = \omega_2 \setminus \mu$, where $\mu$ is least such that $f_\nu^{\xi,\eta} =_{\text{fin}} f_\mu^{\xi,\eta}$, for all $\nu \geq \mu$.

*Case 2:* If the sequence $\{f_\alpha^{\xi,\eta} : \alpha < \omega_2\}$ does not stabilize we let $C^{\xi,\eta}$ be a club in $\omega_2$ such that $f_\alpha^{\xi,\eta} \lneqq_{\text{fin}} f_\beta^{\xi,\eta}$, for all $\alpha, \beta \in C^{\xi,\eta}$ with $\alpha < \beta$. This means that for every such $\alpha$ and $\beta$ the set

$$\{n : f_\alpha(\xi + n) < f_\beta(\xi + n) \leq \eta\}$$

is infinite.

Let $C = \bigcap\{C^{\xi,\eta} : \xi, \eta < \omega_1\}$. Then $C$ is a club in $\omega_2$. We define a coloring $c : [C]^2 \to \omega_1$ by

$$c\{\alpha, \beta\}_< = \max\{\xi : f_\alpha(\xi) \geq f_\beta(\xi)\}.$$

By Chang's conjecture we can find an increasing $\omega_1$ sequence

$$S = \{\alpha_\rho : \rho < \omega_1\}$$

of elements of $C$ such that $c[[S]^2]$ is bounded in $\omega_1$. Let $\xi = \sup(c[[S]^2])+1$. Therefore for every $\rho < \tau < \omega_1$ we have

$$f_{\alpha_\rho} \upharpoonright [\xi, \omega_1) < f_{\alpha_\tau} \upharpoonright [\xi, \omega_1).$$

Now, let $\eta = \sup(\mathrm{ran}(f_{\alpha_1} \upharpoonright [\xi, \xi + \omega)))$. It follows that for every $n$:

$$f_{\alpha_0}(\xi + n) < f_{\alpha_1}(\xi + n) \leq \eta.$$

Since $\alpha_0, \alpha_1 \in C^{\xi,\eta}$ it follows that $C^{\xi,\eta}$ was defined using Case 2. Therefore the sequence $\{f^{\xi,\eta}_{\alpha_\rho} : \rho < \omega_1\}$ is $\leq$-increasing and $f_{\alpha_\rho} \neq_{\text{fin}} f_{\alpha_\tau}$, for all $\rho < \tau$. For each $\rho < \omega_1$ let $n_\rho$ be the least such that $f^{\xi,\eta}_{\alpha_\rho}(\xi + n_\rho) < f^{\xi,\eta}_{\alpha_{\rho+1}}(\xi + n_\rho)$. Then there is a integer $n$ such that $X = \{\rho < \omega_1 : n_\rho = n\}$ is uncountable. It follows that the sequence $\{f_{\alpha_\rho}(\xi + n) : \rho \in X\}$ is strictly increasing. On the other hand it is included in $\eta$ which is countable, a contradiction. □

Therefore, in order to add a strong $\omega_2$-chain in $(\omega_1^{\omega_1}, <_{\text{fin}})$ we need to assume that Chang's conjecture does not hold. In fact, we will assume that there is an increasing function $g : \omega_1 \to \omega_1$ such that

1. $g(\xi)$ is indecomposable, for all $\xi < \omega_1$,
2. $\text{o.t.}(M \cap \omega_2) < g(\delta_M)$, for all $M \in \mathcal{E}^2_0$.

It is easy to add such a function by a preliminary forcing. For instance, we can add by countable conditions an increasing function $g$ which dominates all the canonical functions $c_\alpha$, for $\alpha < \omega_2$, and such that $g(\xi)$ is indecomposable, for all $\xi$. Moreover, we may assume that $g$ is definable in the structure $(H(\aleph_2), \in, \trianglelefteq)$ and so it belongs to $M$, for all $M \in \mathcal{E}^2$.

Our plan is to add an $\omega_2$-chain $\{f_\alpha : \alpha < \omega_2\}$ in $(\omega_1^{\omega_1}, <_{\text{fin}})$ below this function $g$. We can view this chain as a single function $f : \omega_2 \times \omega_1 \to \omega_1$. We want to use conditions of the form $p = (f_p, \mathcal{M}_p)$, where $f_p : A_p \times F_p \to \omega_1$ for some finite $A_p \subseteq \omega_2$ and $F_p \subseteq \omega_1$, and $\mathcal{M}_p \in \mathbb{M}$ is a side condition. Suppose $\alpha, \beta \in A_p$ with $\alpha < \beta$, and $M \in \pi_0(\mathcal{M}_p)$. Then $M$ should localize the disagreement of $f_\alpha$ and $f_\beta$, i.e. $p$ should force that the finite set $\{\xi : f_\alpha(\xi) \geq f_\beta(\xi)\}$ is contained in $M$. This means that if $\xi \in \omega_1 \setminus M$ then $p$ makes the commitment that $f_\alpha(\xi) < f_\beta(\xi)$. Moreover, for every $\eta \in (\alpha, \beta) \cap M$ we should have that $f_\alpha(\xi) < f_\eta(\xi) < f_\beta(\xi)$. Therefore, $p$ imposes that $f_\beta(\xi) \geq f_\alpha(\xi) + \text{o.t.}([\alpha, \beta) \cap M)$. This motivates the definition of the distance function below. Before defining the distance function we need to prove some general properties of side conditions. For a set of ordinals $X$ we let $\overline{X}$ denote the closure of $X$ in the order topology.

*Fact* 3.2 Suppose $P, Q \in \mathcal{E}_0^2$ and $\delta_P \leq \delta_Q$.

1. If $\gamma \in P \cap Q \cap \omega_2$ then $P \cap \gamma \subseteq Q \cap \gamma$.
2. If $P$ and $Q$ are lim-compatible and $\gamma \in \overline{P \cap \omega_2} \cap \overline{Q \cap \omega_2}$ then $P \cap \gamma \subseteq Q \cap \gamma$.

*Proof* (1) For each $\alpha < \omega_2$ let $e_\alpha$ be the $\trianglelefteq$-least injection from $\alpha$ to $\omega_1$. Then $P \cap \gamma = e_\gamma^{-1}[\delta_P]$ and $Q \cap \gamma = e_\gamma^{-1}[\delta_Q]$. Since $\delta_P \leq \delta_Q$ we have that $P \cap \gamma \subseteq Q \cap \gamma$.
(2) If $\gamma \in P \cap Q$ this is (1). Suppose $\gamma$ is a limit point of either $P$ or $Q$ then it is also the limit point of the other. Since $P$ and $Q$ are lim-compatible we have that $\gamma \in \lim(P \cap Q)$. Then $P \cap \gamma = \bigcup\{e_\alpha^{-1}[\delta_P] : \alpha \in P \cap Q\}$ and $Q \cap \gamma = \bigcup\{e_\alpha^{-1}[\delta_Q] : \alpha \in P \cap Q\}$ Since $\delta_P \leq \delta_Q$ we conclude that $P \cap \gamma \subseteq Q \cap \gamma$. □

*Fact* 3.3 Suppose $p \in \mathbb{M}$ and $P, Q \in \pi_0(\mathcal{M}_p)$. If $\delta_P < \delta_Q$ and $P \subseteq Q$ then $P \in Q$.

*Proof* If there is no uncountable model in the interval $(P, Q)_p$, then $P \in Q$ by transitivity. Otherwise, let $S$ be the $\in^*$-largest uncountable model below $Q$ and we proceed by $\in^*$-induction. First note that $S \in Q$ by transitivity and if we let $Q^* = Q \cap S$ then $\delta_{Q^*} = \delta_Q$. Since $P \subseteq S$, we have that $P \subseteq Q^*$ and so $Q^*$ is $\in^*$-above $P$. By the inductive assumption we have $P \in Q^* \subseteq Q$, as desired. □

**Definition 3.4** Let $p = \mathcal{M}_p \in \mathbb{M}^*$, $\alpha, \beta \in \omega_2$ and let $\xi$ be a countable ordinal. Then the binary relation $L_{p,\xi}(\alpha, \beta)$ holds if there is a $P \in \mathcal{M}_p$, with $\delta_P \leq \xi$, such that $\alpha, \beta \in \overline{P \cap \omega_2}$. In this case we will say that $\alpha$ and $\beta$ are *$p, \xi$-linked*.

**Definition 3.5** Let $p = \mathcal{M}_p \in \mathbb{M}^*$ and $\xi < \omega_1$. We let $C_{p,\xi}$ be the transitive closure of the relation $L_{p,\xi}$. If $C_{p,\xi}(\alpha, \beta)$ holds we say that $\alpha$ and $\beta$ are *$p, \xi$-connected*. If $\alpha < \beta$ and $\alpha$ and $\beta$ are $p, \xi$-connected we write $\alpha <_{p,\xi} \beta$.

From Fact 3.2(2) we now have the following.

*Fact* 3.6 Suppose $p = \mathcal{M}_p \in \mathbb{M}^*$ and $\xi < \omega_1$.

1. Suppose $\alpha < \beta < \gamma$ are ordinal in $\omega_2$. If $L_{p,\xi}(\alpha,\gamma)$ and $L_{p,\xi}(\beta,\gamma)$ hold, then so does $L_{p,\xi}(\alpha,\beta)$.
2. If $\alpha <_{p,\xi} \beta$ then there is a sequence $\alpha = \gamma_0 < \gamma_1 < \ldots < \gamma_n = \beta$ such that $L_{p,\xi}(\gamma_i, \gamma_{i+1})$ holds, for all $i < n$.

We now present some properties of the relation $<_{p,\xi}$, in order to define the distance function we will use in the definition of the main forcing.

*Fact* 3.7 Let $p = \mathcal{M}_p \in \mathbb{M}^*$ and $\xi < \omega_1$. Suppose $\alpha < \beta < \gamma < \omega_2$. Then

1. if $\alpha <_{p,\xi} \beta$ and $\beta <_{p,\xi} \gamma$, then $\alpha <_{p,\xi} \gamma$,
2. if $\alpha <_{p,\xi} \gamma$ and $\beta <_{p,\xi} \gamma$, then $\alpha <_{p,\xi} \beta$.

*Proof* Part (1) follows directly from the definition of the relation $<_{p,\xi}$. To prove (2) let $\alpha = \gamma_0 < \ldots < \gamma_n = \gamma$ witness the $p,\xi$-connection between $\alpha$ and $\gamma$ and let $\beta = \delta_0 < \ldots < \delta_l = \gamma$ witness the $p,\xi$-connection between $\beta$ and $\gamma$. We have that $L_{p,\xi}(\gamma_i, \gamma_{i+1})$ holds, for all $i < n$, and $L_{p,\xi}(\delta_j, \delta_{j+1})$ holds, for all $j < l$. We prove that $\alpha$ and $\beta$ are $p,\xi$-connected by induction on $n + l$. If $n = l = 1$ this is simply Fact 3.6(1). Let now $n, l > 1$. Assume for concreteness that $\delta_{l-1} \leq \gamma_{n-1}$. By Fact 3.6(1) $L_{p,\xi}(\delta_{l-1}, \gamma_{n-1})$ holds; so $\alpha <_{p,\xi} \gamma_{n-1}$ and $\beta <_{p,\xi} \gamma_{n-1}$. Now, by the inductive assumption we conclude that $\alpha$ and $\beta$ are $p,\xi$-connected, i.e. $\alpha <_{p,\xi} \beta$. The case $\gamma_{n-1} < \delta_{l-1}$ is treated similarly. □

The above lemma shows in (1) that the relation $<_{p,\xi}$ is transitive and in (2) that the set $(\omega_2, <_{p,\xi})$ has a tree structure. Since for every $M \in \mathcal{E}^2_0$ if $\delta_M \leq \xi$ then o.t.$(M \cap \omega_2) < g(\xi)$ and $g(\xi)$ is indecomposable we conclude that the height of $(\omega_2, <_{p,\xi})$ is at most $g(\xi)$. For every $\alpha <_{p,\xi} \beta$ we let $(\alpha,\beta)_{p,\xi} = \{\eta : \alpha <_{p,\xi} \eta <_{p,\xi} \beta\}$. We define similarly $[\alpha,\beta)_{p,\xi}$ and $(\alpha,\beta]_{p,\xi}$ and $[\alpha,\beta]_{p,\xi}$. If $0 <_{p,\xi} \beta$, i.e. $\beta$ belongs to some $M \in \mathcal{M}_p$ with $\delta_M \leq \xi$ we write $(\beta)_{p,\xi}$ for the interval $[0,\beta)_{p,\xi}$. Thus, $(\beta)_{p,\xi}$ is simply the set of predecessors of $\beta$ in $<_{p,\xi}$. If $\beta$ does not belong to $\overline{M \cap \omega_2}$ for any $M \in \mathcal{M}_p$ with $\delta_M \leq \xi$ we leave $(\beta)_{p,\xi}$ undefined. Note that when defined $(\beta)_{p,\xi}$ is a closed subset of $\beta$ in the ordinal topology.

*Fact* 3.8 Let $p \in \mathbb{M}^*$, $M \in \mathcal{M}_p$, $\xi \in M \cap \omega_1$ and $\beta \in M \cap \omega_2$. Then $(\beta)_{p,\xi} \subseteq M$. Moreover, if we let $p^* = p \cap M$ then $(\beta)_{p,\xi} = (\beta)_{p^*,\xi}$.

*Proof* Let $\alpha <_{p,\xi} \beta$ and fix a sequence $\alpha = \gamma_0 < \gamma_1 < \ldots < \gamma_n = \beta$ such that $L_{p,\xi}(\gamma_i, \gamma_{i+1})$ holds, for all $i < n$. We proceed by induction on $n$. Suppose first $n = 1$ and let $P$ witness that $\alpha$ and $\beta$ are $p,\xi$-linked. Since $\delta_P < \delta_M$ we have by Fact 3.2 that $P \cap \beta \subseteq M$ and by Fact 3.3 that $P \cap M \in M$. Therefore $\alpha,\beta \in \overline{P \cap M \cap \omega_2} \subseteq M$ and so $P \cap M$

witnesses that $\alpha$ and $\beta$ are $p^*,\xi$-linked. Consider now the case $n > 1$. By the same argument as in the case $n = 1$ we know that $\gamma_{n-1}$ and $\beta$ are $p^*,\xi$-linked and then by the inductive hypothesis we conclude that $\alpha$ and $\beta$ are $p^*,\xi$-connected. □

*Fact* 3.9 Let $p \in \mathbb{M}^*$, $M \in \mathcal{M}_p$, $\beta \in \omega_2 \setminus M$ and $\xi \in M \cap \omega_1$. If $(\beta)_{p,\xi} \cap M$ is non empty then it has a largest element, say $\eta$. Moreover, there is $Q \in \mathcal{M}_p \setminus M$ with $\delta_Q \leq \xi$ such that $\eta = \sup(Q \cap M \cap \omega_2)$.

*Proof* Assume $(\beta)_{p,\xi} \cap M$ is non empty and let $\eta$ be its supremum. Note that $\eta$ is a limit ordinal. Since $(\beta)_{p,\xi}$ is a closed subset of $\beta$ in the order topology we know that either $\eta <_{p,\xi} \beta$ or $\eta = \beta$. By Fact 3.8 $(\beta)_{p,\xi} \cap M = (\beta)_{p,\xi} \cap \eta = (\eta)_{p,\xi}$. For every $\rho \in (\eta)_{p,\xi}$ there is some $P \in \mathcal{M}_p \cap M$ with $\delta_P \leq \xi$ such that $\rho \in \overline{P \cap \omega_2}$. Since $\mathcal{M}_p \cap M$ is finite there is such $P$ with $\eta \in \overline{P \cap \omega_2}$. Since $P \in M$ it follows that $\overline{P} \subseteq M$, so $\eta \in M$ and therefore $\eta < \beta$. Finally, since $\eta$ and $\beta$ are $p,\xi$-connected, there is a chain $\eta = \gamma_0 < \gamma_1 < \ldots < \gamma_n = \beta$ such that $\gamma_i$ and $\gamma_{i+1}$ are $p,\xi$-linked, for all $i$. Let $Q$ witness that $\eta = \gamma_0$ and $\gamma_1$ are $p,\xi$-linked. Then $\delta_Q \leq \xi$ and $\eta = \sup(Q \cap M \cap \omega_2)$. Since $\gamma_1 \in \overline{Q \cap \omega_2} \setminus M$ it follows that $Q \notin M$. Therefore, $Q$ is as required. □

We are now ready to define the distance function.

**Definition 3.10** Let $p = \mathcal{M}_p \in \mathbb{M}^*$, $\alpha,\beta \in \omega_2$, and $\xi \in \omega_1$. If $\alpha <_{p,\xi} \beta$ we define the $p,\xi$-distance of $\alpha$ and $\beta$ as

$$d_{p,\xi}(\alpha,\beta) = \text{o.t.}([\alpha,\beta)_{p,\xi}).$$

Otherwise we leave $d_{p,\xi}(\alpha,\beta)$ undefined.

*Remark* 3.11 Notice that for every $p$ and $\xi$ the function $d_{p,\xi}$ is additive, i.e. if $\alpha <_{p,\xi} \beta <_{p,\xi} \gamma$ then

$$d_{p,\xi}(\alpha,\gamma) = d_{p,\xi}(\alpha,\beta) + d_{p,\xi}(\beta,\gamma).$$

Moreover, we have that $d_{p,\xi}(\alpha,\beta) < g(\xi)$, for every $\alpha <_{p,\xi} \beta$.

We can now define the notion of forcing which adds an $\omega_2$ chain in $(\omega_1^{\omega_1}, <_{\text{fin}})$ below the function $g$.

**Definition 3.12** Let $\mathbb{M}_3^*$ be the forcing notion whose elements are pairs $p = (f_p, \mathcal{M}_p)$, where $f_p$ is a partial function from $\omega_2 \times \omega_1$ to $\omega_1$, $\text{dom}(f_p)$ is of the form $A_p \times F_p$ where $0 \in A_p \in [\omega_2]^{<\omega}$, $F_p \in [\omega_1]^{<\omega}$, $\mathcal{M}_p \in \mathbb{M}^*$, and for every $\alpha,\beta \in A_p$ with $\alpha < \beta$, every $\xi \in F_p$ and $M \in \mathcal{M}_p$:

1. $f_p(\alpha,\xi) < g(\xi)$,

2. if $\alpha <_{p,\xi} \beta$ then $f_p(\alpha, \xi) + d_{p,\xi}(\alpha, \beta) \leq f_p(\beta, \xi)$,

We let $q \leq p$ if $f_p \subseteq f_q$, $\mathcal{M}_p \subseteq \mathcal{M}_q$ and for every $\alpha, \beta \in A_p$ and $\xi \in F_q \setminus F_p$ if $\alpha < \beta$ then $f_q(\alpha, \xi) < f_q(\beta, \xi)$.

We first show that for any $\alpha < \omega_2$ and $\xi < \omega_1$ any condition $p \in \mathbb{M}_3^*$ can be extended to a condition $q$ such that $\alpha \in A_q$ and $\xi \in F_q$.

**Lemma 3.13** *Let $p \in \mathbb{M}_3^*$ and $\delta \in \omega_2 \setminus A_p$. Then there is a condition $q \leq p$ such that $\delta \in A_q$.*

*Proof* We let $\mathcal{M}_q = \mathcal{M}_p$, $A_q = A_p \cup \{\delta\}$ and $F_q = F_p$. On $A_p \times F_p$ we let $f_q$ be equal to $f_p$. We need to define $f_q(\delta, \xi)$, for $\xi \in F_p$. Consider one such $\xi$. If $\delta$ does not belong to $\overline{M \cap \omega_2}$, for any $M \in \mathcal{M}_q$ with $\delta_M \leq \xi$, we can define $f_q(\delta, \xi)$ arbitrarily. Otherwise, we need to ensure that if $\alpha \in A_p$ and $\alpha <_{p,\xi} \delta$ then

$$f_{p,\xi}(\alpha, \xi) + d_{p,\xi}(\alpha, \delta) \leq f_q(\delta, \xi).$$

Similarly, if $\beta \in A_p$ and $\delta <_{p,\xi} \beta$ we have to ensure that

$$f_q(\delta, \xi) + d_{p,\xi}(\delta, \beta) \leq f_p(\beta, \xi).$$

By the additivity of $d_{p,\xi}$ we know that if $\alpha <_{p,\xi} \delta <_{p,\xi} \beta$ then $d_{p,\xi}(\alpha, \beta) = d_{p,\xi}(\alpha, \delta) + d_{p,\xi}(\delta, \beta)$. Since $p$ is a condition we know that if $\alpha, \beta \in A_p$ then $f_p(\beta, \xi) \geq f_p(\alpha, \xi) + d_{p,\xi}(\alpha, \beta)$. Let $\alpha^*$ be the largest element of $A_p \cap (\delta)_{p,\xi}$. We can then simply define $f_q(\delta, \xi)$ by

$$f_q(\delta, \xi) = f_p(\alpha^*, \xi) + d_{p,\xi}(\alpha^*, \delta).$$

It is straightforward to check that the $q$ thus defined is a condition. □

**Lemma 3.14** *Let $p \in \mathbb{M}_3^*$ and $\xi \in \omega_1 \setminus F_p$. Then there is a condition $q \leq p$ such that $\xi \in F_q$.*

*Proof* We let $\mathcal{M}_q = \mathcal{M}_p$, $A_q = A_p$ and $F_q = F_p \cup \{\xi\}$. Then we need to extend $f_p$ to $A_q \times \{\xi\}$. Notice that we now have the following commitments. Suppose $\alpha, \beta \in A_p$ and $\alpha < \beta$, then we need to ensure that $f_q(\alpha, \xi) < f_q(\beta, \xi)$ in order for $q$ to be an extension of $p$. If in addition $\alpha <_{p,\xi} \beta$ then we need to ensure that

$$f_q(\alpha, \xi) + d_{q,\xi}(\alpha, \beta) \leq f_q(\beta, \xi)$$

in order for $q$ to satisfy (2) of Definition 3.12. We define $f_q(\beta, \xi)$ by induction on $\beta \in A_q$ as follows. We let $f_q(0, \xi) = 0$. For $\beta > 0$ we let $f_q(\beta, \xi)$ be the maximum of the following set:

$$\{f_q(\alpha, \xi) + 1 : \alpha \in (A_q \cap \beta) \setminus (\beta)_{q,\xi}\} \cup \{f_q(\alpha, \xi) + d_{q,\xi}(\alpha, \beta) : \alpha \in A_q \cap (\beta)_{q,\xi}\}.$$

It is easy to see that $f_q(\beta,\xi) < g(\xi)$, for all $\beta \in A_q$, and that $q$ is a condition extending $p$. □

In order to prove strong properness of $\mathbb{M}_3^*$ we need to restrict to a relative club subset of $\mathcal{E}^2$ of elementary submodels of $H(\aleph_2)$ which are the restriction to $H(\aleph_2)$ of an elementary submodel of $H(2^{\aleph_1 +})$.

**Definition 3.15** Let $\mathcal{D}^2$ be the set of all $M \in \mathcal{E}^2$ such that $M = M^* \cap H(\aleph_2)$, for some $M^* \prec H(2^{\aleph_1 +})$. We let $\mathcal{D}_0^2 = \mathcal{D}^2 \cap \mathcal{E}_0^2$ and $\mathcal{D}_1^2 = \mathcal{D}^2 \cap \mathcal{E}_1^2$.

We split the proof that $\mathbb{M}_3^*$ is $\mathcal{D}^2$-strongly proper in two lemmas.

**Lemma 3.16** *Let $p \in \mathbb{M}_3^*$ and $M \in \mathcal{M}_p \cap \mathcal{D}_0^2$. Then $p$ is an $(M, \mathbb{M}_3^*)$-strongly generic condition.*

*Proof* Given $r \leq p$ we need to find a condition $r \mid M \in M$ such that every $q \leq r \mid M$ which is in $M$ is compatible with $r$. By Lemma 3.14 we may assume that $\sup(P) \in A_r$, for every $P \in \mathcal{M}_r$. The idea is to choose $r \mid M$ which has the same type as $r$ over some suitably chosen parameters in $M$. Let $D = \{\delta_P : P \in \mathcal{M}_r\} \cap M$. Since $M \in \mathcal{D}_0^2$ there is $M^* \prec H(2^{\aleph_1 +})$ such that $M = M^* \cap H(\aleph_2)$. By elementary of $M^*$, we can find in $M$ an $\in$-chain $\mathcal{M}_{r^*} \in \mathbb{M}^*$ extending $\mathcal{M}_r \cap M$, a finite set $A_{r^*} \subseteq \omega_2$ and an order preserving bijection $\pi : A_r \to A_{r^*}$ such:

1. $\pi$ is the identity function on $A_r \cap M$,
2. if $\alpha, \beta \in A_r$ then, for every $\xi \in D$,

$$d_{r^*,\xi}(\pi(\alpha), \pi(\beta)) = d_{r,\xi}(\alpha, \beta).$$

By Lemma 3.13 we can extend $f_r \restriction (A_r \cap M) \times (F_r \cap M)$ to a function $f_{r^*} : A_{r^*} \times (F_r \cap M) \to \omega_1$ such that $(f_{r^*}, \mathcal{M}_{r^*})$ is a condition in $\mathbb{M}_3^*$. Finally, we set $r \mid M = r^*$.

Suppose now $q \leq r \mid M$ and $q \in M$. We need to find a common extension $s$ of $q$ and $r$. We define $\mathcal{M}_s$ to be the closure under intersection of $\mathcal{M}_r \cup \mathcal{M}_q$. Indeed Lemma 2.17 shows that $\mathcal{M}_s \in \mathbb{M}^*$. We first compute the distance function $d_{s,\xi}$ in terms of $d_{r,\xi}$ and $d_{q,\xi}$, for $\xi < \omega_1$. First notice that the new models which are obtained by closing $\mathcal{M}_q \cup \mathcal{M}_r$ under intersection do not create new links and therefore do not influence the computation of the distance function.

Now, consider an ordinal $\xi < \omega_1$. If $\xi \geq \delta_M$ then all ordinals in $M \cap \omega_2$ are pairwise $r,\xi$-linked. The countable models of $\mathcal{M}_q \setminus \mathcal{M}_r$ are all included in $M$ so they do not add any new $s,\xi$-links. It follows that in this case $d_{s,\xi} = d_{r,\xi}$. Consider now an ordinal $\xi < \delta_M$. By Fact 3.8

if $\beta \in M$ then $(\beta)_{s,\xi} = (\beta)_{q,\xi}$. If $\beta \notin M$ then, by Fact 3.9 there is a $\eta \in A_r \cap M$ such that $(\beta)_{s,\xi} \cap M = (\eta)_{q,\xi}$. Let $\xi^* = \max(D \cap (\xi + 1))$. Then, again by Fact 3.9, $\eta$ and $\beta$ are $r, \xi^*$-connected and

$$d_{s,\xi}(\alpha,\beta) = d_{q,\xi}(\alpha,\eta) + d_{r,\xi^*}(\eta,\beta).$$

Let $A_s = A_q \cup A_r$ and $F_s = F_q \cup F_r$. Our next goal is to define an extension, call it $f_s$, of $f_q \cup f_r$ on $A_s \times F_s$. It remains to define $f_s$ on

$$((A_q \setminus A_r) \times (F_r \setminus F_q)) \cup ((A_r \setminus A_q) \times (F_q \setminus F_r)).$$

*Case 1*: Consider first $\xi \in F_r \setminus F_q$ and let us define $f_s$ on $(A_q \setminus A_r) \times \{\xi\}$. We already know that $d_{s,\xi} = d_{r,\xi}$, so we need to ensure that if $\alpha, \beta \in A_s$ and $\alpha <_{s,\xi} \beta$ then

$$f_s(\alpha,\xi) + d_{r,\xi}(\alpha,\beta) \leq f_s(\beta,\xi).$$

Notice that all the ordinals of $A_q$ are $r, \xi$-linked as witnessed by $M$ so then we will also have that for every $\alpha, \beta \in A_q$, if $\alpha < \beta$ then $f_s(\alpha,\xi) < f_s(\beta,\xi)$. In order to define $f_s(\alpha,\xi)$, for $\alpha \in A_q$, let $\alpha^*$ be the maximal element of $(\alpha)_{r,\xi} \cap A_r$ and let $f_s(\alpha,\xi) = f_r(\alpha^*,\xi) + d_{r,\xi}(\alpha^*,\alpha)$. It is straightforward to check that (2) of Definition 3.12 is satisfied in this case.

*Case 2*: Consider now some $\xi \in F_q \setminus F_r$. What we have to arrange is that $f_s(\alpha,\xi) < f_s(\beta,\xi)$, for every $\alpha, \beta \in A_r$ with $\alpha < \beta$. Moreover, for every $\alpha, \beta \in A_s$ with $\alpha <_{s,\xi} \beta$ we have to arrange that

$$f_s(\alpha,\xi) + d_{s,\xi}(\alpha,\beta) \leq f_s(\beta,\xi).$$

We define $f_s$ on $(A_r \setminus A_q) \times \{\xi\}$ by setting

$$f_s(\beta,\xi) = f_q(\pi(\beta),\xi).$$

First, we show that the function $\alpha \mapsto f_s(\alpha,\xi)$ is order preserving on $A_r$. To see this observe that, since $q \leq r^* = r \mid M$ and $\xi \notin F_{r^*}$, the function $\alpha \mapsto f_q(\alpha,\xi)$ is strictly order preserving on $A_{r^*}$. Moreover, $\pi$ is order preserving and the identity on $A_r \cap M = A_r \cap A_q$.

Assume now $\alpha, \beta \in A_s$ and $\alpha <_{s,\xi} \beta$. If $\alpha, \beta \in A_q$ then, since $q$ is a condition, $f_s(\beta,\xi) \geq f_s(\alpha,\xi) + d_{q,\xi}(\alpha,\beta)$. On the other hand, we know that $d_{s,\xi}(\alpha,\beta) = d_{q,\xi}(\alpha,\beta)$, so we have the required inequality in this case. By Fact 3.8 the case $\alpha \in A_r \setminus A_q$ and $\beta \in A_q$ cannot happen.

Suppose $\alpha \in A_q$ and $\beta \in A_r \setminus A_q$. Let $\xi^* = \max(D \cap (\xi + 1))$. By Fact 3.9 there is $\eta \in A_r \cap M$ such that

$$d_{s,\xi}(\alpha, \beta) = d_{s,\xi}(\alpha, \eta) + d_{r,\xi^*}(\eta, \beta).$$

By property (2) of $\pi$ we have that $d_{r^*,\xi^*}(\eta, \pi(\beta)) = d_{r,\xi^*}(\eta, \beta)$. Since $q$ extends $r^*$ it follows that $d_{q,\xi^*}(\eta, \pi(\beta)) \geq d_{r^*,\xi^*}(\eta, \pi(\beta))$. Moreover, $q$ is a condition and so:

$$f_q(\pi(\beta), \xi) \geq f_q(\alpha, \xi) + d_{q,\xi}(\alpha, \pi(\beta)) \geq f_q(\alpha, \xi) + d_{q,\xi^*}(\alpha, \pi(\beta)).$$

Therefore,

$$f_q(\pi(\beta), \xi) \geq f_q(\alpha, \xi) + d_{s,\xi}(\alpha, \beta).$$

The final case is when $\alpha, \beta \in A_r \setminus A_q$ and $\alpha <_{s,\xi} \beta$. Note that in this case, $\alpha$ and $\beta$ are already $r, \xi$-connected, in fact, they are $r, \xi^*$-connected, where as before $\xi^* = \max(D \cap (\xi + 1))$. By property (2) of $\pi$ we have that $\pi(\alpha)$ and $\pi(\beta)$ are $r^*, \xi^*$-connected and

$$d_{r^*,\xi^*}(\pi(\alpha), \pi(\beta) = d_{r,\xi}(\alpha, \beta).$$

Since $\xi^* \leq \xi$ and $q$ extends $r^*$ we have that

$$d_{q,\xi}(\pi(\alpha), \pi(\beta)) \geq d_{r^*,\xi^*}(\pi(\alpha), \pi(\beta)).$$

Since $q$ is a condition we have

$$f_q(\pi(\beta), \xi) \geq f_q(\pi(\alpha), \xi) + d_{q,\xi}(\pi(\alpha), \pi(\beta)).$$

Since $d_{s,\xi}(\pi(\alpha), \pi(\beta)) = d_{q,\xi}(\pi(\alpha), \pi(\beta))$ we have $f_s(\beta, \xi) \geq f_s(\alpha, \xi) + d_{s,\xi}(\alpha, \beta)$, as required.

It follows that $s$ is a condition which extends $q$ and $r$. This completes the proof of Lemma 3.16. □

**Lemma 3.17** *Let $p \in \mathbb{M}_3^*$ and $M \in \pi_1(\mathcal{M}_p)$. Then $p$ is $(M, \mathbb{M}_3^*)$-strongly generic.*

*Proof* Let $r \leq p$. We need to find a condition $r \mid M \in M$ such that any $q \leq r \mid M$ in $M$ is compatible with $r$. We simply set

$$r \mid M = (f_r \restriction (A_r \times F_r) \cap M, \mathcal{M}_p \cap M).$$

We need to show that if $q \leq r \mid M$ is in $M$, then there is a condition $s \leq q, r$. Thanks to Lemma 2.17 we just need to define $f_s$, since we already know that $\mathcal{M}_r \cup \mathcal{M}_q$ is an $\in$-chain and belongs to $\mathbb{M}^*$. Since $\omega_1 \subseteq M$ we have that $F_r \subseteq M$ so we only need to define an extension $f_s$ on $A_r \setminus A_q \times F_q \setminus F_r$. We know that $M \cap \omega_2$ is an initial segment of

$\omega_2$ so all the elements of $A_r \setminus A_q = A_r \setminus M$ are above all the ordinals of $A_q$. Given an ordinal $\xi \in F_q \setminus F_r$ we define $f_s(\beta, \xi)$, for $\beta \in A_r \setminus A_q$ by induction. We set:

$$f_s(\beta,\xi) = \max(\{f_s(\alpha,\xi)+1 : \alpha \in A_r \cap \beta\} \cup \{f_s(\alpha,\xi)+d_{s,\xi}(\alpha,\beta) : \alpha <_{s,\xi} \beta\}.$$

It is easy to check that $(f_s, \mathcal{M}_s)$ is a condition which extends both $q$ and $r$. □

**Corollary 3.18** *The forcing $\mathbb{M}_3^*$ is $\mathcal{D}^2$-strongly proper. Hence it preserves $\omega_1$ and $\omega_2$.*

We have shown that for every $\alpha < \omega_2$ and $\xi < \omega_1$ the set

$$D_{\alpha,\xi} = \{p \in \mathbb{M}_3^* : \alpha \in A_p, \xi \in F_p\}$$

is dense in $\mathbb{M}_3$. If $G$ is a $V$-generic filter in $\mathbb{M}_3^*$ we let

$$f_G = \bigcup \{f_p : p \in G\}.$$

It follows that $f_G : \omega_2 \times \omega_1 \to \omega_1$. For $\alpha < \omega_2$ we define $f_\alpha : \omega_1 \to \omega_1$ by letting $f_\alpha(\xi) = f_G(\alpha, \xi)$, for all $\xi$. It follows that the sequence $(f_\alpha : \alpha < \omega_2)$ is an increasing $\omega_2$-chain in $(\omega_1^{\omega_1}, <_{\text{fin}})$. We have thus completed the proof of the following.

**Theorem 3.19** *There is a $\mathcal{D}^2$-strongly proper forcing which adds an $\omega_2$ chain in $(\omega_1^{\omega_1}, <_{\text{fin}})$.*

## 4 Thin very tall superatomic Boolean algebras

A Boolean algebra $\mathcal{B}$ is called *superatomic* (sBa) iff every homomorphic image of $\mathcal{B}$ is atomic. In particular, $\mathcal{B}$ is an sBa iff its Stone space $S(\mathcal{B})$ is scattered. A very useful tool for studying scattered spaces is the Cantor-Bendixson derivative $A^{(\alpha)}$ of a set $A \subseteq S(\mathcal{B})$, defined by induction on $\alpha$ as follows. Let $A^{(0)} = A$, $A^{(\alpha+1)}$ is the set of limit points of $A^{(\alpha)}$, and $A^{(\lambda)} = \bigcap\{A^{(\alpha)} : \alpha < \lambda\}$, if $\lambda$ is a limit ordinal. Then $S(\mathcal{B})$ is scattered iff for $S(\mathcal{B})^{(\alpha)} = \emptyset$, for some $\alpha$.

When this notion is transferred to the Boolean algebra $\mathcal{B}$, we arrive at a sequence of ideals $I_\alpha$, which we refer to as the Cantor-Bendixson ideals, defined by induction on $\alpha$ as follows. Let $I_0 = \{0\}$. Given $I_\alpha$ let $I_{\alpha+1}$ be generated by $I_\alpha$ together with all $b \in \mathcal{B}$ such that $b/I_\alpha$, is an atom in $\mathcal{B}/I_\alpha$. If $\alpha$ is a limit ordinal, let $I_\alpha = \bigcup\{I_\xi : \xi < \alpha\}$. Then $\mathcal{B}$ is an sBa iff some $I_\alpha = \mathcal{B}$, for some $\alpha$.

The height of an sBa $\mathcal{B}$, ht($\mathcal{B}$), is the least ordinal $\alpha$ such that $I_\alpha = \mathcal{B}$. For $\alpha < \text{ht}\mathcal{B}$ let $\text{wd}_\alpha(\mathcal{B})$ be the cardinality of the set of atoms in $\mathcal{B}/I_\alpha$. The *cardinal sequence* of $\mathcal{B}$ is the sequence $(\text{wd}_\alpha(\mathcal{B}) : \alpha < \text{ht}(\mathcal{B}))$. We say that $\mathcal{B}$ is *$\kappa$-thin-very tall* if $\text{ht}(\mathcal{B}) = \kappa^{++}$ and $\text{wd}_\alpha(\mathcal{B}) = \kappa$, for all $\alpha < \kappa^{++}$. If $\kappa = \omega$ we simply say that $\mathcal{B}$ is *thin very tall.*

Baumgartner and Shelah [2] constructed a forcing notion which adds a thin very tall sBa. This is achieved in two steps. First they adjoin by a $\sigma$-closed $\aleph_2$-cc forcing a function $f : [\omega_2]^2 \to [\omega_2]^{\leq\omega}$ with some special properties. Such a function is called a $\Delta$-function. In the second step they use a $\Delta$-function to define a ccc forcing notion which adds a thin very tall sBa. The purpose of this section is to show how this can be achieved directly by using generalizes side conditions. The following concept from [2] was made explicit by Bagaria in [1].

**Definition 4.1** Given a cardinal sequence $\theta = \langle\kappa_\alpha : \alpha < \lambda\rangle$, where each $\kappa_\alpha$ is an infinite cardinal, we say that a structure $(T, \leq, i)$ is a *$\theta$-poset* if $<$ is a partial ordering on $T$ and the following hold:

1. $T = \bigcup\{T_\alpha : \alpha < \lambda\}$, where each $T_\alpha$ is of the form $\{\alpha\} \times Y_\alpha$, and $Y_\alpha$ is a set of cardinality $\kappa_\alpha$.
2. If $s \in T_\alpha$, $t \in T_\beta$ and $s < t$, then $\alpha < \beta$.
3. For every $\alpha < \beta < \lambda$, if $t \in T_\beta$ then the set $\{s \in T_\alpha : s < t\}$ is infinite.
4. $i$ is a function from $[T]^2$ to $[T]^{<\omega}$ with the following properties:
   a. If $u \in i\{s, t\}$, then $u \leq s, t$
   b. If $u \leq s, t$, then there exists $v \in i\{s, t\}$ such that $u \leq v$.

We let $\Omega(\lambda)$ denote the sequence of length $\lambda$ with all entries equal to $\omega$. The following is implicitly due to Baumgartner (see [1] for a proof).

*Fact* 4.2 Let $\theta = \langle\kappa_\alpha : \alpha < \lambda\rangle$ be a sequence of cardinals. If there exists a $\theta$-poset, then there exists an sBa whose cardinal sequence is $\theta$.

We now define a forcing notion which adds an $\Omega(\omega_2)$-poset. If $x \in \omega_2 \times \omega$ is of the form $(\alpha, n)$ then we denote $\alpha$ by $\alpha_x$ and $n$ by $n_x$.

**Definition 4.3** Let $\mathbb{M}_4$ be the forcing notion whose elements are tuples $p = (x_p, \leq_p, i_p, \mathcal{M}_p)$, where $x_p$ is a finite subset of $\omega_2 \times \omega$, $\leq_p$ is a partial ordering on $x_p$, $i_p : [x_p]^2 \to [x_p]^{<\omega}$, $\mathcal{M}_p \in \mathbb{M}$ and the following hold:

1. if $s, t \in x_p$ and $s <_p t$ then $\alpha_s < \alpha_t$,
2. if $u \in i_p\{s, t\}$ then $u \leq_p s, t$,
3. for every $u \leq_p s, t$ there is $v \in i_p\{s, t\}$ such that $u \leq_p v$,
4. for every $s, t \in x_p$ and $M \in \mathcal{M}_p$ if $s, t \in M$ then $i_p\{s, t\} \in M$.

We let $q \leq p$ if and only if $x_q \supseteq x_p$, $\leq_q \restriction x_p = \leq_p$, $i_q \restriction [x_p]^2 = i_p$ and $\mathcal{M}_p \subseteq \mathcal{M}_q$.

We first observe that a version of Lemma 1.8 holds for $\mathbb{M}_4$.

**Lemma 4.4** *Let $M \in \mathcal{E}^2$ and let $p \in \mathbb{M}_4 \cap M$. Then there is a new condition, which we will call $p^M$, that is the smallest element of $\mathbb{M}_4$ extending $p$ such that $M \in \mathcal{M}_{p^M}$.*

*Proof* If $M \in \mathcal{E}^2_1$ then simply let $p^M = (x_p, \leq_p, i_p, \mathcal{M}_p \cup \{M\})$. If $M \in \mathcal{E}^2_0$, then, as in Lemma 1.8, we let $\mathcal{M}_{p^M}$ be the closure of $\mathcal{M}_p \cup \{M\}$ under intersection and let $p^M = (x_p, \leq_p, i_p, \mathcal{M}_{p^M})$. We need to check that condition (4) of Definition 4.3 is satisfied. Since $p \in M$ we have that $x_p \subseteq M$. In the case $M \in \mathcal{E}^2_1$ the only new model in $\mathcal{M}_{p^M}$ is $M$ so condition (4) holds for $p^M$ since it holds for $p$. In the case $M \in \mathcal{E}^2_0$ there are also models of the form $N \cap M$, where $N \in \pi_1(\mathcal{M}_p)$. However, condition (4) holds for both $N$ and $M$ and so it holds for their intersection. □

Next, we show that $\mathbb{M}_4$ is $\mathcal{E}^2$-proper. We split this in two parts.

**Lemma 4.5** $\mathbb{M}_4$ *is $\mathcal{E}^2_0$-proper.*

*Proof* Let $\theta$ be a sufficiently large regular cardinal and let $M^*$ be a countable elementary submodel of $H(\theta)$ containing all the relevant objects. Then $M = M^* \cap H(\omega_2)$ belongs to $\mathcal{E}^2_0$. Suppose $p \in \mathbb{M}_4 \cap M$. Let $p^M$ be the condition defined in Lemma 4.4, i.e. $p^M = (x_p, \leq_p, i_p, \mathcal{M}_{p^M})$, where $\mathcal{M}_{p^M}$ is the closure of $\mathcal{M}_p \cup \{M\}$ under intersection. We show that $p^M$ is $(M^*, \mathbb{M}_4)$-generic. Let $D \in M^*$ be a dense subset of $\mathbb{M}_4$ and $r \leq p^M$. We need to find a condition $q \in D \cap M^*$ which is compatible with $r$. Note that we may assume that $r \in D$. We define a condition $r \mid M$ as follows. First let $x_{r|M} = x_r \cap M$ and then let $\leq_{r|M} = \leq_r \restriction x_{r|M}$ and $i_{r|M} = i_r \restriction [x_{r|M}]^2$. Condition (4) of Definition 4.3 guarantees that if $s, t \in x_{r|M}$ then $i_r\{s, t\} \subseteq M$. Finally, let $\mathcal{M}_{r|M} = \mathcal{M}_r \cap M$. It follows that $r \mid M = (x_{r|M}, i_{r|M}, i_{r|M}, \mathcal{M}_{r|M})$ belongs to $\mathbb{M}_4 \cap M$. By elementarity of $M^*$ in $H(\theta)$ there is a condition $q \in D \cap M^*$ extending $r \mid M$ such that $(x_q \setminus x_{r|M}) \cap N = \emptyset$, for all $N \in \pi_0(\mathcal{M}_{r|M})$.

*Claim* 4.6 $q$ and $r$ are compatible.

*Proof* We define a condition $s$ as follows. We set $x_s = x_q \cup x_r$ and we let $\leq_s$ be the transitive closure of $\leq_q \cup \leq_r$, i.e. if $u \in x_q \setminus x_r$, $v \in x_r \setminus x_q$ and $t \in x_{r|M}$ are such that $u \leq_q t$ and $t \leq_r v$, then we let $u \leq_s v$.

Similarly, if $v \leq_r t$ and $t \leq_q u$ we let $v \leq_s u$. We let $\mathcal{M}_s$ be the closure under intersection of $\mathcal{M}_q \cup \mathcal{M}_r$. It remains to define $i_s$. For $z \in x_r$ let $A_z = \{t \in x_{r|M} : t \leq_r z\}$ and for $z \in x_q$ let $B_z = \{t \in x_{r|M} : t \leq_q z\}$. We let

$$i_s\{u, v\} = \begin{cases} i_q\{u, v\} & \text{if } u, v \in x_q, \\ i_r\{u, v\} & \text{if } u, v \in x_r, \\ \bigcup_{t \in A_v} i_q\{u, t\} \cup \bigcup_{t \in B_u} i_r\{t, v\} & \text{if } u \in x_q \setminus x_r \text{ and } v \in x_r \setminus x_q. \end{cases}$$

We now need to check property (4) of Definition 4.3, i.e. for every $u, v \in x_s$ and $P \in \mathcal{M}_s$, if $u, v \in P$ then $i_s\{u, v\} \in P$. First of all notice that we only need to show the above property for $P \in \mathcal{M}_q \cup \mathcal{M}_r$, because the other models in $\mathcal{M}_s$ are obtained by intersection and, if (4) holds for $u, v$ and $P$ and also for $u, v$ and $Q$, it also holds for $u, v$ and $P \cap Q$.

*Case 1*: $u, v \in x_q$ and $P \in \mathcal{M}_r$. If $P \in \mathcal{M}_{r|M}$ then $P \in \mathcal{M}_q$ and then (4) holds since $q$ is a condition and $i_s\{u, v\} = i_q\{u, v\}$. Suppose now $P \in \mathcal{M}_r \setminus M$ and $\delta_R < \delta_M$. Then, by Fact 3.3, $P \cap M \in \mathcal{M}_{r|M}$ and so $P \cap M \in \mathcal{M}_q$, therefore (4) of Definition 4.3 holds again. Finally, if $\delta_R \geq \delta_M$ then, by Fact 3.2, $P \cap M \cap \omega_2$ is an initial segment of $M \cap \omega_2$. We know that $i_q\{u, v\} \in M$ and for every $w \in i_q\{u, v\}$ $\alpha_w \leq \min(\alpha_u, \alpha_v)$. Therefore, we have that $i_q\{u, v\} \in P \cap M$. Since $i_s\{u, v\} = i_q\{u, v\}$, we conclude that (4) holds in this case.

*Case 2*: $u, v \in x_r$ and $P \in \mathcal{M}_q$. If $u, v \in x_{r|M}$ then $u, v \in x_q$ and again, since $q$ is a condition and $i_s\{u, v\} = i_q\{u, v\}$, we know that $i_s\{u, v\} \in P$. Suppose now that $u$ and $v$ are not both in $M$. If $P \in \mathcal{E}^2_0$ then $P \subseteq M$ and so we cannot have $u, v \in P$. If $P \in \mathcal{E}^2_1$ we know that $P \cap \omega_2$ is an initial segment of $\omega_2$. Moreover, if $w \in i_s\{u, v\}$ then $\alpha_w \leq \min(\alpha_u, \alpha_v)$ and so if $u, v \in P$ we also have that $i_s\{u, v\} \in P$, so (4) of Definition 4.3 holds again.

*Case 3*: $u \in x_q \setminus x_r$ and $v \in x_r \setminus x_q$. If $P \in \mathcal{E}^2_1$ then $P \cap \omega_2$ is an initial segment of $\omega_2$. Moreover, as before, we have that $\alpha_w \leq \min(\alpha_u, \alpha_v)$, for every $w \in i_s\{u, v\}$. Therefore, $i_s\{u, v\} \in P$. Suppose now $P \in \mathcal{E}^2_0$. If $P \in \pi_0(\mathcal{M}_q)$ then $P \subseteq M$ so $v \notin P$. Now assume $P \in \pi_0(\mathcal{M}_r)$. If $\delta_R < \delta_M$ then by Fact 3.3, $P \cap M \in \mathcal{M}_{r|M}$. However, the condition $q$ is chosen so that $(x_q \setminus x_{r|M}) \cap N$, for all $N \in \pi_0(\mathcal{M}_{r|M})$, therefore in this case $u \notin P$. Assume now $\delta_R \geq \delta_M$. Then, by Fact 3.2, we have that $P \cap M \cap \omega_2$ is an initial segment of $M \cap \omega_2$. Consider first some $t \in A_v$. Then $t \in M$ and $i_q\{u, t\} \in M$. Moreover, $\alpha_w \leq \min(\alpha_u, \alpha_t)$, for every $w \in i_q\{u, t\}$. Since $P \cap M \cap \omega_2$ is an initial segment of $M \cap \omega_2$, it follows that $\alpha_w \in P \cap M$, for every $w \in i_q\{u, t\}$. This implies that $i_q\{u, t\} \in P \cap M$. Finally, consider some $t \in B_u$. Then $t \in M$ and, since $u \in P \cap M$, $\alpha_t \le \alpha_u$ and $P \cap M \cap \omega_2$ is an initial segment of $M \cap \omega_2$, we have that $\alpha_t \in P \cap M$ and so $t \in P \cap M$. Now, since $r$ is a condition, $P \in \mathcal{M}_r$ and $t, v \in P$, we have that $i_r\{t, v\} \in P$, so (4) of Definition 4.3 holds in this case as well.

It follows that $s$ is a condition which extends both $q$ and $r$. This completes the proof of Claim 4.6 and Lemma 4.5. □

□

**Lemma 4.7** *$\mathbb{M}_4$ is $\mathcal{E}_1^2$-proper.*

*Proof* Let $\theta$ be a sufficiently large regular cardinal and $M^*$ an elementary submodel of $H(\theta)$ containing all the relevant objects such that $M = M^* \cap H(\omega_2)$ belongs to $\mathcal{E}_1^2$. Fix $p \in M \cap \mathbb{M}_4$. Let $p^M$ be as in Lemma 4.4. We claim that $p^M$ is $(M^*, \mathbb{M}_4)$-generic. In order to verify this consider a dense subset $D$ of $\mathbb{M}_4$ which belongs to $M^*$ and a condition $r \le p^M$. We need to find a condition $q \in D \cap M^*$ which is compatible with $r$. By extending $r$ if necessary we may assume it belongs to $D$. Let $r \mid M = (x_{r|M}, i_{r|M}, i_{r|M}, \mathcal{M}_{r|M})$ be as in Lemma 4.5. By elementarity of $M^*$ in $H(\theta)$, we can find $q \le r \mid M$, in $D \cap M$, such that $(x_q \setminus x_{r|M}) \cap N = \emptyset$, for all $N \in \mathcal{M}_{r|M}$, and if $u \in x_{r|M}$ and $v \in x_q \setminus x_{r|M}$ then $\alpha_u < \alpha_v$.

*Claim* 4.8 $q$ and $r$ are compatible.

*Proof* We define a condition $s$ as follows. We set $x_s = x_q \cup x_r$, $\le_s = \le_q \cup \le_r$ and $\mathcal{M}_s = \mathcal{M}_q \cup \mathcal{M}_r$. Note that $\le_s$ is a partial order and $\mathcal{M}_s \in \mathbb{M}$. It remains to define $i_s$. We let

$$i_s\{u, v\} = \begin{cases} i_q\{u, v\} & \text{if } u, v \in x_q, \\ i_r\{u, v\} & \text{if } u, v \in x_r, \\ \{z \in x_{r|M} : z \le_q u \text{ and } z \le_r v\} & \text{if } u \in x_q \setminus x_r,\ v \in x_r \setminus x_q. \end{cases}$$

We need to check (4) of Definition 4.3. So, suppose $u, v \in x_s$, $P \in \mathcal{M}_s$ and $u, v \in P$. We need to show that $i_s\{u, v\} \in P$.

*Case 1*: $u, v \in x_q$ and $P \in \mathcal{M}_r$. If $P \cap M \in \mathcal{M}_{r|M}$ this follows from the fact that $q$ is a condition and $\mathcal{M}_{r|M} \subseteq \mathcal{M}_q$. If $P \cap M \notin \mathcal{M}_{r|M}$ then $M \subseteq P$ and, since $i_q\{u, v\} \in M$, it follows that $i_s\{u, v\} \in P$.

*Case 2*: $u, v \in x_r$ and $P \in \mathcal{M}_q$. If $u, v \in M$ then $u, v \in x_q$, so $i_s\{u, v\} \in P$ follows from the fact that $q$ is a condition. Assume now, for concreteness, that $v \notin M$. If $P \in \mathcal{M}_q$ then $v \notin P$ and if $P \in \mathcal{M}_r$ then (4) of Definition 4.3 follows from the fact that $r$ is a condition.

*Case 3*: $u \in x_q \setminus x_r$ and $v \in x_r \setminus x_q$. If $P \in \mathcal{M}_q$ then $v \notin P$. If $P \in \mathcal{M}_r$ then either $P \cap M \in \mathcal{M}_{r|M}$ and then, by the choice of $q$, we have that $u \notin P$. Otherwise $M \subseteq P$ and in this case $x_{r|M} \subseteq P$. Since $i_s\{u, v\} \subseteq x_{r|M}$ it follows that (4) of Definition 4.3 holds in this case as well. □

This completes the proof of Lemma 4.7. □

**Corollary 4.9** *The forcing $\mathbb{M}_4$ is $\mathcal{E}^2$-proper. Hence it preserves $\omega_1$ and $\omega_2$.*

It is easy to see that the set

$$D_{\alpha,n} = \{p \in \mathbb{M}_4 : (\alpha, n) \in x_p\}$$

is dense in $\mathbb{M}_4$, for every $\alpha \in \omega_2$ and $n \in \omega$. Moreover, given $t \in \omega_2 \times \omega$, $\eta < \alpha_t$ and $n < \omega$, one verifies easily that the set

$$E_{t,\eta,n} = \{p : t \in x_p \text{ and } |\{i : (\eta, i) \in x_p \text{ and } (\eta, i) \leq_p t\}| \geq n\}$$

is dense. Then if $G$ is $V$-generic filter on $\mathbb{M}_4$ let

$$\leq_G = \bigcup\{\leq_p : p \in G\} \qquad \text{and} \qquad i_G = \bigcup\{i_p : p \in G\}.$$

It follows that $(\omega_2 \times \omega, \leq_G, i_G)$ is an $\Omega(\omega_2)$-poset in $V[G]$. We have therefore proved the following.

**Theorem 4.10** *There is an $\mathcal{E}^2$-proper forcing notion which adds an $\Omega(\omega_2)$-poset.*

## References

[1] J. Bagaria. Locally-generic Boolean algebras and cardinal sequences. *Algebra Universalis*, 47(3):283–302, 2002.

[2] James E. Baumgartner and Saharon Shelah. Remarks on superatomic Boolean algebras. *Ann. Pure Appl. Logic*, 33(2):109–129, 1987.

[3] Sy David Friedman. Forcing with side conditions. Preprint, 2004.

[4] Piotr Koszmider. On the existence of strong chains in $\mathcal{P}(\omega_1)$/Fin. *J. Symbolic Logic*, 63(3):1055–1062, 1998.

[5] Piotr Koszmider. On strong chains of uncountable functions. *Israel J. Math.*, 118:289–315, 2000.

[6] William J. Mitchell. Notes on a proof of Koszmider. Preprint, http://www.math.ufl.edu/ wjm/papers/koszmider.pdf, 2003.

[7] William J. Mitchell. Adding closed unbounded subsets of $\omega_2$ with finite forcing. *Notre Dame J. Formal Logic*, 46(3):357–371, 2005.

[8] William J. Mitchell. $I[\omega_2]$ can be the nonstationary ideal on Cof($\omega_1$). *Trans. Amer. Math. Soc.*, 361(2):561–601, 2009.

[9] Itay Neeman. Forcing with side conditions. slides, http://www.math.ucla.edu/ ineeman/fwsc.pdf/, 2011.

[10] Saharon Shelah. On long increasing chains modulo flat ideals. *Math. Logic Quarterly*, 56(4):397–399, 2010.

# 12

# Set theory and von Neumann algebras

Asger Törnquist and Martino Lupini

The fifteenth Appalachian Set Theory workshop was held at Carnegie Mellon University in Pittsburgh on March 3, 2012. The lecturer was Asger Törnquist. As a graduate student Martino Lupini assisted in writing this chapter, which is based on the workshop lectures.

## Introduction

The aim of the lectures is to give a brief introduction to the area of von Neumann algebras to a typical set theorist. The *ideal* intended reader is a person in the field of (descriptive) set theory, who works with group actions and equivalence relations, and who is familiar with the rudiments of ergodic theory, and perhaps also orbit equivalence. This should not intimidate readers with a different background: Most notions we use in these notes will be defined. The reader *is* assumed to know a small amount of functional analysis. For those who feel a need to brush up on this, we recommend consulting [Ped89].

What is the motivation for giving these lectures, you ask. The answer is two-fold: On the one hand, there is a strong connection between (non-singular) group actions, countable Borel equivalence relations and von Neumann algebras, as we will see in Lecture 3 below. In the past decade, the knowledge about this connection has exploded, in large part due to the work of Sorin Popa and his many collaborators. Von Neumann algebraic techniques have lead to many discoveries that are also of significance for the actions and equivalence relations themselves, for instance, of new cocycle superrigidity theorems. On the other hand, the increased understanding of the connection between objects of ergodic theory, and their related von Neumann algebras, has also made it possible to construct large families of non-isomorphic von Neumann algebras, which in turn has made it possible to prove *non-classification* type results for the isomorphism relation for various types of von Neumann algebras (and in particular, factors).

For these reasons, it seems profitable that (descriptive) set theorists should know more about von Neumann algebras, and we hope that these lectures can serve as a starting point for those who want to start learning about this wonderful area of mathematics.

Before moving on to the mathematics, a warning or three: Theorems and Lemmas below that are not attributed are *not* due to the authors, and their origin can usually be deduced by perusing the surrounding text. Mistakes, however, are entirely due to the authors (specifically, the first author). On the other hand, it is often implicitly assumed below that Hilbert spaces are separable, and if a result is false *without* this assumption, then it is only a mistake if it also is false *with* this assumption.

## Lecture 1

*Basic definitions, examples, and the double commutant theorem.*

In the following, $H$ will denote a separable Hilbert space with inner product $\langle\cdot,\cdot\rangle$ and norm $\|\cdot\|$, while $B(H)$ will denote the linear space of bounded linear operators on $H$. Unless otherwise specified, $H$ is also assumed to be infinite-dimensional. Define on $B(H)$ the *operator norm*

$$\|T\| = \sup\{\|T\xi\| \mid \xi \in H, \|\xi\| \leq 1\}.$$

Endowed with this norm, $B(H)$ is a Banach space, which is nonseparable, unless $H$ is finite dimensional. Define the *weak (operator) topology* on $B(H)$ as the topology induced by the family of complex valued functions on $B(H)$

$$T \mapsto \langle T\xi, \eta\rangle,$$

where $\xi$ and $\eta$ range over $H$. The *strong (operator) topology* on $B(H)$ is instead defined as the topology induced by the family of functions

$$T \mapsto \|T\xi\|,$$

where $\xi$ ranges over $H$. Both of these topologies are Polish on $B^1(H)$, where $B^1(H)$ denotes the unit ball of $B(H)$ with respect to the operator norm. The Borel structure induced by either of these topologies coincide (see exercise 1.12 below), and $B(H)$ is a standard Borel space with this Borel structure. (Note though that $B(H)$ is Polish in neither of these topologies.) The weak topology is weaker than the strong topology, which in turn is weaker than the norm topology. If $T \in B(H)$, denote by $T^*$ the *adjoint* of $T$, i.e. unique element of $B(H)$ such that for all $\xi, \eta \in H$,

$$\langle T\xi, \eta \rangle = \langle \xi, T^*\eta \rangle .$$

Endowed with this operation, $B(H)$ turns out to be a Banach $*$-algebra, which furthermore satisfies the "C*-axiom",

$$\|T^*T\| = \|T\|^2 ,$$

that is, $B(H)$ is a C*-algebra (in the abstract sense). A subset of $B(H)$ will be called *self-adjoint* if it contains the adjoint of any of its elements. A self-adjoint subalgebra of $B(H)$ will also be called a $*$-*subalgebra.*

**Definition 1.1**

1. A *C*-algebra* is a norm closed $*$-subalgebra of $B(H)$ (for some $H$).
2. A *von Neumann algebra* is a weakly closed self-adjoint subalgebra of $B(H)$ (for some $H$) containing the identity operator $I$.

Since the weak topology is weaker than the norm topology, a von Neumann algebra is, in particular, a C*-algebra. However, the "interesting" von Neumann algebras all turn out to be non-separable with respect to the norm topology. This should be taken as an indication that it would not be fruitful to regard von Neumann algebras simply as a particular kind of C*-algebras; rather, von Neumann algebras have their own, to some degree separate, theory.[1]

The usual catch-phrase that people attach to definition 1.1 to underscore the difference between C*-algebras and von Neumann algebras is that *C*-algebra theory is non-commutative topology* (of locally compact spaces, presumably), while *von Neumann algebra theory is non-commutative measure theory.* This grows out of the observation that all commutative C*-algebras are isomorphic to the $*$-algebra $C_0(X)$ of all complex-valued continuous functions on some locally compact $X$

[1] This is not to suggest that no knowledge about C*-algebras will be useful when studying von Neumann algebras.

(with the sup-norm), whereas all commutative von Neumann algebras are isomorphic to $L^\infty(Y,\nu)$ for some $\sigma$-finite measure space $(Y,\nu)$ (see example 1.4.(1) below). The reader is encouraged to ponder what the locally compact space $X$ would look if we wanted to realize the isomorphism $C(X) \simeq L^\infty(Y,\nu)$, where $Y = \mathbb{N}$ with the counting measure, or even $Y = [0,1]$ with Lebesgue measure. (This should help convince the skeptic why von Neumann algebras merit having their own special theory.)

*Remark* 1.2 Von Neumann algebras were introduced by Murray and von Neumann in a series of papers [MvN36, MvN37, vN40, MvN43] in the 1930s and 1940s, where the basic theory was developed. In older references, von Neumann algebras are often called *W*-algebras.*

*Exercise* 1.3 Show that the map $T \mapsto T^*$ is weakly continuous, but not strongly continuous. (*Hint*: Consider the unilateral shift.) Show that the map $(S,T) \mapsto ST$ is separately continuous (with respect to the weak or strong topology), but not jointly continuous. Conclude that if $A \subseteq B(H)$ is a $*$-subalgebra of $B(H)$, then the weak closure of $A$ is a $*$-subalgebra of $B(H)$.

The reader may also want to verify that the map $(S,T) \mapsto ST$ is continuous on $B^1(H)$ w.r.t. the strong topology, but that this fails for the weak topology.

**Examples 1.4**

1. Let $(M,\mu)$ be a $\sigma$-finite standard measure space. Each $f \in L^\infty(X,\mu)$ gives rise to a bounded operator $m_f$ on $L^2(X,\mu)$, defined by

$$\left(m_f(\psi)\right)(x) = f(x)\psi(x).$$

   The set $\left\{m_f \mid f \in L^\infty(X,\mu)\right\}$ is an abelian von Neumann algebra, which may be seen to be a maximal abelian subalgebra of $B\left(L^2(M,\mu)\right)$ (see Exercise 1.13). It can be shown that any abelian von Neumann algebra looks like this, see [Bla06, III.1.5.18].
2. $B(H)$ is a von Neumann algebra. In particular, when $H$ has finite dimension $n$, $B(H)$ is the algebra $M_n(\mathbb{C})$ of $n \times n$ matrices over the complex numbers.
3. Suppose that $\Gamma$ is a countable discrete group and consider the unitary operators on $L^2(\Gamma)$ defined by

$$\left(U_\gamma\psi\right)(\delta) = \psi\left(\gamma^{-1}\delta\right).$$

Define $\mathcal{A}$ to be the self-adjoint subalgebra of $B\left(L^2(\Gamma)\right)$ generated by $\left\{U_\gamma \mid \gamma \in \Gamma\right\}$. Observe that an element of $\mathcal{A}$ can be written as

$$\sum_{\gamma\in\Gamma} a_\gamma U_\gamma,$$

where $\left(a_\gamma\right)_{\gamma\in\Gamma}$ is a $\Gamma$ -sequence of elements of $\mathbb{C}$ such that

$$\left\{\gamma \in \Gamma \middle| a_\gamma \neq 0\right\}$$

is finite. The weak closure of $\mathcal{A}$ is a von Neumann algebra, called the **group von Neumann algebra** $L(\Gamma)$ of $\Gamma$. For each $\delta \in \Gamma$ define

$$e_\delta(\gamma) = \begin{cases} 1 & \text{if } \gamma = \delta, \\ 0 & \text{otherwise.} \end{cases}$$

For $x = \sum_{\gamma\in\Gamma} a_\gamma U_\gamma \in \mathcal{A}$, define $\tau : \mathcal{A} \to \mathbb{C}$ by

$$\tau(\sum_{\gamma\in\Gamma} a_\gamma U_\gamma) = \left\langle\left(\sum_{\gamma\in\Gamma} a_\gamma U_\gamma\right) e_1, e_1\right\rangle = a_1,$$

where 1 is the identity element of $\Gamma$. Clearly, $\tau$ is a linear functional on $\mathcal{A}$, which is positive ($\tau(x^*x) \geq 0$) and $\tau(I) = 1$ (that is, $\tau$ is a **state** on $\mathcal{A}$). A direct calculation shows that if $y = \sum_{\delta\in\Gamma} b_\delta U_\delta \in \mathcal{A}$ is another finite sum, then

$$\left\langle\left(\sum_{\gamma\in\Gamma} a_\gamma U_\gamma\right)\left(\sum_{\delta\in\Gamma} b_\delta U_\delta\right) e_1, e_1\right\rangle = \sum_{\gamma,\delta\in\Gamma,\gamma\delta=1} a_\gamma b_\delta.$$

Thus $\tau(xy) = \tau(yx)$, i.e., $\tau$ is a *trace* on $\mathcal{A}$. Of course, the formula $\tau(x) = \langle xe_1, e_1\rangle$ makes sense for any operator in $B(L^2(\Gamma))$ and defines a state on $B(L^2(\Gamma))$, and the reader may verify (using that composition is separately continuous) that $\tau$ satisfies the trace property $\tau(xy) = \tau(yx)$ for any $x, y$ in the weak closure of $\mathcal{A}$. Note that $\tau$ is weakly continuous.

**Definition 1.5** For $X \subseteq B(H)$, define the **commutant** of $X$ to be the set

$$X' = \{T \in B(H) : (\forall S \in X) TS = ST\}.$$

Note that (by separate continuity of composition), $X'$ is both weakly and strongly closed. Furthermore, if $X$ is self-adjoint, then so is $X'$, and thus $X'$ is a von Neumann algebra when $X$ is self-adjoint. In particular, the **double commutant** $X'' = (X')'$ is a von Neumann algebra containing $X$.

The next theorem, known as *double commutant theorem*, is a cornerstone of the basic theory of von Neumann algebras. It is due to von Neumann.

**Theorem 1.6** *Let $A \subseteq B(H)$ be a self-adjoint subalgebra of $B(H)$ which contains the identity operator. Then $A$ is strongly dense in $A''$.*

**Corollary 1.7** *If $A$ is a self-adjoint subalgebra of $B(H)$ containing $I$, then the weak and strong closure of $A$ coincide with $A''$, i.e. $\overline{A}^{so} = \overline{A}^{wo} = A''$.*

The corollary follows from the double commutant theorem by noting that $\overline{A}^{so} \subseteq \overline{A}^{wo} \subseteq A''$. To prove the double commutant theorem, we start with the following:

**Lemma 1.8** *Let $A$ be a selfadjoint subalgebra of $B(H)$ which contains $I$, and let $T_0 \in A''$. Then for any $\xi \in H$ and $\varepsilon > 0$ there is $S \in A$ such that $\|(T_0 - S)\xi\| < \varepsilon$.*

*Proof* Define $p \in B(H)$ to be the orthogonal projection onto the closure of the subspace $A\xi = \{T\xi : T \in A\}$ (which contains $\xi$). We claim that $p \in A'$. To see this, note that if $\eta \in \ker(p)$ and $S \in A$, then for every $T \in A$ we have

$$\langle S\eta, T\xi \rangle = \langle \eta, S^*T\xi \rangle = 0,$$

whence $\ker(p)$ is $A$-invariant. It follows that $p \in A'$, since for any $T \in A$ and $\eta \in H$ we have

$$pT\eta = pTp\eta + pT(1-p)\eta = pTp\eta = Tp\eta.$$

Since $T_0 \in A''$ we therefore have $T_0\xi = T_0 p\xi = pT_0\xi \in \operatorname{ran}(p)$, and so there is some $S \in A$ such that $\|S\xi - T_0\xi\| < \varepsilon$, as required. □

*Proof of Theorem 1.6* We begin by defining some notation. Fix $n \in \mathbb{N}$. If $(T_{i,j})_{1 \le i,j \le n}$, where $T_{i,j} \in B(H)$, is a "matrix of operators", then an operator in $B(H^n)$ is defined by "matrix multiplication", i.e.,

$$(S_{(T_{i,j})}(\eta_1, \ldots, \eta_n))_k = \sum_{j=1}^{n} T_{k,j}\eta_j.$$

(Note that every operator in $B(H^n)$ has this form.) For $T \in B(H)$, we let $\mathrm{diag}(T)$ be the operator

$$\mathrm{diag}(T)(\xi_1, \dots, \xi_n) = (T\xi_1, \dots, T\xi_n),$$

corresponding to the matrix of operators which has $T$ on the diagonal, and is zero elsewhere.

Now fix $\xi_1, \dots, \xi_n \in H$, $\varepsilon > 0$ and $T_0 \in A''$, and let $\vec{\xi} = (\xi_1, \dots, \xi_n)$. Note that $\mathrm{diag}(A) = \{\mathrm{diag}(T) : T \in A\}$ is a self-adjoint subalgebra of $B(H^n)$ containing $I$. It suffices to show that $\mathrm{diag}(T_0) \in \mathrm{diag}(A)''$ since then by Lemma 1.8 there is $S \in A$ such that $\|(\mathrm{diag}(S) - \mathrm{diag}(T_0))\vec{\xi}\|_{H^n} < \varepsilon$. Thus $\|S\xi_k - T_0\xi_k\| < \varepsilon$ for all $k \leq n$, which shows that $A$ is strongly dense in $A''$.

To see that $\mathrm{diag}(T_0) \in \mathrm{diag}(A)''$, simply note that $S \in \mathrm{diag}(A)'$ precisely when it has the form $S_{(T_{i,j})}$ where $T_{i,j} \in A'$ for all $i, j$. Thus $\mathrm{diag}(T_0) \in \mathrm{diag}(A)''$. □

*Remark* 1.9 If $X \subseteq B(H)$ is self-adjoint and contains $I$, then it follows from the double commutant theorem that $X''$ is the smallest von Neumann algebra containing $X$, i.e., $X''$ is the von Neumann algebra generated by $X$.

An element $u$ of $B(H)$ is called **partial isometry** if $u^*u$ and (hence) $uu^*$ are orthogonal projections. The first one is called the **support projection** of $u$ and the latter the **range projection** of $u$.

Recall the **polar decomposition** theorem for bounded operators: If $T \in B(H)$, then there is a partial isometry $u$ such that $T = u|T|$, where $|T| = (T^*T)^{\frac{1}{2}}$. In this case, the support projection of $u$ is the orthogonal projection onto $\ker(T)^{\perp}$, while the range projection of $u$ is the orthogonal projection onto $\overline{\mathrm{ran}(T)}$. It is easy to see that $u$ restricted to $\mathrm{ran}(u^*u)$ is an isometry onto $\mathrm{ran}(uu^*)$. The polar decomposition of an operator is unique.

*Exercise* 1.10 Let $M \subseteq B(H)$ be a von Neumann algebra. Let $T \in M$, and let $T = u|T|$ be the polar decomposition of $T$. Prove that $|T| \in M$ and $u \in M$. (*Hint*: To show $|T| \in M$, you may want to recall how the existence of the square root is proved, see e.g. [Ped89]. To show that $u \in M$, use the polar decomposition theorem.)

*Exercise* 1.11 Show that every operator in $B(H^n)$ has the form $S_{(T_{i,j})}$ for some matrix of operators $(T_{i,j})$ in $B(H)$. (*Hint*: For a given operator $S \in B(H^n)$, consider the operators $p_jSp_i$ where $p_i$ is the projection onto the $i$'th coordinate.)

*Exercise* 1.12 Show that the Borel structure on $B^1(H)$ induced by the weak and strong topologies coincide. Conclude that the Borel structure induced by the weak and strong topologies on $B(H)$ coincide, and is standard.

*Exercise* 1.13 (See example 1.4.1.) Let $(X, \mu)$ be a $\sigma$-finite measure space. Show that $L^\infty(X, \mu)$ is a maximal Abelian $*$-subalgebra of the algebra $B(L^2(X, \mu))$ (when identified with the set of multiplication operators $\{m_f : f \in L^\infty(X, \mu)\}$). Conclude that $L^\infty(X, \mu)$ is strongly (and weakly) closed, and so is a von Neumann algebra. (*Hint*: Take $T \in L^\infty(X, \mu)'$, and argue that the function $T(1)$ is essentially bounded.)

# Lecture 2

*Comparison theory of projections, type classification, direct integral decomposition, and connections to the theory of unitary group representations.*

An element $p \in B(H)$ is called a **projection** (or more precisely, an *orthogonal* projection) if

$$p^2 = p^* = p.$$

The reader may easily verify that this is equivalent to (the more geometric definition) $p^2 = p$ and $\ker(p)^\perp = \operatorname{ran}(p)$. (*Warning*: From now on, when we write "projection" we will always mean an orthogonal projection. This is also the convention in most of the literature.) For projections $p, q \in B(H)$, write $p \leq q$ if $\operatorname{ran}(p) \subseteq \operatorname{ran}(q)$. We will say that $p$ is a **subprojection** of $q$. (One may more generally define $T \leq S$ iff $S - T$ is a positive operator for any $T, S \in B(H)$. This definition agrees with our definition of $\leq$ on projections.)

Unlike their C*-algebra brethren[2], von Neumann algebras always have *many* projections. The key fact is this: A von Neumann algebra $M$ contains all spectral projections (in the sense of the spectral theorem) of any normal operator $T \in M$. It follows from this that a von Neumann algebra is generated by its projections (see exercise 2.12 below). It is therefore natural to try to build a structure theory of von Neumann algebras around an analysis of projections.

[2] There are examples of C*-algebras with no non-trivial projections!

**Definition 2.1** Let $M$ be a von Neumann algebra. We let $P(M)$ be the set of projections in $M$. For $p, q \in P(M)$, we say that $p$ and $q$ are **Murray-von Neumann equivalent** (or simply **equivalent**) if there is a partial isometry $u \in M$ such that $u^*u = p$ and $uu^* = q$. We write $p \sim q$. We will say that $p$ is **subordinate** to $q$, written $p \precsim q$, if $p$ is equivalent to a subprojections of $q$ (i.e., there is $p' \leq q$ such that $p \sim p'$).

**Example 2.2** In $B(H)$, $p \precsim q$ iff the range of $p$ has dimension smaller than or equal to the range of $q$. Thus, the ordering on $P(B(H))/\sim$ is linear, and isomorphic to $\{0, 1, ..., n\}$ if $\dim H = n$ and to $\mathbb{N} \cup \{\infty\}$ if $H$ is infinite dimensional.

The previous example highlights why it must be emphasized that the definition of $\sim$ is "local" to the von Neumann algebra $M$ (i.e., that we require $u \in M$ in the definition). Otherwise, $\sim$ would only measure the dimension of the range of the projection $p$, which does not depend on $M$ in any way. However, the idea that $\sim$ and $\precsim$ are somehow related to dimension *relative to M* is essentially correct (though the precise details are subtle). For instance, in $L(\mathbb{F}_n)$ (where $\mathbb{F}_n$ is the free group on $n > 1$ generators) it turns out that $P(L(\mathbb{F}_n))/\sim$ ordered linearly by $\precsim$, and is order-isomorphic to $[0, 1]$. So in some sense, we need a continuous range to measure dimension in $L(\mathbb{F}_n)$. More about this later.

It is clear that $\precsim$ is a transitive relation with $I$ being a maximal element, and 0 being the minimal. Furthermore, we have:

**Proposition 2.3** ("Schröder-Bernstein for projections") *If $p \precsim q$ and $q \precsim p$, then $p \sim q$.*

*Exercise* 2.4 Prove Proposition 2.3.

**Definition 2.5** Let $M$ be a von Neumann algebra. A projection $p \in M$ is said to be

- **finite** if it is not equivalent to a proper subprojection of itself;
- **infinite** if it is not finite;
- **purely infinite** if it has no nonzero finite subprojections;
- **semifinite** if it is not finite, but is the supremum of an increasing family of finite projections.
- **minimal** if it is non-zero, and has no proper non-zero subprojections.

A von Neumann algebra $M$ will be called finite (infinite, purely infinite, or semifinite) if the identity $I \in M$ is finite (respectively, infinite, purely infinite, or semifinite).

**Proposition 2.6** *Let $M$ be a von Neumann algebra. Then the following are equivalent:*

1. *The* center $\mathcal{Z}(M) = M \cap M'$ *of $M$ consists of scalar multiples of the identity $I$, (i.e., $\mathcal{Z}(M) = \mathbb{C}I$).*
2. $P(M)/\sim$ *is linearly ordered by* $\precsim$.

The proof is outlined in exercise 2.16.

**Definition 2.7** A von Neumann algebra $M$ is called a *factor* if (1) (and therefore (2)) of the previous proposition holds.

It is not hard to see that $M_n(\mathbb{C})$ and $B(H)$ are factors. Another source of examples are the group von Neumann algebras $L(\Gamma)$, when $\Gamma$ is an *infinite conjugacy class* (or *i.c.c.*) group, meaning that the conjugacy class of each $\gamma \in \Gamma \setminus \{1\}$ is infinite.

The next theorem shows that factors constitute the building blocks of von Neumann algebras, as all (separably acting, say) von Neumann algebras can be decomposed into a generalized direct sum (i.e., integral) of factors. This naturally shifts the focus of the theory to factors, rather than general von Neumann algebras. Quite often, a general theorem that can be proven for factors can then be extended to all von Neumann algebras using the direct integral decomposition.

**Theorem 2.8** *Let $M$ be a von Neumann algebra. Then there is a standard $\sigma$-finite measure space $(X, \mu)$, a Borel field $(H_x)_{x \in X}$ of Hilbert spaces and a Borel field $(M_x)_{x \in X}$ of von Neumann algebras in $B(H_x)$ such that:*

1. *$H$ is isomorphic to $\int_X H_x d\mu(x)$.*
2. *Identifying $H$ and $\int_X H_x d\mu(x)$, we have*
    - a. *$M_x$ is a factor for all $x \in X$;*
    - b. $M = \int_X M_x d\mu(x)$;
    - c. $\mathcal{Z}(M) = L^\infty(X, \mu)$.

*Moreover, this decomposition is essentially unique.*

The proof is rather involved; it is given in full detail in [Nie80] (see also [Bla06, Dix81]). Some comments about the theorem are in place, however. A Borel field of Hilbert spaces is nothing but a standard Borel space $X$ with a partition $X = \bigsqcup_{n=0,1,2,\ldots,\mathbb{N}} X_n$, and the vectors in $H$ are Borel functions $f : X \to \ell^2(\mathbb{N})$, where for $x \in X_n$ we have that $f(x)$ is in the space generated by the first $n$ standard basis vectors of $\ell^2(\mathbb{N})$, and the inner product is given by $\langle f, g\rangle = \int \langle f(x), g(x)\rangle d\mu(x)$. Equality $M = \int M_x d\mu(x)$ means that every operator $T \in M$ can be written as $T = \int T_x d\mu(x)$ (i.e., $(Tf) = T_x(f(x))$ $\mu$-a.e.), where $T_x \in M_x$, and $x \mapsto T_x$ is Borel (in the obvious sense) w.r.t. the $\sigma$-algebra generated by the weakly open sets.

**Definition 2.9** (Type classification of factors) Let $M$ be a factor. We say that $M$ is

(a) type $\mathrm{I}_n$ if it is finite and $P(M)/\sim$ is order isomorphic to

$$n = \{0, 1, \dots, n-1\}$$

with the usual order;

(b) type $\mathrm{I}_\infty$ if it is infinite and $P(M)/\sim$ is order isomorphic to $\mathbb{N} \cup \{\infty\}$ (where $n < \infty$ for all $n \in \mathbb{N}$);

(c) type $\mathrm{II}_1$ if it is finite and $P(M)/\sim$ is order isomorphic to $[0, 1]$;

(d) type $\mathrm{II}_\infty$ if it is semifinite and $P(M)/\sim$ is order isomorphic to $[0, \infty]$;

(e) type III if it is purely infinite and $P(M)/\sim$ is order isomorphic to $\{0, \infty\}$.

The reader may object that $[0, 1]$ and $[0, \infty]$ are order-isomorphic, and so are $\{0, 1\}$ and $\{0, \infty\}$. The intention is that the $\infty$ indicates that the $\lesssim$-maximal projection is infinite, and so the notation contains additional information. It is clear that any factor $M$ can be at most one of the types, but even more so, the list is in fact exhaustive and complete:

**Theorem 2.10** *Every factor is either type* $\mathrm{I}_n$, $\mathrm{I}_\infty$, $\mathrm{II}_1$ *and* $\mathrm{II}_\infty$ *or type* III. *Moreover, there is at least one factor of each type.*

That the types list all possibilities is not too hard to see. That there is an example of each type is much harder. (We will see examples of all types in the last lectures.)

Up to isomorphism, the algebra $M_n(\mathbb{C})$ of $n \times n$ complex matrices is the only type $I_n$ von Neumann algebra, while $B(H)$ for $H$ infinite dimensional (and separable!) is the only type $I_\infty$ factor. Such an easy description of the isomorphism classes of type II and type III factors is not possible: As we will see below, the type II and III factors cannot even be classified up to isomorphism by countable structures!

The cornerstone of the theory of $\mathrm{II}_1$ factors (or, more generally, finite von Neumann algebras) is the existence of a (unique) faithful trace, which is continuous on the unit ball. We have already seen an example of a trace when we discussed $L(\Gamma)$. The trace on a $\mathrm{II}_1$ factor is an invaluable technical tool when working with projections in a $\mathrm{II}_1$ factor.

**Theorem 2.11** (Murray-von Neumann)

*(A) Every* $\mathrm{II}_1$ *factor* $M$ *has a faithful normal* **trace**, *i.e., a positive linear functional* $\tau : M \to \mathbb{C}$ *which is*

1. **normal**, *meaning that* $\tau$ *is weakly continuous on the unit ball of* $M$*;*
2. **faithful**, *meaning that for* $x \in M$, $x = 0$ *iff* $\tau(x^*x) = 0$*;*
3. **tracial**, *meaning that* $\tau(xy) = \tau(yx)$ *for all* $x, y \in M$.

*The trace is unique up to a scalar multiple.*

*(B) Every* $\mathrm{II}_\infty$ *factor admits a faithful normal semifinite trace defined on the set* $M^+$ *of positive operators in* $M$*. That is, there is an additive map* $\tau : M^+ \to [0, \infty]$ *which satisfy* $\tau(rx) = r\tau(x)$ *for all* $r > 0$*, which is continuous (in the natural sense) on the unit ball of* $M^+$ *with respect to the weak topology, and which is faithful and tracial. The semifinite trace on a* $\mathrm{II}_\infty$ *factor is unique up to multiplication by a scalar.*

The trace gives us a measure of the size of projections, much like the dimension function $p \mapsto \dim(\mathrm{ran}(p))$ does on the projections in $B(H)$. However, for a $\mathrm{II}_1$ factor $M$ we have that $\tau(P(M)) = [0, \tau(I)]$, and so the projections on a $\mathrm{II}_1$ factor have "continuous dimension" (understood locally in the $\mathrm{II}_1$ factor, of course). Because of the trace property, two projections $p, q \in P(M)$ are equivalent iff $\tau(p) = \tau(q)$. The existence of a trace characterizes the finite factors: $M$ is finite iff there is a trace as above on $M$.

*Exercise* 2.12 Let $M$ be a von Neumann algebra, and let $T \in M$ be a normal operator. Show that all the *spectral projections*, i.e., projections in the Abelian von Neumann algebra generated by $T$, belong to $M$. Conclude that $M$ is generated by its projections, i.e., $M = P(M)''$. (*Hint*: Any operator can be written $T = \Re(T) + i\Re(-iT)$, where $\Re(T) = \frac{1}{2}(T + T^*)$.)

*Exercise* 2.13 Prove that if $p$ and $q$ are finite projections in a factor $M$ and $p \sim q$, then $1 - p \sim 1 - q$. Conclude that there is a unitary operator $U \in M$ such that $U^*pU = q$. (*Hint*: Quickly dispense with the case that $M$ is infinite.)

*Exercise* 2.14 If $H$ is infinite dimensional, then $B(H)$ does not admit a trace which is weakly continuous on $B^1(H)$.

The next two exercises are harder.

*Exercise* 2.15 Any two elements $p, q$ of $P(M)$ have inf and sup in $P(M)$ with respect to the relation $\leq$. Denoting these by $p \wedge q$ and $p \vee q$ respectively, one has

$$p \vee q - p \sim q - p \wedge q.$$

This is known as *Kaplanski's identity.* (*Hint:* Consider the domain and range $(1 - q)p$. A proof can be found in [KR97, Theorem 6.1.7].)

*Exercise* 2.16 Prove Proposition 2.6. (A proof can be found in [KR97, Theorem 6.2.6].)

## Applications to unitary representations.

To illustrate the usefulness of the notions of factors and comparison of projections, we briefly turn to study unitary representation of countable discrete groups. What is said in this section applies almost without change to unitary representations of locally compact groups and representations of C*-algebras as well.

**Definition 2.17** Let $\Gamma$ be a countable discrete group. A **unitary representation** of $\Gamma$ on $H$ is a homomorphism $\pi : \Gamma \to U(H)$, where $U(H)$ is the group of unitary operators in $B(H)$.

For notational convenience, we will write $\pi_\gamma$ for $\pi(\gamma)$. Let $\pi$ and $\sigma$ be unitary representations of $\Gamma$ on Hilbert spaces $H_0, H_1$ respectively. We define

$$R_{\pi,\sigma} = \{T \in L(H_0, H_1) : (\forall \gamma \in \Gamma) T\pi_\gamma = \sigma_\gamma T\},$$

where $L(H_0, H_1)$ is the set of bounded linear maps $H_0 \to H_1$. A map $T \in R_{\pi,\sigma}$ is called an *intertwiner* of $\pi$ and $\sigma$. We say that $\pi$ and $\sigma$ are **disjoint**, written $\pi \perp \sigma$, if $R_{\pi,\sigma} = \{0\}$. We let $R_\pi = R_{\pi,\pi}$. Note that $R_\pi = \{\pi_\gamma : \gamma \in \Gamma\}'$, thus $R_\pi$ is a von Neumann algebra.

If $p \in P(R_\pi)$, then $\mathrm{ran}(p)$ is a $\pi$-invariant closed subspace in $H$; conversely, a projection onto any closed $\pi$-invariant subspace must be in $R_\pi$. Note further that if $p, q \in P(R_\pi)$ and $p \sim q$ (in $R_\pi$), then $\pi \restriction \mathrm{ran}(p)$ and $\pi \restriction \mathrm{ran}(q)$ are isomorphic. Thus $R_\pi$ gives us useful information about the invariant subspaces of $\pi$. For instance, the reader may easily verify that $\pi$ is irreducible (i.e., has no non-trivial invariant subspaces) iff $R_\pi = \mathbb{C}I$. More generally, if $p \in P(R_\pi)$ is a minimal non-zero projection (i.e., having no proper non-zero subprojections in $P(R_\pi)$), then $\pi \restriction \mathrm{ran}(p)$ is irreducible.

It is tempting to hope that any unitary representation can be written as a direct sum, or direct integral, of irreducible unitary representations. It can, but the decomposition is very badly behaved (and non-unique) unless $\Gamma$ is abelian by finite. A better behaved decomposition theory is based around the following notion:

**Definition 2.18** A unitary representation $\pi$ of $\Gamma$ is called a **factor representation** (or sometimes a **primary** representation) if $R_\pi$ is a factor. A representation $\pi$ is said to be type $\mathrm{I}_n$, $\mathrm{I}_\infty$, $\mathrm{II}_1$, $\mathrm{II}_\infty$ or III according to what type $R_\pi$ is. Similarly, it is called finite, infinite, semifinite or purely infinite according to what $R_\pi$ is.

The decomposition theory for unitary representation can now be obtained from the decomposition of von Neumann algebras into factors.

**Theorem 2.19** (Mackey) *Let $\pi$ be a unitary representation of $\Gamma$ on a separable Hilbert space. Then there is a standard $\sigma$-finite measure space $(X, \mu)$, a Borel field of Hilbert spaces $(H_x)_{x\in X}$, and a Borel field $(\pi_x)_{x\in X}$ of factor representations of $\Gamma$ such that*

$$\pi \simeq \int_X \pi_x d\mu(x)$$

*and if $x \neq y$, then $\pi_x \perp \pi_y$. Furthermore, this decomposition of $\pi$ is essentially unique.*

*Exercise* 2.20 Show that a type $\mathrm{I}_n$ factor representation ($n \in \mathbb{N} \cup \{\infty\}$) is a direct sum of $n$ irreducible representations.

For type II and III factor representations, nothing like this is true, since there are no minimal non-zero projections. In the case of type III factors, the restriction of $\pi$ to any invariant non-zero closed subspace (i.e., what may be called a “piece” of the representation) is isomorphic to the whole representation. In the case when $\pi$ is type II, every piece of $\pi$ may be subdivided into $n$ smaller pieces that are all isomorphic to each other, but unlike the type III case, the trace (and the associated notion of dimension) gives us a sense of the size of the pieces of $\pi$. (So, in the type III everyone can get as much $\pi$ as they want, in the type II case we can divide the $\pi$ into however many albeit small pieces we like, and in the type I case there is a minimal size of the pieces of $\pi$ (and maximal number of pieces, too).)

We close the section with mentioning the following recent theorem. It solves an old problem of Effros, who asked if the conjugacy relation $\simeq$

for unitary representation of a fixed countable discrete group $\Gamma$ is Borel (in the space $\mathrm{Rep}(\Gamma, H) = \{\pi \in U(H)^\Gamma : \pi \text{ is a homomorphism}\}$).

**Theorem 2.21** (Hjorth-Törnquist, 2011) *Let $H$ be an infinite dimensional separable Hilbert space and let $\Gamma$ be a countable discrete group. The conjugacy relation in* $\mathrm{Rep}(\Gamma, H)$ *is $F_{\sigma\delta}$.*

The proof uses only classical results that have been introduced already above. To give a rough sketch of what happens in the proof, fix $\pi, \sigma \in \mathrm{Rep}(\Gamma, H)$. The idea is that it is an $F_{\sigma\delta}$ statement to say that there are pieces (i.e., projections $p \in P(R_\pi)$) arbitrarily close to $I$ (in the weak topology, say) such that $\pi \restriction \mathrm{ran}(p)$ is isomorphic to a sub-representation of $\sigma$. Call this statement $S_0(\pi, \sigma)$. One first proves that for factor representations $\pi$ and $\sigma$, it holds that $\pi \simeq \sigma$ iff $S_0(\pi, \sigma)$ and $S_0(\sigma, \pi)$. This is trivial in the purely infinite (type III) case, while in finite and semifinite case the trace gives us a way of proving that if pieces of $\pi$ closer and closer to $I$ can fit into $\sigma$, $\pi$ itself can fit into $\sigma$ (and conversely). It then follows that $\sigma \simeq \pi$. The proof is finished by using Theorem 2.19.

The details can be found in [HT12]. The theorem applies more generally to representations of locally compact second countable groups and separable C*-algebras.

# Lecture 3

*The group-measure space construction, the von Neumann algebra of a non-singular countable Borel equivalence relation, orbit equivalence, von Neumann equivalence, and Cartan subalgebras.*

## The group-measure space construction

Let $(X, \mu)$ be a standard $\sigma$-finite measure space, and let $\Gamma$ be a countable discrete group. A Borel action $\sigma : \Gamma \curvearrowright X$ is said to be **non-singular** (w.r.t. $\mu$) if for all $\gamma \in \Gamma$ we have that $\sigma_\gamma \mu \approx \mu$ (i.e., $\mu(A) = 0$ iff $\mu(\sigma_\gamma^{-1}(A)) = 0$), and that $\sigma$ is **measure preserving** (w.r.t. $\mu$) if $\sigma_\gamma \mu = \mu$ for all $\gamma \in \Gamma$ (i.e., $\mu(A) = \mu(\sigma_\gamma^{-1}(A))$ for all measurable $A \subseteq X$). The action is **ergodic** if any invariant measurable set is either null or conull; it is **a.e.free** (or *essentially* free) if for any $\gamma \in \Gamma \setminus \{1\}$ we have that

$$\mu(\{x \in X : \sigma_\gamma(x) = x\}) = 0.$$

The action $\sigma$ induces an **orbit equivalence relation**, denoted $E_\sigma$, on the space $X$, which is defined by

$$xE_\sigma y \iff (\exists\gamma \in \Gamma)\sigma_\gamma(x) = y.$$

Note that $E_\sigma$ is a Borel subset of $X \times X$.

To a non-singular action $\sigma : \Gamma \curvearrowright (X, \mu)$ as above there is an associated von Neumann algebra. The description is slightly easier in the case when $\sigma$ is measure preserving, so we will assume that this is the case. In this case the construction is also closely parallel to the construction of the group von Neumann algebra.

Let $\nu$ be the counting measure on $\Gamma$. Give $\Gamma \times X$ the product measure $\nu\times\mu$, and define for each $\gamma \in \Gamma$ a unitary operator $U_\gamma$ on $L^2(\Gamma\times X, \nu\times\mu)$ by

$$\left(U_\gamma\psi\right)(\delta, x) = \psi\left(\gamma^{-1}\delta, \sigma_{\gamma^{-1}}(x)\right).$$

Further, define for each $f \in L^\infty(X, \mu)$ an operator $m_f \in B(L^2(\Gamma \times X))$ by

$$\left(m_f\psi\right)(\delta, x) = f(x)\,\psi(\delta, x).$$

Define

$$L^\infty(X,\mu) \rtimes_\sigma \Gamma := \left(\{m_f : f \in L^\infty(X,\mu)\} \cup \{U_\gamma : \gamma \in \Gamma\}\right)''.$$

The von Neumann algebra $L^\infty(X,\mu) \rtimes_\sigma \Gamma$ is called the **group-measure space** von Neumann algebra of the action $\sigma$. It is an example of W*-crossed product. It is worth noting that if $X$ consists of a single point then we get the group von Neumann algebra of $\Gamma$.

The group $\Gamma$ acts on $L^\infty(X,\mu)$ by $(\hat\sigma_\gamma(f))(x) = f(\sigma_{\gamma^{-1}}(x))$; note that $\hat\sigma$ is a $*$-automorphism of $L^\infty(X,\mu)$ for every $\gamma \in \Gamma$. An easy calculation shows that $U_\gamma m_f = m_{\hat\sigma_\gamma(f)}U_\gamma$, and so $U_\gamma m_f U_\gamma^* = m_{\hat\sigma_\gamma(f)}$. It follows that

$$\mathcal{A} = \{\sum_{\gamma\in F} m_{f_\gamma}U_\gamma : F \subseteq \Gamma \text{ is finite} \wedge f_\gamma \in L^\infty(X,\mu)\}$$

is a $*$-algebra, which is the smallest $*$-algebra containing the operators $U_\gamma$ and $m_f$, and so $\mathcal{A}$ is dense in $L^\infty(X,\mu)\rtimes_\sigma\Gamma$ by the double commutant theorem.

Assume now that $\mu$ is a *finite* measure. Then

$$e_1(\gamma, x) = \begin{cases} 1 & \text{if } \gamma = 1, \\ 0 & \text{otherwise,} \end{cases}$$

defines an element of $L^2(\Gamma\times X)$, and we can define a state (positive linear functional) on $B(L^2(\Gamma \times X))$ by $\tau(T) = \langle Te_1, e_1 \rangle$. If $x = \sum_{\gamma \in F} m_{f_\gamma} U_\gamma \in \mathcal{A}$, then we have

$$\tau(x) = \left\langle \left( \sum_{\gamma \in F} m_{f_\gamma} U_\gamma \right) e_1, e_1 \right\rangle = \int_X f_1 \, d\mu,$$

and one may easily verify that for $x, y \in \mathcal{A}$ we have $\tau(xy) = \tau(yx)$. Using that multiplication is separately weakly continuous in $B(L^2(\Gamma, X))$ we see that the trace property extends to the weak closure of $\mathcal{A}$. Thus $\tau$ is a weakly continuous trace on $L^\infty(X, \mu) \rtimes_\sigma \Gamma$ with $\tau(I)$ finite, from which it follows that $L^\infty(X, \mu) \rtimes_\sigma \Gamma$ is a finite von Neumann algebra. Taking $f = \chi_A$ to be the characteristic function of some measurable $A \subseteq X$, we see that $\tau(m_f) = \mu(A)$. So if $\mu$ is non-atomic we have $\tau(P(L^\infty(X, \mu) \rtimes_\sigma \Gamma)) = [0, \tau(I)]$.

One may ask if $L^\infty(X, \mu) \rtimes_\sigma \Gamma$ can be a factor. The above analysis shows that *if* it is a factor *and* $\mu$ is finite and non-atomic, then $L^\infty(X, \mu) \rtimes_\sigma \Gamma$ must be a type $\mathrm{II}_1$ factor. Though far from providing an exhaustive answer to this question, the following is a key result in this direction:

**Theorem 3.1** *If $\sigma$ is a.e. free and ergodic, then $L^\infty(X, \mu) \rtimes_\sigma \Gamma$ is a factor.*

The reader may find it amusing to verify that if we take $\sigma$ to be the action of $\mathbb{Z}/n\mathbb{Z}$ on itself by translation (which preserves the counting measure, and is free and ergodic), then the corresponding group-measure space factor is just $M_n(\mathbb{C})$.

A question of central importance is to understand the relationship between the action $\sigma$ and $L^\infty(X, \mu) \rtimes_\sigma \Gamma$. Which properties of the action, if any, are reflected in the group-measure space von Neumann algebra? This is far from completely understood, and a vast body of literature addressing various special cases exists. (In general, the more "rigid" a group is, the more information about the action and the group will be encoded into the group-measure space algebra.)

**Definition 3.2** Let $\sigma : \Gamma \curvearrowright (X, \mu)$ and $\pi : \Gamma \curvearrowright (Y, \nu)$ be measure preserving actions on standard $\sigma$-finite measure spaces. We say that $\sigma$ and $\pi$ are

1. **conjugate** if there is a non-singular Borel bijection $T : X \to Y$ such that for all $\gamma \in \Gamma$ we have $T\sigma_\gamma(x) = \pi_\gamma T(x)$ for almost all $x \in X$.

2. **orbit equivalent** if there is a non-singular Borel bijection $T : X \to Y$ such that for almost all $x, y \in X$ we have
$$xE_\sigma y \iff T(x)E_\pi T(y).$$
3. **von Neumann equivalent** (or **W*-equivalent**) if $L^\infty(X,\mu) \rtimes_\sigma \Gamma$ and $L^\infty(Y,\nu) \rtimes_\pi \Gamma$ are isomorphic.

It is clear that conjugacy implies von Neumann equivalence and orbit equivalence, but little else can be said immediately. We will see below (exercise 3.16) that when the actions $\sigma$ and $\pi$ are free, then orbit equivalence implies von Neumann equivalence.

Let $M = L^\infty(X,\mu) \rtimes_\sigma \Gamma$. A special role is played by the Abelian subalgebra generated by the operators $m_f$ (which we identify with $L^\infty(X,\mu)$ in the obvious way). It can be seen that $\sigma$ is a.e. free iff $L^\infty(X,\mu)$ is a maximal Abelian subalgebra. The subalgebra $L^\infty(X,\mu)$ has the property that the unitary normalizer
$$\{U \in U(M) : UL^\infty(X,\mu)U^* \subseteq L^\infty(X,\mu)\}$$
generates $M$; we say that $L^\infty(X,\mu)$ is a **regular** subalgebra of $M$. So when $\sigma$ is a.e. free, then $L^\infty(X,\mu)$ is a **Cartan subalgebra** of $M$, i.e., a maximal Abelian regular subalgebra (see exercise 3.16 below).

Finally, we define the crossed product when $\sigma$ is non-singular, but not measure preserving. In this case the outcome will be a purely infinite von Neumann algebra. It is still the case that $\hat{\sigma}$ acts on $L^\infty(X,\mu)$ by $*$-automorphisms, and $L^\infty(X,\mu)$ is represented as multiplication operators on $H = L^2(X,\mu)$. Consider then the Hilbert space $L^2(\Gamma,\nu,H)$ of $L^2$ functions with values in $H$. On this Hilbert space we define the operators
$$\left((\lambda_\gamma\psi)(\delta)\right)(x) = \psi(\gamma^{-1}\delta)(x)$$
and
$$(\pi_f(\psi)(\gamma))(x) = (\hat{\sigma}_\gamma f)(x)(\psi(\gamma))(x).$$
We then let $L^\infty(X,\mu)\rtimes_\sigma\Gamma$ be the von Neumann algebra generated by this family of operators. This definition also hints at how one may go about defining the crossed product even more generally: Instead of $L^\infty(X,\mu)$, consider an arbitrary von Neumann algebra $N$ acting on a Hilbert space $H$ and let $\hat{\sigma} : \Gamma \curvearrowright N$ be an action on $N$ by $*$-automorphisms. The formulas above still define bounded operators on $L^2(\Gamma,\nu,H)$, and they generate a von Neumann algebra that we denote by $N \rtimes_{\hat{\sigma}} \Gamma$.

## The von Neumann algebra of a non-singular countable Borel equivalence relation.

In two highly influential papers [FM77a, FM77b], Feldman and Moore developed a string of results related to countable Borel equivalence relations and von Neumann algebras. In the first paper they study countable non-singular Borel equivalence relations and their cohomology; in the second paper they construct a von Neumann algebra $M(E)$ directly from a non-singular countable Borel equivalence relation $E$, and study its properties.

The present section is dedicated to the construction of $M(E)$. It can best be described as the construction of a "matrix algebra" over the equivalence relation. The details that are not provided below can for the most part be found in [FM77b].[3]

Before we can define $M(E)$, we need a few facts from the first paper [FM77a]. The first of these is by now widely known:

**Theorem 3.3** ([FM77a]) *If $E$ is a countable Borel equivalence relation on a standard Borel space $X$, then there is a countable group $\Gamma$ of Borel automorphisms of $X$ which induce $E$. (That is, $E = E_\sigma$ for some Borel action $\sigma : \Gamma \curvearrowright X$.)*

Now fix a countable Borel equivalence relation $E$ on a $\sigma$-finite standard measure space $(X, \mu)$. On $E$, which is a Borel subset of $X^2$, we define two Borel measures

$$\mu_*(A) = \int |A_x| d\mu(x)$$

and

$$\mu^*(A) = \int |A^y| d\mu(y)$$

for each Borel $A \subseteq E$. Here, as usual, $A_x = \{y \in X : (x, y) \in A\}$ and $A^y = \{x : (x, y) \in A\}$. One now has the following:

**Theorem 3.4** ([FM77a]) *Let $E$ and $(X, \mu)$ be as above. Then the following are equivalent:*

1. *$\mu_*$ and $\mu^*$ are absolutely equivalent, i.e., $\mu_* \approx \mu^*$.*
2. *There is a countable group of non-singular Borel automorphisms of $X$ which induce $E$.*

[3] Both [FM77a] and [FM77b] are extremely well written and are warmly recommended.

3. *Any Borel automorphism whose graph is contained in E preserves the measure class of* $\mu$.

We will say that $E$ is **non-singular** (w.r.t. $\mu$) if one (and all) of the conditions in Theorem 3.4 holds. When this is the case, let $D(x, y) = \frac{d\mu_*}{d\mu^*}(x, y)$ be the Radon-Nikodym derivative. From now on, we will always assume that $E$ is non-singular.

*Exercise* 3.5 Prove that if $f \in L^\infty(E, \mu_*)$, then

$$\int f d\mu_* = \int \left( \sum_{y \in [x]_E} f(x, y) \right) d\mu(x),$$

and similarly that for $f \in L^\infty(E, \mu^*)$ we have

$$\int f d\mu^* = \int \left( \sum_{x \in [y]_E} f(x, y) \right) d\mu(y).$$

**Definition 3.6** A function $a \in L^\infty(E, \mu^*)$ is called **left finite** if there is $n \in \mathbb{N}$ such that for $\mu^*$-almost all $(x, y) \in E$ we have

$$|\{z : a(x, z) \neq 0\}| + |\{z : a(z, y) \neq 0\}| \leq n.$$

If $a$ and $b$ are left finite we define the product $ab$ in analogy to matrix multiplication,

$$(ab)(x, y) = \sum_{z \in [x]_E} a(x, z) b(z, y),$$

and we also define the "adjoint matrix" $a^*$ in the natural way, $a^*(x, y) = \overline{a(y, x)}$ (complex conjugation). The following is easily verified:

**Lemma 3.7** *The left finite functions are stable under sum, scalar multiplication, product and adjoint, and so they form a* $*$*-algebra.*

For each left finite $a \in L^\infty(E, \mu^*)$ we can define an operator $L_a$ on $L^2(E, \mu^*)$ by

$$(L_a \psi)(x, y) = \sum_{z \in [x]_E} a(x, z) \psi(z, y).$$

**Lemma 3.8** *Every operator of the form* $L_a$ *is bounded when* $a$ *is left finite. Moreover, for all* $a, b \in L^\infty(E, \mu^*)$ *left finite we have* $L_a L_b = L_{ab}$ *and* $L_a^* = L_{a^*}$. *Thus* $\mathcal{A} = \{L_a : a \text{ is left finite}\}$ *forms a* $*$*-subalgebra of* $B(L^2(E, \mu^*))$.

*Proof* Fix a left finite function $a$ and $n$ such that

$$|\{z : a(x,z) \neq 0\}| + |\{z : a(z,y) \neq 0\}| \leq n$$

for almost all $(x,y) \in E$. We will show that $\|L_a\| \leq n\|a\|_\infty$. For $\psi \in L^2(E,\mu^*)$ we have

$$\|(L_a\psi)\|^2 = \int \left( \sum_{x\in[y]_E} |(L_a\psi)(x,y)|^2 \right) d\mu(y)$$

$$= \int \left( \sum_{x\in[y]_E} |\sum_{z\in[y]_E} a(x,z)\psi(z,y)|^2 \right) d\mu(y). \quad (12.1)$$

Fix a typical $y$. For $x \in [y]_E$ fixed, we get from the Cauchy-Schwarz inequality and left finiteness of $a$ that

$$\left| \sum_{z\in[y]_E} a(x,z)\psi(z,y) \right|^2 \leq n \sum_{z\in[y]_E} |a(x,z)\psi(z,y)|^2$$

$$\leq n\|a\|_\infty^2 \sum_{z\in[y]_E, a(x,z)\neq 0} |\psi(z,y)|^2.$$

Further,

$$\sum_{x\in[y]_E} \sum_{z\in[y]_E, a(x,z)\neq 0} |\psi(z,y)|^2 = \sum_{z\in[y]_E} \sum_{x\in[y]_E, a(x,z)\neq 0} |\psi(z,y)|^2$$

$$\leq n \sum_{z\in[y]_E} |\psi(z,y)|^2,$$

where the last inequality follows since $a$ is left finite. Combining this with (12.1) we get

$$\|(L_a\psi)\|^2 \leq n^2\|a\|_\infty^2 \int \left( \sum_{z\in[y]_E} |\psi(z,y)|^2 \right) d\mu(y) = n^2\|a\|_\infty^2\|\psi\|^2,$$

as required. The remaining claims are left for the reader to verify. □

**Definition 3.9** We define

$$M(E) = \{L_a : a \text{ is left finite}\}''$$

and call this the von Neumann algebra of the equivalence relation $E$.

*Exercise* 3.10 Show that if on $n = \{0, 1 \ldots, n-1\}$ we take $E = n \times n$, then $M(E) \simeq M_n(\mathbb{C})$. Also, describe $M(E)$ with other choices of $E \subseteq n \times n$.

It is easy to see that $M(E)$ only depends on $E$ up to orbit equivalence, i.e., if $F$ is a non-singular countable Borel equivalence relation on $(Y, \nu)$ and there is a non-singular Borel bijection $T$ such that $xEy \iff T(x)FT(y)$ a.e., then $M(F) \simeq M(E)$ as von Neumann algebras.

Following what is standard notation, we denote by $[E]$ the group of all non-singular bijections $\phi : X \to X$ which satisfy $xE\phi(x)$ a.e., and we let $[[E]]$ denote the semigroup of all *partial* $\phi : A \to B$, where $A, B \subseteq X$ are Borel sets, which satisfy $xE\phi(x)$ for a.a. $x \in \mathrm{dom}(\phi)$. Given $\phi \in [[E]]$ and a function $f \in L^\infty(X, \mu)$, a left finite function $a_{\phi,f}$ is defined by

$$a_{\phi,f}(x, y) = \begin{cases} f(x) & \text{if } y = \phi(x), \\ 0 & \text{otherwise.} \end{cases}$$

When $f = 1$, the constant 1 function, then we will write $a_\phi$ for $a_{\phi,1}$. A direct calculation shows that

$$a_{\phi_0,f}a_{\phi_1,g} = a_{\phi_1\circ\phi_0,(g\circ\phi_0)f}. \tag{12.2}$$

In particular, letting $\Delta(x) = x$ for all $x \in X$, we have that $a_{\Delta,f}a_{\Delta,g} = a_{\Delta,fg}$. It follows that $f \mapsto L_{a_{\Delta,f}}$ provides an embedding of $L^\infty(X, \mu)$ into $M(E)$. That is, $L^\infty(X, \mu)$ is naturally identified with the "diagonal matrices" in $M(E)$. When there is no danger of confusion we will therefore write $f$ for $L_{a_{\Delta,f}}$ and write $L^\infty(X, \mu)$ for $\{L_{a_{\Delta,f}} : f \in L^\infty(X, \mu)\}$.

We assume from now on that $E$ is *aperiodic*, i.e., that all $E$ classes are infinite. It is then clear from Theorem 3.3 that one can find a sequence $(\phi_n)_{n\in\mathbb{N}_0}$ in $[E]$ whose graphs are pairwise disjoint and $E = \bigsqcup_{n\in\mathbb{N}_0} \mathrm{graph}(\phi_n)$. It is practical to always assume that $\phi_0 = \Delta$. The following lemma provides a useful standard form for the operators in $M(E)$.

**Lemma 3.11** *Let $(\phi_n)_{n\in\mathbb{N}_0}$ be a sequence in $[E]$ with pairwise disjoint graphs whose union is $E$. Any element $x \in M(E)$ can be written uniquely as*

$$x = \sum_{n=0}^{\infty} f_n L_{a_{\phi_n}}$$

*for some sequence $(f_n)_{n\in\mathbb{N}_0}$ in $L^\infty(X, \mu)$ and with convergence in the* *weak topology. In particular,*

$$M(E) = \left(L^\infty(X,\mu) \cup \{L_{a_{\phi_n}} : n \in \mathbb{N}_0\}\right)''. \tag{12.3}$$

The proof, which we will skip, can be done by hand and is not that hard. The reader should be warned, though, that the Lemma does not provide any information about *which* sequences $(f_n)$ define operators in $M(E)$ in this way. It does allow us to prove two key results about $M(E)$.

**Theorem 3.12** *$L^\infty(X,\mu)$ is a Cartan subalgebra of $M(E)$.*

*Proof* We first prove that $L^\infty(X,\mu)$ is a maximal Abelian subalgebra. For this, suppose $T \in L^\infty(X,\mu)' \cap M(E)$. Using the previous lemma, write $T = \sum_{n=0}^\infty f_n L_{a_{\phi_n}}$. Let $g \in L^\infty(X,\mu)$. Using that $gT = Tg$ and (12.2) above we get

$$\sum_{n=0}^\infty g f_n L_{a_{\phi_n}} = \sum_{n=0}^\infty f_n L_{a_{\phi_n}} g = \sum_{n=0}^\infty f_n (g \circ \phi_n) L_{a_{\phi_n}},$$

and so from the uniqueness of the expansion we have $f_n(g - g \circ \phi_n) = 0$. Thus if $f_n(x) \neq 0$ it follows that $g(x) = g(\phi_n(x))$ for *any* $g \in L^\infty(X,\mu)$, which means that $\phi_n(x) = x$ whenever $f_n(x) \neq 0$. But this shows that $T = f_0 L_{a_{\phi_0}} = f_0 L_{a_\Delta}$, that is, $T \in L^\infty(X,\mu)$.

To see that $L^\infty(X,\mu)$ is regular in $M(E)$, observe that the normalizer of $L^\infty(X,\mu)$ in $M(E)$ contains the unitary elements of $L^\infty(X,\mu)$ as well as all $L_{a_\phi}$ for $\phi \in [E]$, and so generates $M(E)$. □

**Theorem 3.13** *$M(E)$ is a factor iff $E$ is ergodic.*

*Proof* Suppose first that $M(E)$ is a factor, and let $f \in L^\infty(X,\mu)$ be an $E$-invariant function. It follows from (12.3) that $f \in \mathcal{Z}(M(E))$, thus $f$ is a constant multiple of 1. Conversely, if $E$ is ergodic, let $x \in \mathcal{Z}(M(E))$. From the previous theorem we know that $\mathcal{Z}(M(E)) \subseteq L^\infty(X,\mu)$, and so $x = f$ for some $f \in L^\infty(X,\mu)$. Since $f$ is central we have that $fL_{a_\phi} = L_{a_\phi} f$ for all $\phi \in [E]$, and so $f \circ \phi = f$ for all $\phi \in E$. Whence $f$ is $E$ invariant, and therefore constant, which shows that $\mathcal{Z}(M(E)) = \mathbb{C}1$. □

With a little more effort one can go on to prove the following interesting theorem:

**Theorem 3.14** *Let $E$ and $F$ be non-singular countable Borel equivalence relations on standard $\sigma$-finite measure spaces $(X,\mu)$ and $(Y,\nu)$, respectively. Then $E$ is orbit equivalent to $F$ if and only if the inclusions $L^\infty(X,\mu) \subseteq M(E)$ and $L^\infty(Y,\nu) \subseteq M(F)$ are isomorphic, i.e., there is an isomorphism of $M(E)$ and $M(F)$ which maps $L^\infty(X,\mu)$ onto $L^\infty(Y,\nu)$.*

This illustrates that if we want to understand the relationship between $E$ and $M(E)$ we will want to understand the nature of $L^\infty(X,\mu)$ as a subalgebra of $M(E)$.

*Exercise* 3.15 Assume that $\mu(X) < \infty$ and that $\mu$ is $E$-invariant (which means that $\mu$ is invariant under all elements of $[E]$). Show that $\mathbf{1}_{\mathrm{graph}(\Delta)} \in L^2(E,\mu^*)$ and that $\tau(x) = \langle x\mathbf{1}_{\mathrm{graph}(\Delta)}, \mathbf{1}_{\mathrm{graph}(\Delta)}\rangle$ defines a trace on $M(E)$. Conclude that when $\mu$ is non-atomic and $E$ is $\mu$-ergodic, $M(E)$ is a type $\mathrm{II}_1$ factor.

*Exercise* 3.16 Let $\Gamma$ be a countable discrete group and let $\sigma : \Gamma \curvearrowright (X,\mu)$ be a measure preserving a.e. free (!) action. Show that $L^\infty(X,\mu)\rtimes_\sigma \Gamma$ is isomorphic to $M(E_\sigma)$, and in fact that the inclusions $L^\infty(X,\mu) \subseteq L^\infty(X,\mu) \rtimes_\sigma \Gamma$ and $L^\infty(X,\mu) \subseteq M(E_\sigma)$ are isomorphic. Conclude that $L^\infty(X)$ is a Cartan subalgebra of $L^\infty(X,\mu) \rtimes_\sigma \Gamma$, and that $L^\infty(X,\mu) \rtimes_\sigma \Gamma$ is a factor when $\sigma$ is a.e. free and ergodic. (*Hint*: Consider the map $\Gamma \times X \to E$ defined by $(\gamma, x) \mapsto (x, \sigma_\gamma(x))$.)

*Exercise* 3.17 Call a measurable function $b : E \to \mathbb{C}$ **right finite** if $D(x,y)^{-\frac{1}{2}}b(x,y)$ is left finite (where $D$ is the Radon-Nikodym derivative of $\mu_*$ w.r.t. $\mu^*$). Show that

$$R_b(\psi)(x,y) = \sum_{z\in[x]_E} \psi(x,z)b(z,y)$$

defines a bounded operator on $L^2(E,\mu^*)$, and that $\{R_b : b \text{ is right finite}\}$ is a $*$-algebra contained in $M(E)'$.

The next exercise is somewhat harder.

*Exercise* 3.18 With notation as in the previous exercise, show that $M(E)' = \{R_b : b \text{ is right finite}\}''$.

# Lecture 4

*Hyperfinite von Neumann algebras, ITPFI factors, Connes' classification of hyperfinite factors and Krieger's theorems, non-classification results via descriptive set theory, and rigidity.*

This lecture, the last, is dedicated to giving an overview of developments in the field of von Neumann algebras in the last 40 years. In other words, if we had a whole semester's worth of lectures, these are some of the topics that would be covered in detail.

## Hyperfinite von Neumann algebras.

**Definition 4.1** A von Neumann algebra $M$ is called

1. **finite dimensional** if it is isomorphic to $M_{n_1}(\mathbb{C}) \oplus \cdots \oplus M_{n_k}(\mathbb{C})$.
2. **hyperfinite** if there is an increasing sequence $(M_i)$ of finite dimensional sub-algebras of $M$ such that $\bigcup M_i$ is dense in $M$.

Hyperfiniteness is equivalent to a number of other conditions on a von Neumann algebra, among them *amenability* and *injectivity* (neither of which we define here).

The class of hyperfinite von Neumann algebras and factors is very rich, and their theory has been developed further than any other general class. A useful source of examples comes from infinite *tensor products* of finite von Neumann algebras.

## Finite tensor products.

Let $H_1$ and $H_2$ be (complex) Hilbert spaces. We will denote by $H_1 \odot H_2$ the *algebraic* tensor product of $H_1$ and $H_2$. Recall that this means that we have a map $H_1 \times H_2 \to H_1 \odot H_2 : (\xi, \eta) \mapsto \xi \otimes \eta$ with the property that any bilinear map $\rho : H_1 \times H_2 \to E$, where $E$ is some vector space over $\mathbb{C}$, has the form $\rho(\xi, \eta) = \hat{\rho}(\xi \otimes \eta)$, for some linear map $\hat{\rho} : H_1 \odot H_2 \to E$. The elements $\xi \otimes \eta$ are called (elementary) tensors, and the tensors generate $H_1 \odot H_2$.

There is a unique inner product $\langle \cdot, \cdot \rangle$ on $H_1 \odot H_2$ that satisfies

$$\langle \xi_1 \otimes \eta_1, \xi_2 \otimes \eta_2 \rangle = \langle \xi_1, \xi_2 \rangle_{H_1} \langle \eta_1, \eta_2 \rangle_{H_2}.$$

The tensor product of the Hilbert spaces $H_1$ and $H_2$ is the completion of $H_1 \odot H_2$ w.r.t. the norm induced by $\langle \cdot, \cdot \rangle$. It is denoted $H_1 \otimes H_2$.

Let $M_i \subseteq B(H_i)$, $i = 1, 2$, be von Neumann algebras. For $x \in M_1$ and $y \in M_2$ we define an operator $x \otimes y$ on $H_1 \otimes H_2$ by $x \otimes y(\xi \otimes \eta) = (x\xi) \otimes (y\eta)$. It is not hard to see that each such operator is bounded. We let

$$M_1 \otimes M_2 = \{x \otimes y : x \in M_1, y \in M_2\}''$$

and call this the tensor product of $M_1$ and $M_2$. The tensor product of an arbitrary but finite number of Hilbert spaces and von Neumann algebras are defined similarly.

*Exercise* 4.2 Show that $H \simeq H \otimes \mathbb{C}$ for any Hilbert space $\mathbb{C}$. More generally, if $H_1$ and $H_2$ are Hilbert spaces and $\xi \in H_2$ is a unit vector, then $\eta \mapsto \eta \otimes \xi$ is an embedding of $H_1$ into $H_1 \otimes H_2$.

*Exercise* 4.3 Show that $M_n(\mathbb{C})\otimes M_m(\mathbb{C}) \simeq M_{nm}(\mathbb{C})$.

## Infinite tensor products

Let $(H_i)_{i\in\mathbb{N}}$ be a sequence of Hilbert spaces and for each $i \in \mathbb{N}$ let $\xi_i \in H_i$ be a unit vector. (We will call a pair $(H, \xi)$ where $\xi$ is a unit vector in the Hilbert space $H$ a *pointed* Hilbert space. The vector $\xi \in H$ will be called a *base point*.) For each $n \in \mathbb{N}$ we have by exercise 4.2 that $\bigotimes_{i=1}^n H_i$ can be identified naturally with the subspace

$$\{\eta_1\otimes\cdots\otimes\eta_n\otimes\xi_{n+1} : (\forall i \leq n)\eta_i \in H_i\}$$

of $\bigotimes_{i=1}^{n+1} H_i$. Let $H_\infty$ be the inductive limit of the system

$$H_1 \overset{\eta\mapsto\eta\otimes\xi_2}{\hookrightarrow} H_1\otimes H_2 \overset{\eta\mapsto\eta\otimes\xi_3}{\hookrightarrow} H_1\otimes H_2\otimes H_3 \hookrightarrow \cdots$$

which is an inner product space in the natural way; we let

$$\bigotimes_{i=1}^{n} H_i \to \bigotimes_{i=1}^{\infty}(H_i, \xi_i) : \eta \mapsto \eta\otimes\xi_{n+1}\otimes\xi_{n+2}\otimes\cdots$$

denote the canonical embedding. It is clear that $\bigotimes_{i=1}^\infty(H_i, \xi_i)$ is generated by the "elementary tensors" $\eta_1\otimes\eta_2\otimes\cdots$, where $\eta_i \in H_i$, and where $\eta_i = \xi_i$ eventually. We define the **infinite tensor product of the pointed Hilbert spaces** $(H_i, \xi_i)_{i\in\mathbb{N}}$ to be the completion of $H_\infty$. It is denoted $\bigotimes_{i=1}^\infty(H_i, \xi_i)$.

Now let $(M_i)_{i\in\mathbb{N}}$ be a von Neumann algebra acting on $H_i$, for each $i \in \mathbb{N}$. If $x_i \in M_i$, $i \in \mathbb{N}$, is a sequence such that $x_i = I$ eventually, then a bounded operator $\bigotimes_{i\in\mathbb{N}} x_i$ on $\bigotimes_{i=1}^\infty(H_i, \xi_i)$ is defined by requiring that

$$(\bigotimes_{i\in\mathbb{N}} x_i)(\eta_1\otimes\eta_2\otimes\cdots) = x_1\eta_1\otimes x_2\eta_2\otimes\cdots.$$

We define **infinite tensor product of the von Neumann algebras** $(M_i)_{i\in\mathbb{N}}$ w.r.t. the vectors $\xi_i$ as

$$\bigotimes_{i=1}^{\infty}(M_i, \xi_i) = \{\bigotimes_{i\in\mathbb{N}} x_i : x_i \in M_i \text{ and } x_i = I \text{ eventually}\}''.$$

*Warning*: As we will see below, the isomorphism type of the infinite tensor product is *highly* sensitive to the asymptotic behavior of the sequence $(\xi_i)_i$.

**Examples 4.4**

1. Let $(X,\mu)$ and $(Y,\nu)$ be standard probability spaces. It is well-known that $L^2(X,\mu)\otimes L^2(Y,\nu) \simeq L^2(X \times Y, \mu \times \nu)$, and that the isomorphism is given by $f\otimes g \mapsto fg$. It follows that

$$L^\infty(X,\mu)\otimes L^\infty(Y,\nu) \simeq L^\infty(X \times Y, \mu \times \nu),$$

where we represent each $L^\infty$ as multiplication operators on the corresponding $L^2$ space. This is easily generalized: If $(X_i,\mu_i)$ are standard probability spaces and we let $\xi_i = 1$ be the constant 1 function in $L^2(X_i,\mu_i)$, then the tensor product $\bigotimes(L^2(X_i,\mu_i),\xi_i)$ is isomorphic to $L^2(\prod X_i, \prod \mu_i)$, and $\bigotimes(L^\infty(X_i,\mu_i),\xi_i)$ is isomorphic to $L^\infty(\prod X_i, \prod \mu_i)$.

   Now let $\xi \in L^2(Y,\nu)$ be a unit vector (not necessarily 1.) An isometric embedding of $L^2(X,\mu)$ into $L^2(X\times Y, \mu\times\nu)$ is given by $f \mapsto f\xi$, and this corresponds to the embedding $f \mapsto f\otimes\xi$ of $L^2(X,\mu)$ into $L^2(X,\mu)\otimes L^2(Y,\nu)$. Consider also the space $L^2(X \times Y, \mu \times |\xi|^2\nu)$, and the corresponding embedding $f \mapsto f1$, where 1 is the constant 1 function in $L^2(Y, |\xi^2|\nu)$. It is easily verified that the map

$$L^2(X \times Y, \mu \times \nu) \to L^2(X \times Y, \mu \times |\xi|^2\nu) : h \mapsto \frac{h}{\xi}$$

is an isometry which conjugates the two embeddings of $L^2(X,\mu)$ described above.

   Now let $(X_i,\mu_i)_{i\in\mathbb{N}}$ be a sequence of standard probability spaces, and let $\xi_i \in L^2(X_i,\mu_i)$ be a unit vector for each $i \in \mathbb{N}$. Our observations above now show that the tensor product $\bigotimes(L^2(X_i,\mu_i),\xi_i)$ is isomorphic to $L^2(\prod X_i, \prod |\xi_i|^2\mu_i)$ and that $\bigotimes(L^\infty(X_i,\mu_i),\xi_i)$ is isomorphic to $L^\infty(\prod X_i, \prod |\xi_i^2|\mu_i)$. It is of course well-known that the product of infinitely many probability measures is highly sensitive to the asymptotic behavior of the sequence of measures, and so this example indicates that the infinite tensor product of von Neumann algebras also is highly sensitive in this way, for much the same reasons.

2. Let $n_i \in \mathbb{N}$ be an infinite sequence of natural numbers. The easiest way to represent the matrix algebra $M_{n_i}(\mathbb{C})$ is to have it act on itself by matrix multiplication on the left. (If nothing else, this manoeuvre allows us to avoid introducing separate notation for the Hilbert spaces we act on.) Let $\tau : M_{n_i}(\mathbb{C}) \to \mathbb{C}$ be the normalized trace, and let $\langle x,y\rangle_\tau = \tau(y^*x)$ be the associated inner product, which makes $M_{n_i}(\mathbb{C})$ into a Hilbert space. Let $\xi_i \in M_{n_i}(\mathbb{C})$ be a

unit vector. There is no loss in assuming that $\xi_i$ is a diagonal matrix, $\xi_i = \text{diag}(\lambda_{i,1}, \ldots, \lambda_{i,n_i})$. We can now form the infinite tensor product of the matrix algebras $M_{n_i}(\mathbb{C})$ w.r.t. the vectors $\xi_i$ as described above. It turns out that it is a factor. (It can be shown that in general the tensor product (finite or infinite) of factors is again a factor.) A factor of this form is called an **ITPFI factor**, which stands for *infinite tensor product of factors of type* I. It is clear that ITPFI factors are hyperfinite. The sequence $(\lambda_{i,j}^2)$ are called the *eigenvalue sequence* for the ITPFI factor. Those ITPFI factors where for some $n \in \mathbb{N}$ we have $n_i = n$ for all $i \in \mathbb{N}$ are called ITPFI$_n$ factors.

In much the same way as above, we can carry out an analysis of the nature of the inclusions of $M_n(\mathbb{C})$ into $M_n(\mathbb{C}) \otimes M_m(\mathbb{C})$ and how it depends on a choice of $\xi \in M_m(\mathbb{C})$. As we will see below, the isomorphism type of an ITPFI factor depends on the eigenvalue list in a somewhat similar way to how the infinite product of finite measure spaces depends on the measure of the atoms.

3. Certain ITPFI$_2$ factors deserve special mention for their importance in the field. Pride of place goes to the **hyperfinite** II$_1$ **factor**. This factor arises from taking $\xi_i = (\frac{1}{\sqrt{2}}, \frac{1}{\sqrt{2}})$ for all $i \in \mathbb{N}$. The hyperfinite II$_1$ factor is perhaps the most studied and most important of all II$_1$ factors. It is usually denoted $\mathcal{R}$ (or $\mathcal{R}_1$). It is an almost canonical presence, showing up virtually everywhere in the field.

   Another important class of ITPFI$_2$ factors are the **Powers factors**. Let $0 < \lambda < 1$, and let $\xi_i = \left((\frac{\lambda}{1+\lambda})^{\frac{1}{2}}, (1 - \frac{\lambda}{1+\lambda})^{\frac{1}{2}}\right)$ for all $i \in \mathbb{N}$. The resulting ITPFI$_2$ factor is denoted $\mathcal{R}_\lambda$. In 1967, it was shown by Powers in [Pow67] that the family $(\mathcal{R}_\lambda)_{\lambda \in (0,1)}$ constitutes a family of mutually non-isomorphic factors. Previous to that result, it had not been known if there were uncountably many non-isomorphic factors!

A far-reaching study of ITPFI factors was undertaken by Araki and Woods in the paper [AW69], where they introduced a number of invariants and produced new classes of uncountably many non-isomorphic factors.

## Classification

The towering achievement of von Neumann algebra theory from the 1970s is Connes' complete classification of hyperfinite (or, more correctly, injective) factors, [Con76]. The classification divides type III into subcases type III$_\lambda$, $\lambda \in [0, 1]$. Connes showed that there is a unique injective (hyperfinite) factor in each of the classes $\mathrm{II}_1$, $\mathrm{II}_\infty$ and $\mathrm{III}_\lambda$, $0 < \lambda < 1$. In the case $\mathrm{III}_0$, there are many non-isomorphic factors, but they are classified completely by an associated invariant called the *flow of weights* (see e.g. [Con94]). The remaining case, the type $\mathrm{III}_1$ case, proved to be a difficult problem in its own right. It was eventually solved by Haagerup some years later in [Haa87]: Haagerup showed that there is a unique injective factor of type $\mathrm{III}_1$.

Another phenomenal achievement of the period are the results of Krieger in [Kri76]. Krieger showed that every hyperfinite factor is of the form $M(E)$ for some *amenable* non-singular ergodic countable Borel equivalence relation $E$. (Here we use the language of the Feldman-Moore construction (from Lecture 3) to state Krieger's results, though the work of Feldman and Moore postdates Krieger's work.) Krieger moreover showed that for *amenable* ergodic non-singular countable Borel equivalence relations $E$ and $F$, $M(E)$ is isomorphic to $M(F)$ if and only if $E$ and $F$ are orbit equivalent.

*Exercise* 4.5 Let $\mathbb{Z} \curvearrowright 2^{\mathbb{N}}$ be the odometer action (i.e., adding one with carry modulo 2), and let $E_0$ denote the induced equivalence relation. Let $\alpha_i \in (0, 1)$ for all $i \in \mathbb{N}$, and let $\lambda_{i,0} = \alpha_j$ and $\lambda_{i,1} = 1 - \alpha_j$. Give $2^{\mathbb{N}}$ the product measure $\prod_{i=1}^{\mathbb{N}}(\lambda_{i,0}\delta_0 + \lambda_{i,1}\delta_1)$, where $\delta_i$ is the Dirac measure concentrating on $i \in \{0, 1\}$. Show that $E_0$ is measure class preserving, and that $M(E_0)$ is isomorphic to the $\mathrm{ITPFI}_2$ with eigenvalue list $(\lambda_{i,j})$. In particular, if we take $\alpha_i = \frac{1}{2}$ for all $i \in \mathbb{N}$, so that $E_0$ is measure preserving, we get the hyperfinite $\mathrm{II}_1$ factor $\mathcal{R}$. (This exercise may be quite hard.)

## Non-classification

The invariant provided by Connes' classification is a certain non-singular $\mathbb{R}$-action (flow), which is considered up to conjugacy. This is hardly a very simple invariant. It is therefore natural to ask: How difficult is it to classify factors up to isomorphism? It is of course well-known among set theorists that such questions can be fruitfully attacked through the concept of Borel reducibility.[4] We will end these lectures by giving an overview of what is known about the classification problem for factors from this point of view. The reader can find a more detailed survey of these results in [ST09a].

[4] We will assume that the reader is familiar with the notion of Borel reducibility, and related concepts like smooth, non-smooth, classification by countable structures, etc. We refer to [ST09a] for a brief overview of these notions.

Before going on, we remark that the problem of classifying von Neumann algebras and factors acting on a separable Hilbert space fits nicely into the framework of descriptive set theory. Namely, a von Neumann algebra $N \subseteq B(H)$ can be identified with $N \cap B^1(H)$. The unit ball of $B(H)$ is a compact Polish space in the weak topology, and so the space $K(B^1(H))$ of compact subsets of $B^1(H)$ is itself a Polish space. One can now show that the set $\mathrm{vN}(H) = \{N \cap B^1(H) \in K(B^1(H)) :$ $N$ is a von Neumann algebra$\}$ is a Borel set. We think of $\mathrm{vN}(H)$ as the **standard Borel space of separably acting von Neumann algebras**. The subset $\mathcal{F}(H) \subseteq \mathrm{vN}(H)$ of factors turns out to be Borel. (There are other ways of arriving at the space $\mathrm{vN}(H)$, and even a natural choice of Polish topology on it; see [HW98, HW00] for an exhaustive study. The space $\mathrm{vN}(H)$ was introduced and studied by Effros in the papers [Eff65, Eff66], who also proved, among many other things, that $\mathcal{F}(H)$ is Borel.)

The first known non-classification result in the area is due to Woods:

**Theorem 4.6** (Woods, [Woo73]) *The classification of hyperfinite factors is not smooth. More precisely, there is a Borel reduction of $E_0$ to the isomorphism relation of ITPFI$_2$ factors (of type* $\mathrm{III}_0$*).*

This seems to have been the only result of its kind that was known until a few years ago, when the following was proven:

**Theorem 4.7** (Sasyk-Törnquist, [ST09b]) *Let $\Lambda$ be the class of separably acting factors of type* $\mathrm{II}_1$, $\mathrm{II}_\infty$, *or* $\mathrm{III}_\lambda$, $0 \leq \lambda \leq 1$. *Then the isomorphism relation in $\Lambda$ does not admit classification by countable structures.*

Prior to this, it was apparently not even known if the isomorphism relation was non-smooth in any of these classes, except the type $\mathrm{III}_0$ case where it follows from Woods' theorem.[5] In the same paper, the following was also shown:

**Theorem 4.8** (Sasyk-Törnquist, [ST09b]) *Isomorphism of countable graphs is Borel reducible to isomorphism of* $\mathrm{II}_1$ *factors.*

The same construction gives the result for type $\mathrm{II}_\infty$ as well, but the proof falls short of handling the type III cases. It follows from the above

[5] For a long time it was not even known if there were infinitely many non-isomorphic $\mathrm{II}_1$ factors. This problem was solved by McDuff in [McD69].

that isomorphism of factors (of type $\mathrm{II}_1$ and $\mathrm{II}_\infty$) is analytic, and not Borel in the space $\mathrm{vN}(H)$.

The proofs of theorems 4.7 and 4.8 only became possible due to enormous advances in the understanding of the relation between measure preserving countable Borel equivalence relations and their corresponding von Neumann algebra. These advances were spearheaded by Sorin Popa, as well as a number of his collaborators, who in the last two decades have developed a large number of techniques and results that make it possible to analyze group-measure space factors in the presence of certain "rigidity" properties. We refer to [Pop07, Vae10] for an overview of these developments.

None of the factors that were constructed to prove theorems 4.7 and 4.8 are hyperfinite. So what about the classification of hyperfinite type $\mathrm{III}_0$ factors? Can they be classified by countable structures? The answer is again no. This was already shown in [ST09b], by showing that a standard construction of an injective type $\mathrm{III}_0$ factor with a prescribed flow of weights is a Borel construction. Subsequently, this result was improved to show the following non-classification result for $\mathrm{ITPFI}_2$ factors:

**Theorem 4.9** (Sasyk-Törnquist, [ST10]) *The isomorphism relation for* $\mathrm{ITPFI}_2$ *factors is not classifiable by countable structures.*

The proof of this is probably the most elementary and direct of all non-classification results for factors. It only relies on techniques that essentially go back to Araki and Woods in [AW69], as well as straightforward Baire category/turbulence arguments.

We close with a brief discussion of an open problem. The only upper bound known about the classification of von Neumann algebras (and factors) is the following:

**Theorem 4.10** (Sasyk-Törnquist, [ST09b]) *The isomorphism relation in* $\mathrm{vN}(H)$ *is Borel reducible to an orbit equivalence relation induced by a continuous action of the unitary group* $U(\ell^2(\mathbb{N}))$ *on a Polish space.*

The proof is not hard, and it is natural to ask if this upper bound on the complexity of the isomorphism relation for factors is in fact optimal. Given that there currently is no evidence to the contrary, the following conjecture has been made (and stated publicly in many talks):

**Conjecture 4.11** (Törnquist) *The isomorphism relation for separably acting factors is universal (from the point of view of Borel reducibility)* *for orbit equivalence relations induced by a continuous action of the unitary group on a Polish space. In fact, this is already true for isomorphism of* $\mathrm{II}_1$ *factors.*

In other words, the isomorphism relation for $\mathrm{II}_1$ factors attains the upper bound given in Theorem 4.10. A positive solution to this would, in addition to being the ultimate non-classification theorem in the area, give a first example of a "naturally occurring" isomorphism relation which realizes the maximal possible complexity that an orbit equivalence relation induced by the unitary group can have.

## Sources

A great source for learning about von Neumann algebras are the lecture notes of Vaughan Jones [Jon10], freely available online at:

`http://math.berkeley.edu/~vfr/MATH20909/VonNeumann2009.pdf`

Lecture 1 draws its material from [Ped89], [Arv76] and [Sak71]. Lecture 2 is based on [Bla06] and [Nie80], and to a lesser extend on [Dix81] and [Jon10]. In Lecture 3, the discussion of the group measure space von Neumann algebras and crossed products is based on [Bla06] and [Jon10], and the discussion of the von Neumann algebra of a non-singular equivalence relation is based on [FM77b]. The discussion of hyperfinite von Neumann algebras in Lecture 4 draws on [Bla06] and [Con94].

## References

[Arv76] William Arveson. *An invitation to $C^*$-algebras*. Springer-Verlag, New York, 1976. Graduate Texts in Mathematics, No. 39.

[AW69] Huzihiro Araki and E. J. Woods. A classification of factors. *Publ. Res. Inst. Math. Sci. Ser. A*, 4:51–130, 1968/1969.

[Bla06] B. Blackadar. *Operator algebras*, volume 122 of *Encyclopaedia of Mathematical Sciences*. Springer-Verlag, Berlin, 2006. Theory of $C^*$-algebras and von Neumann algebras, Operator Algebras and Non-commutative Geometry, III.

[Con76] A. Connes. Classification of injective factors. Cases $II_1$, $II_\infty$, $III_\lambda$, $\lambda \neq 1$. *Ann. of Math. (2)*, 104(1):73–115, 1976.

[Con94] Alain Connes. *Noncommutative geometry*. Academic Press Inc., San Diego, CA, 1994.

[Dix81] Jacques Dixmier. *von Neumann algebras*, volume 27 of *North-Holland Mathematical Library*. North-Holland Publishing Co., Amsterdam, 1981. With a preface by E. C. Lance, Translated from the second French edition by F. Jellett.

[Eff65] Edward G. Effros. The Borel space of von Neumann algebras on a separable Hilbert space. *Pacific J. Math.*, 15:1153–1164, 1965.

[Eff66] Edward G. Effros. Global structure in von Neumann algebras. *Trans. Amer. Math. Soc.*, 121:434–454, 1966.

[FM77a] Jacob Feldman and Calvin C. Moore. Ergodic equivalence relations, cohomology, and von Neumann algebras. I. *Trans. Amer. Math. Soc.*, 234(2):289–324, 1977.

[FM77b] Jacob Feldman and Calvin C. Moore. Ergodic equivalence relations, cohomology, and von Neumann algebras. II. *Trans. Amer. Math. Soc.*, 234(2):325–359, 1977.

[Haa87] Uffe Haagerup. Connes' bicentralizer problem and uniqueness of the injective factor of type $\mathrm{III}_1$. *Acta Math.*, 158(1-2):95–148, 1987.

[HT12] G. Hjorth and A. Törnquist. The conjugacy relation on unitary representations. *Mathematical Research Letters*, to appear, 2012.

[HW98] Uffe Haagerup and Carl Winsløw. The Effros-Maréchal topology in the space of von Neumann algebras. *Amer. J. Math.*, 120(3):567–617, 1998.

[HW00] Uffe Haagerup and Carl Winsløw. The Effros-Maréchal topology in the space of von Neumann algebras. II. *J. Funct. Anal.*, 171(2):401–431, 2000.

[Jon10] Vaughan Jones. Von Neumann algebras. available on author's webpage, 2010.

[KR97] Richard V. Kadison and John R. Ringrose. *Fundamentals of the theory of operator algebras. Vol. II*, volume 16 of *Graduate Studies in Mathematics*. American Mathematical Society, Providence, RI, 1997. Advanced theory, Corrected reprint of the 1986 original.

[Kri76] Wolfgang Krieger. On ergodic flows and the isomorphism of factors. *Math. Ann.*, 223(1):19–70, 1976.

[McD69] Dusa McDuff. Uncountably many $\mathrm{II}_1$ factors. *Ann. of Math. (2)*, 90:372–377, 1969.

[MvN36] F.J. Murray and J. von Neumann. On rings of operators. *Ann. of Math. (2)*, 37(1):116–229, 1936.

[MvN37] F.J. Murray and J. von Neumann. On rings of operators. II. *Trans. Amer. Math. Soc.*, 41(2):208–248, 1937.

[MvN43] F.J. Murray and J. von Neumann. On rings of operators. IV. *Ann. of Math. (2)*, 44:716–808, 1943.

[Nie80] Ole A. Nielsen. *Direct integral theory*, volume 61 of *Lecture Notes in Pure and Applied Mathematics*. Marcel Dekker Inc., New York, 1980.

[Ped89] Gert K. Pedersen. *Analysis now*, volume 118 of *Graduate Texts in Mathematics*. Springer-Verlag, New York, 1989.

[Pop07] Sorin Popa. Deformation and rigidity for group actions and von Neumann algebras. In *International Congress of Mathematicians. Vol. I*, pages 445–477. Eur. Math. Soc., Zürich, 2007.

[Pow67] Robert T. Powers. Representations of uniformly hyperfinite algebras and their associated von Neumann rings. *Ann. of Math. (2)*, 86:138–171, 1967.

[Sak71] Shôichirô Sakai. *$C^*$-algebras and $W^*$-algebras*. Springer-Verlag, New York, 1971. Ergebnisse der Mathematik und ihrer Grenzgebiete, Band 60.

[ST09a] Román Sasyk and Asger Törnquist. Borel reducibility and classification of von Neumann algebras. *Bull. Symbolic Logic*, 15(2):169–183, 2009.

[ST09b] Roman Sasyk and Asger Törnquist. The classification problem for von Neumann factors. *J. Funct. Anal.*, 256(8):2710–2724, 2009.

[ST10] Román Sasyk and Asger Törnquist. Turbulence and Araki-Woods factors. *J. Funct. Anal.*, 259(9):2238–2252, 2010.

[Vae10] Stefaan Vaes. Rigidity for von Neumann algebras and their invariants. In *Proceedings of the International Congress of Mathematicians. Volume III*, pages 1624–1650, New Delhi, 2010. Hindustan Book Agency.

[vN40] J. von Neumann. On rings of operators. III. *Ann. of Math. (2)*, 41:94–161, 1940.

[Woo73] E. J. Woods. The classification of factors is not smooth. *Canad. J. Math.*, 25:96–102, 1973.

# 13

# The HOD Dichotomy

W. Hugh Woodin[a], Jacob Davis and Daniel Rodríguez

The sixteenth Appalachian Set Theory workshop was held at Cornell University in Ithaca on April 7, 2012. The lecturer was Hugh Woodin. As graduate students Jacob Davis and Daniel Rodríguez assisted in writing this chapter, which is based on the workshop lectures.

## 1 Introduction

This paper provides a more accessible account of some of the material from Woodin [4] and [5]. All unattributed results are due to the first author.

Recall that $0^\#$ is a certain set of natural numbers that codes an elementary embedding $j : L \to L$ such that $j \neq \mathrm{id} \upharpoonright L$. Jensen's covering lemma says that if $0^\#$ does not exist and $A$ is an uncountable set of ordinals, then there exists $B \in L$ such that $A \subseteq B$ and $|A| = |B|$. The conclusion implies that if $\gamma$ is a singular cardinal, then it is a singular cardinal in $L$. It also implies that if $\gamma \geq \omega_2$ and $\gamma$ is a successor cardinal in $L$, then $\mathrm{cf}(\gamma) = |\gamma|$. In particular, if $\beta$ is a singular cardinal, then $(\beta^+)^L = \beta^+$. Intuitively, this says that $L$ is close to $V$. On the other hand, should $0^\#$ exist, if $\gamma$ is an uncountable cardinal, then $\gamma$ is an inaccessible cardinal in $L$. In this case, we could say that $L$ is far from $V$. Thus, the covering lemma has the following corollary, which does not mention $0^\#$.

**Theorem 1.1** (Jensen) *Exactly one of the following holds.*

[a] Research partially supported by NSF grant DMS-0856201.

1. *$L$ is correct about singular cardinals and computes their successors correctly.*
2. *Every uncountable cardinal is inaccessible in $L$.*

Imagine an alternative history in which this $L$ dichotomy was discovered without knowledge of $0^\#$ or more powerful large cardinals. Clearly, (1) is consistent because it holds in $L$. On the other hand, whether or not there is a proper class of inaccessible cardinals in $L$ is absolute to generic extensions. This incomplete evidence might have led set theorists to conjecture that (2) fails. Of course, (2) only holds when $0^\#$ exists but $0^\#$ does not belong to $L$ and $0^\#$ cannot be added by forcing.

Canonical inner models other than $L$ have been defined and shown to satisfy similar covering properties and corresponding dichotomies. Part of what makes them canonical is that they are contained in HOD. In these notes, we will prove a dichotomy theorem of this kind for HOD itself. Towards the formal statement, recall that a cardinal $\delta$ is *extendible* iff for every $\eta > \delta$, there exists $\theta > \eta$ and an elementary embedding $j : V_{\eta+1} \to V_{\theta+1}$ such that $\mathrm{crit}(j) = \delta$ and $j(\delta) > \eta$. The following result expresses the idea that either HOD is close to $V$ or else HOD is far from $V$. We will refer to it as the HOD Dichotomy.

**Theorem 1.2** *Assume that $\delta$ is an extendible cardinal. Then exactly one of the following holds.*

1. *For every singular cardinal $\gamma > \delta$, $\gamma$ is singular in* HOD *and* $(\gamma^+)^{\mathrm{HOD}} = \gamma^+$.
2. *Every regular cardinal greater than $\delta$ is measurable in* HOD.

In this note, we shall prove a dichotomy in which (2) is weakened to hold for all sufficiently large regular cardinals greater than $\delta$; see Corollary 5.7. The full result can be found in [4] Theorem 212.

Notice that we have stated the HOD dichotomy without deriving it from a covering property that involves a "large cardinal missing from HOD". In other words, no analogue of $0^\#$ is mentioned and the alternative history we described for $L$ is what has actually happened in the case of HOD. This leads us to conjecture that (2) fails. One reason is that (2) is absolute between $V$ and its generic extensions by posets that belong to $V_\delta$, which we will show in the next section. There is some evidence for this conjecture. All known large cardinal axioms (which do not contradict the Axiom of Choice) are compatible with $V = \mathrm{HOD}$ and so trivially cannot imply (2). Further, we shall see that the main technique for obtaining independence in set theory (forcing) probably cannot be used to show that (2) is relatively consistent with the existence of an extendible cardinal starting from any known large cardinal hypothesis which is also consistent with the Axiom of Choice. Finally, by definition HOD contains all definable sets of ordinals and this makes it difficult to imagine a meaningful analogue of $0^{\#}$ for HOD.

Besides evidence in favor of this conjecture about HOD, we also have applications. Recall that Kunen proved in ZFC that there is no non-trivial elementary embedding from $V$ to itself. It is a longstanding open question whether this is a theorem of ZF alone. One of our applications is progress on this problem. This and other applications will be listed in Section 7.

## 2 Generic absoluteness

In this section, we establish some basic properties of forcing and HOD, and use them to show that the conjecture about HOD from the previous section is absolute to generic extensions. In other words, if $\mathbb{P}$ is a poset, then clause (2) of Theorem 1.2 holds in $V$ iff it holds in every generic extension by $\mathbb{P}$.

First observe that if $\mathbb{P}$ is a weakly homogeneous (see [1] Theorem 26.12) and ordinal definable poset in $V$, and $G$ is a $V$-generic filter on $\mathbb{P}$, then $\mathrm{HOD}^{V[G]} \subseteq \mathrm{HOD}^{V}$. This is immediate from the basic fact about weakly homogeneous forcing that for all $x_1, \ldots, x_n \in V$ and formula $\varphi(v_1, \ldots, v_n)$, every condition in $\mathbb{P}$ decides $\varphi(\check{x}_1, \ldots, \check{x}_n)$ the same way. We also use here that a class model of ZFC can be identified solely from its sets of ordinals, since each level of its $V$ hierarchy can, using the Axiom of Choice, be encoded by a relation on $|V_\alpha|$ and then recovered by collapsing. We shall use this fact repeatedly.

Let us pause to give an example of the phenomenon we just mentioned in which HOD of the generic extension is properly contained in HOD of the ground model. Let $\mathbb{P}$ be Cohen forcing and $g : \omega \to \omega$ be a Cohen real over $L$. Of course, $g \notin L$. In $L[g]$, let $\mathbb{Q}$ be the Easton poset that forces

$$2^{\omega_n} = \begin{cases} \omega_{n+1} & g(n) = 0 \\ \omega_{n+2} & g(n) = 1. \end{cases}$$

Both $\mathbb{P}$ and $\mathbb{Q}$ are cardinal preserving. Now let $H$ be an $L[g]$-generic filter on $\mathbb{Q}$. Observe that $g \in \mathrm{HOD}^{L[g][H]}$ because it can be read off from $\kappa \mapsto 2^{\kappa}$ in $L[g][H]$. Now let $\lambda$ be a regular cardinal greater than

$|\mathbb{P} * \mathbb{Q}|$. Then $\mathbb{P} * \mathbb{Q} * \text{Coll}(\omega, \lambda)$ and $\text{Coll}(\omega, \lambda)$ have isomorphic Boolean completions, so there exist an $L$-generic filter $J$ on $\text{Coll}(\omega, \lambda)$ and an $L[g][H]$-generic filter $I$ on $\text{Coll}(\omega, \lambda)$ such that $L[J] = L[g][H][I]$. Using the fact that $\text{Coll}(\omega, \lambda)$ is definable and weakly homogeneous we see that

$$L = \text{HOD}^{L[J]} = \text{HOD}^{L[g][H][I]} \subsetneq \text{HOD}^{L[g][H]}$$

where the inequality is witnessed by the Cohen real $g$.

An important fact about forcing which was discovered relatively recently is that if $\delta$ is a regular uncountable cardinal and $\mathbb{P} \in V_\delta$ is a poset, then $V$ is definable from $\mathcal{P}(\delta) \cap V$ in $V[G]$. Towards the precise statement and proof, we make the following definitions.

**Definition 2.1** Let $\delta$ be a regular uncountable cardinal and $N$ be a transitive class model of ZFC. Then

- $N$ has the *$\delta$-covering property* iff for every $\sigma \subseteq N$ with $|\sigma| < \delta$, there exists $\tau \in N$ such that $|\tau| < \delta$ and $\tau \supseteq \sigma$, and
- $N$ has the *$\delta$-approximation property* iff for every cardinal $\kappa$ with $\text{cf}(\kappa) \geq \delta$ and every $\subseteq$-increasing sequence of sets $\langle \tau_\alpha \mid \alpha < \kappa \rangle$ from $N$, $\bigcup \tau_\alpha \in N$.

By Jensen's theorem, $L$ has the $\delta$-covering property in $V$ for every regular $\delta > \omega$ if $0^\#$ does not exist. Next, we show that $V$ has covering and approximation properties in its generic extensions.

**Lemma 2.2** *Let $\delta > \omega$ be regular and let $\mathbb{P}$ be a poset with $|\mathbb{P}| < \delta$. Then $V$ has $\delta$-covering and $\delta$-approximation in $V[G]$ whenever $G$ is a $V$-generic filter on $\mathbb{P}$.*

*Proof* First, we show the covering property. Let $\sigma$ be a name such that $\Vdash \sigma \subset V$ and $|\sigma| < \delta$. By the $\delta$ chain condition, there are fewer than $\delta$ possible values of $|\sigma|$. Let $\gamma < \delta$ be the supremum of these and pick $\dot{f}$ such that $\Vdash \dot{f} : \gamma \twoheadrightarrow \sigma$. To finish this part of the proof, let $\tau$ be the set of possible values for $\dot{f}(\alpha)$ where $\alpha < \gamma$.

Second, we prove the approximation property. Let $p$ force that $\text{cf}(\kappa) \geq \delta$ and $\langle \tau_\alpha \mid \alpha < \kappa \rangle$ is an increasing sequence of sets from $V$. For $\alpha < \kappa$, let $p_\alpha$ decide the value of $\tau_\alpha$. Because $|\mathbb{P}| < \delta \leq \text{cf}(\kappa) \leq \kappa$ there must be some $p_\beta$ that is repeated cofinally often and so determines $\bigcup \tau_\alpha$, thereby forcing the union to belong to $V$. By density, the union is forced to belong to $V$. □

The next theorem is the promised result on the definability of the ground model, which we state somewhat more generally. Part (1) is due to Hamkins and (2) to Laver and Woodin independently.

**Theorem 2.3** *Let $\delta$ be a regular uncountable cardinal. Suppose that $M$ and $N$ are transitive class models of ZFC that satisfy the $\delta$-covering and $\delta$-approximation properties, $\delta^+ = (\delta^+)^N = (\delta^+)^M$, and*

$$N \cap \mathcal{P}(\delta) = M \cap \mathcal{P}(\delta).$$

*(1) Then $M = N$.*
*(2) In particular, $N$ is $\Sigma_2$-definable from $N \cap \mathcal{P}(\delta)$.*

*Proof* For part (1) we show by recursion on ordinals $\gamma$ that for all $A \subseteq \gamma$

$$A \in M \iff A \in N.$$

The case $\gamma \leq \delta$ is clear. By the induction hypothesis, $M$ and $N$ have the same cardinals $\leq \gamma$, and, if $\gamma$ is not a cardinal in these models, then they have the same power set of $\gamma$. Thus, we may assume that $\gamma$ is a cardinal of both $M$ and $N$.

**Case 1** $\mathrm{cf}(\gamma) \geq \delta$

Then, $A \in M$ iff $A \cap \alpha \in M$ for every $\alpha < \gamma$. The forward direction is clear. For the reverse, use the $\delta$-approximation property to see

$$A = \bigcup\{A \cap \alpha \mid \alpha < \gamma\} \in M.$$

The same holds for $N$.

**Case 2** $\gamma > \delta$, $\mathrm{cf}(\gamma) < \delta$ *and* $|A| < \delta$

Define increasing sequences $\langle E_\alpha \mid \alpha < \delta\rangle$ and $\langle F_\alpha \mid \alpha < \delta\rangle$ of subsets of $\gamma$ such that $|E_\alpha|, |F_\alpha| < \delta$, $A \subseteq E_0$, $E_\alpha \subseteq F_\alpha$, $\bigcup_{\alpha<\beta} F_\alpha \subseteq E_\beta$, $E_\alpha \in M$ and $F_\alpha \in N$. For the construction, use the $\delta$-covering property alternately for $M$ and $N$. Then define $E = \bigcup E_\alpha = \bigcup F_\alpha$ and note that $E \in M \cap N$ by $\delta$-approximation property. Let $\theta$ be the order-type of $E$ and $\pi : E \to \theta$ the Mostowski collapse. Then $\pi \in M \cap N$. Also, $\theta < \delta^+ = (\delta^+)^M = (\delta^+)^N$ because $|E| \leq \delta$. By the induction hypothesis,

$$A \in M \iff \pi[A] \in M \iff \pi[A] \in N \iff A \in N.$$

**Case 3** $\gamma > \delta$, $\mathrm{cf}(\gamma) < \delta$ *and* $|A| \geq \delta$

We claim that $A \in M$ iff

(i)$_M$ for every $\alpha < \gamma$, $A \cap \alpha \in M$ and
(ii)$_M$ for every $\sigma \subseteq \gamma$, if $|\sigma| < \delta$ and $\sigma \in M$, then $A \cap \sigma \in M$.

We also claim that $A \in N$ iff (i)$_N$ and (ii)$_N$. The induction hypothesis is that (i)$_M$ iff (i)$_N$ and in case (2) we showed that (ii)$_M$ iff (ii)$_N$, so our claim implies $A \in M$ iff $A \in N$ as desired.

The forward implication of the claim is obvious, so assume (i)$_M$ and (ii)$_M$. Pick $\theta$ with $\text{cf}(\theta) > \gamma$ and the defining formula for $M$ absolute to $V_\theta$. Define an increasing chain $\langle X_\alpha \mid \alpha < \delta \rangle$ of elementary substructures of $V_\theta$ and an increasing chain $\langle Y_\alpha \mid \alpha < \delta \rangle$ of subsets of $V_\theta \cap M$ such that $|X_\alpha|, |Y_\alpha| < \delta$, $A \in X_0$, $\sup(X_0 \cap \gamma) = \gamma$, $X_\alpha \cap N \subseteq Y_\alpha$, $Y_\alpha \in M$ and $\bigcup_{\alpha<\beta}(Y_\alpha \cup X_\alpha) \subseteq X_\beta$. We use Downward Lowenheim-Skolem to obtain $X_\alpha$ and the $\delta$-covering property to obtain $Y_\alpha$. Define $X = \bigcup X_\alpha$ and $Y = \bigcup Y_\alpha$. Then $X \prec V_\theta$ and $Y = X \cap M \prec V_\theta \cap M$. By the $\delta$-approximation property, $Y \in M$. By (ii)$_M$, for every $\alpha < \delta$, $A \cap Y_\alpha \in M$. Again, by the $\delta$-approximation property, $A \cap Y \in M$. Now consider an arbitrary $\alpha \in Y \cap \gamma$ and observe that

- $A \cap \alpha \in Y$ because $A \in X$ so $A \cap \alpha \in X$, and $A \cap \alpha \in M$ by (ii)$_M$; and
- for every $b \in Y$, if $b \cap Y = (A \cap Y) \cap \alpha$, then $Y \models b = A \cap \alpha$, so $b = A \cap \alpha$.

Here we have used (i)$_M$ and $Y \prec V_\theta \cap M$. So the sequence $\langle A \cap \alpha \mid \alpha \in Y \cap \gamma \rangle$ is definable in $M$ from parameters $\gamma$, $Y$ and $A \cap Y$. In particular, this function belongs to $M$. The union of its range is $A$, so $A \in M$.

Part (2) now follows. $A \in N$ iff there is a large regular $\theta$ and a model $M \subseteq V_\theta$ of ZFC-Power Set satisfying $\delta$-covering and $\delta$-approximation in $V_\theta$ such that $M \cap \mathcal{P}(\delta) = N \cap \mathcal{P}(\delta)$ and $A \in M$. This is a $\Sigma_2$ statement. □

We will use the following amazing result. The final equality is not as well known, so we include a proof. Note that $\text{OD}_{\mathcal{A}}$ here denotes the class of all sets that are definable using ordinals and members of $\mathcal{A}$, and $\text{HOD}_{\mathcal{A}}$ is defined correspondingly.

**Theorem 2.4** (Vopěnka) *For every ordinal $\kappa$, there exists $\mathbb{B} \in \text{HOD}$ such that*

$$\text{HOD} \models \mathbb{B} \textit{ is a complete Boolean algebra}$$

*and, for every $a \subseteq \kappa$, there exists a* HOD*-generic filter $G$ on $\mathbb{B}$ such that*

$$\text{HOD}[a] \subseteq \text{HOD}_{\{G\}} = \text{HOD}_{\{a\}} = \text{HOD}[G].$$

*Proof* First define $\mathbb{B}^*$ to be $\mathcal{P}(\mathcal{P}(\kappa)) \cap \mathrm{OD}$ with its Boolean algebra structure. Then $\mathbb{B}^* \in \mathrm{OD}$ and $\mathbb{B}^*$ is OD-complete. Given $a \subseteq \kappa$, we let

$$G^* = \{X \in \mathbb{B}^* \mid a \in X\}$$

and see that $G^*$ is an OD-generic filter on $\mathbb{B}^*$. Fix a definable bijection $\pi$ from $\mathcal{P}(\mathcal{P}(\kappa)) \cap \mathrm{OD}$ to an ordinal. Define $\mathbb{B}$ so that $\pi : \mathbb{B}^* \simeq \mathbb{B}$. Let $G = \pi[G^*]$. Then $G$ is a HOD-generic filter on $\mathbb{B}$. It is straightforward to see that $G \in \mathrm{HOD}_{\{a\}}$ so it remains to see that $\mathrm{HOD}_{\{a\}} \subseteq \mathrm{HOD}[G]$. Let $S \in \mathrm{HOD}_{\{a\}}$; we may assume $S$ is a set of ordinals. Say

$$S = \{\zeta < \theta \mid V_\theta \models \varphi(\zeta, \eta_1, \ldots, \eta_n, a)\}.$$

For each $\zeta < \theta$, let

$$X_\zeta = \{b \subseteq \kappa \mid V_\theta \models \varphi(\zeta, \eta_1, \ldots, \eta_n, b)\}.$$

Then $\zeta \mapsto \pi(X_\zeta)$ belongs to HOD. So $S = \{\zeta \mid \pi(X_\zeta) \in G\}$ belongs to $\mathrm{HOD}[G]$. □

Combining the results in this section, we obtain the following.

**Corollary 2.5** *Let $\mathbb{P} \in \mathrm{OD}$ be a weakly homogeneous poset. Suppose $G$ is a $V$-generic filter on $\mathbb{P}$. Then $\mathrm{HOD}^V$ is a generic extension of $\mathrm{HOD}^{V[G]}$.*

*Proof* Fix $\delta > |\mathbb{P}|$. By Lemma 2.2 and Theorem 2.3, $V$ is definable in $V[G]$ from $A = \mathcal{P}(\delta) \cap V$. In $V$, let $\kappa = |A|$ and $E$ be a binary relation on $\kappa$ such that the Mostowski collapse of $(\kappa, E)$ is $(\mathrm{trcl}(\{A\}, \in)$. Then $V_\gamma \in \mathrm{OD}^{V[G]}_{\{E\}}$ for every $\gamma$, therefore

$$\mathrm{HOD}^V \subseteq \mathrm{HOD}^{V[G]}_{\{E\}}.$$

By Theorem 2.4, we have a $\mathrm{HOD}^{V[G]}$-generic filter $H$ on a Vopěnka algebra so that

$$\mathrm{HOD}^{V[G]}_{\{E\}} = \mathrm{HOD}^{V[G]}[H].$$

Combining all of the above gives

$$\mathrm{HOD}^{V[G]} \subseteq \mathrm{HOD}^V \subseteq \mathrm{HOD}^{V[G]}_{\{E\}} = \mathrm{HOD}^{V[G]}[H].$$

As $\mathrm{HOD}^V$ is nested between $\mathrm{HOD}^{V[G]}$ and a generic extension thereof, it is itself a generic extension of $\mathrm{HOD}^{V[G]}$ (see [1] Theorem 15.43). □

Finally, we discuss again our conjecture that clause (2) of Theorem 1.2 fails. Let us temporarily call this the HOD conjecture although a slightly different statement will get this name later. We wish to see that this conjecture is absolute between $V$ and its generic extensions. Of course, Theorem 1.2 has an extendible cardinal $\delta$ in its hypothesis. We should assume that we are forcing with a poset $\mathbb{P} \in V_\delta$ to assure that if $G$ is a $V$-generic filter on $\mathbb{P}$, then $\delta$ remains extendible in $V[G]$.

To see this, given $\eta > \delta$ limit observe that for each member of $V_\eta^{V[G]}$ we can, by induction on $\eta$ build a name in $V_\eta$ for that member. This is done as usual for nice names by considering maximal antichains, taking advantage of the fact $\mathbb{P}$ is small with respect to $\eta$. Thus $V_\eta[G] = (V_\eta)^{V[G]}$. Now take $j : V_\eta \to V_\theta$ elementary and define $\tilde{j} : V_\eta^{V[G]} \to V_\theta^{V[G]}$ by $j(\tau_G) = j(\tau)_G$. This is a variation on the proof that measurability is preserved by small forcing; see [1] Theorem 21.2 or [3] Theorem 3.

**Corollary 2.6** *The following statement is absolute between $V$ and its generic extensions by posets in $V_\delta$: "$\delta$ is an extendible cardinal and for every singular cardinal $\gamma > \delta$, $\gamma$ is singular in* HOD *and* $(\gamma^+)^{\text{HOD}} = \gamma^+$."

*Proof* If $\mathbb{P}$ is ordinal definable and weakly homogeneous, then it is clear from Corollary 2.5 that the HOD conjecture is absolute between $V$ and $V[G]$. Now consider the general case. Take $\kappa < \delta$ an inaccessible cardinal such that $\mathbb{P} \in V_\kappa$. Let $J$ be a $V[G]$-generic filter on $\text{Coll}(\omega, \kappa)$ and $I$ be a $V$-generic filter on $\text{Coll}(\omega, \kappa)$ such that $V[G][I] = V[J]$. Now $\text{Coll}(\omega, \kappa)$ is ordinal definable and weakly homogeneous so the HOD conjecture is absolute between $V$ and $V[J]$, as well as between $V[G]$ and $V[G][I]$. Therefore, it is absolute between $V$ and $V[G]$. □

## 3 The HOD Conjecture

The official HOD Conjecture is closely related to the conjecture we have been contemplating for two sections. Intuitively, it also says that HOD is not far from $V$, which will turn out to mean that they are close. The HOD Conjecture involves a new concept, which we define first.

**Definition 3.1** Let $\lambda$ be an uncountable regular cardinal. Then $\lambda$ is *$\omega$-strongly measurable in* HOD iff there is $\kappa < \lambda$ such that

1. $(2^\kappa)^{\text{HOD}} < \lambda$ and
2. there is no partition $\langle S_\alpha \mid \alpha < \kappa \rangle$ of $\text{cof}(\omega) \cap \lambda$ into stationary sets such that $\langle S_\alpha \mid \alpha < \kappa \rangle \in \text{HOD}$.

**Lemma 3.2** *Let $\lambda$ be $\omega$-strongly measurable in* HOD*. Then*

$$\text{HOD} \models \lambda \textit{ is a measurable cardinal.}$$

*Proof* We claim that there exists a stationary set $S \subsetneq \mathrm{cof}(\omega) \cap \lambda$ such that $S \in \text{HOD}$ and there is no partition of $S$ into two stationary sets that belong to HOD.

First, let us see how to finish proving the lemma based on the claim. Let $\mathcal{F}$ be the club filter restricted to $S$. That is,

$$\mathcal{F} = \{X \subseteq S \mid \text{there is a club } C \text{ such that } X \supseteq C \cap S\}.$$

Let $\mathcal{G} = \mathcal{F} \cap \text{HOD}$. Clearly, $\mathcal{G} \in \text{HOD}$ and

$$\text{HOD} \models \mathcal{G} \text{ is a } \lambda\text{-complete filter on } \mathcal{P}(S).$$

By the claim,

$$\text{HOD} \models \mathcal{G} \text{ is an ultrafilter on } \mathcal{P}(S).$$

Now we prove the claim by contradiction. Fix a cardinal $\kappa < \lambda$ such that $(2^\kappa)^{\text{HOD}} < \lambda$ and there is no partition $\langle S_\alpha \mid \alpha < \kappa\rangle$ of $\mathrm{cof}(\omega) \cap \lambda$ into stationary sets such that $\langle S_\alpha \mid \alpha < \kappa\rangle \in \text{HOD}$. This allows us to define a subtree $T$ of ${}^{\leq\kappa}2$ with height $\kappa + 1$ and a sequence $\langle S_r \mid r \in T\rangle$ that belongs to HOD such that

1. $S_{\langle\rangle} = \mathrm{cof}(\omega) \cap \lambda$,
2. For every $r \in T$,
   a. $S_r$ is stationary,
   b. $r^\frown\langle 0\rangle$ and $r^\frown\langle 1\rangle$ belong to $T$,
   c. $S_r$ is the disjoint union of $S_{r^\frown\langle 0\rangle}$ and $S_{r^\frown\langle 1\rangle}$, and
   d. if $\mathrm{dom}(r)$ is a limit ordinal, then $S_r = \bigcap\{S_{r\restriction\alpha} \mid \alpha \in \mathrm{dom}(r)\}$.
3. For every limit ordinal $\beta \leq \kappa$ and $r \in {}^{\beta}2 - T$, if $r \restriction \alpha \in T$ for every $\alpha < \beta$, then $\bigcap_{\alpha<\beta} S_{r\restriction\alpha}$ is non-stationary.

First notice that $\mathrm{cof}(\omega) \cap \lambda$ belongs to HOD even though it might mean something else there. Also, $\{S \subsetneq \lambda \mid S \in \text{HOD} \text{ and } S \text{ is stationary}\}$ belongs to HOD even through there might be sets which are stationary in HOD but not actually stationary. In any case, HOD can recognise when a given $S \in \text{HOD}$ is stationary in $V$ and, by the putative failure of the claim, choose a partition of $S$ into two sets which are again stationary in $V$. This choice is done in a uniform way using a wellordering of

$$\{S \subsetneq \lambda \mid S \in \text{HOD} \text{ and } S \text{ is stationary}\}$$

in HOD. This gets us through successor stages of the construction. Suppose that $\beta \le \kappa$ is a limit ordinal and that we have already constructed in HOD $\langle S_r \mid r \in T \cap {}^{<\beta}2\rangle$. By (3) we have recursively maintained that, except for a non-stationary set, $\mathrm{cof}(\omega) \cap \lambda$ equals

$$\bigcup \Big\{ \bigcap \{S_{r\restriction\alpha} \mid \alpha < \beta\} \mid r \text{ is a } \beta\text{-branch of } T \cap {}^{<\beta}2 \text{ and } r \in \mathrm{HOD} \Big\}.$$

Since the club filter over $\lambda$ is $\lambda$-complete and $|{}^{\beta}2|^{\mathrm{HOD}} < \lambda$, there exists at least one such $r$ for which the corresponding intersection is stationary. We put $r \in T \cap {}^{\beta}2$ and define $S_r = \bigcap \{S_{r\restriction\alpha} \mid \alpha < \beta\}$ in this case. That completes the construction. Now take any $r \in T$ with $\mathrm{dom}(r) = \kappa$. Then $S_r$ is the disjoint union of the stationary sets $S_{r\restriction(\alpha+1)} - S_{r\restriction\alpha}$ for $\alpha < \kappa$. This readily contradicts our choice of $\kappa$. □

**Definition 3.3** The HOD *Conjecture* is the statement:
There is a proper class of regular cardinals that are not $\omega$-strongly measurable in HOD.

It turns out that if $\delta$ is an extendible cardinal, then the HOD Conjecture is equivalent to the failure of clause (2) of the dichotomy, Theorem 1.2, which is the conjecture we discussed in the previous two sections. In particular, a model in which the HOD conjecture fails cannot be obtained by forcing. It is clear that if HOD is correct about singular cardinals and computes their successors correctly (clause (1) of Theorem 1.2) then the HOD Conjecture holds, as

$$\{\gamma^+ \mid \gamma \in \mathrm{On} \text{ and } \gamma \text{ is a singular cardinal}\}$$

is a proper class of regular cardinals which are not $\omega$-strongly measurable in HOD.

We close this section with additional remarks on the status of the HOD Conjecture.

(i) It is not known whether more than 3 regular cardinals which are $\omega$-strongly measurable in HOD can exist.
(ii) Suppose $\gamma$ is a singular cardinal, $\mathrm{cof}(\gamma) > \omega$ and $|V_\gamma| = \gamma$. It is not known whether $\gamma^+$ can be $\omega$-strongly measurable in HOD.
(iii) Let $\delta$ be a supercompact cardinal. It is not known whether any regular cardinal above $\delta$ can be $\omega$-strongly measurable in HOD.

# 4 Supercompactness

Recall that a cardinal $\delta$ is $\gamma$-supercompact iff there is a transitive class $M$ and an elementary embedding $j : V \to M$ such that $\mathrm{crit}(j) = \delta$, $j(\delta) > \gamma$ and ${}^{\gamma}M \subseteq M$. Also, $\delta$ is a supercompact cardinal iff $\delta$ is $\gamma$-supercompact cardinal for every $\gamma > \delta$. If $\delta$ is an extendible cardinal, then $\delta$ is supercompact and $\{\alpha < \delta \mid \alpha \text{ is supercompact}\}$ is stationary in $\delta$ (see [2] Theorem 23.7).

There is a standard first-order way to express supercompactness in terms of measures, which we review. First suppose that $j : V \to M$ witnesses that $\delta$ is a $\gamma$-supercompact. Observe that $j[\gamma] \in M$. If we define

$$\mathcal{U} = \{X \subseteq \mathcal{P}_\delta(\gamma) \mid j[\gamma] \in j(X)\},$$

then $\mathcal{U}$ is a $\delta$-complete ultrafilter on $\mathcal{P}_\delta(\gamma)$. Moreover, $\mathcal{U}$ is *normal* in the sense that if $X \in \mathcal{U}$ and $f$ is a choice function for $X$, then there exists $Y \in \mathcal{U}$ and $\alpha < \gamma$ such that $Y \subseteq X$ and $f(\sigma) = \alpha$ for every $\sigma \in Y$. Equivalently, if $\langle X_\alpha \mid \alpha < \gamma\rangle$ is a sequence of sets from $\mathcal{U}$, then the diagonal intersection,

$$\underset{\alpha<\gamma}{\Delta} X_\alpha = \{\sigma \in \mathcal{P}_\delta(\gamma) \mid \sigma \in X_\alpha \text{ for every } \alpha \in \sigma\}$$

also belongs to $\mathcal{U}$. In addition, $\mathcal{U}$ is *fine* in the sense that for every $\alpha < \gamma$,

$$\{\sigma \in \mathcal{P}_\delta(\gamma) \mid \alpha \in \sigma\} \in \mathcal{U}.$$

Suppose, instead, that we are given a $\delta$-complete ultrafilter $\mathcal{U}$ on $\mathcal{P}_\delta(\gamma)$ which is both normal and fine. Then the ultrapower map derived from $\mathcal{U}$ can be shown to witness that $\delta$ is a $\gamma$-supercompact cardinal. We might refer to such an ultrafilter (fine, normal and $\delta$-complete) as a *$\gamma$-supercompactness measure.*

Less well-known is the following characterisation of $\delta$ being supercompact that is more transparently related to extendibility.

**Theorem 4.1** (Magidor) *A cardinal $\delta$ is supercompact iff for all $\kappa > \delta$ and $a \in V_\kappa$, there exist $\bar{\delta} < \bar{\kappa} < \delta$, $\bar{a} \in V_{\bar{\kappa}}$ and an elementary embedding $j : V_{\bar{\kappa}+1} \to V_{\kappa+1}$ such that* $\mathrm{crit}(j) = \bar{\delta}$, *$j(\bar{\delta}) = \delta$ and $j(\bar{a}) = a$.*

*Proof* First we prove the forward direction. Given $\kappa$ and $a$, let $\gamma = |V_{\kappa+1}|$ and $j : V \to M$ witness that $\delta$ is a $\gamma$-supercompact cardinal. Then

$$j \restriction V_{\kappa+1} \in M$$

and witnesses the following sentence in $M$: "There exist $\bar{\delta} < \bar{\kappa} < j(\delta)$, $\bar{a} \in V_{\bar{\kappa}}$ and an elementary embedding $i : V_{\bar{\kappa}+1} \to V_{j(\kappa)+1}$ such that $\mathrm{crit}(i) = \bar{\delta}$, $i(\bar{\delta}) = j(\delta)$ and $i(\bar{a}) = j(a)$." Since $j$ is elementary, we are done.

For the reverse direction, let $\gamma > \delta$ be given. Apply the right side with $\kappa = \gamma + \omega$. (The choice of $a$ is irrelevant.) This yields $\bar{\kappa}$, $\bar{\delta}$ and $j$ as specified. Take $\bar{\gamma}$ such that $j(\bar{\gamma}) = \gamma$. Now $j[\bar{\gamma}] \in V_{\kappa+1}$ so it induces a normal fine ultrafilter $\bar{\mathcal{U}}$ on $\mathcal{P}_{\bar{\delta}}(\bar{\gamma})$. Observe that $\bar{\mathcal{U}} \in V_{\bar{\gamma}+\omega}$, so we can define $\mathcal{U} = j(\bar{\mathcal{U}})$. Then, by elementarity, $V_{\kappa+1}$ believes that $\mathcal{U}$ is a normal fine ultrafilter on $\mathcal{P}_\delta(\gamma)$, and is large enough to bear true witness to such a belief. Thus $\delta$ is $\gamma$-supercompact. □

We will use the Solovay splitting theorem. A proof can be found in [1] Theorem 8.10 or, using generic embeddings, in [1] Lemma 22.27.

**Theorem 4.2** (Solovay) *Let $\gamma$ be a regular uncountable cardinal. Then every stationary subset of $\gamma$ can be partitioned into $\gamma$ many stationary sets.*

We will also make key use of the following theorem, which provides a single set that belongs to every $\gamma$-supercompactness measure on $\mathcal{P}_\delta(\gamma)$. We will refer to this set as the *Solovay set.*

**Theorem 4.3** (Solovay) *Let $\delta$ be supercompact and $\gamma > \delta$ be regular. Then there exists an $X \subseteq \mathcal{P}_\delta(\gamma)$ such that the* sup *function is injective on $X$ and every $\gamma$-supercompactness measure contains $X$.*

*Proof* Let $\langle S_\alpha \mid \alpha < \gamma \rangle$ be a partition of $\gamma \cap \mathrm{cof}(\omega)$ into stationary sets, which exists by Theorem 4.2. For $\beta < \gamma$ such that $\omega < \mathrm{cf}(\beta) < \delta$, let $\sigma_\beta$ be the set of $\alpha < \beta$ such that $S_\alpha$ reflects to $\beta$. In other words,

$$\sigma_\beta = \{\alpha < \beta \mid S_\alpha \cap \beta \text{ is stationary in } \beta\}.$$

Leave $\sigma_\beta$ undefined otherwise. Note that it is not possible to partition $\beta$ into more that $\mathrm{cf}(\beta)$-many stationary sets, as can be seen by considering their restrictions to a club in $\beta$ of order type $\mathrm{cf}(\beta)$, so, $\sigma_\beta \in \mathcal{P}_\delta(\gamma)$. Define $X = \{\sigma_\beta \mid \sup(\sigma_\beta) = \beta\}$. Clearly, the sup function is an injection on $X$ so given $\mathcal{U}$ be a normal fine ultrafilter on $\mathcal{P}_\delta(\gamma)$ it remains to see that $X \in \mathcal{U}$. Let $j : V \to M$ be the embedding associated to $\mathcal{U}$. In fact $\mathcal{U}$ is the corresponding ultrafilter derived from $j$, so what we need to see is that

$$j[\gamma] \in j(X).$$

Let $\beta = \sup(j[\gamma])$ and

$$\langle S^*_\alpha \mid \alpha < j(\gamma)\rangle = j(\langle S_\alpha \mid \alpha < \gamma\rangle).$$

Clearly, $\beta < j(\gamma)$ and $\omega < \mathrm{cf}(\beta) < j(\delta)$, so we are left to show that

$$j[\gamma] = \{\alpha < \beta \mid M \models S^*_\alpha \cap \beta \text{ is stationary in } \beta\}.$$

First we show containment in the forward direction. Consider any $\eta < \gamma$. Then we want $S^*_{j(\eta)} = j(S_\eta)$ to be stationary. Given $C$ a club subset of $\beta$ that belongs to $M$, define $D = \{\alpha < \gamma \mid j(\alpha) \in C\}$. Because $j$ is continuous at ordinals of countable cofinality, $D$ is an $\omega$-club in $\gamma$. But $S_\eta$ contains only ordinals of countable cofinality and is stationary in $\gamma$ so $S_\eta \cap D \neq \emptyset$. Hence $j(S_\eta) \cap C \neq \emptyset$.

For containment in the reverse direction, consider any $\alpha < \beta$ such that, in $M$, $S^*_\alpha \cap \beta$ is stationary in $\beta$. Working in $M$, as $j[\gamma]$ is an $\omega$-club in $\beta$ and $S^*_\alpha$ contains only ordinals of countable cofinality, there exists $\eta < \gamma$ such that $j(\eta) \in S^*_\alpha$. But $j[\gamma]$ is partitioned by the $j[S_\theta]$ for $\theta < \gamma$ so we can take $\theta < \gamma$ such that $j(\eta) \in j[S_\theta] \subseteq j(S_\theta) = S^*_{j(\theta)}$. This means $S^*_\alpha \cap S^*_{j(\theta)} \neq \emptyset$ so $\alpha = j(\theta) \in j[\gamma]$. □

*Remark* 4.4 The proof of Theorem 4.3 can be easily generalised to prove the following. Assume that $j : V \to M$ is a $\gamma$-supercompact embedding, where $\gamma$ is regular. Let $\kappa < \gamma$ be also regular, $\beta = \sup j[\gamma]$ and $\tilde{\beta} = \sup j[\kappa]$. Then given a partition $\langle S_\alpha \mid \alpha < \kappa\rangle$ of $\mathrm{cof}(\omega) \cap \gamma$ into stationary sets, we have that

$$j[\kappa] = \{\alpha \in \tilde{\beta} \mid S^*_\alpha \cap \beta \text{ is stationary in } \beta\},$$

where $\langle S^*_\alpha \mid \alpha < j(\kappa)\rangle = j(\langle S_\alpha \mid \alpha < \kappa\rangle)$.

## 5 Weak extender models

In inner model theory, the word *extender* has taken on a very general meaning as any object that captures the essence of a given large cardinal property. Sometimes ultrafilters or systems of ultrafilters are used. At other times, elementary embeddings or restrictions of elementary embeddings are more relevant. We have already seen two first-order ways to express supercompactness. An easier example is measurability: if $U$ is a normal measure on $\kappa$ and $j : V \to M$ is the corresponding ultrapower map, then $U$ and $j \upharpoonright V_{\kappa+1}$ carry exactly the same information.

Building a canonical inner model with a supercompact cardinal has been a major open problem in set theory for decades. Canonical inner models for measurable cardinals were produced early on. Letting $U$ be a normal measure on $\kappa$ and setting $\bar{U} = U \cap L[U]$, we can see that $\bar{U} \in L[U]$ and $L[U] \models \bar{U}$ is a normal measure on $\kappa$. The general theory of $L[U]$ does not depend on there being measurable cardinals in $V$ but this was an important first step.

**Definition 5.1** A transitive class $N$ model of ZFC is called a *weak extender model for $\delta$ supercompact* iff for every $\gamma > \delta$ there exists a normal fine measure $\mathcal{U}$ on $\mathcal{P}_\delta(\gamma)$ such that

1. $N \cap \mathcal{P}_\delta(\gamma) \in \mathcal{U}$ and
2. $\mathcal{U} \cap N \in N$.

The first condition says that $\mathcal{U}$ *concentrates* on $N$. In the case of the measurable cardinal, which we discussed above, we get the analogous first condition for free because $L[U] \cap \kappa = \kappa \in U$. We might refer to the second condition as saying that $\mathcal{U}$ is *amenable* to $N$.

**Lemma 5.2** *If $N$ is a weak extender model for $\delta$ supercompact, then it has the $\delta$-covering property.*

*Proof* Note that it is enough to prove $\delta$-covering for sets of ordinals. Now, given $\tau \subseteq \gamma$ with $|\tau| < \delta$, let $\mathcal{U}$ be a $\gamma$-supercompactness measure such that $N \cap \mathcal{P}_\delta(\gamma) \in \mathcal{U}$ and $\mathcal{U} \cap N \in N$. By fineness, for each $\alpha < \gamma$, we have that $\{\sigma \in \mathcal{P}_\delta(\gamma) \mid \alpha \in \sigma\} \in \mathcal{U}$. Hence, as $|\tau| < \delta$, by $\delta$-completeness we have $\{\sigma \in \mathcal{P}_\delta(\gamma) \mid \tau \subseteq \sigma\}$ belongs to $\mathcal{U}$. Also as $N \cap \mathcal{P}_\delta(\gamma) \in \mathcal{U}$, there is a $\sigma \in N \cap \mathcal{P}_\delta(\gamma)$ and $\sigma \supseteq \tau$ as desired. □

**Lemma 5.3** *Suppose $N$ is a weak extender model for $\delta$ supercompact and $\gamma > \delta$ is such that $N \models$ '$\gamma$ is a regular cardinal'. Then $|\gamma| = \mathrm{cf}(\gamma)$.*

*Proof* Let $\gamma > \delta$. Of course, $\mathrm{cf}(\gamma) \leq |\gamma|$. Now we prove the reverse inequality. By Lemma 5.2, $N$ satisfies the $\delta$-covering property so, as $N$ believes $\gamma$ is a regular cardinal, we have that $\mathrm{cf}(\gamma) \geq \delta$. Now fix $\mathcal{U}$ a $\gamma$-supercompactness measure, such that $N \cap \mathcal{P}_\delta(\gamma) \in \mathcal{U}$ and $\mathcal{U} \cap N \in N$. As $\gamma$ is a regular cardinal of $N$, we may apply Theorem 4.3 within $N$ and get a Solovay set $X \in N$. So the sup function is an injection on $X$ and $X$ belongs to $\mathcal{U}$. Now fix a club $D \subseteq \gamma$ of order type $\mathrm{cf}(\gamma)$ and define $A = \{\sigma \in \mathcal{P}_\delta(\gamma) \mid \sup(\sigma) \in D\}$.

We first claim that $A \in \mathcal{U}$. Letting $j : V \to M$ be the ultrapower map induced by $\mathcal{U}$, it is enough to show that $j[\gamma] \in j(A)$. Define $\beta = \sup j[\gamma]$. By the definition of $A$, we need to see that $\beta \in j(D)$. Note that $j(D)$ is a club in $j(\gamma)$, and as $D$ is unbounded in $\gamma$ we have that $j[\gamma] \cap j(D)$ is unbounded in $\beta$. Thus $j(D)$ being closed implies $\beta \in j(D)$. Hence $\{\sigma \in X \mid \sup(\sigma) \in D\} \in \mathcal{U}$. Recall that $\mathcal{U}$ is fine, so

$$\gamma = \bigcup \{\sigma \in X | \sup(\sigma) \in D\}.$$

Now, because the sup function is injective on $X$, we have that the cardinality of $\gamma$ is at most $\delta\,|D|$. But the order type of $D$ is $\mathrm{cf}(\gamma)$, so $|\gamma| \leq \delta\ \mathrm{cf}(\gamma)$. Finally remember $\delta \leq \mathrm{cf}(\gamma)$, so $|\gamma| \leq \mathrm{cf}(\gamma)$ which concludes the proof. □

**Corollary 5.4** *Let $N$ be a weak extender model for $\delta$ supercompact and $\gamma > \delta$ be a singular cardinal, then*

1. *$N \models$ '$\gamma$ is singular' and*
2. *$\gamma^+ = (\gamma^+)^N$.*

*Proof* Immediate by Lemma 5.3. □

Next, we characterise the HOD Conjecture in two ways, each of which says HOD is close to $V$ in a certain sense.

**Theorem 5.5** *Let $\delta$ be an extendible cardinal. The following are equivalent.*

1. *The* HOD *Conjecture.*
2. HOD *is a weak extender model for $\delta$ supercompact.*
3. *Every singular cardinal $\gamma > \delta$, is singular in* HOD *and $\gamma^+ = (\gamma^+)^{\mathrm{HOD}}$.*

*Proof* (2) implies (3) is just Lemma 5.4. That (3) implies (1) was shown in the discussion right after the definition of the HOD Conjecture (Definition 3.3). We now prove (1) implies (2).

Given $\zeta > \delta$, we wish to show that there is a $\zeta$-supercompactness measure $\mathcal{U}$ such that $\mathcal{U} \cap \mathrm{HOD} \in \mathrm{HOD}$ and $\mathcal{P}_\delta(\gamma) \cap \mathrm{HOD} \in \mathcal{U}$. For this, take $\gamma > 2^\zeta$, such that $|V_\gamma|^{\mathrm{HOD}} = \gamma$ and fix a regular cardinal $\lambda > 2^\gamma$ such that $\lambda$ is not $\omega$-strongly measurable in HOD. Finally, pick $\eta > \lambda$ such that the defining formula for HOD is absolute for $V_\eta$, whence $\mathrm{HOD}^{V_\eta} = \mathrm{HOD} \cap V_\eta$. As $\delta$ is extendible, there is an elementary embedding $j : V_{\eta+1} \to V_{j(\eta)+1}$ with critical point $\delta$.

*Claim* 5.6 $j[\gamma] \in \mathrm{HOD}^{V_{j(\eta)}}$.

As $\lambda$ is not $\omega$-strongly measurable in HOD and $2^\gamma < \lambda$ (in $V$ and so in HOD) there is a partition $\langle S_\alpha \mid \alpha \in \gamma \rangle$ of $\mathrm{cof}(\omega) \cap \lambda$ into stationary sets such that $\langle S_\alpha \mid \alpha < \gamma\rangle \in \mathrm{HOD}$. Thus $\langle S_\alpha \mid \alpha \in \gamma\rangle \in \mathrm{HOD}^{V_\eta}$. By the elementarity of $j$ we have

$$\langle S^*_\alpha \mid \alpha \in j(\gamma)\rangle = j(\langle S_\alpha \mid \alpha \in \gamma\rangle) \in \mathrm{HOD}^{V_{j(\eta)}}.$$

Let $\beta = \sup j[\lambda]$ and $\tilde{\beta} = \sup j[\gamma]$. By the remark after the proof of Theorem 4.3,

$$j[\gamma] = \{\alpha \in \tilde{\beta} \mid S^*_\alpha \cap \beta \text{ is stationary in } \beta\}$$

This shows that $j[\gamma]$ is OD in $V_{j(\eta)}$. Moreover $V_{j(\eta)}$ is correct about stationarity in $\beta$, thus $j[\gamma] \in \mathrm{HOD}^{V_{j(\eta)}}$. Also note that $j[\zeta] \in \mathrm{HOD}^{V_{j(\eta)}}$.

Now, observe that $\mathrm{HOD}^{V_{j(\eta)}} \subset \mathrm{HOD}$, so we have that $j[\gamma] \in \mathrm{HOD}$. Also $|V_\gamma|^{\mathrm{HOD}} = \gamma$, so we may take $e \in \mathrm{HOD}$ a bijection from $\gamma$ to $V_\gamma^{\mathrm{HOD}}$. Clearly $j(e)[j[\gamma]] = j[V_\gamma \cap \mathrm{HOD}]$ and so $j[V_\gamma \cap \mathrm{HOD}] \in \mathrm{HOD}$. Furthermore, as

$$j \restriction (V_\gamma \cap \mathrm{HOD})$$

is the inverse of the Mostowski collapse, we have that

$$j \restriction (V_\gamma \cap \mathrm{HOD}) \in \mathrm{HOD}.$$

Now, let $\mathcal{U}$ be the ultrafilter on $\mathcal{P}_\delta(\zeta)$ derived from $j$. That is, for $A \subseteq \mathcal{P}_\delta(\zeta)$, $A \in \mathcal{U}$ iff $j[\zeta] \in j(A)$. So,

$$\mathcal{P}_\delta(\zeta) \cap \mathrm{HOD} \in \mathcal{U} \text{ as } j[\zeta] \in \mathrm{HOD}^{V_{j(\eta)}} = j(\mathrm{HOD} \cap V_\eta)$$

$$\mathcal{U} \cap \mathrm{HOD} \in \mathrm{HOD} \text{ as } j \restriction (V_\gamma \cap \mathrm{HOD}) \in \mathrm{HOD} \text{ and } \gamma > 2^\zeta.$$

Thus $\mathcal{U}$ concentrates on HOD and is amenable to HOD as desired. □

As a corollary, we obtain the following version of the HOD Dichotomy, Theorem 1.2.

**Corollary 5.7** *Let $\delta$ be an extendible cardinal. Then exactly one of then following holds.*

1. *For every singular cardinal $\gamma > \delta$, $\gamma$ is singular in* HOD *and* $\gamma^+ = (\gamma^+)^{\mathrm{HOD}}$.
2. *There exists a $\kappa > \delta$ such that every regular $\gamma > \kappa$ is measurable in* HOD.

*Proof* Suppose (2) does not hold, then there are arbitrarily large regular cardinals that are not measurable in HOD. By Lemma 3.2 there are arbitrarily large regular cardinals that are not $\omega$-strongly measurable in HOD. Now by the proof of Theorem 5.5, this implies that HOD is a weak extender model for $\delta$ supercompact. Finally Corollary 5.4 yields (1). □

## 6 Elementary embeddings of weak extender models

We now give more evidence that if $N$ is a weak extender model for $\delta$ supercompact then it is close to $V$. We will prove that if $\delta$ an extendible cardinal, $N$ is a weak extender model for $\delta$ supercompact and $j$ is an elementary embedding between levels of $N$ with $\mathrm{crit}(j) \geq \delta$, then $j \in N$. This implies that if $\delta$ is extendible and the HOD Conjecture holds then there are no elementary embeddings from HOD to HOD with critical point greater or equal $\delta$. This says that a natural analog of $0^\#$ for HOD does not exist. As one would expect from Magidor's characterisation of supercompactness, Theorem 4.1, there is an alternative formulation of "weak extender model for $\delta$ supercompact" in terms of suitable elementary embeddings $j : V_{\bar{\kappa}+1} \to V_{\kappa+1}$ for $\bar{\kappa} < \delta$.

**Theorem 6.1** *Let N be a proper class model of ZFC. Then the following are equivalent:*

1. *N is a weak extender model for $\delta$ supercompact.*
2. *For every $\kappa > \delta$ and $b \in V_\kappa$, there exist two cardinals $\bar{\kappa}$ and $\bar{\delta}$ below $\delta$, $\bar{b} \in V_{\bar{\kappa}}$ and $j : V_{\bar{\kappa}+1} \to V_{\kappa+1}$ such that:*

$$\mathit{crit}(j) = \bar{\delta},\ j(\bar{\delta}) = \delta,\ j(\bar{b}) = b,$$

$$j(N \cap V_{\bar{\kappa}}) = N \cap V_\kappa \ \mathit{and}$$

$$j \restriction (V_{\bar{\kappa}} \cap N) \in N.$$

*Proof (2) implies (1)* Given $\gamma > \delta$, we may assume $\gamma = |V_\gamma|$. Let $\bar{\kappa} = \gamma + \omega$. We obtain $\bar{\kappa}$, $\bar{\delta}$ and $j$ using (2). Take $\bar{\gamma}$ such that $\bar{\kappa} = \bar{\gamma} + \omega$, whence $j(\bar{\gamma}) = \gamma$. Let $\bar{\mathcal{U}}$ be the measure on $\mathcal{P}_{\bar{\delta}}(\bar{\gamma})$ derived form $j$. That is, for $A \in \mathcal{P}_{\bar{\delta}}(\bar{\gamma})$

$$A \in \bar{\mathcal{U}} \iff j[\bar{\gamma}] \in j(A).$$

Define $\mathcal{U} = j(\bar{\mathcal{U}})$. We show that $\mathcal{U}$ is a $\gamma$-supercompactness measure such that $\mathcal{P}_\delta(\gamma) \cap N \in \mathcal{U}$ and $\mathcal{U} \cap N \in N$.

We claim that $P_{\bar{\delta}}(\bar{\gamma}) \cap N \in \bar{\mathcal{U}}$. By (2), we know that $j(N \cap V_{\bar{\kappa}}) = N \cap V_\kappa$. Thus for every $a \in V_{\bar{\kappa}}$ we have

$$j(a \cap N) = j(a) \cap j(N \cap V_{\bar{\kappa}}) = j(a) \cap N.$$

Recalling that $\bar{\kappa} = \bar{\gamma} + \omega$,

$$j\,(P_{\bar{\delta}}(\bar{\gamma}) \cap N) = \mathcal{P}_\delta(\gamma) \cap N.$$

Now, as $j \upharpoonright (N \cap V_{\bar{\kappa}}) \in N$, we have $j[\bar{\gamma}] \in N$, so $j[\bar{\gamma}] \in \mathcal{P}_\delta(\gamma) \cap N = j(P_{\bar{\delta}}(\bar{\gamma}) \cap N)$, which readily implies our claim.

Finally, by elementarity of $j$, we have that $\mathcal{U}$ is a fine and normal measure on $\mathcal{P}_\delta(\gamma)$ and, by the previous claim, $j(N \cap P_{\bar{\delta}}(\bar{\gamma})) \in \mathcal{U}$. It follows that

$$N \cap \mathcal{P}_\delta(\gamma) \in \mathcal{U}.$$

Moreover, as $j \upharpoonright (N \cap V_{\bar{\kappa}}) \in N$, we have that $j(\bar{\mathcal{U}} \cap N) \in N$. Hence

$$j(\bar{\mathcal{U}} \cap N) = j(\bar{\mathcal{U}}) \cap N = \mathcal{U} \cap N \in N.$$

This concludes the first direction. □

*Proof (1) implies (2)* Let $\kappa > \delta$, $b \in V_\kappa$ and fix $\gamma > |V_{\kappa+\omega}|$ such that $|V_\gamma|^N = \gamma$. Fix a $\gamma$-supercompactness measure $\mathcal{U}$ such that $\mathcal{P}_\delta(\gamma) \cap N \in \mathcal{U}$ and $\mathcal{U} \cap N \in N$. Now, fix a bijection $e : \gamma \to V_\gamma^N$ in $N$. We now work in $N$. Define $N_\sigma$ to be the Mostowski collapse of $e[\sigma]$, and

$$Y = \{\sigma \in N \cap \mathcal{P}_\delta(\gamma) \mid N_\sigma = V^N_{\mathrm{otp}(\sigma)}\}.$$

Hence $Y$ is a club of $N \cap \mathcal{P}_\delta(\gamma)$, so it belongs to $\mathcal{U} \cap N$. Thus if $j : V \to M$ is the ultrapower map it follows that $j[\gamma] \in j(Y)$. This implies from the definition of $Y$ that the collapse of $j(e)[j[\gamma]]$ is exactly $V_\gamma \cap j(N \cap V_\delta)$. Note also that $j(e)[j[\gamma]] = j[V_\gamma \cap N]$. Of course $V_\gamma \cap N$ is the collapse of $j[V_\gamma \cap N]$, so

$$V_\gamma \cap N = V_\gamma \cap j(N \cap V_\delta),$$

which implies

$$V_\kappa \cap N = V_\kappa \cap j(N \cap V_\delta).$$

It is clear that for $\sigma \in N$ we have that $e[\sigma] \in V_{\gamma+1} \cap N$. Then by Łos' Theorem we have that $j[V_\gamma \cap N] = j(e)[j[\gamma]] \in j(V_{\gamma+1} \cap N)$. Notice that the collapsing map of $j[V_\gamma \cap N]$ is just the inverse of $j \upharpoonright (V_\gamma \cap N)$, thus

$$j \upharpoonright (V_\kappa \cap N) \in j(V_{\gamma+1} \cap N).$$

Now $M$ being closed under $\gamma$ sequences and $\gamma > |V_{\kappa+\omega}|$ imply $j \upharpoonright (V_{\kappa+1})$ belongs to $M$. Working in $M$ let $i = j \upharpoonright (V_{\kappa+1})$. Now let us prove that the two previous equations imply that $i$ satisfy the conditions of *(2)* relative to $j(\kappa)$, $j(b)$ and $j(N \cap V_{\gamma+1})$ in $M$. Indeed the equations give

$$i \upharpoonright \left(V_\kappa \cap j(N \cap V_{\gamma+1})\right) = i \upharpoonright (V_\kappa \cap N) \in N \cap j(V_{\gamma+1}).$$

Furthermore as $i$ and $j$ agree,

$$i(V_\kappa \cap j(N \cap V_{\gamma+1})) = i(V_\kappa \cap N) = j(N \cap V_{\gamma+1}) \cap V_{j(\kappa)}.$$

Also $j(b) = i(b)$, so by elementarity (2) holds in $V$ with respect to $\kappa$, $b$ and $N$. □

We now prove that if $\delta$ is an extendible cardinal and $N$ is a weak extender model for $\delta$ supercompact, then $N$ sees all elementary embeddings between its levels.

**Theorem 6.2** *Let $\delta$ be an extendible cardinal. Assume that $N$ is a weak extender model for $\delta$ supercompact and $\gamma > \delta$ is a cardinal in $N$. Let*

$$j : H(\gamma^+)^N \to H(j(\gamma)^+)^N$$

*be an elementary embedding with $\delta \leq \mathrm{crit}(j)$ and $j \neq \mathrm{id}$. Then $j \in N$.*

*Proof* Define $b = (j, \gamma)$ and let $\kappa$ be a cardinal much larger than $j(\gamma)$. Now, as $N$ is a weak extender model for $\delta$ supercompact, we may apply Theorem 6.1 to $\kappa$ and $b$. Hence, we get an elementary embedding $\pi : V_{\bar{\kappa}+1} \to V_{\kappa+1}$, two ordinals $\bar{\delta}$, $\bar{\gamma}$ and $\bar{j} \in V_{\bar{\kappa}}$, with the following properties

$$j(N \cap V_{\bar{\kappa}}) = N \cap V_\kappa,\ \pi \restriction (V_{\bar{\kappa}} \cap N) \in N$$

and

$$\mathrm{crit}(\pi) = \bar{\delta},\ \pi(\bar{j}) = j,\ \pi(\bar{\delta}) = \delta,\ \pi(\bar{\gamma}) = \gamma,\ \bar{\kappa} < \delta.$$

Hence, by the elementarity of $\pi$, we have that $\bar{j} : H(\bar{\gamma}^+)^N \to H(\bar{j}(\bar{\gamma})^+)^N$ is an elementary map with $\bar{\delta} \leq \mathrm{crit}(\bar{j})$. Furthermore as $\bar{\kappa}$ is very large above $\bar{j}(\bar{\gamma})$, we have that

$$\pi \restriction \left(H(\bar{j}(\bar{\gamma})^+)^N\right) \in N,$$

hence $\pi \restriction \left(H(\bar{j}(\bar{\gamma})^+)^N\right) \in H(\gamma^+)^N$. Define $\pi^* = j\left(\pi \restriction \left(H(\bar{j}(\bar{\gamma})^+)^N\right)\right)$. Now we wish to show that $\bar{j} \in N$. This will be done by proving that $N$ can actually compute $\bar{j}$. For this, take $\bar{a} \in H(\bar{\gamma}^+)^N$ and $\bar{s} \in H(\bar{j}(\bar{\gamma})^+)^N$. Let $\pi(\bar{a}) = a$ and $\pi(\bar{s}) = s$ then,

$$\begin{aligned} \bar{s} \in \bar{j}(\bar{a}) &\iff s \in j(a) \\ &\iff s \in j(\pi(\bar{a})) \\ &\iff \pi(\bar{s}) \in j(\pi \restriction \left(H(\bar{j}(\bar{\gamma})^+)^N\right)(\bar{a})) \\ &\iff \pi(\bar{s}) \in \pi^*(j(\bar{a})) \\ &\iff \pi \restriction \left(H(\bar{j}(\bar{\gamma})^+)^N\right)(\bar{s}) \in \pi^*(\bar{a}). \end{aligned}$$

Where the last equivalence follows because $\mathrm{crit}(j) > \bar{\kappa}$ and $\bar{\kappa}$ is sufficiently large above $\bar{j}(\bar{\gamma}^+)^N$. Now as $\pi^*$ and $\pi \restriction \left(H(\bar{j}(\bar{\gamma})^+)^N\right)$ are in $N$, $\bar{j} \in N \cap V_{\bar{\kappa}}$. Since $\pi$ stretches $N$ correctly up to rank $\bar{\kappa}$, we conclude $j = \pi(\bar{j}) \in N$ as desired. □

Now we show that, if $\delta$ is an extendible cardinal, then no elementary embedding maps a weak extender model for $\delta$ supercompact to itself. For this we recall the following form of Kunen's theorem.

**Theorem 6.3** (Kunen) *Let $\kappa$ be an ordinal. Then there is no non-trivial elementary embedding*

$$i : V_{\kappa+2} \to V_{\kappa+2}.$$

The proof can be found in [2] Theorem 23.14.

**Theorem 6.4** *Let $N$ be a weak extender model for $\delta$ supercompact. Then there is no elementary embedding $j : N \to N$ with $\delta \leq \mathrm{crit}(j)$ and $j \neq \mathrm{id}$.*

*Proof* Suppose for contradiction that there is such a $j$. Let $\kappa > \delta$ be a fixed point of $j$. Then the restriction of $j$ to $V^N_{\kappa+2}$ is the an elementary embedding $i : V^N_{\kappa+2} \to V^N_{\kappa+2}$ with $\mathrm{crit}(i) \geq \delta$. Theorem 6.2 implies $i \in N$. This contradicts Theorem 6.3 within $N$. □

**Corollary 6.5** *Assume the* HOD *Conjecture. If $\delta$ is an extendible cardinal, then there is no $j : \mathrm{HOD} \to \mathrm{HOD}$ with $\delta \leq crit(j)$ and $j \neq \mathrm{id}$.*

*Proof* Follows from the Theorem 6.4 and Theorem 5.5. □

Finally we give an example $N$ of a weak extender model for $\delta$ supercompact other than $V$. $N$ will be such that there is a nontrivial elementary embedding $j : N \to N$, with $\mathrm{crit}(j) < \delta$. The point of the next example is that actually one can have a weak extender model for $\delta$ supercompact but it lacks structural properties, such as the ones HOD and $L$ possess. Note that this makes Theorem 6.4 actually optimal.

For the example we will use the following fact.

**Lemma 6.6** *Let $\kappa$ be a measurable cardinal, $\mu$ a measure on $\kappa$ and $j : V \to M$ the ultrapower map given by $\mu$. Also let $\nu$ be a $\delta$-complete measure, for some $\delta > \kappa$, $k : V \to N$ the ultrapower map given by $\nu$ and $l : M \to \mathrm{Ult}(M, j(\nu))$ the ultrapower map. Then $k \restriction M = l$.*

*Proof* First, observe that as the critical point of $k$ is above $\kappa$, then $\mu \in N$. Apply $\mu$ to $N$ and let $j' : N \to \mathrm{Ult}(N, \mu)$ be the ultrapower map.

${}^{\delta}N \subset N$ because $\nu$ is $\delta$-complete, so all functions from $\kappa$ to $N$ are in $N$, and this readily implies $j \restriction N = j'$.

Now, for $j(f)(\kappa)$ an element of $M$, we wish to see that $k(j(f)(\kappa)) = l(j(f)(\kappa))$. For simplicity, write $j$ for a restriction of $j$ to a suitable rank-initial segment which can then be treated as an element; likewise for $k$. By elementarity we have that $k(j(f)(\kappa)) = k(j)(k(f))(k(\kappa))$, but $k(j)$ is $j'$ which is the restriction of $j$ to $N$ so,

$$\begin{aligned} k(j(f)(\kappa)) &= j(k(f))(\kappa) \\ &= j(k)(j(f))(\kappa) \\ &= l(j(f))(\kappa) \\ &= l(j(f))(l(\kappa)) \\ &= l(j(f)(\kappa)) \end{aligned}$$

In other words, $k$ restricts to $l$ as desired. □

**Example 6.7** Let $\delta$ be a supercompact cardinal. Then there is $N$ a weak extender model for $\delta$ supercompact , and a nontrivial $j : N \to N$ with $\mathrm{crit}(j) < \delta$.

Let $\kappa < \delta$ be a measurable cardinal and take $\mu$ a measure on $\kappa$. Let

$$V = M_0 \to M_1 \to M_2 \to M_3 \to \cdots \to M_\omega$$

be the internal iteration of $V$ by $\mu$ of length $\omega$. So we have $M_0 = V$, $\kappa_0 = \kappa$; and inductively for naturals $n > 0$ define $\mu_n = i_{n-1,n}(\mu_{n-1})$, $\kappa_n = i_{n-1}(\kappa_{n-1}) = \mathrm{crit}(\mu_n)$ and let $i_{n,n+1} : M_n \to M_{n+1}$ be the map induced by taking the ultrapower of $M_n$ by $\mu_n$. $M_\omega$ is then the direct limit of the system and $i_{n,\omega} : M_n \to M_\omega$ the induced embeddings. $M_\omega$ is well founded and so we identify it with its transitive collapse (see Theorem 19.7 of [1] ). Define $N = M_\omega$.

Now, we show that $N$ is a weak extender model for $\delta$ supercompact. This is equivalent to showing that for unboundedly many $\gamma$ there is a $\gamma$-supercompactness measure that concentrates on $N$ and is amenable to $N$. Note that $i_{0,\omega}(\delta) = \delta$ and that for unboundedly many ordinals $\gamma$, we have that $i_{0,\omega}(\gamma) = \gamma$. Fix such $\gamma$ and let $\mathcal{U}$ be a normal and fine measure on $\mathcal{P}_\delta(\gamma)$. We prove that $\mathcal{U}$ is a suitable measure for $N$. Now, let $\mathcal{W} = i_{0,\omega}(\mathcal{U})$, and $\mathcal{W}_n = i_{0,n}(\mathcal{U})$ (observe that, for each $n$, $\mathcal{W}_n$ is a normal fine measure on $\mathcal{P}_\delta(\gamma)$ in $M_n$). By Lemma 6.6 the map induced by taking $\mathrm{Ult}(V, \mathcal{U})$ restricts to the one given by $\mathrm{Ult}(M_1, \mathcal{W})$. Inductively we have that if $k_n : M_n \to \mathrm{Ult}(V, \mathcal{W}_n)$ is the ultrapower map, then $k_n = k_0 \restriction M_n$. It follows then that as $i_{n,\omega}(\mathcal{W}_n) = \mathcal{W}$ (for each $n$) and $N$ is the direct limit of the initial system, we have that $k = k_0 \upharpoonright N$, where $k : N \to \mathrm{Ult}(N, \mathcal{W})$ is the ultrapower map given by $\mathcal{W}$. Therefore for $A \in N$, $k[\gamma] \in k(A)$ iff $k_0[\gamma] = k_0(A)$, which readily implies $\mathcal{W} = \mathcal{U} \cap N$; in other words $\mathcal{U}$ is amenable to $N$. Also, $\mathcal{U}$ concentrates on $N$ as $N \cap \mathcal{P}_\delta(\gamma) \in \mathcal{W} \subseteq \mathcal{U}$, as desired. Thus $N$ is a weak extender model for $\delta$ supercompact.

Finally, observe that if $j = i_{0,1} \upharpoonright N$ then $j : N \to N$ as $N$ is the $\omega$-th iterate, so we have a nontrivial embedding from $N$ to $N$, the key point here is that $\mathrm{crit}(j) < \delta$.

## 7 Consequences of the HOD Conjecture

We conclude by summarising without proof some results that would follow if the HOD Conjecture were proved to be a theorem of ZFC.

**Theorem 7.1** (ZF) *Assume that ZFC proves the HOD Conjecture. Suppose $\delta$ is an extendible cardinal. Then there is a transitive class $M \subseteq V$ such that:*

1. $M \models \mathrm{ZFC}$
2. *$M$ is $\Sigma_2(a)$-definable for some $a \in V_\delta$*
3. *Every set of ordinals is $< \delta$-generic over $M$*
4. *$M \models$ "$\delta$ is an extendible cardinal"*

The conclusion of the theorem is that there is an inner model M which is both close to V and in which the Axiom of Choice holds. (See [4] Theorem 229 for proof of a stronger result.) This is close to "proving" the Axiom of Choice from large cardinal axioms and suggests the following conjecture.

**Definition 7.2** The *Axiom of Choice Conjecture* asserts in ZF, that if $\delta$ is an extendible cardinal then the Axiom of Choice holds in $V[G]$, where $G$ is $V$-generic for collapsing $V_\delta$ to be countable.

One application of Theorem 7.1 is the following theorem. (See [4] Theorem 228 for a proof.)

**Theorem 7.3** (ZF) *Assume that* ZFC *proves the* HOD *Conjecture. Suppose $\delta$ is an extendible cardinal. Then for all $\lambda > \delta$ there is no nontrivial elementary embedding $j : V_{\lambda+2} \to V_{\lambda+2}$.*

Thus (assuming that ZFC proves the HOD Conjecture) one nearly has a proof of Kunen's Theorem (6.3) without using the Axiom of Choice.

For our final theorem we need a new definition. $L(\mathcal{P}(OR))$ is built in the same way as the usual $L$-hierarchy but allowing the use of all sets of ordinals in definitions. So it is the least model of ZF that contains all sets of ordinals. Note that, under ZF, this is not necessarily the whole of $V$.

**Theorem 7.4** (ZF) *Assume that ZFC proves the HOD Conjecture. Suppose that $\delta$ is an extendible cardinal. Then in $L(\mathcal{P}(OR))$:*

1. *$\delta$ is an extendible cardinal.*
2. *The Axiom of Choice Conjecture holds.*

## References

[1] Thomas Jech, *Set Theory*, Springer-Verlag 2003. Third millennium edition.

[2] Akihiro Kanamori, *The Higher Infinite*, Springer-Verlag 2003. Second edition.

[3] Azriel Lévy & Robert M. Solovay, *Measurable cardinals and the Continuum Hypothesis*, Israel Journal of Mathematics 1967 234-248

[4] W. Hugh Woodin, *Suitable Extender Models 1*, Journal of Mathematical Logic, Vol. 10, Nos. 1 & 2 2010 101-339

[5] W. Hugh Woodin, *Suitable Extender Models 2*, to appear

Printed in the United States
by Baker & Taylor Publisher Services